AF677623

Dorothea Kolossa • Reinhold Haeb-Umbach
Editors

Robust Speech Recognition of Uncertain or Missing Data

Theory and Applications

Springer

Editors
Prof. Dr.-Ing. Dorothea Kolossa
Institute of Communication Acoustics
Ruhr-Universität Bochum
Universitätsstrasse 150
44801 Bochum
Germany
dorothea.kolossa@rub.de

Prof. Dr.-Ing. Reinhold Haeb-Umbach
Department of Communications
Engineering
University of Paderborn
Warburger Strasse 100
33098 Paderborn
Germany
haeb@nt.uni-paderborn.de

ISBN 978-3-642-21316-8 e-ISBN 978-3-642-21317-5
DOI 10.1007/978-3-642-21317-5
Springer Heidelberg Dordrecht London New York

Library of Congress Control Number: 2011932686

ACM codes: I.2.7, G.3

© Springer-Verlag Berlin Heidelberg 2011
This work is subject to copyright. All rights are reserved, whether the whole or part of the material is concerned, specifically the rights of translation, reprinting, reuse of illustrations, recitation, broadcasting, reproduction on microfilm or in any other way, and storage in data banks. Duplication of this publication or parts thereof is permitted only under the provisions of the German Copyright Law of September 9, 1965, in its current version, and permission for use must always be obtained from Springer. Violations are liable to prosecution under the German Copyright Law.
The use of general descriptive names, registered names, trademarks, etc. in this publication does not imply, even in the absence of a specific statement, that such names are exempt from the relevant protective laws and regulations and therefore free for general use.

Cover design: deblik

Printed on acid-free paper

Springer is part of Springer Science+Business Media (www.springer.com)

Preface

Automatic speech recognition suffers from a lack of robustness with respect to noise, reverberation and interfering speech. The growing field of speech recognition in the presence of missing or uncertain input data seeks to ameliorate those problems by using not only a preprocessed speech signal but also an estimate of its reliability, to selectively focus on those segments and features which are most reliable for recognition.

This book presents the state of the art in recognition of uncertain or missing speech data, presenting examples that utilize uncertainty information for noise robustness, for reverberation robustness and for the simultaneous recognition of multiple speech signals, as well as for audiovisual speech recognition.

The editors thank all the authors for their valuable contributions and their cooperation in unifying the layout of the book and the terminology and symbols used. It was a great pleasure working with all of them!

Furthermore, the editors would like to express their gratitude to Ronan Nugent of Springer for his encouragement and support during the creation of this book. We also thank Alexander Krueger and Volker Leutnant for their help with the compilation of the LaTeX document.

Paderborn and Bochum,
February 2011

Reinhold Haeb-Umbach
Dorothea Kolossa

Contents

Abbreviations and Acronyms xvii

1 Introduction 1
Reinhold Haeb-Umbach, Dorothea Kolossa
1.1 Overview of the Book 2
References 5

Part I Theoretical Foundations

2 Uncertainty Decoding and Conditional Bayesian Estimation 9
Reinhold Haeb-Umbach
2.1 Introduction 9
2.2 Speech Recognition Using HMMs 11
2.2.1 Bayesian Classification Rule 11
2.2.2 Presence of Corrupted Features 13
2.2.3 Variants of Uncertainty Decoding 17
2.2.4 Missing-Feature Approaches 18
2.3 Feature Space vs. Acoustic Model-Based Approaches 20
2.4 Posterior Estimation 21
2.4.1 A Priori Model of Clean Speech 24
2.4.2 Observation Model 25
2.4.3 Inference 26
2.5 From Theory to Practice 28
2.6 Conclusions 30
References 31

3 Uncertainty Propagation 35
Ramón Fernandez Astudillo, Dorothea Kolossa
3.1 Uncertainty Propagation 35
3.2 Modeling Uncertainty in the STFT Domain 37
3.3 Piecewise Uncertainty Propagation 39
3.3.1 Uncertainty Propagation Through Linear Transformations 40

3.3.2 Uncertainty Propagation with the Unscented Transform . 41
3.4 Uncertainty Propagation for the Mel Frequency Cepstral Features . 42
3.4.1 Propagation Through the STSA Computation 43
3.4.2 Propagation Through the Mel Filter Bank 44
3.4.3 Propagation Through the Logarithm of the Mel-STSA Analysis 45
3.4.4 Propagation Through the Discrete Cosine Transform 46
3.5 Uncertainty Propagation for the RASTA-PLP Cepstral Features .. 46
3.5.1 Propagation Through the PSD Computation 47
3.5.2 Propagation Through the Bark Filterbank 48
3.5.3 Propagation Through the Logarithm of the Bark-PSD ... 48
3.5.4 Propagation Through RASTA Filter 49
3.5.5 Propagation Through Equal Loudness Pre-emphasis, Power Law of Hearing and Exponential 51
3.5.6 Propagation Through the All-Pole Model and LPCC Computation 51
3.6 Uncertainty Propagation for the ETSI ES 202 050 Standard 52
3.6.1 Propagation of the Complex Gaussian Model into the Time Domain 53
3.6.2 Propagation Through Log-Energy Transformation 55
3.6.3 Propagation Through Blind Equalization 56
3.7 Taking Full Covariances into Consideration 56
3.7.1 Covariances Between Features of the Same Frame 56
3.7.2 Covariance Between Features of Different Frames 57
3.8 Tests and Results.. 57
3.8.1 Monte Carlo Simulation of Uncertainty Propagation 57
3.8.2 Improving the Robustness of ASR with Uncertainty Propagation 61
References 63

Part II Applications: Noise Robustness

4 Front-End, Back-End, and Hybrid Techniques for Noise-Robust Speech Recognition 67
Li Deng
4.1 Introduction 67
4.2 Bayesian Decision Rule with Unreliable Features 69
4.3 Use of Algonquin Model to Compute Feature Uncertainty 71
4.3.1 Algonquin Model of Speech Distortion 71
4.3.2 Step I in Computing Uncertainty: Means 72
4.3.3 Step II in Computing Uncertainty: Variances........... 74
4.3.4 Discussions 75
4.4 Use of a Phase-Sensitive Model for Feature Compensation....... 76
4.4.1 Phase-Sensitive Modeling of Acoustic Distortion — Deterministic Version 77

4.4.2 The Phase-Sensitive Model of Acoustic Distortion — Probabilistic Version ... 80
4.4.3 Feature Compensation Experiments and Lessons Learned 82
4.4.4 Discussions ... 83
4.5 Bayesian Decision Rule with Unreliable Model Parameters ... 84
4.5.1 Bayesian Predictive Classification Rule ... 84
4.5.2 Model Compensation Viewed from the Perspective of the BPC Rule ... 85
4.6 Model and Feature Compensation — A Taxonomy-Oriented Overview ... 86
4.6.1 Model-Domain Compensation ... 86
4.6.2 Feature-Domain Compensation ... 88
4.6.3 Hybrid Compensation Techniques ... 89
4.7 Noise Adaptive Training ... 90
4.7.1 The Basic NAT Scheme and its Performance ... 90
4.7.2 NAT and Related Work — A Brief Overview ... 91
4.8 Summary and Conclusions ... 93
References ... 95

5 Model-Based Approaches to Handling Uncertainty ... 101
M. J. F. Gales
5.1 Introduction ... 101
5.2 General Acoustic Model Adaptation ... 102
5.3 Impact of Noise on Speech ... 104
5.3.1 Static Parameter Mismatch Function ... 104
5.3.2 Dynamic Parameter Mismatch Functions ... 106
5.3.3 Corrupted Speech Distributions ... 106
5.4 Feature Enhancement Approaches ... 107
5.5 Model-Based Noise Compensation ... 110
5.5.1 Vector Taylor Series Compensation ... 111
5.5.2 Sampling-Based Approximations ... 112
5.6 Efficient Model Compensation and Likelihood Calculation ... 113
5.6.1 Compensation Parameter Estimation ... 114
5.6.2 Compensating the Model Parameters ... 115
5.7 Adaptive Training and Noise Estimation ... 116
5.7.1 EM-Based Approaches ... 117
5.7.2 Second-Order Approaches ... 119
5.8 Conclusions and Future Research Directions ... 120
References ... 123

6 Reconstructing Noise-Corrupted Spectrographic Components for Robust Speech Recognition ... 127
Bhiksha Raj and Rita Singh
6.1 Introduction ... 127
6.2 Spectrographic Representation ... 130
6.3 Notation ... 131

6.4 A Note on Estimating Spectrographic Masks ... 132
6.5 Classifier Compensation Methods ... 133
6.5.1 State-Based Imputation ... 134
6.5.2 Marginalization ... 134
6.5.3 Marginalization with Soft Masks ... 135
6.6 Feature Compensation Methods ... 135
6.6.1 Correlation-Based Reconstruction ... 136
6.6.2 Cluster-Based Reconstruction ... 138
6.6.3 Minimum Mean Squared Error Estimation of Spectral Components from Soft Masks ... 141
6.7 Experimental Evaluation ... 144
6.7.1 Experimental Setup ... 144
6.7.2 Recognition Performance with Knowledge of True SNR ... 145
6.7.3 Recognition with Log Spectra ... 145
6.7.4 Effect of Preprocessing ... 146
6.7.5 Recognition with Cepstra ... 147
6.7.6 Effect of Errors in Identifying Unreliable Components ... 148
6.7.7 Experiments with MMSE Estimation from Soft Masks ... 151
6.8 Conclusion and Discussion ... 153
References ... 155

7 Automatic Speech Recognition Using Missing Data Techniques: Handling of Real-World Data ... 157
Jort F. Gemmeke, Maarten Van Segbroeck, Yujun Wang, Bert Cranen, Hugo Van hamme
7.1 Introduction ... 157
7.2 MDT ASR ... 160
7.2.1 Missing Data Techniques ... 160
7.2.2 Gaussian-Dependent Imputation ... 161
7.3 Real-World Data: The SPEECON and SpeechDat-Car Databases ... 162
7.3.1 Isolated Word Test Set ... 162
7.3.2 Training Sets ... 163
7.4 Experimental Setup ... 163
7.4.1 Mask Estimation ... 163
7.4.2 Recognizer Setup ... 165
7.5 MDT and Multi-Condition Training ... 167
7.5.1 Recognition Results ... 167
7.5.2 Discussion ... 170
7.6 MDT and Dereverberation ... 171
7.6.1 Experimental Setup ... 172
7.6.2 Results and Discussion ... 173
7.7 MDT and Feature Enhancement ... 176
7.7.1 Experimental Setup ... 177
7.7.2 Results and Discussion ... 180
7.8 General Discussion and Conclusions ... 182

7.9 Acknowledgements ... 183
References ... 184

8 Conditional Bayesian Estimation Employing a Phase-Sensitive Observation Model for Noise Robust Speech Recognition ... 187
Volker Leutnant and Reinhold Haeb-Umbach
8.1 Introduction ... 188
8.2 Uncertainty Decoding for ASR ... 189
8.2.1 Bayesian Framework of Speech Recognition ... 190
8.2.2 Corrupted Features ... 190
8.3 Conditional Bayesian Estimation of the Feature Posterior ... 193
8.3.1 The A Priori Model ... 194
8.3.2 The Observation Model ... 197
8.3.3 Moments of the Phase Factor Distribution ... 202
8.3.4 Inference ... 205
8.4 Experiments ... 210
8.4.1 Aurora 2 Database ... 210
8.4.2 Aurora 4 Database ... 211
8.4.3 Training of the HMMs ... 211
8.4.4 Training of the A Priori Models ... 212
8.4.5 Experimental Setup ... 212
8.4.6 Experimental Results and Discussion ... 214
8.5 Conclusions ... 218
References ... 220

Part III Applications: Reverberation Robustness

9 Variance Compensation for Recognition of Reverberant Speech with Dereverberation Preprocessing ... 225
Marc Delcroix, Shinji Watanabe, Tomohiro Nakatani
9.1 Introduction ... 226
9.2 Related Work ... 227
9.3 Overview of Recognition System ... 229
9.3.1 Dereverberation Based on Late Reverberation Energy Suppression ... 229
9.3.2 Speech Recognition Based on Dynamic Variance Compensation ... 231
9.3.3 Relation with a Conventional Uncertainty Decoding Approach ... 233
9.4 Proposed Method for Variance Compensation ... 235
9.4.1 Combination of *Static* and *Dynamic* Variance Adaptation 235
9.4.2 Adaptation of Variance Model Parameters ... 237
9.5 Digit Recognition Experiments ... 239
9.5.1 Experimental Settings ... 240
9.5.2 Baseline Results ... 240
9.5.3 Supervised Adaptation ... 242

9.5.4 Unsupervised Adaptation ... 244
9.5.5 Insights about the Proposed Algorithm ... 244
9.6 Experiment on Large Vocabulary Task ... 247
9.6.1 Experimental Settings ... 247
9.6.2 Large Vocabulary Results ... 248
9.7 Conclusion ... 249
Appendix1 ... 250
Appendix2 ... 251
References ... 252

10 A Model-Based Approach to Joint Compensation of Noise and Reverberation for Speech Recognition ... 257
Alexander Krueger and Reinhold Haeb-Umbach
10.1 Introduction ... 258
10.2 Problem Formulation ... 261
10.3 Feature Extraction ... 262
10.4 Bayesian Feature Enhancement Approach ... 264
10.4.1 A Priori Model ... 266
10.4.2 Observation Model ... 268
10.5 Suboptimal Inference ... 276
10.5.1 Feature Enhancement Algorithm ... 277
10.6 Experiments ... 281
10.6.1 Baseline Recognition Results ... 281
10.6.2 Choice of Model Parameters ... 282
10.6.3 Recognition Results for Feature Enhancement ... 285
10.7 Conclusion ... 287
References ... 288

Part IV Applications: Multiple Speakers and Modalities

11 Evidence Modeling for Missing Data Speech Recognition Using Small Microphone Arrays ... 293
Marco Kühne, Roberto Togneri and Sven Nordholm
11.1 Introduction ... 294
11.2 Robust HMM Likelihood Computation Using Evidence Models ... 296
11.3 Microphone Array-Based Evidence Model Parameter Estimation ... 301
11.3.1 BSS Front-End ... 301
11.3.2 Evidence PDF Estimation ... 306
11.4 Speech Recognition Experiments ... 308
11.4.1 Room Simulation ... 308
11.4.2 Model Training ... 309
11.4.3 Performance Measures ... 309
11.4.4 Results ... 310
11.4.5 General Discussion ... 312
11.5 Summary ... 314
References ... 316

12 Recognition of Multiple Speech Sources Using ICA 319
Eugen Hoffmann, Dorothea Kolossa and Reinhold Orglmeister
12.1 Introduction 319
12.2 Blind Source Separation 320
12.2.1 Problem Formulation 321
12.2.2 ICA 322
12.2.3 Permutation Correction 324
12.2.4 Postmasking 326
12.3 Uncertainty Estimation 332
12.3.1 Ideal Uncertainties 333
12.3.2 Masking Error Estimate 333
12.3.3 Uncertainty Propagation 334
12.4 Experiments and Results 334
12.4.1 Recording Conditions 334
12.4.2 Parameter Settings 335
12.4.3 Performance Measures 336
12.4.4 Separation Results 336
12.4.5 Model Training 337
12.4.6 Recognition of Uncertain Data 337
12.4.7 Recognition Results 338
12.5 Conclusions 340
References 341

13 Use of Missing and Unreliable Data for Audiovisual Speech Recognition 345
Alexander Vorwerk, Steffen Zeiler, Dorothea Kolossa, Ramón Fernandez Astudillo, Dennis Lerch
13.1 Introduction 346
13.2 Coupled HMMs in Audiovisual Speech Recognition 346
13.2.1 Two-Stream Coupled HHM for Word Models 347
13.2.2 Missing Features in Coupled HMMs 348
13.3 Audio Features 351
13.4 Video Features 351
13.4.1 Face Recognition 351
13.4.2 Feature Extraction 352
13.5 Audiovisual Speech Recognizer Implementation and Results 356
13.5.1 Face Detection 356
13.5.2 Video Features and Their Uncertainties 358
13.5.3 Audio Features and Their Uncertainties 362
13.5.4 Audiovisual Recognition System 363
13.6 Efficiency Considerations 364
13.6.1 Gaussian Density Computation 364
13.6.2 Conventional Likelihood 366
13.6.3 Uncertainty Decoding 367
13.6.4 Modified Imputation 367

13.6.5 Binary Uncertainties 368
13.6.6 Overview of Uncertain Recognition Techniques 368
13.6.7 Recognition Results 369
13.7 Conclusion 371
References 373

Index 377

List of Contributors

Ramón Fernandez Astudillo
Electronics and Medical Signal Processing Group, Technische Universität Berlin, Germany

Bert Cranen
Department of Linguistics, Radboud University Nijmegen, The Netherlands

Marc Delcroix
NTT Communications Science Laboratories, Kyoto, Japan

Li Deng
Microsoft Research, Redmond, USA

Mark Gales
Cambridge University Engineering Department, Cambridge, UK

Jort F. Gemmeke
Department of Linguistics, Radboud University Nijmegen, The Netherlands

Reinhold Haeb-Umbach
Department of Communications Engineering, University of Paderborn, Germany

Eugen Hoffmann
Electronics and Medical Signal Processing Group, Technische Universität Berlin, Germany

Dorothea Kolossa
Institute of Communication Acoustics, Ruhr-Universität Bochum, Germany

Alexander Krueger
Department of Communications Engineering, University of Paderborn, Germany

Marco Kühne
The University of Western Australia, Crawley, Australia

Dennis Lerch
Electronics and Medical Signal Processing Group, Technische Universität Berlin, Germany

Volker Leutnant
Department of Communications Engineering, University of Paderborn, Germany

Tomohiro Nakatani
NTT Communications Science Laboratories, Kyoto, Japan

Sven Nordholm
Curtin University of Technology, Perth, Australia

Reinhold Orglmeister
Electronics and Medical Signal Processing Group, Technische Universität Berlin, Germany

Bhiksha Raj
Carnegie Mellon University, Pittsburgh, USA

Maarten Van Segbroeck
ESAT Department, Katholieke Universiteit Leuven, Belgium

Rita Singh
Carnegie Mellon University, Pittsburgh, USA

Roberto Togneri
The University of Western Australia, Crawley, Australia

Alexander Vorwerk
Electronics and Medical Signal Processing Group, Technische Universität Berlin, Germany

Hugo Van hamme
ESAT Department, Katholieke Universiteit Leuven, Belgium

Yujun Wang
ESAT Department, Katholieke Universiteit Leuven, Belgium

Shinji Watanabe
NTT Communications Science Laboratories, Kyoto, Japan

Steffen Zeiler
Electronics and Medical Signal Processing Group, Technische Universität Berlin, Germany

Abbreviations and Acronyms

AEC	Acoustic echo canceler
AFE	Advanced front-end
ASR	Automatic speech recognition
AVSR	Audiovisual speech recognition
BPC	Bayesian predictive classification
CHMM	Coupled hidden Markov model
CMLLR	Constrained maximum likelihood linear regression
CMN	Cepstral mean normalization
CMS	Cepstral mean subtraction
D50	Deutlichkeit value
DARPA	Defense Advanced Research Projects Agency
DCT	Discrete cosine transform
DFT	Discrete Fourier transform
DOA	Direction of arrival
DPMC	Data-driven parallel model combination
DVA	Dynamic variance adaptation
ECM	Expectation conditional maximization
EKF	Extended Kalman filter
EKF-$\boldsymbol{\alpha}$	Extended Kalman filter with phase-sensitive observation model
EM	Expectation maximization
ETSI	European Telecommunications Standards Institute
FIR	Finite impulse response
GMM	Gaussian mixture model
GPB1	Generalized pseudo Bayesian of order 1
GPB2	Generalized pseudo Bayesian of order 2
HMM	Hidden Markov model
HTK	Hidden Markov model toolkit
ICA	Independent component analysis
IDFT	Inverse discrete Fourier transform
IEKF	Iterated extended Kalman filter
IID	Independent and identically distributed

IIR	Infinite impulse response
IMM	Interacting multiple model
IVN	Irrelevant variability normalization
JAT	Joint adaptive training
JUD	Joint uncertainty decoding
LDA	Linear discriminant analysis
LDM	Linear dynamic model
LMPSC	Logarithmic mel power spectral coefficient
MDT	Missing data techniques
MFCC	Mel frequency cepstral coefficient
MIDA	Mutual information discriminant analysis
MLLR	Maximum likelihood linear regression
MMSE	Minimum mean squared error
MPE	Minimum phone error
MTFA	Multiplicative transfer function assumption
NAT	Noise-adaptive training
NCMLLR	Noisy constrained maximum likelihood linear regression
PCMLLR	Predictive constrained maximum likelihood linear regression
PDF	Probability density function
PLP	Perceptual linear prediction
PMC	Parallel model combination
POF	Probabilistic optimal filtering
PSD	Power spectral density
RASTA	Relative spectral (filter)
RIR	Room impulse response
RM1	DARPA resource management database
SDVA	Static and dynamic variance adaptation
SFE	Standard front-end
SLDM	Switching linear dynamical model
SNR	Signal-to-noise ratio
STDFT	Short-time discrete Fourier transform
STFT	Short-time Fourier transform
STSA	Short-time spectral amplitude
SVM	Support vector machine
UD	Uncertainty decoding
UKF	Unscented Kalman filter
VAD	Voice activity detection
VTS	Vector Taylor series
WER	Word error rate

Chapter 1
Introduction

Reinhold Haeb-Umbach, Dorothea Kolossa

It has been stated over and over again, and it remains true at the time of this writing: speech recognition by machines is by far not as robust as recognition by humans. While human listeners apparently face little difficulty ignoring reverberation and background noise, and while healthy-of-hearing listeners can even recognize a speaker in babble noise at 0 dB signal-to-noise power ratio, machine performance degrades quite rapidly if the speech signal is degraded by acoustic environmental noise, reverberation, competing speakers or any other kind of distortion.

This lack of robustness has been identified as one if not the major impediment to the ubiquitous use of automatic speech recognition (ASR) technology. Calling it one of the "grand challenges", an expert group of the IEEE Signal Processing Society suggested that future research "*would concentrate on creating and developing systems that would be much more robust against variability and shifts in acoustic environments, reverberation, external noise sources, communication channels (e.g., far-field microphones, cellular phones), speaker characteristics ... and language characteristics*" [1, 2].

The purpose of this book is not to provide a comprehensive overview of the raft of measures that have been proposed to attack the robustness issue. It rather focuses on those approaches that formulate the problem as one of optimal classification in the presence of degraded input data, by considering the true speech features as hidden variables, observable only subject to random distortions which can be described by stochastic processes. These approaches for pattern recognition under observation uncertainty are deeply rooted in statistical pattern recognition and Bayesian estimation theory. They have shown remarkable success in the past and bear the potential for further breakthroughs, as we hope to also illustrate and motivate by this book.

Reinhold Haeb-Umbach
Department of Communications Engineering, University of Paderborn, Warburger Straße 100, 33098 Paderborn, Germany, e-mail: haeb@nt.uni-paderborn.de

Dorothea Kolossa
Institute of Communication Acoustics, Ruhr-Universität Bochum, Universitätsstraße 150, 44801 Bochum, Germany, e-mail: dorothea.kolossa@rub.de

D. Kolossa and R. Haeb-Umbach (eds.), *Robust Speech Recognition of Uncertain or Missing Data: Theory and Applications*, DOI 10.1007/978-3-642-21317-5_1,
© Springer-Verlag Berlin Heidelberg 2011

The following chapters comprise both tutorial-like overview articles and in-depth research reports. While the former may serve as an entry point for newcomers to the field and a source of promising research ideas for the more experienced, the latter give detailed information on the state of the art in robust speech recognition based on missing or uncertain features to cope with noise, reverberation, and multiple speech signals, and to recognize audiovisual input data.

1.1 Overview of the Book

The succeeding 12 chapters are organised in four parts.

In the first, the theoretical background is presented. For that purpose, Chapter 2 by Reinhold Haeb-Umbach introduces the statistical model that is necessary for Bayesian inference in the presence of observation uncertainties. Considering the observed feature vectors as distorted versions of underlying unobservable *clean* ones, optimal classification consists of first estimating the posterior density of the clean features and second of employing this posterior in a modified classification rule. The author argues that the estimation of the posterior is the key to achieving robustness, as optimal point estimates of the clean features and estimation error variances can be obtained from it. Since, in many cases, such estimation error variances are easier to derive in the spectrum domain than in those feature domains most amenable to speech recognition, Chapter 3, authored by Ramón Fernandez Astudillo and Dorothea Kolossa, deals with the strategy of uncertainty propagation, which may be used to obtain an estimation error estimate in a fairly broad range of features better suited for robust speech recognition than the spectrum domain.

The second part deals with noise robustness, and with background noise still posing a major challenge to ubiquitous ASR systems, five chapters are devoted to this subject.

Chapter 4, by Li Deng, gives an overview of uncertainty-based strategies for increased noise robustness, dealing with both front-end and back-end methods. He discusses structured and unstructured methods, where the former assume an analytical model of the distortion while the latter do not. Both classes of approaches have respective strengths and weaknesses, and their hybrids tend to offer the best performance as exemplified in the scheme of noise-adaptive training, which is presented and analyzed in this chapter in light of the author's original contribution to this hybrid scheme. The chapter also devotes a large section to the questions of uncertainty estimation and noise-adaptive training, where possible routes for further development are outlined in some detail.

Mark Gales in Chapter 5 discusses approaches for handling uncertainty in observations with an emphasis on model-based compensation schemes. These methods adjust the acoustic model parameters so that they are more representative of the Hidden Markov Model (HMM) output distributions in the noisy target domain. After a description of the theoretical background, the chapter discusses important practical issues, including noise parameter estimation, efficiency considerations, and the use of adaptive training for multi-style training data.

In Chapter 6, Bhiksha Raj and Rita Singh describe a different set of approaches. Rather than using uncertainties during the decoding process, they use the reliability of spectrum domain features as a binary or real-valued measure to find unreliable segments in the spectrum. These are subsequently replaced by more reliable estimates based on a detailed speech model. Finally, this estimate of spectrum domain features can be used as a more reliable speech estimate, allowing for the use of an unmodified recognition engine in the feature domain of choice.

Another way to utilize feature uncertainties is highlighted by Jort Gemmeke et al. in Chapter 7. The focus here lies on the application of missing data techniques to real-world data obtained in actual noise and reverberation. Using a wide variety of real-world recordings, it is argued that 'reliable' speech features no longer match the acoustic models when trained on non-reverberated, clean speech. Furthermore, it is shown that to regain robustness, multi-condition training and feature enhancement can be combined with new mask estimation techniques to account for the multiple corruption sources by which speech features can be affected.

Chapter 8 by Volker Leutnant and Reinhold Haeb-Umbach gives an in-depth presentation of how to estimate the clean speech feature posterior by Bayesian inference for the recognition of noisy speech. Particular emphasis is placed on a phase-sensitive observation model where the statistical moments of the relative phase between the speech and the noise spectra are computed analytically.

In many applications of automatic speech recognition, hands-free communication is desirable. However, replacing close-talking with distant-talking microphones results in degraded signal quality at the microphones due to acoustic environmental noise and reverberation. The latter is a convolutive distortion caused by reflections of the speech signals on walls and objects. The two chapters of Part 3 of this book discuss techniques which increase the robustness of an ASR system with regard to reverberated speech at its input.

While dereverberation techniques have been successfully employed in voice communications to increase audible quality and ease of listening, as preprocessors to an ASR system they often introduce distortions which may render the recognition accuracy even below the accuracy obtained on unprocessed speech. Marc Delcroix et al. therefore develop a variance compensation scheme in Chapter 9, where the mismatch between the variances of the HMM observation models estimated on nonreverberant training data and those observed on the preprocessed test data are compensated for, resulting in considerable improvement in ASR recognition accuracy.

The subsequent Chapter 10 by Alexander Krueger and Reinhold Haeb-Umbach treats feature vector dereverberation as a Bayesian inference problem and applies the very same techniques as those employed in Chapter 8 for noise robustness. The difference solely resides in the choice of observation model, which relates the observed features to the underlying clean features. Here, a stochastic model describing the effect of noise and reverberation on the log mel spectral feature vectors is presented. A *minimum mean squared error* estimate of the clean, i.e., noise-free and unreverberated, feature vector is computed and forwarded to the ASR decoder.

The fourth and final part of this book deals with the use of multiple channels or modalities for enhanced speech recognition. It is comprised of two chapters that apply uncertainty techniques in conjunction with blind source separation, with the purpose of recognizing simultaneously uttered speech from multiple talkers, and a third which exploits uncertainties for better integration of different modalities to obtain improvements in audiovisual speech recognition.

Aiming at a close coupling of microphone array processing and missing data speech recognition, Chapter 11 by Marco Kühne et al. shows how the so-called evidence probability density function (pdf), i.e., the pdf of the clean speech features, given the observed corrupted data, can be estimated from a blind source separation front-end and how it is to be used in the ASR classifier. Various choices for the parametric form of the pdf are discussed and evaluated.

Chapter 12 by Eugen Hoffmann et al. is also concerned with speech recognition in the presence of multiple speech sources. It first demonstrates that source separation by Independent Component Analysis (ICA) can be improved by nonlinear postprocessing employing a spectrographic mask, and then shows how information about the reliability of the recovered source can be gathered from the estimated masks, which leads to improved speech recognition results when used in a missing data speech recognizer.

The final Chapter 13 deals with improving adaptive stream integration for audiovisual speech recognition by estimating the time-variant reliability of the audio and video feature streams. Alexander Vorwerk et al. demonstrate improved recognition accuracy with both binary and continuous-valued uncertainty information used in a recognizer employing missing data principles.

Thanks to the valuable contributions of all chapter authors, the editors believe that this book provides a good overview and entry point for researchers and practitioners to the field of robust speech recognition by uncertainty decoding and missing data techniques. We hope that our readers will find it a source of inspiring ideas and concepts for their own work and research.

References

1. Baker, J., Deng, L., Glass, J., Khudanpur, S., Lee, C.H., Morgan, N., O'Shaughnessy, D.: Developments and directions in speech recognition and understanding, part 1. Signal Processing Magazine, IEEE **26**(3), 75–80 (2009)
2. Baker, J., Deng, L., Glass, J., Khudanpur, S., Lee, C.H., Morgan, N., O'Shaughnessy, D.: Developments and directions in speech recognition and understanding, part 2. Signal Processing Magazine, IEEE **26**(4), 78–85 (2009)

Part I
Theoretical Foundations

Chapter 2
Uncertainty Decoding and Conditional Bayesian Estimation

Reinhold Haeb-Umbach

Abstract In this contribution classification rules for HMM-based speech recognition in the presence of a mismatch between training and test data are presented. The observed feature vectors are regarded as corrupted versions of underlying and unobservable *clean* feature vectors, which have the same statistics as the training data. Optimal classification then consists of two steps. First, the posterior density of the clean feature vector, given the observed feature vectors, has to be determined, and second, this posterior is employed in a modified classification rule, which accounts for imperfect estimates. We discuss different variants of the classification rule and further elaborate on the estimation of the clean speech feature posterior, using conditional Bayesian estimation. It is shown that this concept is fairly general and can be applied to different scenarios, such as noisy or reverberant speech recognition.

2.1 Introduction

Improving the robustness of state-of-the-art automatic speech recognition (ASR) continues to be an important research area. Current hidden Markov model (HMM) based speech recognition systems are notorious for performing well in matched training and test conditions while quickly degrading in the presence of a mismatch. While such a mismatch may be caused by many factors, probably one of the most studied problems is improving the robustness of a recognizer trained on clean training data to test data being corrupted by acoustic environmental noise.

Huge research efforts have been devoted to overcoming this lack of robustness, and a wealth of methods has been proposed. These can be categorized into either methods that try to compensate the effect of distortions on the features (so-called

Reinhold Haeb-Umbach
Department of Communications Engineering, University of Paderborn, Warburger Straße 100, 33098 Paderborn, Germany, e-mail: haeb@nt.uni-paderborn.de

D. Kolossa and R. Haeb-Umbach (eds.), *Robust Speech Recognition of Uncertain or Missing Data: Theory and Applications*, DOI 10.1007/978-3-642-21317-5_2,
© Springer-Verlag Berlin Heidelberg 2011

front-end methods) or approaches that modify the models used in the recognizer to better match the incoming distorted feature stream (*back-end* methods).

Traditionally, front-end methods aim at obtaining point estimates of the uncorrupted, clean features. Likewise, back-end methods usually try to obtain point estimates of parameters, such as the mean vectors of the observation probabilities. These estimates are then "plugged" into the Bayesian decision rule as if they were perfect estimates. However, more recently the focus has shifted to estimating the features or parameters together with a measure of reliability of the estimate and propagating the uncertainty to the decision rule [4, 9, 17, 20, 26–28, 37, 46, 47]. The underlying rationale is that an estimate is never perfect and that the recognizer can benefit from knowing the estimation error variance by de-emphasizing the contributions of unreliable estimates to the overall decision on the word sequence.

The use of such optimal decision rules is by no means new. How to modify the Bayesian decision rule in the presence of missing or noisy features can be found in many textbooks on pattern recognition; see, e.g., [22]. Two developments, however, are fairly recent: how to modify the decision rule for HMM classifiers and how to obtain reliability information for a given class of distortions.

In this contribution we first show how a suboptimal decision rule for HMM-based speech recognition in the presence of corrupted feature vectors can be derived from the optimal, though infeasible, decoding rule. It will be seen that this decoding rule exploits the temporal correlation between the feature vectors, which is otherwise not used in an HMM-based recognizer due to the predominant conditional independence assumption. Several other decoding rules proposed in the literature, including missing-feature techniques, can be viewed as approximations to this rule.

A key element of the classification rule is the posterior density of the clean feature vector, given the observed and corrupted feature vectors. We show how this posterior can be estimated in principle by using well-known results from the field of optimal filtering. While the principle approach is applicable to various kinds of signal degradation, a crucial element, the observation model, has to be developed specifically for each kind of degradation, such as noise or reverberation.

This contribution is organized as follows: starting from the optimal Bayesian decision rule, we first derive a classification rule for an HMM-based speech recognizer in the presence of corrupted feature vectors, and follow it with a discussion of related uncertainty decoding rules, including missing feature theory. While uncertainty can be treated in either the feature or the model domain, we concentrate here on the feature domain and motivate this in Section 2.3. Next, the estimation of the clean speech feature posterior is discussed, comprising the components a priori model, observation model and inference algorithm. Since an in-depth treatment for a specific problem, such as noisy or reverberant speech recognition, is beyond the scope of this contribution, we refer to other chapters of this book for details. Finally, we briefly discuss the limitations of this approach and thus share perspectives for future research. The chapter ends with some conclusions in Section 2.6.

2.2 Speech Recognition Using HMMs

The Bayesian decision rule is the most fundamental equation governing statistical speech recognition. When HMMs are used as acoustic models, certain approximations, notably the famous conditional independence assumption, have to be made to arrive at a computationally tractable decoding rule. An implicit assumption that is often made is that the acoustic model training has been carried out on feature vectors representative of those to be found on the test data and that the required probability densities governing the decision rule are perfectly known. As in practice this will hardly be the case, we are going to discuss in the following how the classification rule has to be modified accordingly.

2.2.1 Bayesian Classification Rule

Given a sequence of D-dimensional feature vectors of length T, $\mathbf{x}_{1:T} = (\mathbf{x}_1, \ldots, \mathbf{x}_T)$, $\mathbf{x}_t \in \mathbb{R}^D$, the classification task consists of finding that sequence of words $\hat{\mathbf{W}}$ from a given vocabulary which maximizes the posterior probability $P(\mathbf{W}|\mathbf{x}_{1:T})$:

$$\hat{\mathbf{W}} = \underset{\mathbf{W}}{\operatorname{argmax}}\ P(\mathbf{W}|\mathbf{x}_{1:T}). \tag{2.1}$$

Using Bayes' rule for conditional probabilities, and noting that the term $p(\mathbf{x}_{1:T})$ does not influence the maximization, we obtain the classification rule in the well-known form:

$$\hat{\mathbf{W}} = \underset{\mathbf{W}}{\operatorname{argmax}}\ \{P(\mathbf{W}) \cdot p(\mathbf{x}_{1:T}|\mathbf{W})\}. \tag{2.2}$$

$P(\mathbf{W})$ is the a priori probability that the word sequence $\mathbf{W}$ was uttered and is specified by a *language model*. The *acoustic modeling* is concerned with estimating $p(\mathbf{x}_{1:T}|\mathbf{W})$, the probability density of observing the given feature vector sequence under the assumption of the word sequence $\mathbf{W}$. In an HMM-based speech recognizer the sequence of hidden states $q_{1:T} = (q_1, \ldots, q_T)$ underlying the sequence of observations is introduced:

$$p(\mathbf{x}_{1:T}|\mathbf{W}) = \sum_{\{q_{1:T}\}} p(\mathbf{x}_{1:T}|q_{1:T}, \mathbf{W}) \cdot P(q_{1:T}|\mathbf{W}), \tag{2.3}$$

where the summation is carried out over all HMM state sequences corresponding to $\mathbf{W}$.

In the following we will drop the dependence of the state sequence $q_{1:T}$ on $\mathbf{W}$ for notational convenience. However, one should always keep in mind that $q_{1:T} = q_{1:T}(\mathbf{W})$, i.e., that the state sequence depends on the word sequence under investigation.

In order to compute (2.3) recursively, both terms under the sum are factorized:

$$p(\mathbf{x}_{1:T}|q_{1:T}) = \prod_{t=1}^{T} p(\mathbf{x}_t|\mathbf{x}_{1:t-1}, q_{1:T}), \tag{2.4}$$

$$P(q_{1:T}) = \prod_{t=1}^{T} P(q_t|q_{1:t-1}), \tag{2.5}$$

where, for the sake of simplicity, we do not write the first factor separately and $P(q_1|q_{1:0})$ is understood to mean $P(q_1|q_{1:0}) = P(q_1)$, and correspondingly in similar expressions.

The expressions can be simplified by assuming that a feature vector $\mathbf{x}_t$ is independent of preceding and succeeding feature vectors if the state q_t is known (so-called *conditional independence* assumption) and by assuming that the state sequence is a first-order Markov process:

$$p(\mathbf{x}_{1:T}|q_{1:T}) = \prod_{t=1}^{T} p(\mathbf{x}_t|q_t), \tag{2.6}$$

$$P(q_{1:T}) = \prod_{t=1}^{T} P(q_t|q_{t-1}). \tag{2.7}$$

Using (2.6) and (2.7) in (2.3) and then in (2.2) we arrive at the well-known result:

$$\hat{\mathbf{W}} = \underset{\mathbf{W}}{\operatorname{argmax}} \left\{ P(\mathbf{W}) \cdot \sum_{\{q_{1:T}\}} \prod_{t=1}^{T} p(\mathbf{x}_t|q_t) \cdot P(q_t|q_{t-1}) \right\}. \tag{2.8}$$

An approximate value for (2.8) can be computed by the Viterbi algorithm, in which the summation over all possible state sequences is replaced by $\max_{q_{1:T}}$.

From pattern classification it is well known that the Bayesian decision rule minimizes the classification error rate [22]. This, however, is only true if the *true* probability $P(\mathbf{W})$ and the true density $p(\mathbf{x}|\mathbf{W})$, from which the data to be classified are drawn, are known.[1] In practice, however, this will never be the case! The reason is twofold. First, there is often a mismatch between training and test data, or, to phrase it more precisely, the probability distribution from which the test data are drawn is different from that of the training data. The reasons for this can be manifold. For example, the test data can be distorted by additive noise or reverberation, which was not present in the training data.

Second, as the training data set is limited, there will be inevitably an estimation error, i.e., the density $p(\mathbf{x}|\mathbf{W})$ estimated on the training data will not be identical to the true distribution underlying the data.

In [27] it is shown how an optimal decoding rule for ASR can be derived which takes into account the finite precision of the estimated parameters. However, this approach is computationally very expensive, since the number of parameters of the

[1] The use of $\mathbf{x}$ in $p(\mathbf{x}|\mathbf{W})$ instead of $\mathbf{x}_t$ shall indicate that the random variable $\mathbf{x}$ is meant rather than its realization $\mathbf{x}_t$. In most of this contribution, however, we do not make a distinction between the random variable and its realization for notational convenience with hope that the meaning is obvious through the context.

acoustic model can be very large, up to a few million for large vocabulary recognizers. It has therefore been proposed to account for the difference of the statistics of the test data and the acoustic model estimated on the training data in the feature space [17]. Rather than assuming an uncertainty about the acoustic model, we assume that the density from which the feature vectors are drawn is indeed $p(\mathbf{x}|\mathbf{W})$. However, the feature vector $\mathbf{x}$ is not available and only a related feature vector $\mathbf{y}$ is observed. Obviously, this is an equivalent way of representing a mismatch between the acoustic model used by the recognizer and the feature vectors to be recognized. Here we will take this latter approach as it is computationally by far less expensive.

In the following we will present a modified Bayesian decision rule which accounts for this feature space uncertainty.

2.2.2 Presence of Corrupted Features

We assume that the test features $\mathbf{x}_{1:T}$, which are drawn from the acoustic model $p(\mathbf{x}_{1:T}|\mathbf{W})$, are not observable. A corrupted version $\mathbf{y}_{1:T}$ is observed instead. Since $\mathbf{x}_{1:T}$ is not available, the recognition task is stated now as finding the word sequence which maximizes the posterior $p(\mathbf{W}|\mathbf{y}_{1:T})$:

$$\hat{\mathbf{W}} = \underset{\mathbf{W}}{\operatorname{argmax}}\ P(\mathbf{W}|\mathbf{y}_{1:T}). \tag{2.9}$$

As we want to make use of the acoustic model $p(\mathbf{x}_{1:T}|\mathbf{W})$ obtained from the training, the clean speech feature vector sequence is introduced as a hidden variable:

$$\begin{aligned} P(\mathbf{W}|\mathbf{y}_{1:T}) &= \int_{\mathbb{R}^{DT}} P(\mathbf{W}|\mathbf{x}_{1:T},\mathbf{y}_{1:T})p(\mathbf{x}_{1:T}|\mathbf{y}_{1:T})d\mathbf{x}_{1:T} \\ &= \int_{\mathbb{R}^{DT}} P(\mathbf{W}|\mathbf{x}_{1:T})p(\mathbf{x}_{1:T}|\mathbf{y}_{1:T})d\mathbf{x}_{1:T}. \end{aligned} \tag{2.10}$$

Here the integration is over the whole feature space $\mathbb{R}^{DT}$.

In (2.10) we have used the fact that the corrupted feature vector sequence $\mathbf{y}_{1:T}$ does not deliver more information about the word sequence than is present in the clean speech sequence $\mathbf{x}_{1:T}$. Thus, if the density is conditioned on $\mathbf{x}_{1:T}$, the conditioning on $\mathbf{y}_{1:T}$ can be removed. Next, Bayes' rule is applied to $P(\mathbf{W}|\mathbf{x}_{1:T})$ to obtain

$$P(\mathbf{W}|\mathbf{y}_{1:T}) = P(\mathbf{W}) \int_{\mathbb{R}^{DT}} \frac{p(\mathbf{x}_{1:T}|\mathbf{W})}{p(\mathbf{x}_{1:T})} p(\mathbf{x}_{1:T}|\mathbf{y}_{1:T})d\mathbf{x}_{1:T}. \tag{2.11}$$

Note that unlike in eq. (2.2), the prior $p(\mathbf{x}_{1:T})$ cannot be neglected as it is inside the integral.

Introducing the hidden state sequence as in (2.3) and using the assumptions (2.6) and (2.7), eq. (2.11) is written as

$$P(\mathbf{W}|\mathbf{y}_{1:T}) = P(\mathbf{W}) \sum_{\{q_{1:T}\}} \int_{\mathbb{R}^{DT}} \frac{p(\mathbf{x}_{1:T}|\mathbf{y}_{1:T})}{p(\mathbf{x}_{1:T})} \prod_{t=1}^{T} p(\mathbf{x}_t|q_t) P(q_t|q_{t-1}) d\mathbf{x}_{1:T}. \tag{2.12}$$

The conditional and unconditional probability densities of the feature vector sequence can be written as

$$p(\mathbf{x}_{1:T}|\mathbf{y}_{1:T}) = \prod_{t=1}^{T} p(\mathbf{x}_t|\mathbf{x}_{1:t-1}, \mathbf{y}_{1:T}), \tag{2.13}$$

$$p(\mathbf{x}_{1:T}) = \prod_{t=1}^{T} p(\mathbf{x}_t|\mathbf{x}_{1:t-1}). \tag{2.14}$$

To arrive at an expression similar to (2.8) we need to interchange the order of integration and multiplication in (2.12). This can be done if the following approximations are made:

$$p(\mathbf{x}_t|\mathbf{x}_{1:t-1}, \mathbf{y}_{1:T}) = p(\mathbf{x}_t|\mathbf{y}_{1:T}), \tag{2.15}$$

$$p(\mathbf{x}_t|\mathbf{x}_{1:t-1}) = p(\mathbf{x}_t). \tag{2.16}$$

Using this to simplify (2.13) and (2.14) and employing the result in eq. (2.12), we obtain

$$P(\mathbf{W}|\mathbf{y}_{1:T}) = P(\mathbf{W}) \sum_{\{q_{1:T}\}} \int_{\mathbb{R}^{DT}} \prod_{t=1}^{T} \frac{p(\mathbf{x}_t|q_t) p(\mathbf{x}_t|\mathbf{y}_{1:T})}{p(\mathbf{x}_t)} P(q_t|q_{t-1}) d\mathbf{x}_{1:T} \tag{2.17}$$

$$= P(\mathbf{W}) \sum_{\{q_{1:T}\}} \prod_{t=1}^{T} \int_{\mathbb{R}^{D}} \frac{p(\mathbf{x}_t|q_t) p(\mathbf{x}_t|\mathbf{y}_{1:T})}{p(\mathbf{x}_t)} d\mathbf{x}_t \cdot P(q_t|q_{t-1}) \tag{2.18}$$

With this the Bayesian decision rule in the presence of corrupted features becomes

$$\hat{\mathbf{W}} = \underset{\mathbf{W}}{\operatorname{argmax}} \left\{ P(\mathbf{W}) \sum_{\{q_{1:T}\}} \prod_{t=1}^{T} \int_{\mathbb{R}^{D}} \frac{p(\mathbf{x}_t|q_t) p(\mathbf{x}_t|\mathbf{y}_{1:T})}{p(\mathbf{x}_t)} d\mathbf{x}_t \cdot P(q_t|q_{t-1}) \right\}. \tag{2.19}$$

This decoding rule was first published in [28], however, here we have presented a different derivation. A related decoding rule has been presented in [7] in the context of simultaneous missing feature mask estimation and word recognition.

A comparison of (2.19) with (2.8) reveals that its only difference with the classical decoding rule presented in Section 2.2.1 is that the observation likelihood $p(\mathbf{x}_t|q_t)$ has to be replaced by the likelihood

$$p^{(LH)}(\mathbf{y}_{1:T}|q_t) = \int_{\mathbb{R}^{D}} \frac{p(\mathbf{x}_t|q_t) p(\mathbf{x}_t|\mathbf{y}_{1:T})}{p(\mathbf{x}_t)} d\mathbf{x}_t. \tag{2.20}$$

It follows that the same decoding algorithms can be applied to (2.8) and (2.19). The only difference is in the likelihood computation.

The key element of this new likelihood term is the clean feature posterior $p(\mathbf{x}_t|\mathbf{y}_{1:T})$, given all observed corrupted feature vectors. To reduce computational effort and to reduce latency, often only a causal posterior is employed: $p(\mathbf{x}_t|\mathbf{y}_{1:t})$, i.e., for the estimation of the clean speech feature vector at frame t only past observations and the current observation are used.

Computation of the observation probability is the most time-consuming single processing step in a speech recognizer. Certainly, the numerical evaluation of the integral (2.20) would increase the computational burden beyond the limits of practical interest. Fortunately, the integral can be solved analytically if the involved densities are approximated by Gaussians or Gaussian mixture densities.

1. In most of today's speech recognition systems the state conditioned observation probability of the uncorrupted feature is assumed to be a mixture of Gaussians:

$$p(\mathbf{x}_t|q_t = l) = \sum_{j=1}^{J} c_{l,j} \mathcal{N}(\mathbf{x}_t; \boldsymbol{\mu}_{l,j}, \boldsymbol{\Sigma}_{l,j}) \tag{2.21}$$

   where $c_{l,j}$ is the weight, $\boldsymbol{\mu}_{l,j}$ the mean vector and $\boldsymbol{\Sigma}_{l,j}$ the covariance matrix of the j-th of J mixture components of the observation probability of state $q_t = l$.
2. While the a priori probability density $p(\mathbf{x}_t)$ of the unconditional clean speech feature vector can be obtained via (2.21) by a weighted sum of Gaussian mixture densities, we suggest approximating it by a single unimodal Gaussian:

$$p(\mathbf{x}_t) = \mathcal{N}(\mathbf{x}_t; \boldsymbol{\mu}_{\mathbf{x}}, \boldsymbol{\Sigma}_{\mathbf{x}}). \tag{2.22}$$

   We tested this approximation on the Aurora 2 and Aurora 4 databases and found that it is quite valid with certain reservations concerning the log energy component.
3. Further, we assume that the feature posterior is a Gaussian:

$$p(\mathbf{x}_t|\mathbf{y}) = \mathcal{N}(\mathbf{x}_t; \boldsymbol{\mu}_{\mathbf{x}_t|\mathbf{y}}, \boldsymbol{\Sigma}_{\mathbf{x}_t|\mathbf{y}}). \tag{2.23}$$

   Here we used the notation $\mathbf{x}_t|\mathbf{y}$ to represent either causal or non-causal estimation of the parameters of the Gaussian (i.e., either $\mathbf{x}_t|\mathbf{y}_{1:t}$ or $\mathbf{x}_t|\mathbf{y}_{1:T}$). The estimation of the parameters is explained in Section 2.4 below. The use of a Gaussian posterior is the most debatable assumption. Some researchers therefore suggested using a Gaussian mixture model instead [47]. This, however, has strong impact on the computational effort for the likelihood computation in the recognizer, as the number of mixture components of $p^{(LH)}(\mathbf{y}_{1:T}|q_t)$ is equal to the product of the number of densities of the mixtures $p(\mathbf{x}_t|\mathbf{y})$ and $p(\mathbf{x}_t|q_t)$. Here, we therefore prefer to stick to a single Gaussian model for the posterior density.

With these assumptions the integral can be solved analytically [22]:

$$p^{(LH)}(\mathbf{y}|q_t = l) = \sum_{j=1}^{J} c_{l,j} \int_{\mathbb{R}^D} \frac{\mathcal{N}(\mathbf{x}_t; \boldsymbol{\mu}_{l,j}, \boldsymbol{\Sigma}_{l,j}) \cdot \mathcal{N}(\mathbf{x}_t; \boldsymbol{\mu}_{\mathbf{x}_t|\mathbf{y}}, \boldsymbol{\Sigma}_{\mathbf{x}_t|\mathbf{y}})}{\mathcal{N}(\mathbf{x}_t; \boldsymbol{\mu}_{\mathbf{x}}, \boldsymbol{\Sigma}_{\mathbf{x}})} d\mathbf{x}_t$$

$$= \sum_{j=1}^{J} \tilde{c}_{l,j,t} \mathcal{N}\left(\boldsymbol{\mu}_t^{(eq)}; \boldsymbol{\mu}_{l,j}, \boldsymbol{\Sigma}_{l,j} + \boldsymbol{\Sigma}_t^{(eq)}\right) \tag{2.24}$$

where

$$\left(\boldsymbol{\Sigma}_t^{(eq)}\right)^{-1} \boldsymbol{\mu}_t^{(eq)} = \boldsymbol{\Sigma}_{\mathbf{x}_t|\mathbf{y}}^{-1} \boldsymbol{\mu}_{\mathbf{x}_t|\mathbf{y}} - \boldsymbol{\Sigma}_{\mathbf{x}}^{-1} \boldsymbol{\mu}_{\mathbf{x}}, \tag{2.25}$$

$$\left(\boldsymbol{\Sigma}_t^{(eq)}\right)^{-1} = \boldsymbol{\Sigma}_{\mathbf{x}_t|\mathbf{y}}^{-1} - \boldsymbol{\Sigma}_{\mathbf{x}}^{-1}, \tag{2.26}$$

$$\tilde{c}_{l,j,t} = c_{l,j} \frac{\mathcal{N}\left(\mathbf{0}; \boldsymbol{\mu}_{\mathbf{x}_t|\mathbf{y}}, \boldsymbol{\Sigma}_{\mathbf{x}_t|\mathbf{y}}\right)}{\mathcal{N}(\mathbf{0}; \boldsymbol{\mu}_{\mathbf{x}}, \boldsymbol{\Sigma}_{\mathbf{x}}) \cdot \mathcal{N}\left(\mathbf{0}; \boldsymbol{\mu}_t^{(eq)}, \boldsymbol{\Sigma}_t^{(eq)}\right)}. \tag{2.27}$$

Note that the *equivalent* mean $\boldsymbol{\mu}_t^{(eq)}$ and covariance $\boldsymbol{\Sigma}_t^{(eq)}$, as well as the weight $\tilde{c}_{l,j,t}$, are time variant, which is indicated by the chosen notation.

If we assume all Gaussians to have diagonal covariance matrices, we obtain simplified expressions:

$$p^{(LH)}(\mathbf{y}|q_t = l) = \sum_{j=1}^{J} \tilde{c}_{l,j,t} \prod_{d=1}^{D} \mathcal{N}\left(\mu_{t,d}^{(eq)}; \mu_{l,j,d}, \sigma_{l,j,d}^2 + \sigma_{t,d}^{2(eq)}\right) \tag{2.28}$$

where

$$\mu_{t,d}^{(eq)} = \sigma_{t,d}^{2(eq)} \left(\frac{\mu_{(\mathbf{x}_t|\mathbf{y}),d}}{\sigma_{(\mathbf{x}_t|\mathbf{y}),d}^2} - \frac{\mu_{\mathbf{x}_d}}{\sigma_{\mathbf{x}_d}^2}\right), \tag{2.29}$$

$$\frac{1}{\sigma_{t,d}^{2(eq)}} = \frac{1}{\sigma_{(\mathbf{x}_t|\mathbf{y}),d}^2} - \frac{1}{\sigma_{\mathbf{x}_d}^2}, \tag{2.30}$$

$$\tilde{c}_{l,j,t} = c_{l,j} \frac{\prod_{d=1}^{D} \mathcal{N}\left(0; \mu_{(\mathbf{x}_t|\mathbf{y}),d}, \sigma_{(\mathbf{x}_t|\mathbf{y}),d}^2\right)}{\prod_{d=1}^{D} \mathcal{N}\left(0; \mu_{\mathbf{x}_d}, \sigma_{\mathbf{x}_d}^2\right) \mathcal{N}\left(0; \mu_{t,d}^{(eq)}, \sigma_{t,d}^{2(eq)}\right)}. \tag{2.31}$$

Here, the index $(\cdot)_d$ represents the vector component.

Eq. (2.28) states that the originally trained observation probability of the clean feature must be changed by increasing the variance by $\sigma_{t,d}^{2(eq)}$, and that it has to be evaluated at $\mu_{t,d}^{(eq)}$. With uncertainty decoding the feature enhancement thus associates a confidence with each value that it outputs. For example, the confidence is low if the speech is buried by additive noise, while it is high in high signal-to-noise ratio frames and vector components. This varying reliability can be advantageously used in the recognizer. Since a word covers several acoustic frames, the word decision is obtained by placing more emphasis on frames with high reliability and less on corrupted frames.

2.2.3 Variants of Uncertainty Decoding

In the following we show that many of the decoding rules proposed in the past for robust speech recognition can be viewed as approximations to the rule derived in the previous section.

A first approximation is to let the clean feature posterior be only dependent on the current observation, i.e., to replace $p(\mathbf{x}_t|\mathbf{y}_{1:t})$ with $p(\mathbf{x}_t|\mathbf{y}_t)$. The likelihood (2.20) then simplifies to

$$p^{(LH1)}(\mathbf{y}_t|q_t) = \int_{\mathbb{R}^D} \frac{p(\mathbf{x}_t|q_t)p(\mathbf{x}_t|\mathbf{y}_t)}{p(\mathbf{x}_t)} d\mathbf{x}_t. \tag{2.32}$$

Using this in (2.19) we obtain

$$\hat{\mathbf{W}} = \underset{\mathbf{W}}{\operatorname{argmax}} \left\{ P(\mathbf{W}) \sum_{\{q_{1:T}\}} \prod_{t=1}^{T} \int_{\mathbb{R}^D} \frac{p(\mathbf{x}_t|q_t)p(\mathbf{x}_t|\mathbf{y}_t)}{p(\mathbf{x}_t)} d\mathbf{x}_t \cdot P(q_t|q_{t-1}) \right\}. \tag{2.33}$$

This is the result given in [20, 31, 39].

In [17], [47] the denominator $p(\mathbf{x}_t)$ in (2.33) has been neglected:

$$p^{(LH2)}(\mathbf{y}_t|q_t) = \int_{\mathbb{R}^D} p(\mathbf{x}_t|q_t)p(\mathbf{x}_t|\mathbf{y}_t) d\mathbf{x}_t. \tag{2.34}$$

This approximation can be motivated on the grounds that the prior $p(\mathbf{x}_t)$ must always have a larger variance than the posterior $p(\mathbf{x}_t|\mathbf{y}_t)$. If the variance of the prior is much larger, the denominator can be considered constant for the range of values of $\mathbf{x}_t$, where the numerator assumes values larger than zero. However, this argument no longer holds in the presence of strong distortions, for example, if the signal-to-noise ratio (SNR) is very low. Then $p(\mathbf{x}_t|\mathbf{y}_t) = p(\mathbf{x}_t)$ and the use of the approximate decision rule, which neglects the prior, results in artifacts and poor performance [37].

Another popular form is obtained if the right-hand side of (2.32) is multiplied by $p(\mathbf{y}_t)$. Then Bayes' rule for conditional probabilities can be applied to obtain

$$p^{(LH3)}(\mathbf{y}_t|q_t) = \int_{\mathbb{R}^D} \frac{p(\mathbf{x}_t|q_t)p(\mathbf{x}_t|\mathbf{y}_t)}{p(\mathbf{x}_t)} p(\mathbf{y}_t) d\mathbf{x}_t = \int_{\mathbb{R}^D} p(\mathbf{x}_t|q_t)p(\mathbf{y}_t|\mathbf{x}_t) d\mathbf{x}_t \tag{2.35}$$

Using (2.35) in (2.19) we obtain

$$\hat{\mathbf{W}} = \underset{\mathbf{W}}{\operatorname{argmax}} \left\{ P(\mathbf{W}) \sum_{\{q_{1:T}\}} \prod_{t=1}^{T} \int_{\mathbb{R}^D} p(\mathbf{y}_t|\mathbf{x}_t)p(\mathbf{x}_t|q_t) d\mathbf{x}_t \cdot P(q_t|q_{t-1}) \right\}, \tag{2.36}$$

which results in exactly the same decoding method as if (2.32) were used. The formulation (2.35) has for example been used in [36] in the context of environmental noise compensation.

In [37] it has been argued that this decision rule has a fundamental issue at very low SNR, where $p(\mathbf{y}_t|\mathbf{x}_t) = p(\mathbf{y}_t)$. This term, now being independent of $\mathbf{x}_t$, can be taken outside the integral. As a consequence the integral (2.36) evaluates to 1, which means that the frame is ignored in terms of acoustic discrimination. But this is actually the correct behavior! If the current frame is completely distorted it is eliminated from the decoding decision, just as in marginalization in the context of missing feature speech recognition; see the following Section 2.2.4. If, however, the original decoding rule (2.19) is used, which employs the likelihood according to (2.20), the issue is resolved: if successive feature vectors are correlated, then $p(\mathbf{x}_t|\mathbf{y}_{1:T}) \neq p(\mathbf{x}_t)$, even if the t-th observation $\mathbf{y}_t$ is completely corrupted. The neighboring observations are now used to estimate $\mathbf{x}_t$. This provides additional information which would otherwise not be exploited by the recognizer, since the HMM back-end employs the conditional independence assumption, where neighboring frames are deemed independent of the current frame, given the state index. In the context of network speech recognition, where the communication link between front-end and back-end is assumed to be characterized by packet losses or bit errors, it has indeed been shown that the decoding rule (2.19) is clearly superior to (2.33) [28].

In [37] it has further been proposed to use an acoustic model-based approach instead, where a mixture of conditional densities can be used with its components covering different regions of the acoustic space. While a heavily distorted frame may be ignored for some components, for others it may still allow for some degree of discrimination.

Finally, if a point estimate is used, i.e., $p(\mathbf{x}_t|\mathbf{y}) = \delta(\mathbf{x}_t - \hat{\mathbf{x}}_t(\mathbf{y}))$, the likelihood (2.20) reduces to

$$p^{(LH4)}(\mathbf{y}|q_t) = \frac{p(\hat{\mathbf{x}}_t(\mathbf{y})|q_t)}{p(\hat{\mathbf{x}}_t(\mathbf{y}))}. \tag{2.37}$$

As the denominator is independent of the word sequence under investigation, the decision rule reduces to (2.8) with $\mathbf{x}_t$ replaced by its estimate $\hat{\mathbf{x}}_t(\mathbf{y})$. This is no longer an uncertainty decoding rule but the well-known *plug-in* approach, where the estimate is taken as if it were the true value of the clean feature vector.

2.2.4 Missing-Feature Approaches

In missing-feature approaches, also termed missing data approaches, to speech recognition, one attempts to determine which cells of a spectrogram-like time-frequency display of speech are unreliable (termed *missing*) because of degradation due to noise or due to other types of interference. These cells are either ignored in subsequent processing or they are filled in by estimates of their putative values [7, 10, 42].

Let the indices of the components of the feature vector $\mathbf{y}$ be segregated into two mutually exclusive sets $\mathscr{U}$ and $\mathscr{R}$. Let $\mathbf{y}_u$ be a vector constructed from all components of $\mathbf{y}$, whose indices lie in $\mathscr{U}$, and $\mathbf{y}_r$, of indices lying in $\mathscr{R}$. The identification of the index sets $\mathscr{U}$ and $\mathscr{R}$, which correspond to the unreliable (missing) and reliable feature components, for each frame is termed mask estimation, and is not at all an easy task. Many approaches have been proposed to solve this problem [10, 42]. The discussion of mask estimation methods, however, is beyond the scope of this chapter.

There are two major approaches to deal with missing features.

- *Feature vector imputation*: For some feature vector component $d \in \mathscr{U}$, an estimate $\hat{\mathbf{x}}_{t,d}(\mathbf{y})$ is determined from the observed features $\mathbf{y}$. This estimate replaces the missing $\mathbf{y}_{t,d}$ in the feature vector to be decoded. Using our earlier formulation employing the feature posterior $p(\mathbf{x}_t|\mathbf{y})$, imputation means that the following posterior is assumed:

$$p(\mathbf{x}_{t,d}|\mathbf{y}) = \begin{cases} \delta(\mathbf{x}_{t,d} - \mathbf{y}_{t,d}) & d \in \mathscr{R} \\ \delta(\mathbf{x}_{t,d} - \hat{\mathbf{x}}_{t,d}(\mathbf{y})) & d \in \mathscr{U} \end{cases}. \tag{2.38}$$

- *Marginalisation*: Here, the classifier is modified to perform recognition using $\mathbf{y}_r$ alone. In other words, the feature posterior is approximated by

$$p(\mathbf{x}_{t,d}|\mathbf{y}) = \begin{cases} \delta(\mathbf{x}_{t,d} - \mathbf{y}_{t,d}) & d \in \mathscr{R} \\ p(\mathbf{x}_{t,d}) & d \in \mathscr{U} \end{cases}. \tag{2.39}$$

Imputation and marginalisation can thus be thought to provide a coarse approximation of the clean speech feature posterior density. In imputation the feature posterior is replaced by a point estimate. Even if the point estimate is the optimal *minimum mean squared error* (MMSE) estimate it does not necessarily result in optimal performance, as the limited reliability of the estimate is not accounted for. In marginalisation the feature is completely eliminated. However, we have seen in the last section that the correlation between successive frames or across the feature vector can be exploited to avoid the marginalisation and thus elimination of the feature from the recognition.

A binary decision when a feature is deemed reliable or missing incorporates a threshold. For the choice of the threshold one has to strike a balance between accepting the feature as reliable (although it might be somewhat corrupted) or completely dismissing the feature (although the feature might contain some information about the underlying uncorrupted feature). This binary decision is problematic and researchers have therefore proposed using so-called *soft masks*, where each observation $\mathbf{y}_{t,d}$ is assigned a probability $\gamma_{t,d} = P(\text{reliable}|\mathbf{y}_{t,d})$ that it is reliable and dominated by speech rather than by noise [8]. However, heuristics are again required to determine the reliability factor $\gamma_{t,d}$. It can for example be chosen as a function of the instantaneous SNR.

With a soft mask, the imputation technique results in a clean speech posterior as follows:

$$p(\mathbf{x}_{t,d}|\mathbf{y}) = \gamma_{t,d}\delta(\mathbf{x}_{t,d} - \mathbf{y}_{t,d}) + (1-\gamma_{t,d})\delta(\mathbf{x}_{t,d} - \hat{\mathbf{x}}_{t,d}(\mathbf{y})). \tag{2.40}$$

Note that this is again a very specific form of the clean speech posterior!

Various sophistications and improvements have been proposed to this basic missing-feature approach. They include improved marginalization by arguing that in the log mel spectral domain the clean speech feature can never be larger than the observed noisy one. This, however, is also problematic, as this argument disregards the influence of the phase factor between speech and noise, which may cause $\mathbf{x}_t$ to be larger than $\mathbf{y}_t$. For a discussion of the phase factor, see another contribution to this book [34]. Another variant of the missing-feature approach, termed *modified imputation*, is to estimate the imputed value to be a weighted average of the observed feature and an a priori value, such as the mean of the observation probability $p(\mathbf{x}_t|q_t)$, where the weights are obtained from reliability information [30].

Finally, to mention one of the most sophisticated models, a mixture of a Gaussian and a uniform density has been proposed:

$$p(\mathbf{x}_{t,d}|\mathbf{y}) = w\mathcal{N}(\mathbf{x}_{t,d}; \mu_d, \sigma_d^2) + (1-w)\mathscr{F}(\mathbf{x}_{t,d}; a, b), \tag{2.41}$$

where

$$\mathscr{F}(\mathbf{x}_{t,d}; a, b) = \begin{cases} \frac{1}{b-a} & a \le \mathbf{x}_{t,d} < b \\ 0 & \text{else} \end{cases} \tag{2.42}$$

denotes the uniform density and w is a weighting factor. For a discussion of this and related models, see Chapter 11 by Kühne et al. in this book.

2.3 Feature Space vs. Acoustic Model-Based Approaches

As discussed earlier, a mismatch between the acoustic model $p(\mathbf{x}|\mathbf{W})$ and the statistics of the test data $\mathbf{y}$ quickly leads to a degradation of the recognition performance. To increase robustness, i.e., to reduce the mismatch, either the test data have to be modified (so-called feature space or front-end approaches) or the acoustic model has to be adjusted to the new statistics of the test data (so-called acoustic model-based or back-end approaches).

There are pros and cons for either approach. Acoustic model-based methods are in general more powerful as they allow for different compensation parameters for different regions of the acoustic space. Compensation parameters can be chosen differently for each HMM state, or regression classes of HMM states can share the same set of parameters. This degree of freedom seems to be particularly important if the mismatch is due to phonetic differences which are not uniform across the acoustic-phonetic space, such as different speaker characteristics, speaking styles and accents.

Front-end methods, on the other hand, typically carry out a compensation irrespective of the underlying phonetic content, as the phoneme hypothesis is only available in the back-end. To allow for multiple mappings also for feature space methods, the feature space can be subdivided by a data-driven method such as vector quantization and training of a Gaussian mixture model. In this class fall methods such as Codeword-Dependent Cepstral Normalization (CDCN) [1], Probabilistic Optimal Filtering (POF) [40], SPLICE [14] and Joint Uncertainty Decoding (JUD) [36], where piecewise linear mappings between clean features and features distorted by additive noise are used. The mappings are either learned from stereo training data [2, 40], where the same utterance is available in both clean and corrupted form, or by using a model of the distortion which allows us to compute the statistics of the distorted features from clean data according to the model.

Due to their greater flexibility, acoustic model-based approaches to enhancing robustness appear to be superior in terms of recognition performance; see for example the excellent results on the Aurora 2 database reported in [35]. The price to pay, however, is computational cost. The modification of the acoustic model, which may comprise several hundred thousands of parameters, is computationally so expensive that model-based compensation schemes have not yet made large inroads in very large vocabulary tasks. Another potential disadvantage of acoustic model-based approaches is that they have to be compatible with the predominant acoustic modeling paradigm employing hidden Markov models, which may pose restrictions on the choice of statistical models. For example, it is very difficult to exploit direct correlation of successive feature vectors in the back-end in order to reconstruct a corrupted frame, as HMMs employ the conditional independence assumption. In the front-end this can be accomplished more easily, as we will show in the following sections.

Note that various compensation techniques can be employed either in feature or in model space, such as Maximum Likelihood Linear Regression (MLLR) or the Vector Taylor Series (VTS) approach to noise robust speech recognition. A good overview of feature and model space approaches to noise robust speech recognition can be found in [19]. See also Chapter 5 on model-based approaches to handling uncertainty in this book.

In a sense, Uncertainty Decoding as discussed in Section 2.2 can be viewed as reconciling front-end and back-end methods. While the estimation of the compensation parameters is carried out in the feature space, the compensation is actually done in the model space: the parameters are used to modify the likelihood computation, which is part of the back-end processing.

2.4 Posterior Estimation

In this contribution it has been argued that the uncertainty decoding rule (2.19) which employs the clean speech feature posterior density $p(\mathbf{x}_t|\mathbf{y})$ is the most appropriate representation. In Section 2.2.3 we have demonstrated this by noting that a variety of alternatives proposed in the literature can actually be viewed as approximations to this classification rule.

The use of the clean speech feature posterior density $p(\mathbf{x}_t|\mathbf{y})$ is also beneficial for other reasons. First, knowledge of the posterior density enables one to compute an optimal estimate of $\mathbf{x}_t$ with respect to any criterion. For example the MMSE estimate of the clean speech feature vector is the conditional mean $\hat{\mathbf{x}}_t^{(\mathrm{MMSE})} = \int \mathbf{x}_t p(\mathbf{x}_t|\mathbf{y}) d\mathbf{x}_t$. Further, a measure of accuracy of the estimate can be obtained from the posterior. In the case of jointly Gaussian random variables it is well known that the covariance of the posterior $\mathbf{\Sigma}_{\mathbf{x}_t|\mathbf{y}}$ is identical with the covariance matrix of the estimation error $\mathbf{e}_t = (\mathbf{x}_t - \hat{\mathbf{x}}_t^{(\mathrm{MMSE})})$. Thus the feature uncertainty is obtained as a by-product of the estimation of the clean speech feature posterior and, in principle, no heuristics are required to assess the reliability of the estimate. Finally, a major advantage of employing $p(\mathbf{x}_t|\mathbf{y})$ is that considerable literature in the field of optimal filtering is available about how to estimate the posterior from distorted observations; see, e.g., [6, 43]. An excellent overview of the benefits of conditional Bayesian estimation for various speech enhancement tasks can be found in [49].

Conceptually, the posterior can be estimated recursively via the following equations, where we have restricted ourselves to causal processing, i.e., rather than computing $p(\mathbf{x}_t|\mathbf{y}_{1:T})$ we compute $p(\mathbf{x}_t|\mathbf{y}_{1:t})$:

$$p(\mathbf{x}_t|\mathbf{y}_{1:t-1}) = \int_{\mathbb{R}^D} p(\mathbf{x}_t|\mathbf{x}_{t-1}, \mathbf{y}_{1:t-1}) p(\mathbf{x}_{t-1}|\mathbf{y}_{1:t-1}) d\mathbf{x}_{t-1}, \tag{2.43}$$

where

$$\begin{aligned} p(\mathbf{x}_{t-1}|\mathbf{y}_{1:t-1}) &= \frac{p(\mathbf{y}_{1:t-1}|\mathbf{x}_{t-1}) p(\mathbf{x}_{t-1})}{p(\mathbf{y}_{1:t-1})} \\ &= \frac{p(\mathbf{y}_{t-1}|\mathbf{x}_{t-1}, \mathbf{y}_{1:t-2}) p(\mathbf{y}_{1:t-2}|\mathbf{x}_{t-1}) p(\mathbf{x}_{t-1})}{p(\mathbf{y}_{1:t-1})} \\ &= \frac{p(\mathbf{y}_{t-1}|\mathbf{x}_{t-1}, \mathbf{y}_{1:t-2}) p(\mathbf{x}_{t-1}|\mathbf{y}_{1:t-2}) p(\mathbf{y}_{1:t-2})}{p(\mathbf{y}_{1:t-1})} \\ &= \frac{p(\mathbf{y}_{t-1}|\mathbf{x}_{t-1}, \mathbf{y}_{1:t-2}) p(\mathbf{x}_{t-1}|\mathbf{y}_{1:t-2})}{p(\mathbf{y}_{t-1}|\mathbf{y}_{1:t-2})} \\ &= \frac{p(\mathbf{y}_{t-1}|\mathbf{x}_{t-1}, \mathbf{y}_{1:t-2}) p(\mathbf{x}_{t-1}|\mathbf{y}_{1:t-2})}{\int_{\mathbb{R}^D} p(\mathbf{y}_{t-1}|\mathbf{x}_{t-1}, \mathbf{y}_{1:t-2}) p(\mathbf{x}_{t-1}|\mathbf{y}_{1:t-2}) d\mathbf{x}_{t-1}} \\ &\propto p(\mathbf{y}_{t-1}|\mathbf{x}_{t-1}, \mathbf{y}_{1:t-2}) p(\mathbf{x}_{t-1}|\mathbf{y}_{1:t-2}). \end{aligned} \tag{2.44}$$

In the following we will assume that the clean speech feature vector sequence is a first-order Markov process, such that (2.43) can be simplified to

$$p(\mathbf{x}_t|\mathbf{y}_{1:t-1}) = \int_{\mathbb{R}^D} p(\mathbf{x}_t|\mathbf{x}_{t-1}) p(\mathbf{x}_{t-1}|\mathbf{y}_{1:t-1}) d\mathbf{x}_{t-1}. \tag{2.45}$$

Often, conditionally independent and identically distributed (i.i.d.) observations are assumed, i.e., $p(\mathbf{y}_{t-1}|\mathbf{x}_{t-1},\mathbf{y}_{1:t-2}) = p(\mathbf{y}_{t-1}|\mathbf{x}_{t-1})$. Then (2.44) simplifies to

$$p(\mathbf{x}_{t-1}|\mathbf{y}_{1:t-1}) \propto p(\mathbf{y}_{t-1}|\mathbf{x}_{t-1})p(\mathbf{x}_{t-1}|\mathbf{y}_{1:t-2}). \tag{2.46}$$

Thus, for the determination of the posterior, the following issues have to be addressed:

- An appropriate dynamic model of the clean speech feature trajectory $p(\mathbf{x}_t|\mathbf{x}_{t-1})$ to be used in (2.45) has to be established.
- An observation model $p(\mathbf{y}_t|\mathbf{x}_t,\mathbf{y}_{1:t-1})$ to be used in (2.44) – or the simplified version $p(\mathbf{y}_t|\mathbf{x}_t)$ employed in (2.46) – has to be derived for the type of corruption at hand.
- While (2.43) and (2.44) provide the principal way of recursively determining the clean speech feature posterior, in practice it may turn out that the posterior cannot be determined analytically. The implementation may require the storage of the entire (non-Gaussian) pdf which is, in general terms, equivalent to an infinite dimensional vector [43]. Therefore a computationally tractable inference algorithm has to be chosen which computes (an approximation of) $p(\mathbf{x}_t|\mathbf{y}_{1:t})$.

Figure 2.1 illustrates the processing stages. Note that the observation model often employs parameters which have to be estimated from the signal to be recognized. For example, in noisy speech recognition an estimate of the noise has to be determined from the incoming data, and in the case of reverberant speech recognition, parameters of the reverberation model, such as the reverberation time T_{60}, are required.[2]

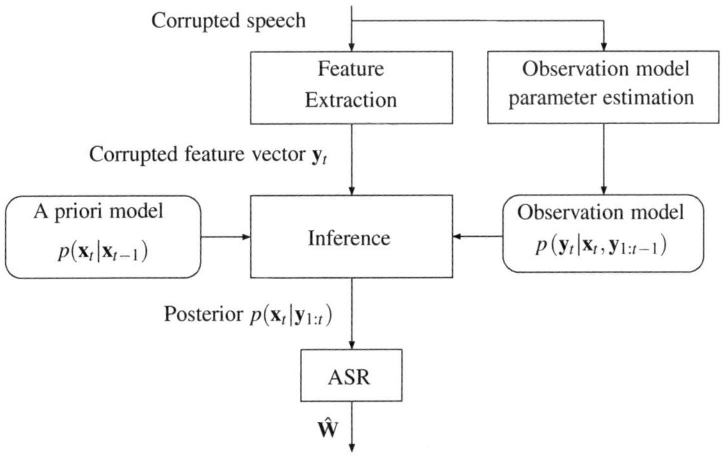

Fig. 2.1: Block diagram of robust speech recognition based on Bayesian inference

[2] This is the time interval, after which the energy in the tail of the impulse response is below 60 dB of the total energy of the impulse response.

In the following we will give an overview of the aforementioned system components.

2.4.1 A Priori Model of Clean Speech

The a priori model $p(\mathbf{x}_t|\mathbf{x}_{t-1})$ captures the statistical properties of clean speech with an approach typically far less complex and having far fewer parameters than the acoustic model $p(\mathbf{x}|q)$ used in the back-end.

The most simple a priori model is a model which assumes the feature vectors $\mathbf{x}_t$, $t = 1,2,\ldots$, to be i.i.d. random variables, such as the Gaussian mixture model (GMM), which is employed in [14, 37, 46] in the context of noise compensation:

$$p(\mathbf{x}) = \sum_{m=1}^{M} w_m p(\mathbf{x}|m), \tag{2.47}$$

where m is the component index.

A more sophisticated a priori model captures correlations between successive feature vectors. In [18] switching linear dynamic models (SLDM) have been proposed for this purpose. A linear dynamic model $\mathbf{x}_t = \mathbf{A}\mathbf{x}_{t-1} + \mathbf{v}_t$ is often used in tracking and control systems. A single dynamic model, however, is too coarse to describe the complex dynamics of the trajectory of a speech feature vector. It is better modeled by an interaction or superposition of multiple such linear dynamic models:

$$\begin{aligned} p(\mathbf{x}_t|\mathbf{x}_{t-1}) &= \sum_{m=1}^{M} p(\mathbf{x}_t|\mathbf{x}_{t-1},\gamma_t = m)P(\gamma_t = m|\mathbf{x}_{t-1}) \\ &\approx \sum_{m=1}^{M} p(\mathbf{x}_t|\mathbf{x}_{t-1},\gamma_t = m) \sum_{n=1}^{M} P(\gamma_t = m|\gamma_{t-1} = n)P(\gamma_{t-1} = n|\mathbf{x}_{t-1}). \end{aligned} \tag{2.48}$$

Here we assumed M interacting stochastic linear prediction models, where γ_t is the hidden state variable which indicates the active model at time t. The use of a zero-th order process, i.e., $P(\gamma_t = m|\gamma_{t-1} = n) = P(\gamma_t = m)$ has also been proposed [50]. The state-conditional processes are assumed to be Gaussian:

$$p(\mathbf{x}_t|\mathbf{x}_{t-1},\gamma_t = m) = \mathcal{N}(\mathbf{x}_t;\mathbf{A}_m\mathbf{x}_{t-1} + \mathbf{b}_m, \mathbf{V}_m); \quad m = 1,\ldots,M \tag{2.49}$$

where $\mathbf{A}_m$, $\mathbf{b}_m$ and $\mathbf{V}_m$ represent the state transition matrix, a bias vector and the prediction error covariance matrix of the m-th model, respectively. The training of the models can be carried out on clean speech training data using standard Expectation-Maximization techniques [18].

We have observed that a first-order a priori model $p(\mathbf{x}_t|\mathbf{x}_{t-1})$ is particularly beneficial if the corruptions are well concentrated in time and can be clearly identified. An example for this is a lossy communication link between the speech capturing device and the speech recognition engine. In the case of a packet loss a complete feature vector is missing, while the received feature vector at the recognizer input is without errors if the packet is received. Then the lost frame can be quite well reconstructed using the correlations between feature vectors as represented by a first-order a priori model [28]. We have also observed large advantages over the zero-th order a priori model (2.47) in the case of feature enhancement for reverberant speech recognition. Here, a feature vector affects many subsequent vectors due to the reverberation. The first-order a priori model then helps discern the correlation of successive feature vectors due to the temporal correlation of speech from that introduced by the reverberation [32]. In the case of noisy speech recognition many authors claim excellent results using a GMM as an a priori model [14, 47] while others advocate for a first-order model [12].

Note that the use of the a priori model has significant impact on the complexity of the inference algorithm in determining the clean speech feature posterior, as will be explained later on.

2.4.2 Observation Model

The observation model relates the clean speech feature vector $\mathbf{x}_t$ with the observed corrupted vector $\mathbf{y}_t$ via a probabilistic model $p(\mathbf{y}_t|\mathbf{x}_t, \mathbf{y}_{1:t-1})$ or $p(\mathbf{y}_t|\mathbf{x}_t)$. Two basic approaches can be discerned. In the first approach the density is obtained by deriving an analytic expression of the impact of the distortion on the computed feature vector. Alternatively, the relationship between $\mathbf{x}_t$ and $\mathbf{y}_t$ can be learned from stereo data. While this latter approach does not require a mathematical analysis of the effect of the distortion, there is the need for stereo data, which often makes this approach unattractive.

There is also a third, hybrid approach, where a model of the corrupting process is used to artificially generate the corrupted data from the clean data and thus to artificially generate stereo training data which are then used to learn a mapping between $\mathbf{x}_t$ and $\mathbf{y}_t$.

While the a priori model and the inference method are quite generic and may be applied to different scenarios, the observation model has to be developed for the specific type of corruption under investigation.

For noise robust speech recognition the most common model is given by [1],

$$\mathbf{y}_t = \mathbf{C}\log\left(\exp(\mathbf{C}^{-1}\mathbf{x}_t) + \exp(\mathbf{C}^{-1}\mathbf{n}_t)\right), \tag{2.50}$$

where we assumed a *Mel Frequency Cepstral Coefficient* (MFCC) feature vector with $\mathbf{C}$ denoting the Discrete Cosine Transform and $\mathbf{C}^{-1}$ its (pseudo)inverse. The functions exp() and log() are to be understood to operate element-wise on their vector argument. This model has been subsequently extended to include the phase factor between the speech and noise spectra [16, 23, 47]. As the model is nonlinear the probability density $p(\mathbf{y}_t|\mathbf{x}_t)$ is non-Gaussian. Since a Gaussian, however, is preferred for mathematical tractability and computational convenience, various approximate Gaussian or Gaussian mixture models have been derived; see [25] for an overview of different approaches.

Observation models have also been derived for speech transmitted over a communication link to a remote speech recognition server, where the link is characterized by bit errors or packet losses [38, 41, 48].

Interestingly, not many analytical observation models have been derived for speech corrupted by reverberation [32, 45]. The model in [32] relates the currently observed reverberant log mel spectral feature vector to the current and past clean speech feature vectors according to

$$p(\mathbf{y}_t|\mathbf{x}_t,\mathbf{y}_{1:t-1}) \approx \mathcal{N}\left(f(\mathbf{x}_{t-L_H:t},\hat{\mathbf{h}}) + \boldsymbol{\mu}_w, \boldsymbol{\Sigma}_w\right) \tag{2.51}$$

where $f(\cdot,\cdot)$ is a nonlinear function of the current and the L_H most recent clean speech feature vectors and $\hat{\mathbf{h}}$ is an estimate of a log mel spectral representation of the room impulse response. The parameters $\boldsymbol{\mu}_w$ and $\boldsymbol{\Sigma}_w$ of the Gaussian have been determined on artificially generated stereo data using the image method [3].

Finally one should mention that the observation model may contain unknown parameters, for example, the noise term $\mathbf{n}_t$ in (2.50) or the room impulse response model $\hat{\mathbf{h}}$ in (2.51). Unlike in the estimation of the parameters of the a priori model, the parameters of the observation model usually cannot be determined off-line up front, but have to be estimated on the very same data on which the inference has to be conducted. The reason is that the parameters are specific to the test data to be recognized.

2.4.3 Inference

The term *inference* denotes the computation of the clean feature posterior according to Equations (2.43) and (2.44). If a zero-th order a priori model is employed, the posterior is obtained by applying Bayes' rule for conditional probabilities and no recursive estimation is required. However, as the observation model is usually nonlinear, the solution is still far from being trivial and various approximations have to be made [15, 21].

If a first-order a priori model is employed and if both the a priori model and the observation model are linear and Gaussian, the function recursions (2.45) and (2.46) become the Kalman filter.

This no longer holds if switching linear dynamic models are used as an a priori model. Actually, the computation of the posterior results in a complexity which increases exponentially with time. This can be easily seen if the posterior density is written as the weighted sum of posterior densities, given individual state sequence histories:

$$p(\mathbf{x}_t|\mathbf{y}_{1:t}) = \sum_{\ell=1}^{M^t} p(\mathbf{x}_t|\mathbf{y}_{1:t},\Gamma_t^{(\ell)})P(\Gamma_t^{(\ell)}|\mathbf{y}_{1:t}). \tag{2.52}$$

Here, $\Gamma_t^{(\ell)}$ denotes the ℓ-th sequence of state variables (state histories) starting at time frame 1 and ending at time t:

$$\Gamma_t^{(\ell)} = \{\gamma_1^{(\ell)}, \gamma_2^{(\ell)}, \ldots, \gamma_t^{(\ell)}\}; \quad \ell = 1, \ldots, M^t. \tag{2.53}$$

Since $\gamma_\tau^{(\ell)} \in \{1, \ldots, M\}$, $1 \leq \tau \leq t$, there are M^t such sequences of length t. In (2.53), $\gamma_\tau^{(\ell)}$ represents the particular value the state variable takes for the ℓ-th state sequence in the τ-th time interval. Obviously the number of terms in the sum, and thus the computational complexity, increases exponentially with time t.

The probability of a particular state sequence, given measurements $\mathbf{y}_{1:t}$, can be computed recursively as follows:

$$\begin{aligned} P(\Gamma_t^{(\ell)}|\mathbf{y}_{1:t}) &= P(\Gamma_t^{(\ell)}|\mathbf{y}_{1:t-1}, \mathbf{y}_t) \\ &\propto p(\mathbf{y}_t|\Gamma_t^{(\ell)}, \mathbf{y}_{1:t-1}) P(\Gamma_t^{(\ell)}|\mathbf{y}_{1:t-1}) \\ &= p(\mathbf{y}_t|\Gamma_t^{(\ell)}, \mathbf{y}_{1:t-1}) P(\gamma_t^{(\ell)}, \Gamma_{t-1}^{(\ell)}|\mathbf{y}_{1:t-1}) \\ &= p(\mathbf{y}_t|\Gamma_t^{(\ell)}, \mathbf{y}_{1:t-1}) P(\gamma_t^{(\ell)}|\Gamma_{t-1}^{(\ell)}, \mathbf{y}_{1:t-1}) P(\Gamma_{t-1}^{(\ell)}|\mathbf{y}_{1:t-1}). \end{aligned} \tag{2.54}$$

The sequence of state variables can be modeled either to be i.i.d. ($P(\gamma_t^{(\ell)}|\Gamma_{t-1}^{(\ell)}, \mathbf{y}_{1:t-1}) = P(\gamma_t^{(\ell)})$) or, what is more common, as a first-order Markov chain:

$$P(\gamma_t^{(\ell)}|\Gamma_{t-1}^{(\ell)}, \mathbf{y}_{1:t-1}) = P(\gamma_t^{(\ell)}|\gamma_{t-1}^{(\ell)}). \tag{2.55}$$

As an exact inference for arbitrarily long observation sequences is not possible due to the exponentially increasing complexity, suboptimal algorithms have been developed [6]. These algorithms achieve constant complexity in time by limiting the number of state sequences to a finite value, e.g., M or M^2.

The state-sequence conditional posterior $p(\mathbf{x}_t|\mathbf{y}_{1:t}, \Gamma_t^{(\ell)})$ in (2.52) can be obtained from a Kalman filter. However, note that the observation model is nonlinear; see, for example, (2.50) and (2.51). A common approach is to use an extended Kalman filter (EKF), where the nonlinearity $\mathbf{y}_t = f(\mathbf{x}_t)$ is developed around the current estimate $\hat{\mathbf{x}}_{t|1:t-1}^{(\ell)}$ by a Taylor series up to the linear term to obtain the predicted observation $\hat{\mathbf{y}}_{t|1:t-1}^{(\ell)} = E[\mathbf{y}_t|\mathbf{x}_{1:t-1}, \Gamma_t^{(\ell)}]$. Alternatively, an unscented Kalman filter (UKF) can be employed [29]. The UKF approximates the posterior $p(\mathbf{x}_t|\mathbf{y}_{1:t}, \Gamma_t^{(\ell)})$ by a Gaussian density, which is represented by a set of deterministically chosen sample points. These sample points completely capture the true mean and covariance of the Gaussian density. When propagated through the nonlinear function, the mean and the covariance of the resulting Gaussian are identical with mean and covariance of the true output density. Moments of higher order, however, are not identical. For noise robust speech recognition it has been reported that superior recognition performance was obtained with the UKF as compared to the EKF [13].

In the case of severe nonlinearities it can be beneficial to iterate the EKF or UKF. For example, in an EKF the observation model is linearized by Taylor series expansion and truncation after the linear term and a first estimate of the state vector $\hat{\mathbf{x}}_{t|1:t}^{(\ell)}$ is obtained. This estimate then serves as the expansion point for a second iteration. Similarly, the UKF can also be iterated. Iteration indeed improves performance for the observation models used in noisy speech recognition [21, 24].

An alternative to the use of an EKF or UKF is the use of a particle filter. It performs sequential Monte Carlo estimation based on a point mass (or "particle") representation of probability densities. They have also been used with some success for robust speech recognition [44, 51]. However, a downside of this approach is that due to the high dimension of the feature vector the number of particles required to achieve a good representation of the densities is large, rendering this approach computationally inefficient.

2.5 From Theory to Practice

In this contribution we have tried to place speech recognition in the presence of corrupted features in a solid statistical framework. A classification rule has been derived, where the clean feature posterior, given the corrupted feature vector sequence, is a key element. It has then been outlined how this posterior can be estimated, making use of results from optimal filtering theory.

However, for the determination of $p(\mathbf{x}_t|\mathbf{y})$, approximations had to be introduced at almost all processing stages to arrive at a computationally tractable solution:

- The dynamic model describing the a priori knowledge of speech feature trajectories is rather coarse, even if switching linear dynamic models are used. Further, the SLDM model parameters are not known and have to be estimated from training data, where the estimation itself may cause its own problems. We carried out Maximum Likelihood estimation employing the EM algorithm. However, ML may not be the most suitable criterion. For example, we observed that the likelihood on the training data and even the likelihood on the test data was only a poor predictor of the speech recognition performance. Furthermore, the likelihood function is certainly nonlinear and multi-modal, making the initialization of the models a critical issue. We have studied various initialization methods [33], including a novel scheme with ideas borrowed from the k-means++ clustering algorithm [5]. For reverberant speech recognition on the Aurora 5 database using the approach outlined in this chapter and described in more detail in Chapter 10, we found that careful selection of initial values of the SLDM parameters can improve performance. However, the random selection of seed points resulted in a variation of the word error rate of as much as $\pm 10\%$ around the mean value!
- The observation models are often extremely nonlinear; see, for example, the observation models for noisy speech recognition and reverberant speech recog-

nition briefly mentioned in Section 2.4.2 and discussed extensively in Chapters 8 and 10 of this book. The models have to be linearized, either analytically via a Taylor series approximation, or statistically using the unscented transform. An indication of the severeness of the nonlinearity is that the iteration of the Taylor series expansion leads to improved recognition performance. A specific complication in the case of noisy speech recognition is the unknown phase factor between the clean speech and the noise spectra. In [11] a case study was conducted, where the observation likelihood was approximated by a Monte Carlo method. While being too computationally demanding for practical use, it demonstrated that significant improvements in the word error rate can be obtained if the observation model is further refined.

- Optimal inference is not feasible due to exponentially increasing complexity and due to the nonlinear observation model. Thus approximation algorithms have to be employed. In [32] three algorithms for inference in the presence of an SLDM a priori model have been compared on a reverberant speech recognition task: The *Generalized Pseudo Bayes Algorithm 1* (GPB1) and *2* (GPB2), and the *Interacting Multiple Model Algorithm* (IMM) [6]. The three approaches differ in the way they restrict the growth of the number of model histories with time: In GPB1 the mixture model is collapsed to a single Gaussian after each iteration, which serves as the starting point for the next iteration. The IMM approach differs from the GPB1 in the prediction step, where for each filter a separate initial mean and covariance matrix are computed. Finally, for the GPB2 approach all possible combinations of two successive states γ_t and γ_{t-1} are considered requiring M^2 Kalman filtering operations per frame, while the operations required by GPB1 and IMM are in the order of M. In [32] it was observed that the word error rate was reduced by 2% to 15% when going from GPB1 to the more complex GPB2.

As a result of all these approximations, the computed posterior $p(\mathbf{x}_t|\mathbf{y})$ is, after all, only an estimate of the true posterior. Furthermore, the quality of the estimate is unknown. One can, however, assume that the estimate of the posterior mean $\hat{\mathbf{x}}_t^{(\mathrm{MMSE})} = E[\mathbf{x}_t|\mathbf{y}]$ is probably more exact than the estimate of the covariance $\boldsymbol{\Sigma}_{\mathbf{x}_t|\mathbf{y}}$. This may be safely concluded when considering the variance of a mean and a variance estimator for Gaussian distributed samples: while the estimation error variance of the mean estimator is proportional to the square of the (true) variance, the variance of the variance estimator is proportional the fourth power of the true variance. It is therefore safer to use the MMSE estimate alone than to carry out full uncertainty decoding, which employs estimates of covariances. Indeed, problems with estimated covariances in the context of uncertainty decoding have been reported in the literature and heuristics have been proposed to overcome them [20, 37].

Thus, despite the elegant probabilistic framework, there is no guarantee that the estimation of the covariance of the estimation error $\mathbf{e} = (\mathbf{x}_t - \hat{\mathbf{x}}_t)$ via the feature posterior $p(\mathbf{x}_t|\mathbf{y})$ works any better than sensible heuristics. Future research should therefore be targeted at improving the estimation of the feature posterior, e.g., by more sophisticated a priori or observation models. Another important research topic is developing methods to assess the quality of the estimated posterior, in particular the ability of the estimated covariance term to describe the true estimation error covariance.

2.6 Conclusions

Bayesian classification and estimation theory provides an elegant framework for deriving algorithms for robust speech recognition. In this contribution we have shown how front-end techniques for robust ASR can take advantage of a priori knowledge about speech features or feature trajectories to guide the inference towards sensible solutions in the presence of corrupted observations. However, various approximations have to be introduced to arrive at mathematically feasible and computationally attractive solutions, whose impact on the final recognition performance is unknown. While this can be considered bad news, the good news is that it leaves room for improvements and future exciting research on the recurring topic of robustness in speech recognition.

Acknowledgements The author wishes to express his gratitude to the following Ph.D. students who made very valuable contributions to the work described in this chapter: Valentin Ion, Alexander Krueger, Volker Leutnant and Stefan Windmann. The work has been supported by the Deutsche Forschungsgemeinschaft (DFG) under contract nos. Ha 3455/2-3 and Ha3455/6-1, by the DFG Research Training Group GK-693 and by the Paderborn Institute for Scientific Computation (PaSCo).

References

1. Acero, A.: Acoustical and environmental robustness in automatic speech recognition. Ph.D. thesis, Carnegie Mellon University (1990)
2. Afifi, M., Cui, X., Gao, Y.: Stereo-based stochastic mapping for robust speech recognition. In: Proc. of IEEE International Conference on Acoustics, Speech and Signal Processing (ICASSP). Honolulu, Hi. (2007)
3. Allen, J.: Image method for efficiently simulating small-room acoustics. J. Acoust. Soc. Am. **65**(4), 943–590 (1979)
4. Arrowood, J., Clements, M.: Using observation uncertainty in HMM decoding. In: Proc. of International Conference on Spoken Language Processing (ICSLP). Denver, Colorado (2002)
5. Arthur, D., Vassilvitskii, S.: k-means++: The advantages of careful seeding. In: Proc. SODA, pp. 1027–1035 (2007)
6. Bar-Shalom, Y., Rong Li, X., Kirubarajan, T.: Estimation with Applications to Tracking and Navigation. John Wiley and Sons, Inc. (2001)
7. Barker, J., Cooke, M., Ellis, D.: Decoding speech in the presence of other sources. Speech Commununication **45**, 5–25 (2005)
8. Barker, J., Josifovski, L., Cooke, M., Green, P.: Soft decisions in missing data techniques for robust automatic speech recognition. In: Proc. of International Conference on Spoken Language Processing (ICSLP), pp. 373–376. Beijing, China (2000)
9. Bernard, A., Alwan, A.: Joint channel decoding – Viterbi recognition for wireless applications. In: Proc. of Eurospeech, Aalborg, Denmark (2001)
10. Cooke, M., Green, P., Josifovski, L., Vizinho, A.: Robust automatic speech recognition with missing and unreliable acoustic data. Speech Commununication **34**(3), 267 – 285 (2001)
11. van Dalen, R., Gales, M.: Asymptotically exact noise-corrupted speech likelihoods. In: Proc. of Annual Conference of the International Speech Communication Association (Interspeech). Makuhari, Japan (2010)
12. Deng, J., Bouchard, M., Yeap, T.H.: Speech feature estimation under the presence of noise with a switching linear dynamical model. In: Proc. of IEEE International Conference on Acoustics, Speech and Signal Processing (ICASSP). Toulouse, France (2006)
13. Deng, J., Bouchard, M., Yeap, T.H.: Noisy speech feature estimation on the Aurora2 database using a switching linear dynamic model. Journal of Multimedia **2**(2), 47–52 (2007)
14. Deng, L., Acero, A., Plumpe, M., Huang, X.: Large vocabulary speech recognition under adverse acoustic environments. In: Proc. of International Conference on Spoken Language Processing (ICSLP). Beijing, China (2000)
15. Deng, L., Droppo, J., Acero, A.: Log-domain speech feature enhancement using sequential map noise estimation and a phase-sensitive model of the acoustic environment. In: Proc. of International Conference on Spoken Language Processing (ICSLP), vol. 1, pp. 192–195. Denver, Co. (2002)
16. Deng, L., Droppo, J., Acero, A.: Enhancement of log Mel power spectra of speech using a phase-sensitive model of the acoustic environment and sequential estimation of the corrupting noise. IEEE Transactions on Speech and Audio Processing **12**(2), 133 – 143 (2004)
17. Deng, L., Droppo, J., Acero, A.: Dynamic compensation of HMM variances using the feature enhancement uncertainty computed from a parametric model of speech distortion. IEEE Trans. Speech and Audio Processing **13**(3), 412–421 (2005)
18. Droppo, J., Acero, A.: Noise robust speech recognition with a switching linear dynamic model. In: Proc. of IEEE International Conference on Acoustics, Speech and Signal Processing (ICASSP). Montreal, Canada (2004)
19. Droppo, J., Acero, A.: Environmental robustness. In: J. Benesty, M. Sondhi, Y. Huang (eds.) Handbook of Speech Processing. Springer, London (2008)
20. Droppo, J., Acero, A., Deng, L.: Uncertainty decoding with SPLICE for noise robust speech recognition. In: Proc. of IEEE International Conference on Acoustics, Speech and Signal Processing (ICASSP). Orlando, Fl. (2002)

21. Droppo, J., Deng, L., Acero, A.: A comparison of three non-linear observation models for noisy speech features. In: Proc. Eurospeech, vol. 1, pp. 681–684. Geneva, Switzerland (2003)
22. Duda, R., Hart, P., Stork, D.: Pattern Classification. John Wiley and Sons (2001)
23. Faubel, F., McDonough, J., Klakow, D.: A phase-averaged model for the relationship between noisy speech, clean speech and noise in the log-mel domain. In: Proc. of Annual Conference of the International Speech Communication Association (Interspeech). Brisbane, Australia (2008)
24. Frey, B.J., Deng, L., Acero, A., Kristjansson, T.T.: ALGONQUIN: Iterating Laplace's method to remove multiple types of acoustic distortion for robust speech recognition. In: Proc. of Eurospeech. Aalborg, Denmark (2001)
25. Gales, M.: Model-based approaches to handling uncertainty. In: Robust Speech Recognition of Uncertain or Missing Data. Springer, London (2011)
26. Haeb-Umbach, R., Ion, V.: Soft features for improved distributed speech recognition over wireless networks. In: Proc. of International Conference on Spoken Language Processing (ICSLP). Jeju Island, Korea (2004)
27. Huo, Q., Lee, C.H.: A Bayesian predictive approach to robust speech recognition. IEEE Trans. Speech and Audio Processing **8**(8), 200–204 (2000)
28. Ion, V., Haeb-Umbach, R.: A novel uncertainty decoding rule with applications to transmission error robust speech recognition. IEEE Transactions on Audio, Speech, and Language Processing **16**, 1047–1060 (2008)
29. Julier, S., Uhlmann, J., Durrant-Whyte, H.: A new method for the nonlinear transformation of means and covariances in filters and estimators. IEEE Transactions on Automatic Control **45**(3), 477–482 (2000)
30. Kolossa, D., Klimas, A., Orglmeister, R.: Separation and robust recognition of noisy, convolutive speech mixtures using time-frequency masking and missing data techniques. In: IEEE Workshop on Applications of Signal Processing to Audio and Acoustics, pp. 82–85 (2005)
31. Kristjansson, T.T., Frey, B.J.: Accounting for uncertainty in observations: A new paradigm for robust automatic speech recognition. In: Proc. of IEEE International Conference on Acoustics, Speech and Signal Processing (ICASSP). Orlando, Fl. (2002)
32. Krueger, A., Haeb-Umbach, R.: Model based feature enhancement for reverberant speech recognition. IEEE Transactions on Audio, Speech, and Language Processing **18**(7), 1692–1707 (2010)
33. Krueger, A., Leutnant, V., Haeb-Umbach, R., Ackermann, M., Bloemer, J.: On the initialisation of dynamic models for speech features. In: Proc. ITG Fachtagung Speech Communication. Bochum, Germany (2010)
34. Leutnant, V., Haeb-Umbach, R.: Conditional Bayesian estimation employing a phase-sensitive observation model for noise robust speech recognition. In: Robust Speech Recognition of Uncertain or Missing Data. Springer, London (2011)
35. Li, J., Deng, L., Yu, D., Gong, Y., Acero, A.: HMM adaptation using a phase-sensitive acoustic distortion model for environment-robust speech recognition. In: Proc. of IEEE International Conference on Acoustics, Speech and Signal Processing (ICASSP), pp. 4069–4072. Las Vegas, Nv. (2008)
36. Liao, H., Gales, M.J.F.: Joint uncertainty decoding for noise robust speech recognition. In: Proc. of Annual Conference of the International Speech Communication Association (Interspeech). Lisbon, Portugal (2005)
37. Liao, H., Gales, M.J.F.: Issues with uncertainty decoding for noise robust speech recognition. Speech Commununication **50**, 265–277 (2008)
38. Lindberg, B., Tan, Z. (eds.): Automatic Speech Recognition on Mobile Devices and over Communication Networks. Springer, London (2008)
39. Morris, A., Barker, J., Bourlard, H.: From missing data to maybe useful data: Soft data modelling for noise robust ASR. Proc. WISP **06** (2001)
40. Neumeyer, L., Weintraub, M.: Probabilistic optimum filtering for robust speech recognition. In: Proc. of IEEE International Conference on Acoustics, Speech and Signal Processing (ICASSP). Adelaide, Australia (1994)

41. Peinado, A.M., Segura, J.C.: Speech Recognition over Digital Channels. John Wiley & Sons Ltd. (2006)
42. Raj, B., Stern, R.: Missing-feature approaches in speech recognition. IEEE Signal Processing Magazine **22**(5), 101–116 (2005)
43. Ristic, B., Arulampalam, S., Gordon, N.: Beyond the Kalman Filter – Particle Filters for Tracking Applications. Artech House (2004)
44. Schmalenstroeer, J., Haeb-Umbach, R.: A comparison of particle filtering variants for speech feature enhancement. In: Proc. of Annual Conference of the International Speech Communication Association (Interspeech). Lisbon, Portugal (2005)
45. Sehr, A., Kellermann, W.: Towards robust distant-talking automatic speech recognition in reverberant environments. In: E. Hänsler, G. Schmidt (eds.) Speech and Audio Processing in Adverse Environments. Springer, London (2008)
46. Stouten, V., Van hamme, H., Wambacq, P.: Accounting for the uncertainty of speech estimates in the context of model-based feature enhancement. In: Proc. of International Conference on Spoken Language Processing (ICSLP). Jeju Island, Korea (2004)
47. Stouten, V., Van hamme, H., Wambacq, P.: Model-based feature enhancement with uncertainty decoding for noise robust ASR. Speech Commununication **48**(11) (2006)
48. Tan, Z.H., Dalsgaard, P., Lindberg, B.: Automatic speech recognition over error-prone wireless networks. Speech Commununication **47**(1-2), 220–242 (2005)
49. Vary, P.: Speech enhancement by conditional estimation: Noise reduction, error concealment & bandwidth extension, what makes the difference? In: Proc. International Workshop on Acoustic Echo and Noise Control (IWAENC) (2008)
50. Windmann, S., Haeb-Umbach, R.: Approaches to iterative speech feature enhancement and recognition. IEEE Transactions on Audio, Speech, and Language Processing **17**(5), 974–984 (2009)
51. Wölfel, M.: Enhanced speech features by single-channel joint compensation of noise and reverberation. IEEE Transactions on Audio, Speech, and Language Processing **17**(2), 312–323 (2009)

Chapter 3
Uncertainty Propagation

Ramón Fernandez Astudillo, Dorothea Kolossa

Abstract While it is often fairly straightforward to estimate the reliability of speech features in the time-frequency domain, this may not be true in other domains more amenable to speech recognition, such as for RASTA-PLP features or those obtained with the ETSI advanced front-end. In such cases, one useful approach is to estimate the uncertainties in the domain where noise reduction preprocessing is carried out, and to subsequently transform the uncertainties, along with the actual features, to the recognition domain. In order to develop suitable approaches, we will first give a short overview of relevant strategies for propagating probability distributions through nonlinearities. Secondly, for some feature domains suitable for robust recognition, we will show possible implementations and sensible approximations of uncertainty propagation and discuss the associated error margins and trade-offs.

3.1 Uncertainty Propagation

In automatic speech recognition (ASR) an incoming speech signal is transformed into a set of observation vectors $\mathbf{x} = \mathbf{x}_1 \cdots \mathbf{x}_L$ which are input to the recognition model. ASR systems which work with uncertain input data replace this observation set with a distribution of possible observed values conditioned on the available information $\mathbf{I}$, $p(\mathbf{x}_1 \cdots \mathbf{x}_L|\mathbf{I})$. This uncertainty represents the lost information that causes the mismatch between the observed speech and the trained ASR model. By combining this distribution with observation uncertainty techniques [24], superior recognition robustness can be attained. There are multiple sources from which

Ramón Fernandez Astudillo
Electronics and Medical Signal Processing Group, TU Berlin, 10587 Berlin, e-mail: ramon@astudillo.com

Dorothea Kolossa
Electronics and Medical Signal Processing Group, TU Berlin, 10587 Berlin, e-mail: dorothea.kolossa@gmx.de

D. Kolossa and R. Haeb-Umbach (eds.), *Robust Speech Recognition of Uncertain or Missing Data: Theory and Applications*, DOI 10.1007/978-3-642-21317-5_3,
© Springer-Verlag Berlin Heidelberg 2011

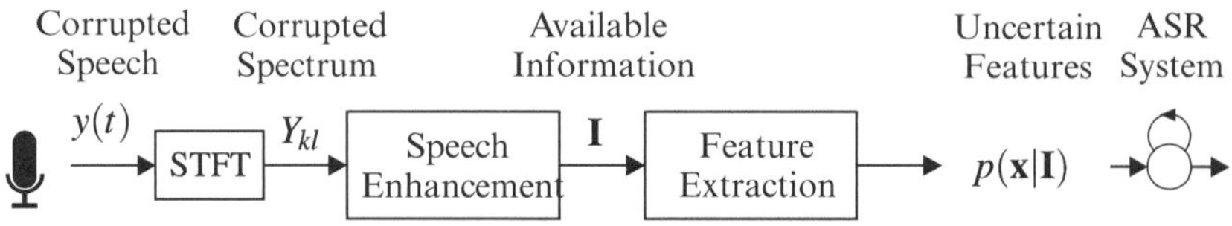

Fig. 3.1: Speech recognition using speech enhancement in short-time Fourier domain and uncertainty propagation. The available information obtained after speech enhancement in the STFT domain $\mathbf{I}$ is transformed into the feature domain to yield a probabilistic or uncertain description of the features $p(\mathbf{x}|\mathbf{I})$.

this distribution over the features can be computed. Uncertainty of observation has been determined directly from speech enhancement techniques in the feature domain, such as SPLICE [10] and model-based feature enhancement (MBFE) [32], or by the use of a state-space noise model [33]. Other models of uncertainty have been obtained by considering the relation between speech and additive noise in the logarithm domain, a step common to many feature extractions. Using this relation, uncertainty has been computed based on parametric models of speech distortion [9, 34] or the residual uncertainty after feature enhancement [5]. Other approaches have exploited different sources of uncertainty such as channel transmission errors [17] or reverberation, as described in detail in Chapter 9.

This chapter discusses a particular set of techniques, which aim to determine $p(\mathbf{x}_1 \cdots \mathbf{x}_L|\mathbf{I})$ by propagating the information attained from preprocessing in the short-time Fourier transform (STFT) domain into the feature domain used for recognition (see Fig. 3.1). The STFT domain is the domain most often used for single-channel and multi-channel speech enhancement. Speech distortion phenomena like additive noise, reverberation, channel distortions and cocktail party effect can be modeled more easily in this domain than in the domains where speech recognition takes place. Some speech characteristics such as vocal tract transfer function and fundamental frequency are also easier to model in this domain. Furthermore, the time domain signal can be completely recovered from the STFT of the signal, which makes this domain particularly suitable for speech enhancement targeted to human listeners.

Despite the positive properties of the STFT domain, it is rarely used for automatic speech recognition. Nonlinear transformations of the STFT such as the mel frequency cepstral coefficients (MFCCs) and the relative spectral filtered perceptual linear prediction features (RASTA-PLPs) are used instead. Such features have a smaller intra- and inter-speaker variability and achieve a more compact and classifiable representation of the acoustic space. The use of STFT domain speech enhancement is therefore usually limited to obtaining a point estimate of the clean speech which is transformed into the domain of recognition features.

Uncertainty propagation techniques attempt to increase the amount of information passed from the speech enhancement step, usually in the STFT domain, to the domain of recognition features by adding a measure of reliability. This idea was first used in the context of source separation by means of independent component analysis (ICA) [21]. For this purpose, the estimated STFT of each speaker was treated as a Gaussian variable. The uncertainty was then used to compensate for the loss of information incurred by time-frequency post-masking of ICA results. This same model was later employed to propagate the uncertainty generated after single-channel speech enhancement, using the estimated noise variance [22]. These models were later extended to the complex Gaussian uncertainty model in [3], which arises naturally from the use of conventional minimum mean square error (MMSE) speech enhancement techniques as a measure of uncertainty [4]. In both cases, the transformation of the uncertain described features into the domain of recognition features was attained by using a piecewise approach combining closed form solutions with pseudo-Monte Carlo approximations.

Similarities can be drawn between these two techniques and the works in [29, 30]. Here, the STFT domain was reconstructed using missing-feature techniques [27], which allows the computation of an MMSE estimate of the unreliable components from the reliable components as well as a variance or measure of uncertainty. This variance was then propagated into the domain of recognition features by using general nonlinear approximators like multilayer perceptrons [29] and regression trees [30], which had to be previously trained using noisy data samples. Other work which can also be related to uncertainty propagation techniques is that in [31]. Here, a noise variance estimate in the STFT domain was obtained with a minimum statistics based noise variance estimator, typically used for speech enhancement. This estimate along with a measure of its reliability was propagated into the MFCC domain using the log-normal assumption [14] and combined with MBFE techniques [31].

This chapter will be centered on the piecewise propagation approach, and it is organized as follows: Section 3.2 deals with the problem of modeling uncertainty in the spectral domain, Sections 3.3 through 3.6 describe the propagation to three feature domains of interest, Section 3.7 considers the implications of using full covariances in the approach, and Section 3.8 introduces some results and draws conclusions.

3.2 Modeling Uncertainty in the STFT Domain

Let $y(t)$ be a corrupted version of a clean speech signal $x(t)$. A time frequency representation of $y(t)$ can be obtained using an N-point STFT as

$$Y_{kl} = \sum_{t'=1}^{N} y\left(t' + (l-1)M\right) h(t') \exp\left(-2\pi j \frac{(t'-1)(k-1)}{N}\right), \tag{3.1}$$

where $t' \in [1 \cdots N]$ is the sample index within the frame, $l \in [1 \cdots L]$ is the frame index, $k \in [1 \cdots K]$ corresponds to the index of each Fourier bin up to half the sampling frequency and j is the imaginary unit. Each frame is shifted by M samples with respect to the previous one and weighted by a window function $h(t')$.

Speech enhancement methods obtain an estimation of the clean signal by applying a time-frequency-dependent gain function G_{kl} to obtain each corresponding estimation $\hat{X}_{kl}$ of the clean Fourier coefficient X_{kl} as

$$\hat{X}_{kl} = G_{kl} \cdot Y_{kl}. \tag{3.2}$$

The gain function can be obtained, for example, from multi-channel speech enhancement such as ICA with post-masking [21], missing-feature techniques [30], or conventional MMSE speech enhancement [11]. Apart from this point estimate, uncertainty propagation algorithms need a reliability measure in the form of a variance, defined as

$$\lambda_{kl} = E\left\{\left|X_{kl} - \hat{X}_{kl}\right|^2\right\}, \tag{3.3}$$

where X_{kl} is the unknown clean Fourier coefficient. When MMSE estimations are used, a measure of reliability can be easily obtained from the variance of the resulting posterior distribution [4, 30]. In cases in which more complex algorithms are used, as in the case of ICA with post-masking [21] and the ETSI advanced front-end [13], approximate measures of reliability can be obtained based on the amount of change inflicted by the speech enhancement algorithm [2, 21].

Once an estimation of the clean Fourier coefficient $\hat{X}_{kl}$ and a corresponding measure of reliability λ_{kl} have been obtained, this information can be used to yield a probabilistic description of each clean Fourier coefficient X prior to propagation (see Fig. 3.2). Although it would also be possible to include the cross-covariance between different Fourier coefficients, this is usually ignored for simplicity. This is a usual simplification also used when employing similar models for speech enhancement [12, 25].

With regard to the type of distributions used to model each uncertain Fourier coefficient, in [21], a Gaussian distribution with variance equal to the uncertainty variance was used as

$$|X_{kl}| = \left|\left|\hat{X}_{kl}\right| + \delta_{kl}\right| \tag{3.4}$$

where δ_{kl} followed the zero mean Gaussian distribution

$$p(\delta_{kl}) = \frac{1}{\sqrt{2\pi\lambda_{kl}}} \exp\left(-\frac{\delta_{kl}^2}{2\lambda_{kl}}\right). \tag{3.5}$$

In [29, 30] the uncertainty was determined using an MMSE estimation based on missing-feature techniques [27]. The posterior of this MMSE estimator resulted in a Gaussian mixture model for each Fourier coefficient amplitude.[1]

The works in [2, 3] present an extension of the Gaussian model in [21], in which the distribution used is circularly symmetric complex Gaussian:

[1] Despite this fact, only the variance was considered for propagation.

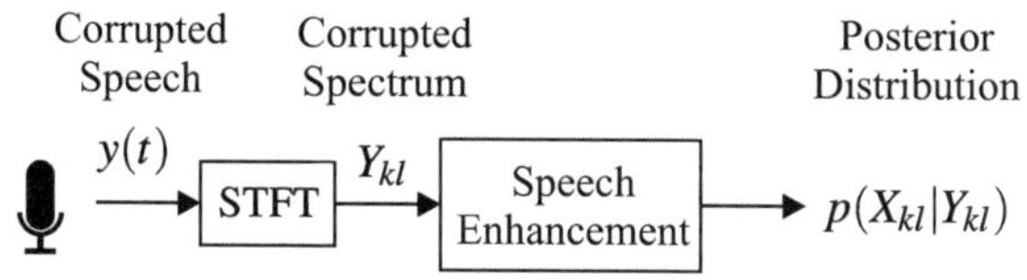

Fig. 3.2: The available information after speech enhancement in the STFT domain is summarized in the form of a posterior distribution

$$p(X_{kl}|Y_{kl}) = \frac{1}{\pi\lambda_{kl}} \exp\left(-\frac{|X_{kl} - \hat{X}_{kl}|^2}{\lambda_{kl}}\right). \tag{3.6}$$

This has the advantage of being an uncertainty model not just of the magnitude but also of the complex-valued Fourier coefficient, which allows propagation back into the time domain. It also arises naturally in conventional MMSE speech enhancement methods like the amplitude, log-amplitude and Wiener estimators, as seen in [4].

The complex Gaussian uncertainty model can be related to the Gaussian uncertainty model used in [21]. Both models originate, in fact, from a non-central Chi distribution uncertainty model for the amplitude with two and one degrees of freedom, respectively [1]. The approaches to uncertainty propagation introduced in the next chapter will, however, be centered on the complex Gaussian model due the particular advantages previously discussed. A detailed discussion of the methods can be found in [1].

3.3 Piecewise Uncertainty Propagation

The problem of propagating uncertainty corresponds to that of transforming a random variable through a nonlinear transformation, in this case the feature extraction. There are many possible methods to compute this transformation; however, there are two characteristics of the particular problem of uncertainty propagation for robust automatic speech recognition that limit the set of useful techniques:

1. ASR systems are expected to work in real time or with reasonable offline execution times. Computational resources are also scarce in some particular cases.
2. Feature extractions encompass linear and nonlinear operations performed jointly on multiple features of the same time frame or combining features from different time frames. Such transformations can therefore become very complex.

Regarding the first characteristic, the fastest solution to a propagation problem is finding a closed-form solution for the pdf or the needed cumulants of the transformed random variable. This implies solving changes of variable, derivatives or integrals; due to the complexity of the feature extraction transformations, such solutions are rarely obtainable. Regarding the second characteristic, black-box approaches like pseudo-Monte Carlo methods or universal nonlinear approximators are not dependent on the complexity of the transformations involved. Such methods suffer, however, from other shortcomings such as the need for training with suitable noise samples, or poor trade-offs between accuracy and speed under certain conditions.

Given this context, neither closed-form nor black-box solutions can provide a general solution for the propagation of uncertainty. There exists, however, the possibility of combining both techniques in a piecewise approach that will provide a set of solutions for some well-known feature extractions. In this approach, the feature extraction will be divided into different steps and the optimal method to propagate the statistical information through each step will be selected. The main disadvantage of this technique is that, even if we want to propagate only first and second order moments, we still need to consider the complete pdfs of the uncertain features in the intermediate steps. On the positive side, we can optimize each process individually and share transformations among different feature extractions.

In [1], the piecewise approach to uncertainty propagation was applied to the mel frequency cepstral and RASTA-LPCC feature extractions, as well as to the additional transformations used in the advanced front-end specified in the ES 202 050 standard of the European Telecommunications Standards Institute (ETSI). It was also demonstrated that for the complex Gaussian model, the resulting uncertain features were accurately described by a Gaussian distribution. The next sections will detail the individual solutions used for these transformations, as well as the general solutions obtainable for linear transformations and via pseudo-Monte Carlo methods. The last section will present some of the experiments on the accuracy of the piecewise approach and its results in robust automatic speech recognition.

3.3.1 Uncertainty Propagation Through Linear Transformations

In general, given a random vector variable $\mathbf{x}$ and a linear transformation defined by the matrix $\mathbf{T}$, the transformed mean and covariance correspond to

$$E\left\{\mathbf{T}\mathbf{x}^T\right\} = \mathbf{T}E\left\{\mathbf{x}\right\}^T \tag{3.7}$$

and

$$\mathrm{Cov}\left\{\mathbf{T}\mathbf{x}^T\right\} = \mathbf{T}\mathrm{Cov}\left\{\mathbf{x}\right\}\mathbf{T}^T. \tag{3.8}$$

Furthermore, if a variable is Gaussian-distributed, it retains this condition when linearly transformed. However, it has to be taken into account that when $\mathbf{T}$ is not diagonal, the linear transformation induces a non-diagonal covariance, which has to be considered in subsequent transformations.

3.3.2 Uncertainty Propagation with the Unscented Transform

Monte Carlo and pseudo-Monte Carlo methods provide a numerical solution to the propagation problem when a closed form solution cannot be found. These methods are based on replacing the continuous distribution by a set of sample points that are representative of the distribution characteristics. Statistical parameters like mean and variance can be simply computed from the transformed samples using conventional estimators. Multiple techniques are available to optimize the PDF sampling process in order to minimize the number of iterations needed and maximize accuracy. In the context of this work, however, most methods still have prohibitive execution time requirements.

The unscented transform [19] is a pseudo-Monte Carlo method that has very low computational requirements for reasonable accuracy, particularly when some conditions about the initial and final uncertainty distributions as well as the nonlinear transformation applied are met. Let

$$\mathbf{y}_l = \mathbf{g}(\mathbf{x}_l) \tag{3.9}$$

be a nonlinear vector-valued function. Let $\mathbf{x}_l$ be a frame containing a size N multivariate random variable of distribution $p(\mathbf{x}_l)$ with mean and covariance matrix equal to $\boldsymbol{\mu}_l^x$ and $\boldsymbol{\Sigma}_l^x$, respectively. The result of transforming $\mathbf{x}_l$ through $\mathbf{g}()$ is the multivariate random variable $\mathbf{y}_l$ of distribution $p(\mathbf{y}_l)$. The unscented transform approximates the first- and second-order moments of $p(\mathbf{y}_l)$ by transforming a set of $2N+1$ so-called sigma points $\mathbf{S} = \{\mathbf{S}_1 \cdots \mathbf{S}_{2N+1}\}$ which capture the characteristics of $p(\mathbf{x}_l)$. The mean and covariance of $\mathbf{y}_l$ can be approximated by using the following weighted averages:

$$E\{\mathbf{y}_l\} \approx \sum_{i=1}^{2N+1} W_i \cdot \mathbf{g}(\mathbf{S}_i), \tag{3.10}$$

$$\mathrm{Cov}\{\mathbf{y}_l\} \approx \sum_{i=2}^{2N+1} W_i \cdot (\mathbf{g}(\mathbf{S}_i) - E\{\mathbf{y}_l\}) \cdot (\mathbf{g}(\mathbf{S}_i) - E\{\mathbf{y}_l\})^T. \tag{3.11}$$

The sigma points $\mathbf{S}$ and the weights are deterministically chosen given the mean and variance of the initial probability distribution as

$$\begin{cases} W_1 = & \frac{\kappa}{(N+\kappa)} \\ \mathbf{S}_1 = & \boldsymbol{\mu}_l^x \\ \mathbf{S}_i = & \boldsymbol{\mu}_l^x + \left(\sqrt{(N+\kappa)\cdot \boldsymbol{\Sigma}_l^x}\right)_i \\ \mathbf{S}_{i+N} = & \boldsymbol{\mu}_l^x - \left(\sqrt{(N+\kappa)\cdot \boldsymbol{\Sigma}_l^x}\right)_i \\ W_i = & \frac{1}{2(N+\kappa)} \quad \text{for} \quad i \in \{2 \cdots N+1\}, \end{cases} \tag{3.12}$$

where $(\ )_i$ corresponds to the i-th row of the matrix square root. The additional parameter κ allows the kurtosis and higher-order cumulants of $p(\mathbf{x}_l)$ to be modeled. If the option chosen is to assume a Gaussian distribution for $p(\mathbf{x}_l)$, this leads to

$$\kappa = 3 - N. \tag{3.13}$$

The unscented transform is not impervious to the curse of dimensionality and suffers from accuracy problems if some conditions are not met. The assumed condition of zero odd moments for the initial distribution $p(\mathbf{x}_l)$ can be problematic in the context of uncertainty propagation.

3.4 Uncertainty Propagation for the Mel Frequency Cepstral Features

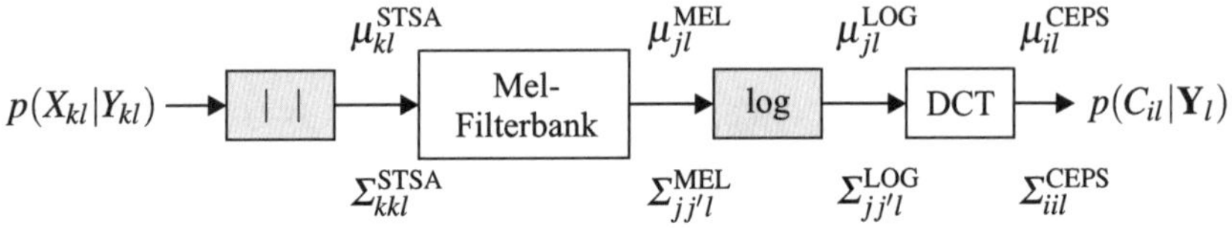

Fig. 3.3: Steps involved in the mel frequency cepstral feature extraction as well as moments being propagated. Shaded blocks denote nonlinear transformations

The mel frequency cepstral coefficients (MFCCs) are computed from the STFT of a signal X_{kl} by the following nonlinear transformation (see Fig. 3.3):

$$C_{il} = \sum_{j=1}^{J} T_{ij} \log\left(\sum_{k=1}^{K} W_{jk}|X_{kl}|^{\alpha}\right), \tag{3.14}$$

where W_{jk} is the mel filter bank weight and

$$T_{ij} = \sqrt{\frac{2}{J}} \cos\left(\frac{\pi i}{J}(j-0.5)\right) \tag{3.15}$$

is the discrete cosine transform (DCT) weight. The exponent α usually takes the values 1 or 2, corresponding to the use of the short-time spectral amplitude (STSA) or the estimated power spectral density (PSD), respectively. This transformation is also very similar to the initial steps of the RASTA-PLP transformation later considered. The case of the STSA will be therefore be discussed here, whereas the case of the PSD will be commented on in the RASTA-PLP section.

3.4.1 Propagation Through the STSA Computation

If we consider the complex Gaussian Fourier uncertainty model introduced in Section 3.1, we have that each Fourier coefficient of the clean signal X_{kl} is circularly symmetric complex Gaussian-distributed. The posterior distribution of each Fourier coefficient given the available information is thus given by Equation (3.6), where $\hat{X}_{kl}$ is the estimated Fourier coefficient of the clean signal given in Equation (3.2) and λ_{kl} is the associated variance or uncertainty of estimation given in Equation (3.3). If we express this uncertain Fourier coefficient in polar coordinates we have

$$X_{kl} = A_{kl}\, e^{j\alpha_{kl}}, \tag{3.16}$$

where j corresponds here to the imaginary unit. The distribution of its amplitude

$$A_{kl} = |X_{kl}| \tag{3.17}$$

can be the obtained by marginalizing the phase in polar coordinates from Equation (3.6) as

$$p(A_{kl}|Y_{kl}) = \int p(X_{kl}|Y_{kl}) d\alpha_{kl} = \int p(A_{kl}, \alpha_{kl}|Y_{kl}) d\alpha_{kl}, \tag{3.18}$$

which leads to the Rice distribution:

$$p(A_{kl}|Y_{kl}) = \frac{2A_{kl}}{\lambda_{kl}} \exp\left(-\frac{A_{kl}^2 + |\hat{X}_{kl}|^2}{\lambda_{kl}}\right) I_0\left(\frac{2A_{kl}|\hat{X}_{kl}|}{\lambda_{kl}}\right), \tag{3.19}$$

where I_0 is the modified Bessel function of order zero obtained by applying [15, Eq. (8.431.5)]. The n-th order moment of a Rice distribution has the following closed-form solution obtained by applying [15, Eqs. 6.631.1, 8.406.3, 8.972.1]:

$$E\{A_{kl}^n|Y_{kl}\} = \Gamma\left(\frac{n}{2}+1\right) (\lambda_{kl})^{\frac{n}{2}}\, \mathrm{L}_{\frac{n}{2}}\left(-\frac{|\hat{X}_{kl}|^2}{\lambda_{kl}}\right), \tag{3.20}$$

where $\Gamma()$ is the gamma function and $\mathrm{L}_{\frac{n}{2}}$ is the Laguerre polynomial. [2]

[2] Note that the upper index in the Laguerre polynomial is omitted since it is always zero.

Given a general closed-form solution of the n-th order moment, the mean of the uncertain STSA can be computed as

$$\mu_{kl}^{\mathrm{STSA}} = E\{A_{kl}|Y_{kl}\} = \Gamma(1.5)\sqrt{\lambda_{kl}}\mathrm{L}_{\frac{1}{2}}\left(-\frac{|\hat{X}_{kl}|^2}{\lambda_{kl}}\right), \tag{3.21}$$

where $\mathrm{L}_{\frac{1}{2}}()$ can be expressed in terms of Bessel functions by combining [15, Eq. 8.972.1] and [28, Eq. 4B-9] as

$$\mathrm{L}_{\frac{1}{2}}(x) = \exp\left(\frac{x}{2}\right)\left[(1-x)I_0\left(-\frac{x}{2}\right) - xI_1\left(-\frac{x}{2}\right)\right]. \tag{3.22}$$

Note the logical similarity between (3.21) and the MMSE-STSA estimator given in [12], since both correspond to the mean amplitude of a complex Gaussian distribution. Note also that for a variance λ_{kl} equal to that of the Wiener filter posterior, both formulas coincide [4].

The variance of the uncertain STSA can be computed in a similar form, using the relation between second-order cumulants and moments [26, Eq. 2.4] as

$$\Sigma_{kkl}^{\mathrm{STSA}} = E\{A_{kl}^2|Y_{kl}\} - E\{A_{kl}|Y_{kl}\}^2, \tag{3.23}$$

where the second-order moment can be obtained from Equation (3.20) using [15, Eq. 8.970.4] as

$$E\{A_{kl}^2|Y_{kl}\} = \lambda_{kl} + |\hat{X}_{kl}|^2, \tag{3.24}$$

leading to

$$\Sigma_{kkl}^{\mathrm{STSA}} = \lambda_{kl} + |\hat{X}_{kl}|^2 - \left(\mu_{kl}^{\mathrm{STSA}}\right)^2. \tag{3.25}$$

The covariance of each STSA frame $\mathbf{A}_l$ is diagonal since each Fourier coefficient is considered statistically independent of the others.

3.4.2 Propagation Through the Mel Filter Bank

The propagation through the mel filter bank transformation is much simpler, since it is a linear transformation. The computation of the mean $\boldsymbol{\mu}_l^{\mathrm{Mel}}$ and covariance $\boldsymbol{\Sigma}_l^{\mathrm{Mel}}$ of the uncertain features after the mel filter bank transformation for each frame can be solved using Equations (3.7) and (3.8) for $\mathbf{T} = \mathbf{W}$, yielding

$$\mu_{jl}^{\mathrm{Mel}} = \sum_{k=1}^{K} W_{jk}\mu_{kl}^{\mathrm{STSA}} \tag{3.26}$$

and

$$\Sigma_{jj'l}^{\text{Mel}} = \sum_{k=1}^{K}\sum_{k'=1}^{K} W_{jk}W_{j'k'}\Sigma_{kk'l}^{\text{STSA}} = \sum_{k=1}^{K} W_{jk}W_{j'k}\Sigma_{kkl}^{\text{STSA}}, \tag{3.27}$$

where the last equality was obtained by using the fact that the STSA covariance is diagonal. It has to be taken into account that the mel filters overlap. Consequently, the filter bank matrix $\mathbf{W}$ is not diagonal and thus $\mathbf{\Sigma}_l^{\text{Mel}}$ is not diagonal either.

3.4.3 Propagation Through the Logarithm of the Mel-STSA Analysis

After the mel filter bank, the uncertain features of each frame follow an unknown distribution (the sum of Rice random variables of different parameters [1]) and are no longer statistically independent. The calculation of a closed-form solution for the resulting PDF seems therefore unfeasible. The propagation through the logarithm domain was already solved in [21] by approximating the mel-STSA uncertain features with the log-normal distribution. This same distribution had also been originally used in [14] to propagate a model of corrupted speech from spectral to logarithm domain and later in [31] for the propagation of noise variance statistics. However, this assumption is suboptimal in the case of the complex Gaussian Fourier uncertainty model. The empirically obtained pdf of the mel-STSA uncertain features is closer to the Gaussian distribution than the log-normal distribution.

The fact that the distribution of the features is no longer known but exhibits a relatively low skewness, together with the fact that, after the mel filter bank transformation, the number of features per frame has been reduced by almost one order of magnitude,[3] favors the use of the pseudo-Monte Carlo method known as the unscented transform, presented in Section 3.3.2.

Indeed, in the particular case of the mel-STSA uncertain features, the unscented transform can provide a more accurate estimate than the log-normal assumption, and is therefore used, yielding

$$\mu_{jl}^{\text{LOG}} \approx \sum_{i=1}^{2N+1} W_i \cdot \log(S_{ji}), \tag{3.28}$$

$$\Sigma_{jj'l}^{\text{LOG}} \approx \sum_{i=2}^{2N+1} W_i \cdot (\log(S_{ji}) - \mu_{jl}^{\text{LOG}}) \cdot (\log(S_{j'i}) - \mu_{j'l}^{\text{LOG}}), \tag{3.29}$$

where each of the $2N+1$ vector sigma points $\mathbf{S}_i$ was obtained using Equations (3.12) and (3.13) with $\boldsymbol{\mu}_l^x = \boldsymbol{\mu}_l^{\text{MEL}}$, $\mathbf{\Sigma}_l^x = \mathbf{\Sigma}_l^{\text{MEL}}$, and N is equal to the number of Bark filter banks J.

[3] Typically, the filter bank compresses each 256–512 frequency bin frame $\mathbf{X}_l$ into a compact representation of 20–25 filter outputs $\mathbf{M}_l$.

3.4.4 Propagation Through the Discrete Cosine Transform

The cosine transform is a linear transformation and therefore can be easily computed by using Equations (3.7) and (3.8) as

$$\mu_{il}^{\text{CEPS}} = \sum_{j=1}^{J} T_{ij}\mu_{jl}^{\text{LOG}} \tag{3.30}$$

and

$$\Sigma_{iil}^{\text{CEPS}} = \sum_{j=1}^{J}\sum_{j'=1}^{J} T_{ij}T_{ij'}\Sigma_{jj'l}^{\text{LOG}}. \tag{3.31}$$

Note that despite the fact that the full covariance after the DCT transformation can be easily obtained, only the values of the diagonal were used in the experiments to reduce the computational load.[4] Note also that the eventual use of cepstral liftering poses no difficulty, since it is also a linear transformation that can be included as a factor in the transform matrix $\mathbf{T}$.

3.5 Uncertainty Propagation for the RASTA-PLP Cepstral Features

The perceptual linear prediction (PLP) features [16] will first be described without uncertainties for clarity. Subsequent Sections 3.5.1 through 3.5.5 will detail each step of the computation. The feature extraction is depicted in detail in Fig. 3.4, and as it can be seen, it shares the initial steps with the MFCC feature extraction:

$$B_{jl} = \sum_{k=1}^{K} V_{jk}|X_{kl}|^{\alpha}, \tag{3.32}$$

where the V_{kl} correspond to the weights of a Bark filter bank, similar to the mel filter bank, and α equals 2. The next transformation is the application of an equal loudness preemphasis ψ_j and the cubic root compression to obtain a perceptually modified power spectrum

$$H_{jl} = \left(\psi_j B_{jl}\right)^{0.33}, \tag{3.33}$$

where ψ_j is a multiplicative gain in the linear domain that only varies with the filter bank index j. The final step of the RASTA-PLP feature extraction is the computation of the autoregressive all-pole model. Its implementation is done via the inverse DFT and the Levinson-Durbin recursion to obtain the LPCs [8, pp. 290–302]. As an

[4] Note that this does not imply that the correlation induced by the mel filter bank can be ignored, since it is necessary for the computation of the diagonal values of the covariance as well.

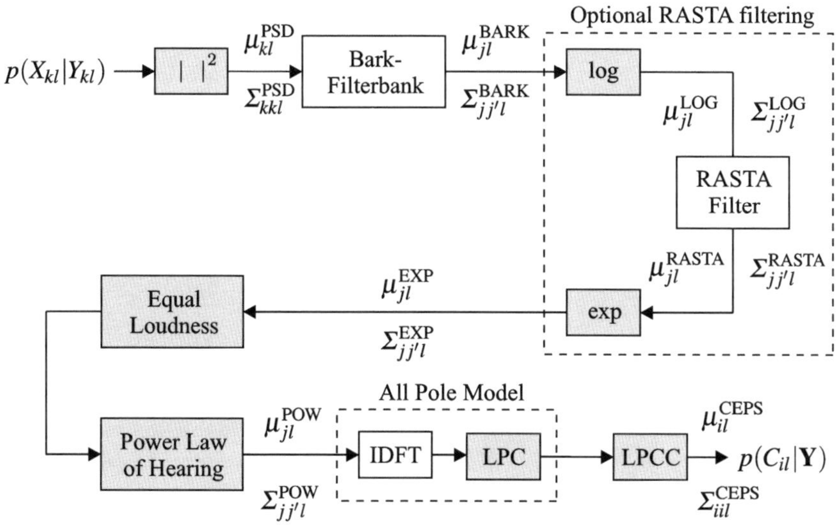

Fig. 3.4: Steps involved in the perceptual linear prediction feature extraction with optional RASTA filtering as well as moments being propagated. Shaded blocks denote nonlinear transformations

optional postprocessing step to improve feature performance, LPC-based cepstral coefficients (LPCCs) can also be computed using [7, Eq. 3].

A possible improvement of the LPCCs is the application of a RelAtive SpecTrAl (RASTA) filter to the logarithmic output of the filter bank [16]. This process implies transforming the output of the filter bank through the logarithm and applying an infinite impulse response (IIR) filter that operates over the time dimension (index l) of the feature matrix. This filter corresponds to the difference equation

$$y_{jl} = \sum_{d=1}^{4} b_d x_{jl-d} - a_1 y_{jl-1}, \tag{3.34}$$

where y_{jl} and y_{jl-1} are the l-th and $(l-1)$-th RASTA filter outputs and $x_{jl} \cdots x_{jl-4}$ correspond to the four previous logarithm domain inputs of $\log(B_{jl})$. The scalars a_1 and $b_0 \cdots b_4$ are the normalized feedforward and feedback coefficients. After this filtering in the logarithm domain, the signal is transformed back using the exponential transformation; and the RASTA-PLP transformation is completed.

3.5.1 Propagation Through the PSD Computation

Since a complex Gaussian distribution has a Rice-distributed amplitude for which the n-th order moments are known, the mean μ_{kl}^{PSD} of the uncertain PSD features can be computed from Equation (3.24) as

$$\mu_{kl}^{\mathrm{PSD}} = E\{A_{kl}^2|Y_{kl}\} = \lambda_{kl} + |\hat{X}_{kl}|^2, \tag{3.35}$$

and the covariance can be computed similarly to the STSA case as

$$\Sigma_{kkl}^{\mathrm{PSD}} = E\{A_{kl}^4|Y_{kl}\} - E\{A_{kl}^2|Y_{kl}\}^2 = 2\lambda_{kl}|\hat{X}_{kl}|^2 + \lambda_{kl}^2, \tag{3.36}$$

where $E\{A_{kl}^4|Y_{kl}\}$ is obtained by solving Equation (3.20) for $n = 4$ using [15, Eq. 8.970.5].

3.5.2 Propagation Through the Bark Filterbank

As in the case of the mel filter bank, the Bark filter bank is a linear transformation and therefore propagating mean and covariance requires only matrix multiplications, giving

$$\mu_{jl}^{\mathrm{BARK}} = \sum_{k=1}^{K} V_{jk}\mu_{kl}^{\mathrm{PSD}} \tag{3.37}$$

and

$$\Sigma_{jj'l}^{\mathrm{BARK}} = \sum_{k=1}^{K}\sum_{k'=1}^{K} V_{jk}V_{j'k'}\Sigma_{kk'l}^{\mathrm{PSD}} = \sum_{k=1}^{K} V_{jk}V_{j'k}\Sigma_{kkl}^{\mathrm{PSD}} \tag{3.38}$$

where V_{jk} was defined in Section 3.5.

3.5.3 Propagation Through the Logarithm of the Bark-PSD

From simulation tests, it can be observed that the distribution of the uncertain Bark-PSD features is more skewed than that of the mel-STSA features. It can also be observed that the log-normality assumption offers a good approximation for the distribution of the filter bank-PSD features regardless of which of the two filter banks is used [1]. The log-normality assumption, also used in other propagation approaches [14, 21, 31], is equivalent to the assumption of the logarithm of the uncertain Bark-PSD features being Gaussian-distributed [18, Chapter 14]. The corresponding statistics in the Bark domain can be then computed by propagating back the Gaussian distribution from the log domain through the exponential transformation and matching the first and second moments, thus obtaining [14, Eqs C. 11 and C. 12]

$$\mu_{jl}^{\mathrm{LOG}} = \log(\mu_{jl}^{\mathrm{Bark}}) - \frac{1}{2}\log\left(\frac{\Sigma_{ijl}^{\mathrm{BARK}}}{\mu_{jl}^{\mathrm{BARK}}\mu_{jl}^{\mathrm{BARK}}} + 1\right) \tag{3.39}$$

and

$$\Sigma_{jj'l}^{\mathrm{LOG}} = \log\left(\frac{\Sigma_{jj'l}^{\mathrm{BARK}}}{\mu_{jl}^{\mathrm{Bark}}\mu_{j'l}^{\mathrm{Bark}}} + 1\right). \tag{3.40}$$

3.5.4 Propagation Through RASTA Filter

As an IIR filter, the RASTA filter is a recursive linear transformation corresponding to the difference equation given in Equation (3.34). Computing the propagation of the mean through this transformation is immediate due to the linearity of the expectation operator. Applying Equation (3.7), we obtain

$$\boldsymbol{\mu}_l^{\mathrm{RASTA}} = E\{\mathbf{y}_l\} = \sum_{d=0}^{4} b_d \boldsymbol{\mu}_l^{\mathrm{LOG}} - a_1 \boldsymbol{\mu}_{l-1}^{\mathrm{RASTA}}, \tag{3.41}$$

where the last term $\boldsymbol{\mu}_{l-1}^{\mathrm{RASTA}}$ was obtained in the previous frame computation.

The computation of the covariance is more complex, however, due to the created temporal correlation between inputs and outputs and the correlation between different features created after the filter bank. If we consider the feature covariance matrix for frame l we have that, by [26, Eq. 2.4],

$$\boldsymbol{\Sigma}_l^{\mathrm{RASTA}} = E\{\mathbf{y}_l\mathbf{y}_l^T\} - E\{\mathbf{y}_l\}E\{\mathbf{y}_l\}^T, \tag{3.42}$$

where the second term can be directly calculated from Equation (3.41) but the first term must be solved. By substituting (3.34) into $E\{\mathbf{y}_l\mathbf{y}_l^T\}$ we obtain

$$E\{\mathbf{y}_l\mathbf{y}_l^T\} = E\left\{\left(\sum_{d=0}^{4} b_d\mathbf{x}_{l-d} - a_1\mathbf{y}_{l-1}\right)\left(\sum_{d=0}^{4} b_d\mathbf{x}_{l-d} - a_1\mathbf{y}_{l-1}\right)^T\right\}. \tag{3.43}$$

Taking into account the linearity of the expectation operator, the distributivity of matrix multiplication, and the properties of the transpose operator, we can further simplify this to

$$\begin{aligned} E\{\mathbf{y}_l\mathbf{y}_l^T\} = &\sum_{d=0}^{4} b_d^2 E\{\mathbf{x}_{l-d}\mathbf{x}_{l-d}^T\} + a_1^2 E\{\mathbf{y}_{l-1}\mathbf{y}_{l-1}^T\} - \sum_{d=0}^{4} b_d a_1 E\{\mathbf{y}_{l-1}\mathbf{x}_{l-d}^T\} \\ &- \sum_{d=0}^{4} b_d a_1 E\{\mathbf{x}_{l-d}\mathbf{y}_{l-1}^T\}, \end{aligned} \tag{3.44}$$

where the last two terms are the transpose of each other and account for the correlation between the inputs $\mathbf{x}_l \ldots \mathbf{x}_{l-4}$ and the shifted output $\mathbf{y}_{l-1}$. Since this is not zero due to the action of the IIR filter, it has to be further calculated. Both terms are computed analogously. If, for example, we take the term $E\{\mathbf{y}_{l-1}\mathbf{x}_{l-d}^T\}$, we can further expand it by inserting Equation (3.34) shifted by one time unit, which leads to

$$\begin{aligned} E\{\mathbf{x}_{l-d}\mathbf{y}_{l-1}^T\} &= E\left\{\mathbf{x}_{l-d}\left(\sum_{p=0}^{4} b_p\mathbf{x}_{l-1-p} - a_1\mathbf{y}_{l-2}\right)^T\right\} \\ &= \sum_{p=0}^{4} b_p E\{\mathbf{x}_{l-d}\mathbf{x}_{l-1-p}^T\} - a_1 E\{\mathbf{x}_{l-d}\mathbf{y}_{l-2}^T\}. \end{aligned} \tag{3.45}$$

Since the inputs are assumed uncorrelated (in time), the first term will only be non-zero if the inputs are the same, that is, if $d = 1 + p$, which simplifies the formula to

$$E\{\mathbf{x}_{l-d}\mathbf{y}_{l-1}^T\} = b_{d-1}E\{\mathbf{x}_{l-d}\mathbf{x}_{l-d}^T\} - a_1 E\{\mathbf{x}_{l-d}\mathbf{y}_{l-2}^T\}. \tag{3.46}$$

The second term of this formula is again an input and output correlation that can be expanded as in Equation (3.45) and simplified the same way. By recursively replacing the time shifted equivalents of Equation (3.34) into the remaining input output correlation we reach a point at which the shifted output is no longer dependent on the inputs, thus obtaining

$$\begin{aligned} E\{\mathbf{x}_{l-d}\mathbf{y}_{l-1}^T\} &= b_{d-1}E\{\mathbf{x}_{l-d}\mathbf{x}_{l-d}^T\} - a_1 b_{d-2}E\{\mathbf{x}_{l-d}\mathbf{x}_{l-d}^T\} + a_1^2 E\{\mathbf{x}_{l-d}\mathbf{y}_{l-3}^T\} \\ &\quad + \sum_{q=1}^{d}(-1)^{q-1}a_1^{q-1}b_{d-q}E\{\mathbf{x}_{l-d}\mathbf{x}_{l-d}^T\}. \end{aligned} \tag{3.47}$$

Substituting this result into Equation (3.44), and substituting it as well as the propagated mean in Equation (3.41) into Equation (3.42), we obtain the following formula for the covariance:

$$\begin{aligned} \boldsymbol{\Sigma}_l^{\mathrm{RASTA}} &= \sum_{d=0}^{4} b_d^2 E\{\mathbf{x}_{l-d}\mathbf{x}_{l-d}^T\} + a_1^2 E\{\mathbf{y}_{l-1}\mathbf{y}_{l-1}^T\} \\ &\quad + 2\sum_{d=0}^{4} b_d \sum_{q=1}^{d}(-1)^q a_1^q b_{d-q}E\{\mathbf{x}_{l-d}\mathbf{x}_{l-d}^T\} - \boldsymbol{\mu}_l^{\mathrm{RASTA}}\left(\boldsymbol{\mu}_l^{\mathrm{RASTA}}\right)^T, \end{aligned} \tag{3.48}$$

where $E\{\mathbf{y}_{l-1}\mathbf{y}_{l-1}^T\}$ is obtained from the previous frame computation. This closed-form solution is a particular case of uncertainty propagation through a generic IIR filter. Such a transformation can always be solved using the conventional matrix solutions in Equations (3.7) and (3.8), as long as the output-output and input-output correlations are taken into consideration and included in the input covariance matrix.

The computation of the appearing correlations can be performed on the fly by solving the linear transformations at each step.

It should also be noted that the RASTA filter induces a correlation between features of different frames, $E\{\mathbf{y}_l\mathbf{y}_{l-d}^T\}$. In order to achieve full propagation of second-order information, this correlation should be taken into account. However, for the sake of simplicity, the created time correlation is ignored in the upcoming uncertainty propagation steps. Experimental results show that this correlation has a low influence.

3.5.5 Propagation Through Equal Loudness Pre-emphasis, Power Law of Hearing and Exponential

Both equal loudness preemphasis and the power law of hearing transformation, given in (3.33), can be performed in the log domain where they can be formulated as the following linear transformation

$$H_{jl} = \exp(0.33\log(\psi_j) + 0.33\log(B_{jl})). \tag{3.49}$$

Since the assumed log-normal distribution of uncertainty in the Bark domain led to Gaussian-distributed uncertainty in the log domain, this changes neither after the RASTA nor after the equal loudness preemphasis and power law of hearing transformations. Consequently, applying the exponential transformation leads to a log-normal distribution. The mean and covariance of this distribution can be computed by combining the solution for the linear transformations, in Equations (3.7) and (3.8), with the log-normal forward propagation through the exponential given in [14, Eqs C. 5 and C. 8]:

$$\mu_{jl}^{\text{EXP}} = \exp\left(0.33\log(\psi_j) + 0.33\mu_{jl}^{\text{RASTA}} + \frac{0.33^2\Sigma_{jjl}^{\text{RASTA}}}{2}\right) \tag{3.50}$$

and

$$\Sigma_{jj'l}^{\text{EXP}} = \mu_{jl}^{\text{EXP}}\mu_{j'l}^{\text{EXP}}\left(\exp\left(0.33^2\Sigma_{jj'l}^{\text{RASTA}}\right) - 1\right). \tag{3.51}$$

3.5.6 Propagation Through the All-Pole Model and LPCC Computation

As discussed at the beginning of this section, the computation of the cepstral coefficients from the all-pole model implies computing the inverse Fourier transformation, implemented in this case with the inverse discrete Fourier transform (IDFT), obtaining the linear prediction coefficients with the Levinson-Durbin recursion and finally computing the cepstra. Although some of these steps, like the inverse DFT, are linear, the most suitable solution here is to use the unscented transform to solve all of these transformations in one single nonlinear transformation $\mathbf{g}$. This is advantageous because the dimensionality of the features is low enough and experimental results show that the asymmetry of the pdf is relatively low. The computation of the LPCC mean and covariance can be performed using the corresponding unscented transform Equations (3.10) and (3.11) with $\boldsymbol{\mu}_l^x = \boldsymbol{\mu}_l^{\mathrm{EXP}}$ and $\boldsymbol{\Sigma}_l^x = \boldsymbol{\Sigma}_l^{\mathrm{EXP}}$

3.6 Uncertainty Propagation for the ETSI ES 202 050 Standard

As an example of a state-of-the-art robust feature extraction, the propagation problem was solved for the advanced front-end of the ETSI standard. This standard uses a mel cepstrum based feature extraction with PSD features instead of the STSA features. Mel-PSD features do, however, behave identically to the Bark-PSD features explained in Section 3.5 and thus the uncertainty propagation approach explained in this section remains valid when using Equations (3.35), (3.36), (3.37), (3.38), (3.39), (3.40), (3.30) and (3.31) and taking into account the change between Bark and mel matrices. In addition to these transformations already introduced, the advanced front-end introduces two additional transformations: the log-energy and a blind equalization step.

The log-energy is used as a measure of energy obtained directly from the time domain raw data. This is computed from each time domain frame of the STFT (see Equation (3.1))

$$E_l = \log\left(\sum_{t'=1}^{N} x\left(t' + (l-1)M\right)^2\right), \tag{3.52}$$

where $t' \in [1 \cdots N]$ is the sample index within the frame and M is the frame shift.

Blind equalization (BE) can be considered as a noise suppression scheme in the cepstral domain similar to cepstral mean subtraction. The particular implementation described here forms part of the ETSI standard for robust speech recognition [13, Section 5.4]. This BE implementation is based on the minimization of a distortion measure between the actual signal and a given reference of the flat spectrum cepstra C_i^{ref}. Each equalized cepstral coefficient $\tilde{C}_{il}$ is obtained as

$$\tilde{C}_{il} = C_{il} - b_{il} \tag{3.53}$$

where b_{il} is a bias correction computed recursively from[5]

$$b_{il} = (1-\mu_l)b_{il-1} + \mu_l\left(C_{il-1} - C_i^{\mathrm{ref}}\right) \tag{3.54}$$

[5] Note that the ETSI 202 050 manual and the C code implementation differ in the multiplying factor of the first summand of Equation (3.54). Here, the implementation of the C code was used as the reference.

and where the step size factor μ_l is obtained online using the previous frame energy information by applying the following rule:

$$\mu_l = \begin{cases} \nu & \text{if } \frac{E_{l-1}-211}{64} >= 1 \\ \nu \cdot \frac{E_{l-1}-211}{64} & \text{if } 0 < \frac{E_{l-1}-211}{64} < 1 \\ 0 & \text{if } \frac{E_{l-1}-211}{64} =< 0, \end{cases} \tag{3.55}$$

with $\nu = 0.0087890625$ a fixed parameter given in [23].

3.6.1 Propagation of the Complex Gaussian Model into the Time Domain

Given the uncertain Fourier spectrum of a signal $\mathbf{X}$, where each X_{kl} is statistically independent of the others and complex Gaussian-distributed according to Equation (3.6), it is possible to transform $\mathbf{X}$ back to the time domain with the inverse short time Fourier transform ISTFT, thus obtaining a Gaussian-distributed uncertain time domain signal $x(t)$. In order to do this, it is first necessary to apply the IDFT to obtain, from each frame of uncertain Fourier coefficients $X_{1l} \dots X_{Kl}$, the corresponding time-domain frame $x_{1l} \dots x_{Nl}$, where K denotes the number of bins up to half the sampling frequency and $N = 2K-2$ is the size of the frame.[6] In order to apply the IDFT, it is first necessary to obtain the whole spectrum (N bins), including the frequency bins at half the sampling frequency. This can be obtained as

$$\mathbf{X}'_l = [X_{1l} \dots X_{Kl}, X^*_{K-1l} \dots X^*_{2l}], \tag{3.56}$$

where X^*_{kl} denotes the conjugate of X_{kl}. Since the complex conjugate only implies multiplying the imaginary component of the Fourier coefficient by minus 1, we have that

$$E\{X^*_{kl}\} = E\{X_{kl}\}^*, \tag{3.57}$$

whereas the variance of the conjugate remains unchanged. Taking this into account, the mean of the whole spectrum frame corresponds to

$$E\{\mathbf{X}'_l\} = \hat{\mathbf{X}}'_l = [\hat{X}_{1l} \dots \hat{X}_{Kl}, \hat{X}^*_{K-1l} \dots \hat{X}^*_{2l}]. \tag{3.58}$$

The correlations between real and imaginary components of the complex-valued coefficients can be obtained using the definition of complex Gaussian distribution (see Equation (3.6)). Considering the real X^R_{kl} and imaginary X^I_{kl} components of the spectrum separately, the resulting diagonals of their correlation matrices yield

[6] In the particular case of the ETSI standard STFT, $K = 129$, $N = 256$. Note also that the ETSI STFT uses zero padding. Since this is a trivial operation, it is omitted here for clarity.

$$E\{X_{kl}^{\prime R} X_{kl}^{\prime R}\} = \mathrm{Var}\{X_{kl}^{\prime R}\} + E\{X_{kl}^{\prime R}\}^2 = \lambda_{kl}/2 + \left(\hat{X}_{kl}^R\right)^2 \tag{3.59}$$

and

$$E\{X_{kl}^{\prime I} X_{kl}^{\prime I}\} = \mathrm{Var}\{X_{kl}^{\prime I}\} + E\{X_{kl}^{\prime I}\}^2 = \lambda_{kl}/2 + \left(\hat{X}_{kl}^I\right)^2 \tag{3.60}$$

for $k \in [1 \cdots N]$. In addition to these correlations, the appearing correlations between the original Fourier coefficients and their conjugates (excluding the DC component) have to be considered. These correspond to

$$E\{X_{kl}^{\prime R} X_{k'l}^{\prime R}\} = \lambda_{kl}/2 + \left(\hat{X}_{kl}^R\right)^2 \tag{3.61}$$

and

$$E\{X_{kl}^{\prime I} X_{k'l}^{\prime I}\} = -\lambda_{kl}/2 - \left(\hat{X}_{kl}^I\right)^2 \tag{3.62}$$

with $k \in [2 \cdots N]$ and $k' = N - k + 2$. Taking this into account, we consider the IDFT defined as

$$x_{t'l} = \mathrm{Re}\left\{ \frac{1}{N} \sum_{k=1}^{N} X_{kl}' \exp\left(2\pi j \frac{(t'-1)(k-1)}{N} \right) \right\}, \tag{3.63}$$

where $t' \in [1 \cdots N]$ is the sample index within the frame.

This transformation can be expressed as the following matrix operation[7]:

$$\mathbf{x}_l = \mathrm{Re}\left\{ \mathbf{F} \mathbf{X}_l^T \right\}, \tag{3.64}$$

where

$$F_{kt'} = F_{kt'}^R + jF_{kt'}^I = \frac{1}{N} \exp\left(2\pi j \frac{(t'-1)(k-1)}{N} \right) \tag{3.65}$$

is the inverse Fourier matrix expressed in Cartesian and polar coordinates, which may also be extended to include the inverse window function if one is used.

Since X_{kl}' can be decomposed into its real and imaginary components,

$$X_{kl}' = X_{kl}^{\prime R} + iX_{kl}^{\prime I}, \tag{3.66}$$

Equation (3.64) can be expressed as the following linear transformation involving real-valued matrices:

$$\mathbf{x}_l = \mathrm{Re}\left\{ \mathbf{F} \mathbf{X}_l^T \right\} = \mathbf{F}^R \left(\mathbf{X}_l^{\prime R} \right)^T - \mathbf{F}^I \left(\mathbf{X}_l^{\prime I} \right)^T \tag{3.67}$$

where $F_{kt}^R, F_{kt}^I \in \mathbb{R}$, and by the definition of complex Gaussian distribution, $\mathbf{X}_l^{\prime R}$ and $\mathbf{X}_l^{\prime I}$ are multivariate Gaussian distributed with correlations given by Equations (3.59), (3.60), (3.61) and (3.62). Consequently, $\mathbf{x}_l$ will be multivariate Gaussian-distributed with mean

[7] Re{} operates element-wise over the vector.

$$E\{\mathbf{x}_l\} = \mathbf{F}^R \left(\hat{\mathbf{X}}_l'^R\right)^T - \mathbf{F}^I \left(\hat{\mathbf{X}}_l'^I\right)^T \tag{3.68}$$

and correlation

$$E\{\mathbf{x}_l\mathbf{x}_l^T\} = \mathbf{F}^R E\left\{\mathbf{X}_l'^R(\mathbf{X}_l'^R)^T\right\}\left(\mathbf{F}^R\right)^T + \mathbf{F}^I E\left\{\mathbf{X}_l'^I(\mathbf{X}_l'^I)^T\right\}\left(\mathbf{F}^I\right)^T, \tag{3.69}$$

where for this last formulation, the statistical independence of real and imaginary components has been used, and the correlation matrices in Equations (3.59), (3.60), (3.61) and (3.62) have been expressed in matrix form.

Once each frame of the uncertain Fourier spectrum $\mathbf{X}_l$ has been transformed into an uncertain time domain frame $\mathbf{x}_l$, the uncertain time domain signal can be recovered using the overlap and add method (see [6, Eq. 3]).

3.6.2 Propagation Through Log-Energy Transformation

The computation of the log-energy is attained from each time domain frame of the STFT as given by Equation (3.52). As shown in the last section, for the complex Gaussian uncertainty model, each frame of time-domain elements $x_{1l}\cdots x_{Nl}$ is multivariate Gaussian-distributed with a covariance matrix given by Equation (3.69). Consequently, the mean of the squared time domain elements $x_{1l}^2\cdots x_{Nl}^2$ will correspond to the diagonal of Equation (3.69). Furthermore, given the formula which relates the fourth-order cross-cumulant and the raw moments of a random variable [26, Eq. 2.7], and provided that a multivariate Gaussian distribution has a fourth-order cross-cumulant equal to zero, we have

$$E\{x_{t'l}^2 x_{t''l}^2\} = E\{x_{t'l}^2\}E\{x_{t''l}^2\} + 2E\{x_{t'l}x_{t''l}\}^2 - 2E\{x_{t'l}\}^2E\{x_{t''l}\}^2, \tag{3.70}$$

which enables us to compute the full covariance of the squared Gaussian variables from the mean and covariance given by Equations (3.68) and (3.69). To complete the transformation in Equation (3.52), the logarithm of the sum of $x_{1l}^2\cdots x_{Nl}^2$ has to be computed. The sum is a linear transformation and thus can be computed with Equations (3.7) and (3.8), whereas the logarithm has to be approximated. For this purpose, we are using the log-normality assumption specified by Equations (3.39) and (3.40). Combining the formulas for both mean and covariance results in

$$\Sigma_l^{\mathrm{LogE}} = \log\left(\sum_{t'=1}^{N}\sum_{t''=1}^{N} E\{x_{t'l}^2 x_{t''l}^2\}\right) - 2\log\left(\sum_{t'=1}^{N} E\{x_{t'l}^2\}\right) \tag{3.71}$$

and

$$\mu_l^{\mathrm{LogE}} = \log\left(\sum_{t'=1}^{N} E\{x_{t'l}^2\}\right) - \frac{1}{2}\Sigma_l^{\mathrm{LogE}}, \tag{3.72}$$

where in this case, for the sake of simplicity, Equation (3.40) has been re-expressed in terms of mean and correlation instead of mean and covariance using [26, Eq. 2.4].

3.6.3 Propagation Through Blind Equalization

The propagation of uncertainty through the blind equalization transformation, given by Equations (3.53), (3.54) and (3.55), can be rather complex if we consider that it includes the product of two random variables, the step-factor μ_l, obtained from a nonlinear transformation of the log-energy of the previous frame E_{l-1} (see Equations (3.54) and (3.55)), and the Gaussian-distributed cepstral coefficient C_{il}. As will be shown in the experimental setup, this propagation problem can be simplified by considering the log-energy as deterministic and using the mean of the log energy $\mu_{l-1}^{\mathrm{LogE}}$ instead of E_{l-1}. In this case, the propagation problem can be decoupled into two conventional linear propagation problems solvable using Equations (3.7) and (3.8) on (3.53) and (3.54) individually. The temporal correlations induced between features of different frames are also ignored here.

3.7 Taking Full Covariances into Consideration

3.7.1 Covariances Between Features of the Same Frame

One of the main limiting factors when using uncertainty propagation for robust ASR is the consumption of computation resources. An ASR system can be implemented in multiple types of devices, from embedded systems to distributed systems spanning across several servers. Such systems have different costs for the different operations involved in the uncertainty propagation algorithms and therefore determining the exact computational cost is not possible in general.

One easy modification, which provided a sensible reduction of computational time, was ignoring the correlation created between the features after the filter bank transformations (see Sections 3.4.2, 3.5.2). The assumption of diagonal covariances after the filter bank reduces the number of operations needed for the computation of the covariances in most transformations. In the case of the mel, DCT, Bark, log-Bark, RASTA, exponential and power law of hearing transformations, the number of operations is divided by the dimension of the mel/Bark features (number of filters) J.[8] Furthermore, it also greatly simplifies the unscented transform algorithm (see Section 3.3.2) since, apart from reducing the number of computations, the matrix

[8] The value of J varies with the implementation but is usually around 20.

square root needed for the determination of the sigma points turns into a conventional square root.

The assumption of diagonal covariance can be applied not only after the mel/Bark filter banks, but also after the IDFT transformation when computing the log-energy (see Section 3.6.2).

The effect of both assumptions will be analyzed in the experimental setup in Section 3.8.

3.7.2 Covariance Between Features of Different Frames

Apart from the RASTA filter, there are various linear transformations typically used in feature extractions that combine features of different time frames. Many such transformations can be solved by applying Equations (3.7) and (3.8), as, for example, the computations of the delta and acceleration features and cepstral mean subtraction. Nevertheless, if a previous transformation (e.g., a RASTA filter) has induced a correlation between features of different frames, this should be included in the covariance matrix. For the typical output of RASTA filtering, for example, the uncertain feature distribution is described by a $J \cdot L$ element mean matrix and L intra-frame covariance matrices of size J^2. If all inter-frame time correlations are to be considered, an $(L \cdot J)^2$ matrix is needed. This matrix can be optimized, taking into account that the time correlations only span nearby frames, but it remains, nevertheless, a very significant increase in computational cost. Inter-frame correlation is therefore ignored in delta, acceleration and cepstral mean subtraction transformations. As will be seen in Section 3.8.2, this does not much affect the accuracy of the uncertainty propagation algorithms.

3.8 Tests and Results

3.8.1 Monte Carlo Simulation of Uncertainty Propagation

For the assumed complex Gaussian model of uncertainty, given by Equation (3.6), it is very simple to generate random samples of the distribution. This allows for an approximation of the true uncertain feature distributions by using a Monte Carlo simulation. For this purpose a simulation experiment with additive noise at different SNR levels was carried out. For each case, samples of the resulting STFT matrix of complex Gaussian-distributed values were drawn and transformed through the different feature extractions. The resulting feature matrices of samples were used to compute different statistics. The kurtosis was used as a measure of Gaussianity to determine average and extreme cases. Two cases were analyzed. Uncertainty

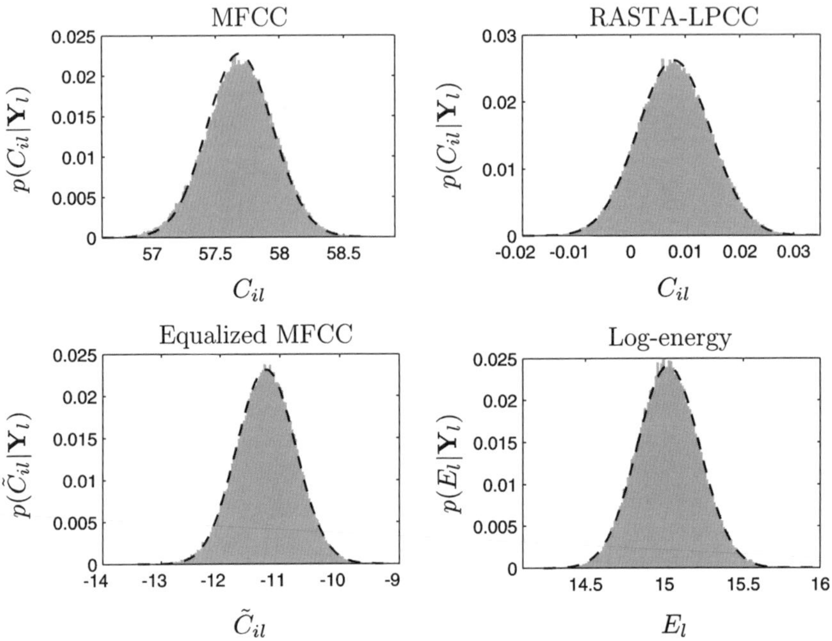

Fig. 3.5: Comparison between uncertainty distributions computed using a Monte Carlo simulation (solid gray) with the assumed uncertainty distributions (dashed) computed from the propagated first- and second-order moments with full covariance (F). From left to right, top to bottom: MFCC, RASTA-LPCC, equalized MFCC according to the ETSI-AFE standard and log-energy

propagation considering the full covariance after the mel filter bank, labeled F, and uncertainty propagation with diagonal covariances after the filter bank, labeled D.[9]

Figure 3.5 shows the uncertain feature distributions representative of the average case. The true distribution of the features obtained through the Monte Carlo simulation (shaded) are compared with the assumed distributions (dashed). The parameters of the assumed distributions are computed from the first- and second-order moments computed as indicated in Section 3.3 using full covariances. The results show that the hypothesis of Gaussianity of the three explored feature types holds reasonably well in all scenarios.

To complete the analysis, first- and second-order moments obtained through the Monte Carlo simulation were compared with the ones obtained with the proposed formulas. For the estimation of the mean, the Monte Carlo and proposed propagation estimations were almost always indistinguishable. The accuracy of the mean estimation was also not affected by the use of full (F) or diagonal covariances (D). The error in the propagated covariance was rather low when using full covariance in all scenarios. When using diagonal covariances, the error increased, particularly for certain features, but the estimated variances remained approximately proportional

[9] See [1] for a detailed explanation of the experiments.

to the true variances. Figure 3.6 displays the variance of the feature component with the highest absolute variance estimation error, comparing Monte Carlo (gray), full covariance (F, solid black) and diagonal covariance (D, dashed) solutions. The error of variance propagation is most noticeable in the case of RASTA-LPCC features propagated with diagonal covariance (D), although this error is only exhibited by the higher-order cepstral coefficients. The use of full covariance (F) provides an accurate computation of the first- and second-order moments.

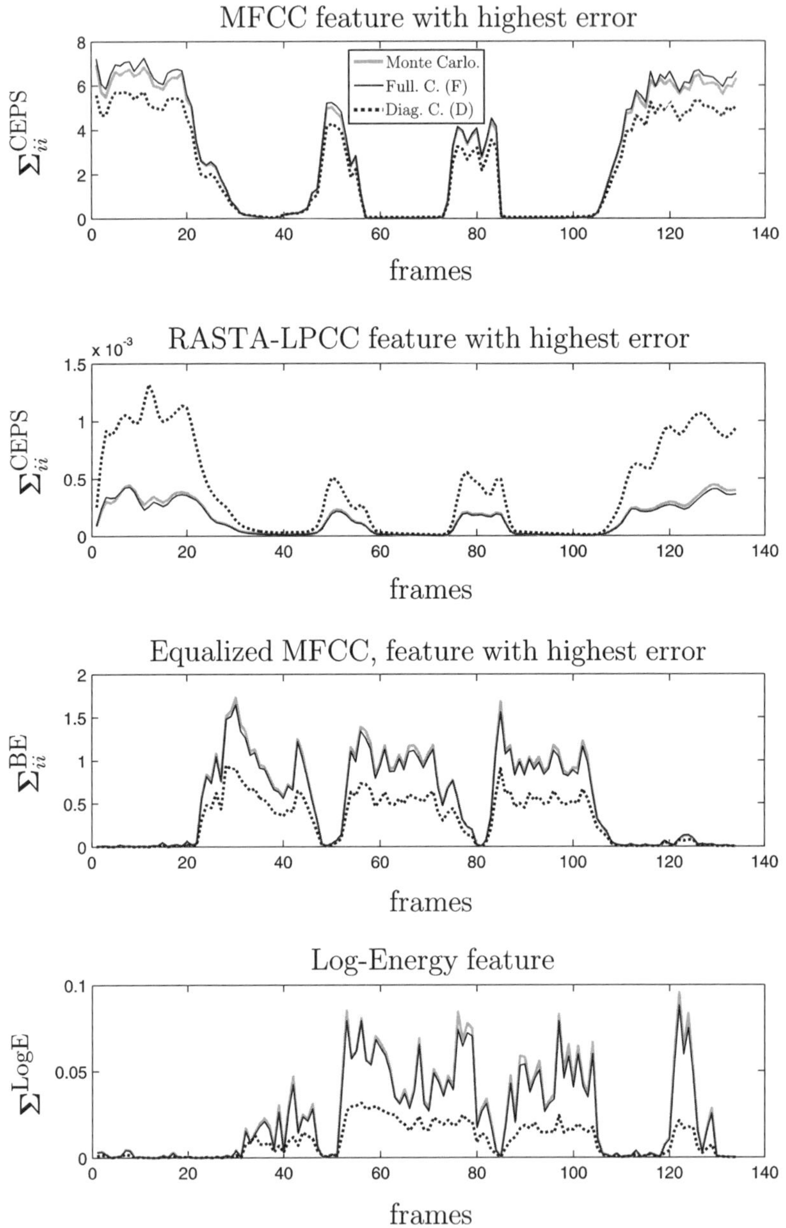

Fig. 3.6: Comparison between propagated variance computed using a Monte Carlo simulation (solid gray) with the covariance computed with the proposed uncertainty propagation algorithms using diagonal (D, dotted black) and full covariance (F, solid black). Only the feature with the highest absolute variance estimation error, when comparing the Monte Carlo simulation with the propagated solution using full covariances, is displayed. From top to bottom: MFCC, RASTA-LPCC, equalized MFCC according to the ETSI-AFE standard and log-energy

Table 3.1: Word error rates [%] for the advanced front-end baseline (AFE), and its combination with uncertainty propagation and modified imputation (MI) or uncertainty decoding (UD). Use of diagonal or full covariance by propagation marked as (D) or (F), respectively. The best results are displayed in bold

Non Reverberant					
SNR	AFE	+MI+D	+UD+D	+MI+F	+UD+F
∞	0.51	0.65	0.54	0.70	0.54
15	2.85	2.65	2.66	2.74	**2.62**
10	5.97	**5.52**	5.65	5.62	5.65
05	14.60	**13.22**	13.56	13.50	13.49
00	34.32	**31.55**	32.83	31.80	32.64
Hands Free Office					
SNR	AFE	+MI+D	+UD+D	+MI+F	+UD+F
∞	5.85	3.94	4.86	**3.86**	4.72
15	11.06	7.80	9.38	**7.46**	9.10
10	17.54	13.45	15.47	**13.01**	15.04
05	30.29	25.10	27.90	**24.47**	27.40
00	51.36	45.89	48.69	**45.15**	48.33
Hands Free Living Room					
SNR	AFE	+MI+D	+UD+D	+MI+F	+UD+F
∞	14.27	**11.82**	13.22	11.97	12.94
15	20.54	16.39	18.72	**16.19**	18.41
10	28.60	23.56	26.40	**23.12**	25.97
05	42.73	37.16	40.34	**36.70**	39.88
00	62.32	57.40	60.17	**56.78**	59.84

3.8.2 Improving the Robustness of ASR with Uncertainty Propagation

The usefulness of the proposed propagation algorithms for robust ASR has been confirmed by positive results in a variety of ASR tasks employing single and multiple channel speech enhancement algorithms in the STFT domain [1, 4, 20]. The results displayed here correspond to the improvement of the European Telecommunications Standards Institute (ETSI) advanced front-end (AFE) in the Aurora 5 test environment [13], originally described in [2]. The highly optimized AFE combines multiple noise suppression approaches to achieve very robust features. At certain steps of the algorithm, speech is enhanced through techniques which need a good noise suppression to work properly. Since this might not always be the case, the uncertainty is set as proportional to the rate of change those steps inflict.

The generated uncertainty is then propagated using the algorithms presented in the previous sections and employed in the domain of recognition features to achieve more robust speech recognition. In order to integrate the uncertainty with the ASR system, uncertainty decoding [10], labeled UD, and modified imputation, labeled MI, [21] (see Chapter 12) are used. In addition to this, the two cases considered in the Monte Carlo simulation, diagonal (D) and full (F) covariances, are also considered here.

The results displayed in Table 3.1 show how the use of uncertainty decoding achieves a general reduction in the word error rates (WER) of the standard's baseline, particularly for the reverberant environments for which the standard is not optimized. Furthermore, the use of diagonal covariances during propagation has only a slightly negative or even a positive effect on the recognition rates. It has to be taken into account that the computational cost of the uncertainty propagation algorithms for diagonal covariance is close to twice that of the conventional MFCC feature extraction. The computational cost of this step is also very small compared with the previous noise suppression steps. The only noticeable increase in costs of the use of uncertainty propagation is the need for transmitting 28 rather than the 14 features usually computed at the terminal side of the standard.

References

1. Astudillo, R.F.: Integration of short-time Fourier domain speech enhancement and observation uncertainty techniques for robust automatic speech recognition. Ph.D. thesis, Technical University Berlin (2010)
2. Astudillo, R.F., Kolossa, D., Mandelartz, P., Orglmeister, R.: An uncertainty propagation approach to robust ASR using the ETSI advanced front-end. IEEE Journal of Selected Topics in Signal Processing **4**, 824 833 (2010)
3. Astudillo, R.F., Kolossa, D., Orglmeister, R.: Propagation of statistical information through non-linear feature extractions for robust speech recognition. In: Proc. MaxEnt 2007 (2007)
4. Astudillo, R.F., Kolossa, D., Orglmeister, R.: Accounting for the uncertainty of speech estimates in the complex domain for minimum mean square error speech enhancement. In: Proc. Interspeech (2009)
5. Benítez, M.C., Segura, J.C., Torre, A., Ramírez, J., Rubio, A.: Including uncertainty of speech observations in robust speech recognition. In: Proc. International Conference on Acoustics Speech and Signal Processing (ICASSP), pp. 137–140 (2004)
6. Cohen, I., Berdugo, B.: Speech enhancement for non-stationary noise environments. Signal Processing **81**(11), 2403 – 2418 (2001)
7. Davis, S., Mermelstein, P.: Comparison of parametric representations for monosyllabic word recognition in continuously spoken sentences. IEEE Trans. Acoust., Speech, Signal Processing **28** **(4)**(2), 357– 366 (1980)
8. Deller, J.R., Hansen, J.H.L., Proakis, J.G.: Discrete-Time Processing of Speech Signals. Prentice-Hall, Inc. (1987)
9. Deng, L., Droppo, J., Acero, A.: Exploiting variances in robust feature extraction based on a parametric model of speech distortion. In: Proc. International Conference on Spoken Language Processing (ICSLP) (2002)
10. Droppo, J., Acero, A., Deng, L.: Uncertainty decoding with SPLICE for noise robust speech recognition. In: Proc. IEEE International Conference on Acoustics, Speech, and Signal Processing (ICASSP)., vol. 1, pp. 57–60 (2002)
11. Ephraim, Y., Cohen, I.: Recent Advancements in Speech Enhancement, pp. 1–22. CRC Press (May 17, 2004)
12. Ephraim, Y., Malah, D.: Speech enhancement using a minimum-mean square error short-time spectral amplitude estimator. IEEE Trans. Acoust., Speech, Signal Processing **32**(6), 1109–1121 (1984)
13. ETSI: ETSI standard document, "Speech Processing, Transmission and Quality Aspects (STQ); Distributed speech recognition; Front-end feature extraction algorithm; Compression algorithms, ETSI ES 202 050 v1.1.5" (January 2007)
14. Gales, M.J.F.: Model-based technique for noise robust speech recognition. Ph.D. thesis, Gonville and Caius College (1995)
15. Gradshteyn, I.S., Ryzhik, I.: Table of Integrals, Series and Products. Elsevier (2007)
16. Hermansky, H., Morgan, N.: RASTA processing of speech. IEEE Trans. on Speech and Audio Processing **2**(4), 578–589 (1994). DOI 10.1109/89.326616
17. Ion, V., Haeb-Umbach, R.: Improved source modeling and predictive classification for channel robust speech recognition. In: Proc. Interspeech (2006)
18. Johnson, N.L.: Continuous Univariate Distributions, Vol. 1. Wiley Interscience (1970)
19. Julier, S., Uhlmann, J.: A general method for approximating nonlinear transformations of probability distributions. Tech. rep., Dept. of Engineering Science, University of Oxford, Oxford, UK (1996)
20. Kolossa, D., Astudillo, R.F., Hoffmann, E., Orglmeister, R.: Independent component analysis and time-frequency masking for speech recognition in multi-talker conditions. EURASIP Journal on Audio, Speech, and Music Processing (2010)
21. Kolossa, D., Klimas, A., Orglmeister, R.: Separation and robust recognition of noisy, convolutive speech mixtures using time-frequency masking and missing data techniques. In: Proc. Workshop on Applications of Signal Processing to Audio and Acoustics (WASPAA), pp. 82–85 (2005)

22. Kolossa, D., Sawada, H., Astudillo, R.F., Orglmeister, R., Makino, S.: Recognition of convolutive speech mixtures by missing feature techniques for ICA. In: Proc. Asilomar Conference on Signals, Systems, and Computers, pp. 1397–1401 (2006)
23. Kuroiwa, S., Tsuge, S., Ren, F.: Blind equalization via minimization of VQ distortion for ETSI standard DSR front-end. In: Proc. International Conference on Natural Language Processing and Knowledge Engineering (NLP-KE), pp. 585–590 (2003). DOI 10.1109/NLPKE.2003.1275974
24. Liao, H., Gales, M.: Issues with uncertainty decoding for noise robust automatic speech recognition. Speech Communication **50**(4), 265 – 277 (2008). DOI DOI:10.1016/j.specom.2007.10.004
25. McAulay, R., Malpass, M.: Speech enhancement using a soft-decision noise suppression filter. IEEE Trans. Acoust., Speech, Signal Processing **28**(2), 137–145 (1980)
26. Nikias, C.L., Petropulu, A.P.: Higher-Order Spectra Analysis: A Nonlinear Signal Processing Framework. Prentice Hall Signal Processing Series (1993)
27. Raj, B., Stern, R.: Reconstruction of missing features for robust speech recognition. Speech Communication **43**(5), 275–296 (2004)
28. Rice, S.O.: Mathematical Analysis of Random Noise, vol. 23. Bell Telephone Labs Inc. (1944)
29. Srinivasan, S., Wang, D.: A supervised learning approach to uncertainty decoding for robust speech recognition. In: Proc. IEEE International Conference on Acoustics, Speech and Signal Processing (ICASSP), vol. 1, pp. I –I (2006)
30. Srinivasan, S., Wang, D.: Transforming binary uncertainties for robust speech recognition. IEEE Trans. Audio, Speech and Language Processing **15**(7), 2130–2140 (2007)
31. Stouten, V., Van hamme, H., Wambacq, P.: Application of minimum statistics and minima controlled recursive averaging methods to estimate a cepstral noise model for robust ASR. In: Proc. IEEE Int. Conf. Acoustics, Speech, Signal Processing (ICASSP), vol. 1, pp. I–I (2006). DOI 10.1109/ICASSP.2006.1660133
32. Stouten, V., Van hamme, H., Wambacq, W.: Model based feature enhancement with uncertainty decoding for noise robust ASR. Speech Communication. **48(11)**, 1502–1514 (2006)
33. Windmann, S., Haeb-Umbach, R.: Parameter estimation of a state-space model of noise for robust speech recognition. Audio, Speech, and Language Processing, IEEE Transactions on **17**(8), 1577 –1590 (2009)
34. Yoma, N., McInnes, F., Jack, M.: Improving performance of spectral subtraction in speech recognition using a model for additive noise. IEEE Trans. Speech, Audio Processing **6 (6)**, 579–582 (1998)

Part II
Applications: Noise Robustness

Chapter 4
Front-End, Back-End, and Hybrid Techniques for Noise-Robust Speech Recognition

Li Deng

Abstract Noise robustness has long been an active area of research that captures significant interest from speech recognition researchers and developers. In this chapter, with a focus on the problem of uncertainty handling in robust speech recognition, we use the Bayesian framework as a common thread for connecting, analyzing, and categorizing a number of popular approaches to the solutions pursued in the recent past. The topics covered in this chapter include 1) Bayesian decision rules with unreliable features and unreliable model parameters; 2) principled ways of computing feature uncertainty using structured speech distortion models; 3) use of a phase factor in an advanced speech distortion model for feature compensation; 4) a novel perspective on model compensation as a special implementation of the general Bayesian predictive classification rule capitalizing on model parameter uncertainty; 5) taxonomy of noise compensation techniques using two distinct axes, feature vs. model domain and structured vs. unstructured transformation; and 6) noise-adaptive training as a hybrid feature-model compensation framework and its various forms of extension.

4.1 Introduction

Noise-robust speech recognition has been an active area of research for many years, and is still a vigorous research area with many practical applications today, e.g., [1, 7–10, 17, 18, 24, 39, 59, 66, 80, 85]. There are numerous challenges to building a speech recognition system that is robust to environmental noise. Noise is unpredictable, time-varying, and has a variety of properties. Not only is accurate noise estimation itself a difficult problem, but, even given an accurate model of noise, nonlinear interactions between clean speech and noise in generating noise-distorted speech (often parameterized in the log power spectrum or cepstrum) also give rise to high complexity in decoding speech, with a high degree of imperfection.

Microsoft Research, One Microsoft Way, Redmond, WA 98052
deng@microsoft.com

D. Kolossa and R. Haeb-Umbach (eds.), *Robust Speech Recognition of Uncertain or Missing Data: Theory and Applications*, DOI 10.1007/978-3-642-21317-5_4,
© Springer-Verlag Berlin Heidelberg 2011

Standard noise robustness methods can be divided into the broad categories of feature compensation and model compensation. Feature compensation is also called feature enhancement or front-end denoising, where the effect of noise is removed from the observed noisy speech features without using speech decoding results and without changing the parameters of the acoustic model (e.g., HMM). On the other hand, in model compensation the acoustic model parameters can be modified to incorporate the effects of noise, where each component of the model can be adapted to account for how the noise affects its mean and variance. While typically achieving higher performance than feature compensation, model compensation often incurs significantly greater computational cost with straightforward implementation (unless drastic approximations are made, such as the use of sub-space techniques, e.g., [64]).

In the past decade, with the earliest work published in the same year [5, 19, 32, 57], a new approach to robust ASR has emerged that is aimed at propagating the uncertainty in the acoustic features due to either the noise effect itself or the residual noise after feature compensation into the decoding process of speech recognition. These techniques provide, at the frame level, dynamic compensation of HMM variances as well as its means, based on the estimate of uncertainty caused by imperfect feature enhancement, and incorporate the compensated parameters into the decoding process. The goal is to achieve recognition performance that is comparable to model compensation techniques, with computational cost similar to feature compensation. This way of handling uncertainty in speech feature data, the theme of this book, appears to strike a balance between computational cost and noise-robust speech recognition accuracy. Several key issues related to the scheme of the above uncertainty decoding have been addressed by a number of researchers, with various new approaches proposed and developed in [28, 43, 50, 51, 65–67, 82].

Given the large number of noise-robust speech recognition techniques developed over the past two decades, this chapter provides a selective overview of them and focuses on the topics that have particular relevance for the future development of noise-robust speech recognition technology. Section 2 starts by introducing the Bayesian perspective as a common thread that connects the remaining topics presented in this chapter, and provides a theoretical background for treating uncertain data. A concrete example is given in Section 3 on how data uncertainty produced in feature compensation can be computed in a principled way. The Algonquin model of speech distortion is used in the example. A more detailed model of speech distortion and mismatch than Algonquin, which makes use of the non-uniform distribution of phase factors between clean speech and mixing noise (due to many data points in filter banks while computing cepstral features), is presented and discussed in Section 4. Feature compensation experiments using this phase-sensitive model and insights gained from these experiments are also provided. In Section 5, we shift the discussion from feature-domain compensation and the associated uncertainty to their model-domain counterparts. Uncertainty in model parameters is studied within the Bayesian framework, where both feature and model uncertainties are integrated into a most general form of Bayesian predictive classification rule. Importantly, in this general framework, model compensation can be viewed as a special realization of the Bayesian predictive classification rule. Then, in Section 6, a taxonomy of a multitude of model compensation techniques is provided based on the structured vs. unstructured transformation of the model parameters. This same axis of structured vs. unstructured transformation in the feature domain is also used to categorize a multitude of feature compensation techniques. Finally, in Section 7, we discuss hybrid feature-model compensation techniques and, in particular, its important member, noise-adaptive training.

4.2 Bayesian Decision Rule with Unreliable Features

In this section, we start from the Bayesian framework to account for uncertainty of the features that form the input into speech recognition systems. This framework provides the theoretical basis for "uncertainty decoding", which manifests itself in various forms in the literature. In Section 5 of this chapter, we apply the same framework to account for uncertainty of model parameters, providing the basis for model compensation.

The standard Bayesian decision rule for speech recognition, with fixed model parameter set Λ, gives

$$\hat{\mathbf{W}} = \underset{\mathbf{W}}{\operatorname{argmax}}\, p(\mathbf{y}|\mathbf{W},\Lambda)P(\mathbf{W}), \tag{4.1}$$

where $P(\mathbf{W})$ is the prior probability that the speaker utters a word sequence $\mathbf{W}$, and $p(\mathbf{y}|\Lambda,\mathbf{W})$ is the probability that the speaker produces the acoustic feature sequence $\mathbf{y} = [\mathbf{y}_1, \mathbf{y}_2, ..., \mathbf{y}_t, ...\mathbf{y}_T]$ when $\mathbf{W}$ is the intended word sequence. The computation of probability $P(\mathbf{y}|\Lambda,\mathbf{W})$ uses deterministic parameters, denoted by Λ, in the speech model.

Using the rule of total probability and conditional independence, we have

$$p(\mathbf{W},\mathbf{y}) = \int p(\mathbf{W},\mathbf{y},\mathbf{x})d\mathbf{x} = \int p(\mathbf{W}|\mathbf{x})p(\mathbf{x}|\mathbf{y})d\mathbf{x}. \tag{4.2}$$

Thus Eq. (4.1) becomes

$$\hat{\mathbf{W}} = \underset{\mathbf{W}}{\operatorname{argmax}} \int p(\mathbf{W}|\mathbf{x},\Lambda)p(\mathbf{x}|\mathbf{y})d\mathbf{x}, \tag{4.3}$$

where the conditional $p(\mathbf{x}|\mathbf{y})$ represents the effect of feature compensation from noisy speech $\mathbf{y}$ to enhanced speech $\mathbf{x}$, and $p(\mathbf{W}|\mathbf{x},\Lambda)$ is the objective function for the decoding problem of speech recognition with input feature $\mathbf{x}$.

Applying Bayes' rule on $p(\mathbf{W}|\mathbf{x})$, we obtain from Eq. (4.3)

$$\hat{\mathbf{W}} = \underset{\mathbf{W}}{\operatorname{argmax}} \int \frac{p(\mathbf{x}|\mathbf{W},\Lambda)}{p(\mathbf{x})} p(\mathbf{x}|\mathbf{y},\Lambda)d\mathbf{x}P(\mathbf{W}). \tag{4.4}$$

Since the prior speech distribution of $p(\mathbf{x})$ is sufficiently broad, with its variance being significantly larger than the posterior (an assumption which was also used in [32]), we can simplify the rule by assuming it is approximately constant over the range of $p(\mathbf{x})$ values of interest. Thus, Eq. (4.4) becomes

$$\hat{\mathbf{W}} \approx \underset{\mathbf{W}}{\operatorname{argmax}} \int p(\mathbf{x}|\mathbf{W},\Lambda)p(\mathbf{x}|\mathbf{y},\Lambda)d\mathbf{x}P(\mathbf{W}), \tag{4.5}$$

which is the uncertainty decoding rule used in [19, 28, 50]. It was pointed out in [66] that under low-SNR conditions the above assumption no longer holds. This observation accounts for poor uncertainty decoding results at low SNR.

One main advantage of the approximate uncertainty decoding rule of Eq. (4.5) is the simplicity in its incorporation into the recognizer's decision rule. This simplicity arises from the fact that a product of two Gaussians remains a Gaussian. One major improvement over the conventional uncertainty decoding rule of Eq. (4.5) is the exploitation of temporal correlations in this rule. The work of [51] explicitly models such correlations in the context of uncertainty decoding. A similar motivation for explicitly exploiting the temporal correlation is presented in [20, 27] in the context of feature compensation, which gives the mean values of the distribution $p(\mathbf{x}|\mathbf{y})$. On the other hand, appending differential parameters to the static ones in the implementation of the uncertainty decoding rule implicitly represents the temporal correlations, as carried out in the work of [28, 50]. Another major improvement comes from a series of work reported in [65–67], where the authors developed "joint uncertainty decoding" in which the acoustic space is subdivided into regions and the joint density of clean and noisy speech is estimated using stereo data.

One of most practical issues in uncertainty decoding is the computation of feature uncertainty from $p(\mathbf{x}|\mathbf{y})$, or equivalently $p(\mathbf{y}|\mathbf{x}) = \alpha \frac{p(\mathbf{x}|\mathbf{y})}{p(\mathbf{x})}$, in Eq. (4.4). In [32], the SPLICE model, originally developed in [17, 18] with numerous further improvements [3, 18, 22, 25, 30, 33, 87], was used to compute feature uncertainty based on some approximation of $p(\mathbf{y}|\mathbf{x})$ (although to a lesser degree than the assumption of flat $p(\mathbf{x})$). Feature uncertainty is expressed directly as a function of the SPLICE parameters determined separately using the techniques described in [17, 25]. In [50, 51], feature uncertainty was determined by detailed analysis in an interesting scenario of distributed speech recognition where bit errors or lost speech data packets on IP connections are encountered.

Finally, in [19, 28], feature uncertainty was computed using a parametric statistical model of acoustic distortion, sometimes called the Algonquin model [37, 38], a probabilistic extension of the commonly used deterministic VTS (Vector Taylor Series) model [2, 56, 71]. Because the approach developed in this work has generality, we will review it in some detail in the following section, and point out further potential of this approach.

4.3 Use of Algonquin Model to Compute Feature Uncertainty

In this section, we use the log-spectral features to represent all acoustic data, be it clean speech, noise, noisy speech, or enhanced speech. We provide an iterative solution, where each iteration contains a closed form solution, for the computation of the "uncertainty" expressed as the MMSE estimate and its variance. It is important to point out that when the complex Fourier transform is taken as the feature to represent the acoustic data, a rather different type of analysis and solution can be used [6].

4.3.1 Algonquin Model of Speech Distortion

The Algonquin algorithm [37, 38, 75] was originally developed as a feature compensation technique, and was extended for use in uncertainty decoding in [19, 28]. Underlying the Algonquin algorithm is the Algonquin model which characterizes the relationship among clean speech, corrupting noise, and noisy speech in the log-spectral or the cepstral domain with an SNR-independent modeling-error residual. Without such a modeling residual, the Algonquin model would be the same as the standard VTS model of speech distortion [2, 56, 71]. The residual captures crude properties of the modeling error in deriving the VTS model. The modeling error includes notably the ignorance of phases between clean speech and noise vectors, both in the log domain involving multiple filter bank channels, in producing the noisy speech vector; see the detailed analysis in Section II of [27]. We now provide an overview of the Algonquin model.

Let $\mathbf{y}, \mathbf{x}$, and $\mathbf{n}$ be single-frame vectors of log mel filter energies for noisy speech, clean speech, and additive noise, respectively. These quantities are shown in [27] to satisfy the following relationship when the phase relationship between clean speech and noise is considered:

$$
\begin{aligned}
\mathbf{y} &= \mathbf{x} + \log\left[(\mathbf{1}+e^{\mathbf{n}-\mathbf{x}})[\mathbf{1}+\frac{2\,\boldsymbol{\lambda} e^{\frac{\mathbf{n}-\mathbf{x}}{2}}}{(\mathbf{1}+e^{\mathbf{n}-\mathbf{x}})}]\right] \\
&\approx \mathbf{x} + \log(\mathbf{1}+e^{\mathbf{n}-\mathbf{x}}) + \frac{\boldsymbol{\lambda}}{\cosh(\frac{\mathbf{n}-\mathbf{x}}{2})},
\end{aligned}
\tag{4.6}
$$

where $\boldsymbol{\lambda}$ is the inner product of the clean speech and noise vectors of mel filter energies in the linear domain, and the last step of approximation uses the assumption that $\boldsymbol{\lambda} \ll \cosh(\frac{\mathbf{n}-\mathbf{x}}{2})$.

In order to simplify the complicated evaluation of the small prediction residual (Eq. (4.6)) of

$$
\mathbf{r} = \frac{\boldsymbol{\lambda}}{\cosh(\frac{\mathbf{n}-\mathbf{x}}{2})},
\tag{4.7}
$$

an "ignorance" modeling approach is taken to model it as a zero mean, Gaussian random vector. This thus gives a probabilistic model of

$$\mathbf{y} = \mathbf{x} + \mathbf{g}(\mathbf{n} - \mathbf{x}) + \mathbf{r}, \tag{4.8}$$

where $\mathbf{g}(\mathbf{z}) = \log\big(\mathbf{1} + e^{\mathbf{z}}\big)$, and $\mathbf{r} \sim \mathcal{N}(\mathbf{r}; \mathbf{0}, \boldsymbol{\Psi})$.

The Gaussian assumption for the residual $\mathbf{r}$ in Eq. (4.7) allows straightforward computation of the conditional likelihood of the noisy speech vector according to

$$p(\mathbf{y}|\mathbf{x},\mathbf{n}) = \mathcal{N}[\mathbf{y}; \mathbf{x} + \mathbf{g}(\mathbf{n} - \mathbf{x}), \boldsymbol{\Psi}]. \tag{4.9}$$

We call Eq. (4.9) the Algonquin model, originally developed in [37], where the parameter $\boldsymbol{\Psi}$ is fixed and independent of SNR. This model was later extended to the phase-sensitive model by making $\boldsymbol{\Psi}$ dependent on SNR using more detailed knowledge of the property of speech and noise mixing.

The Algonquin model is a central element of the Algonquin algorithm, which uses prior Gaussian mixture models for both clean speech, $p(\mathbf{x})$, and noise, $p(\mathbf{n})$, in the log domain for computing an MMSE estimate of clean speech $\mathbf{x}$. In deriving the estimate, a variational algorithm is used to obtain an approximate posterior. Multiple point Taylor linearization is used also, one at each of the Gaussian mean vectors in the Gaussian mixture model of clean speech. The Algonquin algorithm in its original form, which uses multiple points of Taylor series expansion, was compared carefully with the single-point expansion method for log-spectral feature enhancement [20, 27] in internal evaluation (unpublished). The single-point expansion method requires more iterations to converge but overall is much more efficient than the multiple-point expansion method of Algonquin and does not require the use of variational inference. In terms of performance, the single-point expansion method produces better results, especially after the dynamic prior is introduced to model the temporal correlation in feature enhancement [27].

4.3.2 Step I in Computing Uncertainty: Means

As with the Algonquin model assumption, the following Gaussian-mixture distribution is used as the prior model for clean speech:

$$p(\mathbf{x}_t) = \sum_{m=1}^{M} c_m p(\mathbf{x}_t|m) = \sum_{m=1}^{M} c_m \mathcal{N}(\mathbf{x}_t; \boldsymbol{\mu}_m^x, \boldsymbol{\Sigma}_m^x).$$

For simplicity in presentation without loss of generality, the prior model for noise is assumed to be a time-varying delta function instead of a Gaussian mixture as in the original Algonquin model:

$$p(\mathbf{n}_t) = \delta(\mathbf{n}_t - \bar{\mathbf{n}}_t), \tag{4.10}$$

where $\bar{\mathbf{n}}_t$ is assumed known, and can be determined by any noise tracking algorithm, e.g., [25, 26, 58].

The MMSE estimate is computed as the expected value of the posterior probability $p(\mathbf{x}|\mathbf{y})$:

$$\hat{\mathbf{x}} = E[\mathbf{x}|\mathbf{y}] = \int \mathbf{x}p(\mathbf{x}|\mathbf{y})d\mathbf{x}. \tag{4.11}$$

Using Bayes' rule and using the prior speech and noise models just described, this MMSE estimate becomes

$$\begin{aligned}\hat{\mathbf{x}} &= \frac{\int \mathbf{x}p(\mathbf{y}|\mathbf{x})p(\mathbf{x})d\mathbf{x}}{p(\mathbf{y})} \\ &= \frac{\sum_{m=1}^{M} c_m \int\int \mathbf{x}p(\mathbf{n})p(\mathbf{x}|m)p(\mathbf{y}|\mathbf{x},\mathbf{n})d\mathbf{x}d\mathbf{n}}{p(\mathbf{y})} \\ &= \frac{\sum_{m=1}^{M} c_m \int \mathbf{x}p(\mathbf{x}|m)p(\mathbf{y}|\mathbf{x},\bar{\mathbf{n}})d\mathbf{x}}{p(\mathbf{y})}.\end{aligned} \tag{4.12}$$

Substituting the parametric acoustic distortion model of Eq. (4.9) into Eq. (4.12) and carrying out the needed integration in an analytical form via the use of iterative Taylor series approximation (truncation to the first order), we have approximated the evaluation of the MMSE estimate in Eq. (4.12) using the following iterative procedure. First, train and fix all parameters in the clean speech model: $c_m, \boldsymbol{\mu}_m^x$, and $\boldsymbol{\Sigma}_m^x$. Then, compute the noise estimate and the weighting matrices:

$$\begin{aligned}\mathbf{W}_1(m) &= (\boldsymbol{\Sigma}_m^x + \boldsymbol{\Psi})^{-1}\boldsymbol{\Psi}, \\ \mathbf{W}_2(m) &= \mathbf{I} - \mathbf{W}_1(m).\end{aligned} \tag{4.13}$$

Next, fix the total number, J, of intra-frame iterations. For each frame $t = 2,3,...,T$ in a noisy utterance $\mathbf{y}_t$, set iteration number $j = 1$, and initialize the clean speech estimate with

$$\hat{\mathbf{x}}_t^{(1)} = \underset{\boldsymbol{\mu}_m^x}{\operatorname{argmax}} \mathcal{N}[\mathbf{y}_t; \boldsymbol{\mu}_m^x + \mathbf{g}(\bar{\mathbf{n}}_t - \boldsymbol{\mu}_m^x), \boldsymbol{\Psi}]. \tag{4.14}$$

Then, execute the following steps for each time frame (and then sequentially over time frames):

- Step 1: Compute

$$\gamma_t^{(j)}(m) = \frac{c_m \mathcal{N}(\mathbf{y}_t; \boldsymbol{\mu}_m^x + \mathbf{g}^{(j)}, \boldsymbol{\Sigma}_m^x + \boldsymbol{\Psi})}{\sum_{m=1}^{M} c_m \mathcal{N}(\mathbf{y}_t; \boldsymbol{\mu}_m^x + \mathbf{g}^{(j)}, \boldsymbol{\Sigma}_m^x + \boldsymbol{\Psi})},$$

  where $\mathbf{g}^{(j)} = \log\left(\mathbf{1} + e^{\bar{\mathbf{n}}_t - \hat{\mathbf{x}}_t^{(j)}}\right)$.
- Step 2: Update the MMSE estimate:

$$\hat{\mathbf{x}}_t^{(j+1)} = \sum_m \gamma_t^{(j)}(m)\left[\mathbf{W}_1(m)\boldsymbol{\mu}_m^x + \mathbf{W}_2(m)(\mathbf{y}_t - \mathbf{g}^{(j)})\right]. \tag{4.15}$$

- Step 3: If $j < J$, increment j by 1, and continue the iteration by returning to Step 1. If $j = J$, then increment t by 1 and start the algorithm again by resetting $j = 1$ to process the next time frame until the end of the utterance $t = T$.

The expectation of the enhanced speech feature vector is obtained as the final iteration of the estimate above for each time frame:

$$\boldsymbol{\mu}_{\hat{\mathbf{x}}_t} = \hat{\mathbf{x}}_t^{(J)}. \tag{4.16}$$

4.3.3 Step II in Computing Uncertainty: Variances

Given the expectation for the enhanced speech feature computed as just described, the variance of the enhanced speech feature can now be computed according to

$$\boldsymbol{\Sigma}_{\hat{\mathbf{x}}_t} = E[\mathbf{x}_t\mathbf{x}_t^T|\mathbf{y}_t] - \boldsymbol{\mu}_{\hat{\mathbf{x}}_t}\boldsymbol{\mu}_{\hat{\mathbf{x}}_t}^T, \tag{4.17}$$

where the second-order moment is

$$\begin{aligned} E[\mathbf{x}_t\mathbf{x}_t^T|\mathbf{y}_t] &= \int \mathbf{x}_t\mathbf{x}_t^T p(\mathbf{x}_t|\mathbf{y}_t,\bar{\mathbf{n}}_t)d\mathbf{x}_t \\ &= \frac{\int \mathbf{x}_t\mathbf{x}_t^T p(\mathbf{x}_t)p(\mathbf{y}_t|\mathbf{x}_t,\bar{\mathbf{n}}_t)d\mathbf{x}_t}{p(\mathbf{y}_t)} \\ &= \frac{\sum_{m=1}^M c_m \overbrace{\int \mathbf{x}_t\mathbf{x}_t^T p(\mathbf{x}_t|m)p(\mathbf{y}_t|\mathbf{x}_t,\bar{\mathbf{n}}_t)d\mathbf{x}_t}^{I_m(\mathbf{y}_t)}}{p(\mathbf{y}_t)}. \end{aligned} \tag{4.18}$$

After using the zero-th order Taylor series to approximate the nonlinear function $\mathbf{g}(\bar{\mathbf{n}}_t - \mathbf{x}_t)$ by $\mathbf{g}_0(\bar{\mathbf{n}}_t - \mathbf{x}_0)$, the integral in Eq. (4.18) becomes

$$\begin{aligned} I_m(\mathbf{y}_t) &\approx \int \mathbf{x}_t\mathbf{x}_t^T \mathscr{N}(\mathbf{x}_t;\boldsymbol{\mu}_m^x,\boldsymbol{\Sigma}_m^x)\mathscr{N}(\mathbf{y}_t;\mathbf{x}_t+\mathbf{g}_0,\boldsymbol{\Psi})d\mathbf{x}_t \\ &= \int \mathbf{x}_t\mathbf{x}_t^T \mathscr{N}(\mathbf{x}_t;\boldsymbol{\mu}_m^x,\boldsymbol{\Sigma}_m^x)\mathscr{N}(\mathbf{x}_t;\mathbf{y}_t-\mathbf{g}_0,\boldsymbol{\Psi})d\mathbf{x}_t \\ &= \int \mathbf{x}_t\mathbf{x}_t^T \mathscr{N}\left[\mathbf{x}_t;\boldsymbol{\theta}_m(t),(\boldsymbol{\Sigma}_m^x+\boldsymbol{\Psi})^{-1}\boldsymbol{\Sigma}_m^x\boldsymbol{\Psi}\right]d\mathbf{x}_t N_m(\mathbf{y}_t) \\ &= \left[(\boldsymbol{\Sigma}_m^x+\boldsymbol{\Psi})^{-1}\boldsymbol{\Sigma}_m^x\boldsymbol{\Psi}+\boldsymbol{\theta}_m\boldsymbol{\theta}_m^T\right]N_m(\mathbf{y}_t) \end{aligned} \tag{4.19}$$

where

$$\begin{aligned} \boldsymbol{\theta}_m(t) &= (\boldsymbol{\Sigma}_m^x+\boldsymbol{\Psi})^{-1}\left[\boldsymbol{\Psi}\boldsymbol{\mu}_m^x+\boldsymbol{\Sigma}_m^x(\mathbf{y}_t-\mathbf{g}_0)\right], \\ N_m(\mathbf{y}_t) &= \mathscr{N}\left[\mathbf{y}_t-\mathbf{g}_0;\boldsymbol{\mu}_m^x,\boldsymbol{\Sigma}_m^x+\boldsymbol{\Psi}\right] = \mathscr{N}\left[\mathbf{y}_t;\boldsymbol{\mu}_m^x+\mathbf{g}_0,\boldsymbol{\Sigma}_m^x+\boldsymbol{\Psi}\right]. \end{aligned}$$

Substituting the result of Eq. (4.19) into Eq. (4.18), we obtain

$$E[\mathbf{x}_t\mathbf{x}_t^T|\mathbf{y}_t] = \sum_{m=1}^{M} \eta_m(\mathbf{y}_t)\left[(\boldsymbol{\Sigma}_m^x+\boldsymbol{\Psi})^{-1}\boldsymbol{\Sigma}_m^x\boldsymbol{\Psi}+\boldsymbol{\theta}_m(t)\boldsymbol{\theta}_m^T(t)\right], \tag{4.20}$$

where

$$\eta_m(\mathbf{y}_t) = \frac{c_m N_m(\mathbf{y}_t)}{\sum_{m=1}^{M} c_m N_m(\mathbf{y}_t)},$$

and where we used the result that $p(\mathbf{y}_t) = \sum_{m=1}^{M} c_m N_m(\mathbf{y}_t)$ for the denominator.

Equation (4.17) then gives the estimate of the variance for the enhanced feature. In the implementation, an iterative procedure is also used to estimate the variance, for the same purpose of reducing errors caused by the approximation of $\mathbf{g}(\bar{\mathbf{n}}-\mathbf{x})$ by $\mathbf{g}_0(\bar{\mathbf{n}}-\mathbf{x}_0)$. For each iteration, the variance estimate takes the final form of

$$\begin{aligned}\boldsymbol{\Sigma}_{\hat{\mathbf{x}}_t} = & \sum_{m=1}^{M} \eta_m(\mathbf{y}_t)\left[(\boldsymbol{\Sigma}_m^x+\boldsymbol{\Psi})^{-1}\boldsymbol{\Sigma}_m^x\boldsymbol{\Psi}+\boldsymbol{\theta}_m(t)\boldsymbol{\theta}_m^T(t)\right] \\ & - \left[\sum_m \gamma_t(m)\left(\mathbf{W}_1(m)\boldsymbol{\mu}_m^x+\mathbf{W}_2(m)(\mathbf{y}_t-\mathbf{g}_0)\right)\right] \cdot \\ & \left[\sum_m \gamma_t(m)\left(\mathbf{W}_1(m)\boldsymbol{\mu}_m^x+\mathbf{W}_2(m)(\mathbf{y}_t-\mathbf{g}_0)\right)\right]^T\end{aligned} \tag{4.21}$$

after combining Eqs. (4.17), (4.20), and (4.15). Note that the weights $\gamma_t(m)$ above in the form of posterior probability are computed for each of the iterations.

4.3.4 Discussions

As can be seen from Eqs. (4.21) and (4.16), the first two moments of the probabilistic feature enhancement, characterized by $p(\mathbf{x}|\mathbf{y})$, is dynamic or time-varying on a frame-by-frame basis. This is very powerful, and is difficult to achieve using the common model compensation techniques. To the best of our knowledge, the only other technique that also achieves frame-by-frame compensation with practical success is variable-parameter HMMs [84, 85], but this benefit comes with a higher computation cost.

Characterization of feature uncertainty is typically done using the first two moments of $p(\mathbf{x}|\mathbf{y})$ as shown above. The approach we took as shown above permits the computation of any higher-order moments. But how to incorporate such more detailed information about uncertainty into the decoding rule of speech recognizers based on Gaussian mixture HMMs in a computationally efficient way is an open problem.

One extension of the uncertainty computation technique discussed above is to remove the temporal conditional independence assumption, which has been discussed in some detail in [27, 51]. Two solutions are offered in [27] and [51] both demonstrating clear performance improvement after incorporating the improved uncertainty estimates.

In [65, 67], the authors presented an interesting analysis demonstrating problematic issues with front-end or feature uncertainty decoding schemes, e.g.,[5, 19, 32, 50, 51, 57]. The crux of the problem is the often-found acoustic regions where at low SNR no discriminative information is retained since only a single set of compensation parameters is propagated from the front-end processor to the recognizer's decoder. They developed an improved scheme, called joint uncertainty decoding (JUD), where the model-based concept is embedded into uncertainty decoding. Specifically, instead of linking feature components to the recognizer components, they associate each feature component with a set of recognition model components. Introducing the model component sets provides discriminative information at the otherwise non-discriminative acoustic regions in the original front-end uncertainty decoding without use of any discriminative method. How to exploit some concepts from well-established powerful discriminative techniques (e.g., [44]) to further improve joint uncertainty decoding, or simply to improve the more efficient front-end uncertainty decoding, is an interesting research topic.

Another possible extension of the uncertainty computation technique discussed in this section is to use more advanced models of speech distortion than the Algonquin model. A series of such models have been developed, exploiting the SNR dependency of the residual error in the Algonquin model but based on more detailed analysis than the Algonquin model of the phase relationship between clean speech and the mixing noise. We will take up this topic in the next section, with a review on how these phase-sensitive models have been used for fixed-point feature compensation, which has yet to be extended to derive feature compensation uncertainty with these models.

4.4 Use of a Phase-Sensitive Model for Feature Compensation

Traditionally, the interaction model for environmental distortion ignores the phase asynchrony between the clean speech and the mixing noise [1, 71]; it is known as the VTS-model. This type of crude model has been improved over the past several years to achieve higher fidelity that removes the earlier simplifying assumption by including random phase asynchrony. As discussed in the preceding section, the Algonquin model is a simple kind of extension that lumps all modeling errors, including the phase effect, into a zero mean, fixed-variance Gaussian residual in an "ignorant" manner [19, 27, 37, 38]. Phase sensitive models developed subsequently make further improvements over the Algonquin model, resulting in an SNR-dependent residual component. A series of work on the phase-sensitive models, including their successful applications to both feature compensation and model compensation, can be found in [21, 23, 31, 61, 62, 75, 81]. In this section, we limit our review on this part of the literature to feature compensation only.

4.4.1 Phase-Sensitive Modeling of Acoustic Distortion — Deterministic Version

To introduce the background and for simplicity, we derive the phase-sensitive model in the log filter bank domain. (This can be easily extended to the cepstral domain.) Using the discrete-time, linear system model for acoustic distortion in the time domain, we have the well-known relationship among noisy speech ($y(t)$), clean speech ($x(t)$), additive noise ($n(t)$), and the impulse response of the linear distortion channel ($h(t)$):

$$y(t) = x(t) * h(t) + n(t).$$

In the frequency domain, the equivalent relationship is

$$Y[k] = X[k]H[k] + N[k], \tag{4.22}$$

where k is the frequency bin index in the DFT given a fixed-length time window, and $H(k)$ is the (frequency domain) transfer function of the linear channel.

The power spectrum of the noisy speech can then be obtained from the DFT in Eq. (4.22) by

$$\begin{aligned} |Y[k]|^2 &= |X[k]H[k] + N[k]|^2 \\ &= |X[k]|^2|H[k]|^2 + |N[k]|^2 + (X[k]H[k])(N[k])^* \\ &\quad + (X[k]H[k])^*N[k] \\ &= |X[k]|^2|H[k]|^2 + |N[k]|^2 + 2|X[k]||H[k]||N[k]|\cos\theta_k, \end{aligned} \tag{4.23}$$

where θ_k denotes the (random) phase angle between the two complex variables $N[k]$ and $(X[k]H[k])$. Equation (4.23) incorporates the phase relationship between the (linearly filtered) clean speech and the additive corrupting noise in the speech distortion process. It is noted that in the traditional, phase-insensitive models for acoustic distortion, the last term in Eq. (4.23) has been assumed to be zero. This is correct only in the expected sense. The phase-sensitive model presented here based on Eq. (4.23) with non-zero instantaneous values in the last term removes this common but unrealistic assumption.

After applying a set of mel scale filters (L in total) to the spectrum $|Y[k]|^2$ in the frequency domain, where the l-th filter is characterized by the transfer function $W_k^{(l)} \geq 0$ (where $\sum_k W_k^{(l)} = 1$), we obtain a total of L mel filter bank energies of

$$\begin{aligned} \sum_k W_k^{(l)}|Y[k]|^2 = \sum_k W_k^{(l)}|X[k]|^2|H[k]|^2 + \sum_k W_k^{(l)}|N[k]|^2 \\ + 2\sum_k W_k^{(l)}|X[k]||H[k]||N[k]|\cos\theta_k, \end{aligned} \tag{4.24}$$

with $l = 1, 2, ..., L$.

Denoting the various filter bank energies in Eq. (4.24) by

$$|\tilde{Y}^{(l)}|^2 = \sum_k W_k^{(l)} |Y[k]|^2,$$
$$|\tilde{X}^{(l)}|^2 = \sum_k W_k^{(l)} |X[k]|^2,$$
$$|\tilde{N}^{(l)}|^2 = \sum_k W_k^{(l)} |N[k]|^2, \tag{4.25}$$

and

$$|\tilde{H}^{(l)}|^2 = \frac{\sum_k W_k^{(l)} |X[k]|^2 |H[k]|^2}{|\tilde{X}^{(l)}|^2},$$

we simplify Eq. (4.24) to

$$|\tilde{Y}^{(l)}|^2 = |\tilde{X}^{(l)}|^2 |\tilde{H}^{(l)}|^2 + |\tilde{N}^{(l)}|^2 + 2\alpha^{(l)} |\tilde{X}^{(l)}||\tilde{H}^{(l)}||\tilde{N}^{(l)}|, \tag{4.26}$$

where we define the "phase factor" as

$$\alpha^{(l)} \equiv \frac{\sum_k W_k^{(l)} |X[k]||H[k]||N[k]| \cos\theta_k}{|\tilde{X}^{(l)}||\tilde{H}^{(l)}||\tilde{N}^{(l)}|}. \tag{4.27}$$

Since $\cos\theta_k \leq 1$, we have

$$|\alpha^{(l)}| \leq \frac{\sum_k W_k^{(l)} |X[k]||H[k]||N[k]|}{|\tilde{X}^{(l)}||\tilde{H}^{(l)}||\tilde{N}^{(l)}|}.$$

The right-hand side is the normalized inner product of vectors $\bar{N}$ and $\bar{X}^H$, with elements $\bar{N}_k \equiv \sqrt{W_k^{(l)}} |\tilde{N}^{(l)}(k)|$ and $\bar{X}_k^H \equiv \sqrt{W_k^{(l)}} |\tilde{X}^{(l)}(k)||\tilde{H}^{(l)}(k)|$. Hence

$$|\alpha^{(l)}| \leq \frac{< \bar{N}, \bar{X}^H >}{|\bar{N}||\bar{X}^H|} \leq 1.$$

Further, we define the log mel filter bank energy (log spectrum) vectors

$$\mathbf{y} = \begin{bmatrix} \log|\tilde{Y}^{(1)}|^2 \\ \log|\tilde{Y}^{(2)}|^2 \\ \vdots \\ \log|\tilde{Y}^{(l)}|^2 \\ \vdots \\ \log|\tilde{Y}^{(L)}|^2 \end{bmatrix}, \mathbf{x} = \begin{bmatrix} \log|\tilde{X}^{(1)}|^2 \\ \log|\tilde{X}^{(2)}|^2 \\ \vdots \\ \log|\tilde{X}^{(l)}|^2 \\ \vdots \\ \log|\tilde{X}^{(L)}|^2 \end{bmatrix},$$

$$\mathbf{n} = \begin{bmatrix} \log|\tilde{N}^{(1)}|^2 \\ \log|\tilde{N}^{(2)}|^2 \\ \vdots \\ \log|\tilde{N}^{(l)}|^2 \\ \vdots \\ \log|\tilde{N}^{(L)}|^2 \end{bmatrix}, \mathbf{h} = \begin{bmatrix} \log|\tilde{H}^{(1)}|^2 \\ \log|\tilde{H}^{(2)}|^2 \\ \vdots \\ \log|\tilde{H}^{(l)}|^2 \\ \vdots \\ \log|\tilde{H}^{(L)}|^2 \end{bmatrix}, \tag{4.28}$$

and the vector of phase factors

$$\boldsymbol{\alpha} = \begin{bmatrix} \alpha^{(1)} \\ \alpha^{(2)} \\ \vdots \\ \alpha^{(l)} \\ \vdots \\ \alpha^{(L)} \end{bmatrix}.$$

Then, we rewrite Eq. (4.26) as

$$\begin{aligned} e^{\mathbf{y}} &= e^{\mathbf{x}} \bullet e^{\mathbf{h}} + e^{\mathbf{n}} + 2\,\boldsymbol{\alpha} \bullet e^{\mathbf{x}/2} \bullet e^{\mathbf{h}/2} \bullet e^{\mathbf{n}/2} \\ &= e^{\mathbf{x}+\mathbf{h}} + e^{\mathbf{n}} + 2\,\boldsymbol{\alpha} \bullet e^{(\mathbf{x}+\mathbf{h}+\mathbf{n})/2}, \end{aligned} \tag{4.29}$$

where the $\bullet$ operation for two vectors denotes element-wise product, and each exponentiation of a vector above is also an element-wise operation. To obtain the log mel filter bank energy for noisy speech, we apply the log operation on both sides of Eq. (4.29):

$$\begin{aligned} \mathbf{y} &= \log\left[e^{\mathbf{x}+\mathbf{h}} \bullet (\mathbf{1} + e^{\mathbf{n}-\mathbf{x}-\mathbf{h}} + 2\boldsymbol{\alpha} \bullet e^{\frac{\mathbf{x}+\mathbf{h}+\mathbf{n}}{2} - \mathbf{x} - \mathbf{h}})\right] \\ &= \mathbf{x} + \mathbf{h} + \log[\mathbf{1} + e^{\mathbf{n}-\mathbf{x}-\mathbf{h}} + 2\boldsymbol{\alpha} \bullet e^{\frac{\mathbf{n}-\mathbf{x}-\mathbf{h}}{2}}] \\ &\equiv \mathbf{y}(\mathbf{x}, \mathbf{n}, \mathbf{h}, \boldsymbol{\alpha}). \end{aligned} \tag{4.30}$$

From Eq. (4.30), the phase factor (vector) $\boldsymbol{\alpha}$ can be solved as a function of the remaining variables:

$$\begin{aligned}
\boldsymbol{\alpha} &= \frac{e^{\mathbf{y}-\mathbf{x}-\mathbf{h}} - e^{\mathbf{n}-\mathbf{x}-\mathbf{h}} - \mathbf{1}}{2e^{\frac{\mathbf{n}-\mathbf{x}-\mathbf{h}}{2}}} \\
&= 0.5(e^{\mathbf{y}-\frac{\mathbf{n}+\mathbf{x}+\mathbf{h}}{2}} - e^{\frac{\mathbf{n}-\mathbf{x}-\mathbf{h}}{2}} - e^{-\frac{\mathbf{n}-\mathbf{x}-\mathbf{h}}{2}}) \\
&\equiv \boldsymbol{\alpha}(\mathbf{x},\mathbf{n},\mathbf{h},\mathbf{y}). \qquad (4.31)
\end{aligned}$$

Equation (4.30) or Eq. (4.31) constitutes the (deterministic) version of the phase-sensitive model for acoustic distortion due to additive noise in the log-spectral domain.

4.4.2 The Phase-Sensitive Model of Acoustic Distortion — Probabilistic Version

We now use the nonlinear relationship between the phase factor $\boldsymbol{\alpha}$ and the log-domain signal quantities of $\mathbf{x},\mathbf{n},\mathbf{h}$, and $\mathbf{y}$, as derived above and shown in Eqs. (4.30) or (4.31), as the basis to develop a probabilistic phase-sensitive model for the acoustic environment. The outcome of a probabilistic model for the acoustic environment is explicit determination of the conditional probability, $p(\mathbf{y}|\mathbf{x},\mathbf{n},\mathbf{h})$, of noisy speech observations ($\mathbf{y}$) given all other variables $\mathbf{x},\mathbf{n}$, and $\mathbf{h}$. This conditional probability is what is required in the Bayesian network model to specify the conditional dependency. This conditional probability is also required for deriving an optimal estimate of clean speech, which was carried out in [23].

To determine the form of $p(\mathbf{y}|\mathbf{x},\mathbf{n},\mathbf{h})$, we first need to assume a form of the statistical distribution for the phase factor $\boldsymbol{\alpha} = \{\alpha^{(l)}, l = 1,2,...,L\}$. To accomplish this, we note that the angle θ_k between the complex variables of $N[k]$ and $(X[k]H[k])$ is uniformly distributed over $(-\pi,\pi)$. This amounts to the maximal degree of randomness in mixing speech and noise, and has been empirically observed to be correct.

Then, from the definition of $\alpha^{(l)}$ in Eq. (4.27), it can be shown that the phase factor $\alpha^{(l)}$ for each mel filter l can be approximated by a (weighted) sum of a number of independent, zero mean random variables $\cos(\theta_k)$ distributed (non-uniformly but symmetrically) over $(-1,1)$, where the total number of terms equals the number of DFT bins (with a non-zero gain) allocated to the mel filter. When the number of terms becomes large, as is typical for high-frequency filters, the central limit theorem postulates that $\alpha^{(l)}$ will be approximately Gaussian. The law of large numbers further postulates that the Gaussian distribution will have a zero mean since each term of $\cos(\theta_k)$ has a zero mean.

Thus, the statistical distribution for the phase factor can be reasonably assumed to be a zero mean Gaussian:

$$p(\alpha^{(l)}) = \mathscr{N}(\alpha^{(l)}; 0, \Sigma_\alpha^{(l)}),$$

where the filter-dependent variance $\Sigma_\alpha^{(l)}$ is estimated from a set of training data. Since noise and (channel-distorted) clean speech are mixed independently for each

DFT bin, we can also reasonably assume that the different components of the phase factor $\boldsymbol{\alpha}$ are uncorrelated. Thus, we have the multivariate Gaussian distribution of

$$p(\boldsymbol{\alpha}) = \mathcal{N}(\boldsymbol{\alpha}; \mathbf{0}, \boldsymbol{\Sigma}_\alpha), \tag{4.32}$$

where $\boldsymbol{\Sigma}_\alpha$ is a diagonal covariance matrix.

Given $p(\boldsymbol{\alpha})$, we are in a position to derive an appropriate form for $p_y(\mathbf{y}|\mathbf{x},\mathbf{n},\mathbf{h})$. To do so, we first fix the values of $\mathbf{x},\mathbf{n}$, and $\mathbf{h}$, treating them as constants. We then view Eq. (4.30) as a (monotonic) nonlinear transformation from random variables $\boldsymbol{\alpha}$ to $\mathbf{y}$. Using the well-known result from probability theory on determining the pdf for functions of random variables, we have

$$p_y(\mathbf{y}|\mathbf{x},\mathbf{n},\mathbf{h}) = |J_{\boldsymbol{\alpha}}(\mathbf{y})|\; p_\alpha(\boldsymbol{\alpha}|\mathbf{x},\mathbf{n},\mathbf{h}), \tag{4.33}$$

where $J_{\boldsymbol{\alpha}}(\mathbf{y}) = \frac{1}{\frac{\partial \mathbf{y}}{\partial \boldsymbol{\alpha}}}$ is the Jacobian of the nonlinear transformation.

The diagonal elements of the Jacobian can be computed using Eq. (4.30) and then using Eq. (4.29) by

$$\begin{aligned}\operatorname{diag}\left(\frac{\partial \mathbf{y}}{\partial \boldsymbol{\alpha}}\right) &= \frac{2e^{\frac{\mathbf{n}-\mathbf{x}-\mathbf{h}}{2}}}{\mathbf{1} + e^{\mathbf{n}-\mathbf{x}-\mathbf{h}} + 2\boldsymbol{\alpha} \bullet e^{\frac{\mathbf{n}-\mathbf{x}-\mathbf{h}}{2}}} \\ &= \frac{2e^{\frac{\mathbf{n}+\mathbf{x}+\mathbf{h}}{2}}}{e^{\mathbf{x}+\mathbf{h}} + e^{\mathbf{n}} + 2\boldsymbol{\alpha} \bullet e^{\frac{\mathbf{n}+\mathbf{x}+\mathbf{h}}{2}}} \\ &= 2\, e^{\frac{\mathbf{n}+\mathbf{x}+\mathbf{h}}{2} - \mathbf{y}}.\end{aligned} \tag{4.34}$$

The determinant of the diagonal matrix of Eq. (4.34) is then the product of all the diagonal elements.

Also, the Gaussian assumption for $\boldsymbol{\alpha}$ gives

$$p(\boldsymbol{\alpha}|\mathbf{x},\mathbf{n},\mathbf{h}) = p\left[\boldsymbol{\alpha}(\mathbf{x},\mathbf{n},\mathbf{h},\mathbf{y})\right] = \mathcal{N}\left[\boldsymbol{\alpha}(\mathbf{x},\mathbf{n},\mathbf{h},\mathbf{y}); \mathbf{0}, \boldsymbol{\Sigma}_\alpha\right]. \tag{4.35}$$

Substituting Eqs. (4.34) and (4.35) into Eq. (4.33), we establish the following probabilistic model of the acoustic environment:

$$\begin{aligned}p_y(\mathbf{y}|\mathbf{x},\mathbf{n},\mathbf{h}) &= \frac{1}{2}\ \left|\operatorname{diag}\left(e^{\mathbf{y}-\frac{\mathbf{n}+\mathbf{x}+\mathbf{h}}{2}}\right)\right| \\ &\mathcal{N}\left[\frac{1}{2}(e^{\mathbf{y}-\frac{\mathbf{n}+\mathbf{x}+\mathbf{h}}{2}} - e^{\frac{\mathbf{n}-\mathbf{x}-\mathbf{h}}{2}} - e^{-\frac{\mathbf{n}-\mathbf{x}-\mathbf{h}}{2}}); \mathbf{0}, \boldsymbol{\Sigma}_\alpha\right].\end{aligned} \tag{4.36}$$

Because $\boldsymbol{\alpha}$ is the inner product (proportional to cosine of the phase) of the mel filter vectors of noise and clean speech characterizing their phase relationship, a Gaussian distribution on it makes the distortion model of Eq. (4.36) phase-sensitive.

4.4.3 Feature Compensation Experiments and Lessons Learned

The phase-sensitive model expressed in the form of Eq. (4.6) is used directly in the MMSE estimate of clean speech, which gives the compensated cepstral features that feed into the speech recognizer. Deng et al. [23] provides details of the MMSE estimate derivation using first-order Taylor series expansion. Second-order Taylor series expansion can also be used to give a somewhat more accurate MMSE estimate without incurring more computation (unpublished).

As reported in [23], a diagnostic experiment was carried out to assess the role of phase asynchrony in feature enhancement for noise-robust speech recognition. To eliminate the factor of noise power estimation inaccuracy, phase-removed true noise power is used since in the Aurora 2 task the true noise's waveforms are made readily available [46]. Table 4.1 lists the percent accuracy in the Aurora 2 standard task of digit recognition (as a function of the feature enhancement algorithm iterations using the phase-sensitive model; see the algorithm in [23]). Clean HMMs (simple back-end) as provided by the Aurora 2 task are used for recognizing enhanced features.

Table 4.1: Percent accurate digit recognition rate for the Aurora 2 task as a function of the feature enhancement algorithm iteration number using the phase-sensitive model. Phase-removed true noise features (noise power spectra) are used in this diagnostic experiment as the n-layer variables

Itrs	1	2	4	7	12
SetA	94.12	96.75	97.96	98.11	98.12
SetB	94.80	97.29	98.10	98.48	98.55
SetC	91.00	94.50	96.50	97.86	98.00
Ave.	93.77	96.52	97.72	98.21	**98.27**

When the phase information is removed, how much does the performance suffer? To examine this issue, several spectral subtraction methods are used where the same phase-removed true noise features are used as in Table 1. After careful tuning of the spectral subtraction parameter of the floor value, the best accuracy is 96% (see detailed results in Table 4.2), significantly below the accuracy of 98% obtained with the use of the phase-sensitive model.

However, when instead of the true noise power, the estimated noise power is used (with the algorithm for noise power estimation described in [23]), improvement of recognition accuracy from the use of the phase-insensitive model to the use of the phase-sensitive model becomes much smaller, from 84.80% to 85.74%; see detailed results in Table 4.3.

What could be the reason for the drastic difference between the performance improvements (from the phase-insensitive to phase-sensitive models) with and without noise estimation errors? Let us examine Eq. (4.26). It is clear that the third, phase-related term and the second, noise-power term are added to contribute to the power

Table 4.2: Performance (percent accurate) for the Aurora 2 task using four versions of spectral subtraction (SS) with the same phase-removed true noise features as in Table 1

Floor	e^{-20}	e^{-10}	e^{-5}	e^{-3}	e^{-2}
SS1	93.57	94.26	**95.90**	92.18	90.00
SS2	12.50	44.00	65.46	88.69	84.44
SS3	88.52	89.26	93.19	90.75	88.00
SS4	10.00	42.50	63.08	87.41	84.26

Table 4.3: Right column: percent accurate digit recognition rates for the Aurora 2 task using noise estimation and phase-sensitive feature enhancement algorithms, both described in [23]. Left column: The baseline results obtained with the phase-insensitive model

	Baseline (no phase)	Enhanced (with phase)
SetA	85.66	86.39
SetB	86.15	86.30
SetC	80.40	83.35
Ave.	**84.80**	**85.74**

of noisy speech. If the estimation error in the second, noise-power term is comparable to the entire third term, then the addition of the third term would not be very meaningful in accounting for the power of noisy speech. This is the most likely explanation for the huge performance improvement when true noise power is used (Tables 4.1 and 4.2) and the relatively mild improvement when noise power estimation contains errors (Table 4.3). The analysis above shows the critical role of noise power estimation in enabling the effectiveness of the phase-sensitive model of environmental distortion.

4.4.4 Discussions

In this section, we described one of the most advanced acoustic distortion models and its application to model-based or structured feature compensation. In terms of the degree of model sophistication, the phase-sensitive model is superior to the Algonquin model, which in turn is superior to the standard deterministic VTS model. For example, as a special case, when the variance from the phase factor is assumed to be independent of SNR, the phase-sensitive model becomes reduced to the Algonquin model. And as the phase factor is eliminated, the model becomes further reduced to the VTS model.

When any of these acoustic distortion models is used for feature enhancement, we call the resulting techniques model-based or structured feature compensation.

This contrasts with the feature normalization techniques widely in use in front-end design for speech recognition, where no such acoustic distortion model is exploited. Examples of the latter include feature moment normalization and cepstral time smoothing. One main difference between these two classes of feature compensation techniques is that the feature normalization methods do not provide any mechanism to determine feature uncertainty. The model-based feature enhancement methods, however, are equipped with this mechanism. In the preceding section, we illustrated how the Algonquin model was used for computing uncertainty in the enhanced features and then for uncertainty decoding. Recently, work has been reported [60] on the use of phase-sensitive models for computing feature uncertainty and then for uncertainty decoding. In this approach, which is detailed in Chapter 8 of this book, the authors propose a phase-sensitive distortion model, where the phase factor α is no longer modeled as a Gaussian and where the moments of α are computed analytically. This distortion model together with an a priori model of clean speech is used to compute the feature posterior by Bayesian inference. The obtained feature posterior is then used for uncertainty decoding.

Finally, we remark that model-based feature compensation also contrasts with the different class of noise-robust speech recognition techniques which we call model-domain compensation, where speech recognition model parameters (e.g., means and variances of the HMM) are modified. We will present model-domain compensation in the next section in the context of studying uncertainty in the model parameter space instead of in the feature space discussed so far.

4.5 Bayesian Decision Rule with Unreliable Model Parameters

4.5.1 Bayesian Predictive Classification Rule

Effective exploitation of uncertainty is a key ingredient in nearly all branches of statistical pattern recognition. In the previous sections, we discussed the uncertainty in the feature domain and examined how the feature-domain uncertainty can be propagated to the back-end speech decoder. In this section, we turn to the more authentic Bayesian framework, which accounts for uncertainty in model parameters, providing a theoretical basis for understanding model compensation.

In the already successful applications of HMM-based speech recognition and speaker verification, uncertainty in the HMM parameter values has been represented by their statistical distributions (e.g., [49, 52, 53]). The motivation of this model-space Bayesian approach has been the widely varied speech properties due to many different sources, including speakers (both intra-speaker and inter-speaker) and acoustic environments, across and possibly within training and test data. In order to take advantage of the model parameter uncertainty, the decision rule for recognition or decoding has been improved from the conventional plug-in MAP rule to the Bayesian predictive classification (BPC) rule [49]:

$$\hat{\mathbf{W}} = \underset{\mathbf{W}}{\operatorname{argmax}} \left[\int_{\Lambda \in \Omega} p(\mathbf{y}|\Lambda,\mathbf{W}) p(\Lambda|\phi,\mathbf{W}) d\Lambda \right] P(\mathbf{W}), \tag{4.37}$$

where ϕ is the set of (deterministic) hyper-parameter characterizing the distribution of the random model parameters, Ω denotes all possible values that the random parameters Λ can take, and the integral becomes the desired acoustic score. How to simplify the integration in Eq. (4.37) to enable robust speech recognition in the model space can be found in [53], which, unfortunately, is not as simple as propagating the uncertainty in the feature space to the effect of only modifying the deterministic HMM variance values as discussed in the preceding sections.

It is possible to combine the effects of model parameter uncertainty and feature uncertainty into a single, comprehensive decision rule. Taking the integrals in both the feature and the model-parameter domains, we have

$$\hat{\mathbf{W}} = \underset{\mathbf{W}}{\operatorname{argmax}} P(\mathbf{W}) \int_{\Lambda \in \Omega} \int_{\mathbf{x}} \frac{p(\mathbf{x}|\mathbf{W},\Lambda)}{p(\mathbf{x})} p(\mathbf{x}|\mathbf{y},\Lambda) p(\Lambda|\phi,\mathbf{x}) d\mathbf{x} d\Lambda. \tag{4.38}$$

4.5.2 Model Compensation Viewed from the Perspective of the BPC Rule

As discussed in [41], there are a large number of model compensation or model-based techniques developed in the past to handle uncertainty in noise-robust speech recognition. It is interesting to place such techniques in the context of the implementation of the BPC rule exemplified in Eq. (4.38).

Let us first simplify Eq. (4.38) by disregarding feature uncertainty; that is, assume the input features have a zero variance value:

$$p(\mathbf{x}|\mathbf{y},\Lambda) = \delta(\mathbf{y} = \mathbf{x}). \tag{4.39}$$

This then gives

$$\begin{aligned} \hat{\mathbf{W}} &= \underset{\mathbf{W}}{\operatorname{argmax}} P(\mathbf{W}) \int_{\Lambda \in \Omega} \frac{p(\mathbf{y}|\mathbf{W},\Lambda)}{p(\mathbf{y})} p(\Lambda|\phi,\mathbf{y}) d\Lambda \\ &= \underset{\mathbf{W}}{\operatorname{argmax}} \int_{\Lambda \in \Omega} p(\mathbf{W}|\mathbf{y},\Lambda) p(\Lambda|\phi,\mathbf{y}) d\Lambda. \end{aligned} \tag{4.40}$$

We further assume that the model-parameter distribution, $p(\Lambda|\phi,\mathbf{y})$, is sharp. It is obvious then that the mode of $p(\Lambda|\phi,\mathbf{y})$ will be at the parameter set which matches best the noisy speech input $\mathbf{y}$. We denote this set of model parameters by $\Lambda(\mathbf{y})$, which is the goal that all the model compensation techniques are searching for. So under the assumption of

$$p(\Lambda|\phi,\mathbf{y}) = \delta(\Lambda = \Lambda(\mathbf{y})), \tag{4.41}$$

we further simplify Eq. (4.40) to

$$\hat{\mathbf{W}} = \underset{\mathbf{W}}{\operatorname{argmax}}\, p(\mathbf{W}|\mathbf{y}, \Lambda(\mathbf{y})), \tag{4.42}$$

which gives rise to the decision rule of all model compensation techniques published in the literature, where $\Lambda(\mathbf{y})$ is at the mode of the distribution $p(\Lambda|\phi,\mathbf{y})$, or

$$\Lambda(\mathbf{y}) = \underset{\Lambda}{\operatorname{argmax}}\, p(\Lambda|\phi,\mathbf{y}). \tag{4.43}$$

From the above discussion, we can view the decision rule of Eq. (4.40) as a generalized form of model compensation. An implementation of this general form of decision rule and the associated training procedure was carried out to compensate for acoustic variations using generic linear transformations [53]. Structured models for speech and noise interactions such as VTS, Algonquin, or phase-sensitive models have not been explored using the generalized form of model compensation of Eq. (4.40).

Another insight gained from the discussion above is that the generalized form of model compensation (Eq. (4.40)), and further, the generalized form (Eq. (4.38)) that includes both model and feature compensation require active feedback from the speech decoding results to update both speech and noise parameters. Fertile research can be carried out in this direction.

4.6 Model and Feature Compensation — A Taxonomy-Oriented Overview

We now return to the traditional techniques for model-domain compensation where model parameter uncertainty is not considered and where the model parameters are updated as fixed values. We provide an overview aimed at categorizing a large number of existing techniques in model compensation and at clarifying some related nomenclature which sometimes causes confusion in the literature. A similar overview is provided for the counterpart of feature-domain compensation.

4.6.1 Model-Domain Compensation

Many model compensation techniques have been developed by speech recognition researchers over the past 20 years or so. They are used to update the model parameters, e.g., HMM means and variances, contrasting the feature compensation techniques that reduce the noise and other distortions from the speech feature vectors. Model compensation is also called a model-based or model-domain approach, and the goal is to determine from a set of "adaptation" data the model parameter set $\Lambda(\mathbf{y})$, which has the desirable property of Eq. (4.43), so as to implement the decision rule described by Eq. (4.42).

The techniques of model compensation can be broadly classified into two main categories, depending on two different approaches. In the first category, one uses unstructured, generic transformations to convert the model parameters. The transformations are typically linear, and often, a set of linear transformations are used. The techniques are general, applicable not only to noise compensation but also to other types of acoustic variations, notably speaker adaptation. They involve many parameters and thus require a reasonably large amount of data to estimate them. This category of techniques is also called model adaptation or adaptive scheme, with typical algorithms of

- Maximum Likelihood Linear Regression (MLLR) [40];
- Maximum A Posteriori (MAP) [59];
- constrained MLLR [40];
- noisy constrained MLLR [55]; and
- multi-style training [17, 34].

Multi-style training is classified into this category of unstructured model compensation because there is structured knowledge built into the training, where instead of using a single noise condition, many types of noisy speech are included in the training data. The hope is that one of the types will appear in the deployment condition. Multi-style training updates the model parameters generally in a less efficient and less structured way than does the linear transformation approach. It requires much more training data as a result, and it often produces the updated distributions in HMMs that are unnecessarily broad, making the trained model weakly discriminative.

In the second category of the model compensation techniques, structured transformations are used, which are generally nonlinear and which take into account the way the noisy speech features (e.g., log spectra or cepstra) are produced from the mixing speech and noise. These techniques are sometimes called predictive scheme or structured model adaptation, and the structured transformations are sometimes called mismatch function, interaction model, or acoustic distortion model. In contrast to the unstructured techniques, the structured methods make use of physical knowledge, or a "model", as an approximation to the physics of the speech and noise mixing process. As such, they are not applicable to other types of acoustic variation compensation such as speaker adaptation. Common techniques in the category of structured model compensation include

- Parallel Model Combination (PMC, with log-normal approximation) [39];
- Vector Taylor Series (VTS) [2, 64];
- phase-sensitive model compensation [61–63, 78].

Note that the technique reported in [78] uses linear SPLINE interpolation to approximate the phase-sensitive model presented in Section 4. The division above is based on the type of mismatch or distortion functions in use. Alternative and further divisions can be made based on the kind of approximation exploited to represent the mismatch functions.

One main practical advantage of the structured model compensation techniques over the unstructured counterpart is the much smaller number of free parameters that need to be estimated. However, these parameters, e.g., those in the noise and channel models, are harder to estimate than the generic linear transformation parameters. Typically, second-order approaches are needed to estimate the variance parameters effectively [62, 63]. In addition, the mismatch functions may be inaccurate, and the computational complexity is higher as well.

4.6.2 Feature-Domain Compensation

The same scheme for classifying the model compensation techniques based on structured vs. unstructured methods can be used for feature compensation. The latter is also called feature enhancement or denoising in the literature. The techniques in the category of structured feature compensation involve the use of the same or similar structured transformations as those discussed earlier for model compensation. Because of the use of the distortion or interaction model of speech and noise mixing, structured feature compensation is often called model-based feature enhancement or compensation in the literature. Note that here "model" refers to the distortion model or structured transformation, with examples in Eqs. (4.6), (4.8) and (4.30) ([71, 82]), and it may sometimes also refer to the use of Gaussian mixture models for clean speech. On the other hand, the "model" in model-based compensation refers to the model used for speech recognition, or HMM (e.g., [41]). Some of the commonly used techniques in the category of structured feature-domain compensation have been discussed in Section 3 (the part with mean calculation) and Section 4. Here is a brief summary with some more examples:

- VTS [71];
- Algonquin [20, 37];
- Phase-sensitive modeling for feature enhancement [23, 31, 81].

All these structured enhancement techniques use the estimate of Minimum Mean Squared Error (MMSE) for clean speech in the cepstral or log-spectral domains. In these domains, nonlinearity comes into play. The log domain is believed to be a better one than the linear spectral domain because it is closer to what the back-end HMM speech recognizer receives (see [85] for experimental evidence and related discussions).

In the unstructured category of feature-domain compensation, the techniques developed make no use of structured knowledge of how speech and noise mix expressed in the log domain. They either use stereo data to learn the impact of noise on speech in the log domain implicitly (e.g., SPLICE), or operate in the linear spectral domain, where the mixing process is much simpler (e.g., spectral subtraction). They also include several popular feature normalization methods. Examples of commonly used unstructured feature compensation techniques include

- SPLICE and its extensions, known as stochastic vector mapping [3, 17, 18, 33, 47, 87];
- spectral subtraction [12];
- Wiener filtering combined with the HMM or trajectory-HMM methods [35, 76, 77];
- MMSE estimator on Fourier spectral amplitude [36];
- MMSE estimator on cepstra [85];
- cepstral mean, variance, and histogram normalization (cf., review in [34]); and
- RASTA, FDLP (frequency-domain linear prediction), and other types of modulation spectra (e.g., [72]).

4.6.3 Hybrid Compensation Techniques

We provide a summary in Table 4.4, with entries of classes F1, F2, M1, and M2, of the two-way classification for each of the feature-domain and model-domain compensation techniques discussed so far in this section. The feature and model techniques can be combined to form hybrid techniques, shown as H1 and H2 in Table 4.4.

	Feature Domain	**Model Domain**	**Hybrid**
Un-structured	Class F1	Class M1	Class H1
Structured	Class F2	Class M2	Class H2

Table 4.4: A summary and classification of noise-robust speech recognition techniques

As a summary, some typical examples in each of the classes of techniques in Table 4.4 are provided below again, including the two classes of hybrid techniques to be discussed:

- Class F1: SPLICE, spectral subtraction, Wiener filter, HMM, MMSE, MMSE-Cepstra, CMN (cepstral mean normalization), CVN (cepstral variance normalization), CHN (cepstral histogram normalization), RASTA, modulation spectra;
- Class F2: VTS, Algonquin, phase model;
- Class M1: MLLR, MAP, C-MLLR, N-CMLLR, multi-style training;
- Class M2: PMC, VTS, phase model;
- Class H1: NAT-SS (noise-adaptive training with spectral subtraction), NAT-SPLICE (noise-adaptive training with SPLICE), JAT (or NAT-LR), IVN (irrelevant variability normalization); and
- Class H2: NAT-VTS, UD, JUD.

Examples of structured hybrid techniques are the various uncertainty decoding (UD) techniques discussed in earlier sections in this chapter, where the structured transformation as provided by the Algonquin model was used to compute the uncertainty in the feature that is subsequently propagated to the HMM decoder. A more comprehensive discussion on uncertainty propagation and decoding is provided in Chapter 3 and other chapters in this book.

The best example of unstructured hybrid techniques is noise-adaptive training (NAT) and its various extensions. Due to its excellent performance and the recent renewed interest in this scheme, we will devote the next section to this important topic.

4.7 Noise Adaptive Training

Noise adaptive training (NAT) is a hybrid strategy of feature compensation and model compensation. The part of feature compensation can be in any form of noise reduction or feature enhancement (structured H2 with F2, or unstructured H1 with F1). The part of model compensation, however, takes the specific form of multi-style (re)training operating on the feature-compensated training data. This original scheme of NAT, first reported in [17], which demonstrated its surprisingly high performance, has formed one of the two standard paradigms (i.e., the multi-style acoustic model of denoised features) in the Aurora experimental framework for the evaluation of speech recognition systems under noisy conditions. Hence, the effectiveness of NAT published in [17] has been verified by at least hundreds of additional experiments with all sorts of feature enhancement techniques and databases worldwide. In addition, the original scheme of NAT has been further developed in various directions during the last decade. In this section, we will provide an overview of these developments.

4.7.1 The Basic NAT Scheme and its Performance

Let us first review and demonstrate the effectiveness of NAT in Figure 4.1 (with data extracted and reorganized from [17]), where the word error rate produced by the author using a standard HMM system (with a noisy 5K-vocabulary Wall Street Journal task) is shown as a function of SNR with added noise. Five sets of results are shown. The results labeled as "Noisy-Noisy matched" (in green) were produced by adding the same noise samples to the training and test data at each SNR level. And the noise used was stationary. In this way, the perfect matching condition was created artificially in the noisy speech domain. This is the condition that perfect model compensation schemes are striving for, and conventional wisdom posits that this sets the upper bound for the system performance. However, when the NAT scheme was developed and applied with two relatively simple unstructured feature compensation techniques, spectral subtraction (SS) and SPLICE, both re-trained HMMs (NAT-SS and NAT-SPLICE) outperform the noisy-matched system under almost all SNR conditions (except the clean condition with slight degradation). And NAT-SPLICE did better than NAT-SS. This finding has been verified by many types of noise and SNR levels with SPLICE and SS (some of them were published in [17]). Other feature compensation techniques discussed in this chapter were also used with success as a component of the NAT training. More recently, this kind of multi-style training in NAT was also successfully applied to multilingual speech recognition [68].

There have been some discussions among speech recognition researchers as to whether the model or the feature domain is more appropriate for noise compensation. The results discussed above demonstrate that while feature compensation alone may not clearly outperform model compensation (with some possible exception, e.g., [13]), a simple hybrid such as NAT is already sufficient to beat it. The fact that NAT with a more sophisticated type of feature compensation (e.g., SPLICE) outperforms a simpler type (e.g., spectral subtraction), as shown in Figure 4.1, points to the importance of developing higher-quality feature compensation techniques.

In practical deployment, however, the basic paradigm of NAT discussed above is difficult to realize because it requires knowledge of the exact noise characteristics (e.g., noise type and level). Also, if the noise characteristics are numerous, and especially if they are time-varying, NAT cannot be easily carried out in advance even for small systems. More recent developments of the NAT framework have offered possible solutions to this problem, which we will review next.

4.7.2 NAT and Related Work — A Brief Overview

The basic NAT scheme just discussed can be viewed as a way of estimating "clean" speech model parameters, which have been assumed to be available and accurate in all the noise compensation and speech recognition techniques presented so far. However, this assumption rarely holds in practice, since the training data used for building large systems typically contain mixed clean and noisy speech. On the one hand, this situation is similar to speaker adaptive training (SAT) [4], which deals with speaker variation caused by mixed speakers in the training set. On the other hand, the NAT problem differs from SAT in that there is the golden "target" for feature adaptation or compensation, which is truly clean speech. In SAT, there is no such predefined golden "sheep" speaker as the adaptation target.

In the basic NAT scheme of [17], the feature compensation components (SPLICE and SS) have fixed parameters during NAT training. Only the "pseudo-clean" HMM parameters are learned with the objective function, which is independent of SPLICE and SS parameters. And the optimization criterion is maximum likelihood, optimized via EM. This basic scheme has been extended in [48, 54] in two ways. First, the unstructured feature compensation components (SPLICE and SS) are improved to the structured compensation technique of VTS, changing from the unstructured hybrid technique NAT-SPLICE or NAT-SS to the structured hybrid technique NAT-VTS. Second, the free parameters in the feature compensator VTS are subject to

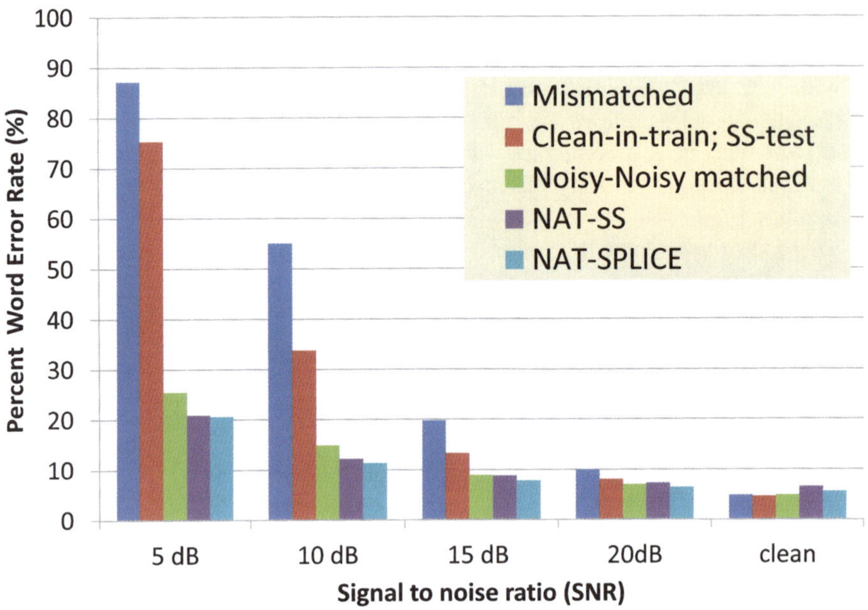

Fig. 4.1: Word error rates for a noisy WSJ task demonstrating that NAT outperforms the best noisy matched condition

joint training with the HMM parameters. These free parameters include the noise means and variances as part of the VTS distortion model. The use of EM for maximum likelihood estimation remains the same. In the work of both [48, 54], the motivation comes from two key insights provided in [17]:

1. Any feature compensation technique inevitably introduces, however small they may be, undesirable residual errors. These errors cause a special kind of model mismatch if no retraining is done; and
2. No absolute clean speech data in training are available, and hence the ultimate models built after the use of any feature compensation technique should settle at best with "pseudo-clean" speech models.

The more recent work of [54] differs from [48] in the way to train the HMM and VTR noise model parameters. One uses iteratively separate steps in the training while insisting on the mapping from noisy speech to "pseudo-clean" speech [48]. The other uses single-step optimization, disregarding the true distribution of the clean speech, matching the best to the adaptation scheme performed at runtime, and achieving somewhat better recognition accuracy on the same task [54]. In this regard, the NAT-VTS version of [54] is very similar to SATin spirit.

Another interesting extension of NAT is called Joint Adaptation Training (JAT) [66], where linear regression instead of VTS or SPLICE/SS is used to represent the feature transformation. As in [48] and [54], the adaptive transformation in JAT is parameterized, and its parameters are jointly trained with the HMM parameters by the same kind of maximum likelihood criterion as in [17, 48, 54]. Since the feature transformation is linear regression (LR) in the JAT technique, by the nomenclature established in this chapter, JAT can also be rightfully called NAT-LR.

Another set of extension and generalization of the basic NAT scheme is called Irrelevant Variability Normalization (IVN) (e.g., [80, 88]), which is a general framework designed not only for noise compensation but also for other types of acoustic mismatch, and it encompasses SAT as a special case. A specific version of IVN [48] designed for noise compensation using VTS as the feature enhancement method, that gives rise to NAT-VTS in our NAT nomenclature, was discussed earlier. In [88], a total of six feature compensation functions are presented, extending the original SPLICE and SS mapping functions proposed in the basic NAT framework [17] in a systematic manner. Among these feature compensation functions are SPLICE, clusters of linear regressions, and clusters of bias removal mappings. And again, joint training of all parameters in the feature compensation functions and in the HMM is carried out in an iterative, two-step fashion. It is noted that in the original NAT-SPLICE of [17], feature compensation is accomplished utterance by utterance in the training set using SPLICE, whose parameters were trained separately from the HMM training for pseudo-clean speech. In the decoding phase, the same SPLICE technique is applied to the test input features. Differently, in IVN of [80, 88], the "environment" variable has to be detected, a process called "acoustic sniffing", during the joint HMM and feature compensation parameters' training. The same "acoustic sniffing" is needed in the decoding phase also.

Additional improvements of IVN over the basic NAT framework of [17] are the use of MAP instead of maximum likelihood training [88], and the use of sequential estimation of the feature compensation parameters [80].

Finally, we point out that the mechanism presented in [53] for compensating for extraneous or "irrelevant" variability of spontaneous speech is very similar to that of IVN discussed above. In both techniques, the "condition" or "environment" variable is used to denote a set of discrete unknown or hidden factors underlying the extraneous variations in the observed acoustic data. And they both use a joint training strategy for both the HMM and the transformation parameters. The main difference is that the transformation from the distorted domain to the "canonical" domain is done on the HMM mean vectors in [53] and on the features in IVN.

4.8 Summary and Conclusions

Handling uncertainty in both data and model is a general problem in pattern recognition, and it has special relevance to noise-robust speech recognition, where the uncertainty with both types abounds. In this chapter, selected topics in noise-robust speech recognition have been presented with in-depth discussion, framed using the Bayesian perspective and centered on the theme of uncertainty treatment.

Since noise robustness is a vast subject, a number of other relevant topics — for example, estimation of noise and channel characteristics, long-term reverberant distortion, fast-changing non-stationary noise tracking, single-microphone speaker separation, and voice activity detection in noisy environments — have not been included in the overview and discussion in this chapter. Also, robust speech recognition performance figures have not been systematically provided in this chapter for comparisons of the techniques presented. The references included in this chapter and other chapters in this book should fill in most of the missing topics, as well as the information about performance comparisons.

Despite significant progress in noise-robust speech recognition over the past two decades or so, many of which have been discussed in this chapter, the problem is far from being solved. Further research in this area is required to enable a sufficiently high level of accuracy in real-world speech recognition applications encompassing a full range of acoustic conditions. Here, a brief discussion is offered from the author's perspective on the expected future research activities in the area of noise-robust speech recognition in the relatively long term.

First, better acoustic modeling than the current ones for speech and noise and the interactions between them is needed. To illustrate this need, we use a simple example here. Estimation of clean speech and mixing noise from noisy speech is a mathematically ill-defined problem, with one known and two unknown variables and one constraint. The only way to have sensible solutions is to impose and exploit prior knowledge as constraints. The hallmark of the Bayesian framework is its formalization of the use of prior knowledge in a mathematically rigorous manner. As elaborated in [9, 10], some powerful sources of prior knowledge in machine speech recognition, especially under noisy environments, come from human speech perception and production. Computational models with appropriate complexity are the first step in exploiting such prior knowledge; e.g., [14–16, 69, 73]. In addition, algorithms extracting key insights from human speech perception and production knowledge and the corresponding computational models are needed to benefit robust speech recognition, e.g., [11, 75]. An example of the benefit can be found in a solution to the multi-talker speech separation and recognition problem, a most difficult kind in robust speech recognition, where the interfering "noise" is another speaker's speech. Powerful graphical modeling and related algorithms, coupled with the Algonquin and phase-sensitive interaction models (see Sections 4.3.1 and 4.4), as well as the use of speech and noise dynamics as part of prior knowledge built into the Bayesian network, offer recognition accuracy superior to that of humans [45]. Given that embedding only very crude dynamic characteristics of speech and noise gives promising results, exploitation of more insightful and relevant knowledge is expected to be fruitful research.

Second, related to the above discussion on better acoustic modeling for speech and noise interactions, we need to develop better techniques to characterize and exploit the effects of noise on speech in a wide variety of front-end features. All the work reviewed in this chapter assumes the use of log spectra or cepstra (e.g.,

MFCC), based on which all nonlinear structured acoustic distortion models are developed and approximated. However, state-of-the-art front-end features may not always be strict cepstra. Tandem features [72], PLP features, some discriminatively derived features [74], and mean-normalized MFCCs often perform better than plain MFCCs. But these features make it difficult to derive the structured distortion models to enable structured feature and model compensation (classes of F2 and M2 in Table 4.4 as well as NAT-VTR). Further, an emerging technology developed from machine learning, deep learning (e.g., [42]), which has just started entering the field of speech recognition [29, 70, 86], proposes a fundamentally different way of feature extraction from the largely handcrafted features such as MFCC. The layer-by-layer feature extraction strategy in deep learning provides the opportunity to automatically derive powerful features from the primitive raw data, e.g., speech waveform or Fourier transform. Given that in the linear domains of waveform and linear spectrum speech and noise interaction models become much simpler than in MFCC, but that there are some special difficulties involved in using the primitive feature of waveform in speech recognition (e.g., [79]), how to do noise robust speech recognition in the deep learning framework will require complete rethinking of the HMM framework, which we have been so familiar with as the implicit assumption in this and many other chapters in this book.

Third, as elaborated in Section 4.7, the hybrid strategy of feature and model compensation exemplified by NAT, JAT, and IVN is a powerful framework that can handle not only noise-induced uncertainty but also other types of variations extraneous to phonetic distinction. Better integrated algorithm design, improved joint optimization of model and transformation parameters, more effective tracking techniques for time-varying distortion factors, and clever use of metadata as labels for the otherwise hidden "distortion condition" variables that are frequently available from specific speech applications are all fruitful research directions.

References

1. A. Acero: Acoustical and Environmental Robustness in Automatic Speech Recognition. Kluwer Academic Publishers (1993)
2. A. Acero, L. Deng, T. Kristjansson, and J. Zhang: HMM adaptation using vector Taylor series for noisy speech recognition. In: Proc. ICSLP, vol.3, pp. 869-872 (2000)
3. M. Afify, X. Cui, and Y. Gao: Stereo-based stochastic mapping for robust speech recognition. In: Proc. ICASSP (2007)
4. T. Anastasakos, J. McDonough, R. Schwartz, and J. Makhoul: A compact model for speaker-adaptive training. In: Proc. ICSLP (1996)
5. J. Arrowood and M. Clements: Using observation uncertainty in HMM decoding. In: Proc. ICSLP, Denver, Colorado (2002)
6. R. F. Astudillo, D. Kolossa, and R. Orglmeister: Accounting for the uncertainty of speech estimates in the complex domain for minimum mean squared error speech enhancement. In: Proc. Interspeech (2009)
7. H. Attias, Li Deng, Alex Acero, and John Platt: A new method for speech denoising and robust speech recognition using probabilistic models for clean speech and for noise. In: Proc. of the Eurospeech Conference (2001)

8. H. Attias, J. Platt, Alex Acero, and Li Deng: Speech denoising and dereverberation using probabilistic models. In: Proc. NIPS (2000)
9. J. Baker, Li Deng, Jim Glass, S. Khudanpur, C.-H. Lee, N. Morgan, and D. O'Shaughnessy: Research developments and directions in speech recognition and understanding. IEEE Signal Processing Magazine, vol. 26, no. 3, pp. 75-80 (2009)
10. J. Baker, Li Deng, S. Khudanpur, C.-H. Lee, J. Glass, N. Morgan, and D. O'Shaughnessy: Updated MINDS report on speech recognition and understanding. IEEE Signal Processing Magazine, vol. 26, no. 4 (2009)
11. J. Bilmes and C. Bartels: Graphical model architectures for speech recognition. IEEE Signal Processing Magazine, vol. 22, no. 5, pp. 89-100 (2005)
12. S.F. Boll: Suppression of acoustic noise in speech using spectral subtraction. IEEE Trans. on Acoustics, Speech, and Signal Processing, 27:113-120 (1979)
13. K. Demuynck, X. Zhang, D. Van Compernolle, and H. Van hamme: Feature versus model based noise robustness. In: Proc. Interspeech (2010)
14. L. Deng: Computational models for auditory speech processing. In: Computational Models of Speech Pattern Processing, (NATO ASI Series), pp. 67-77, Springer Verlag (1999)
15. L. Deng: Computational models for speech production. Computational Models of Speech Pattern Processing, (NATO ASI Series), pp. 199-213, Springer Verlag (1999)
16. L. Deng, D. Yu, and A. Acero: Structured speech modeling. IEEE Trans. on Audio, Speech and Language Processing (Special Issue on Rich Transcription), vol. 14, No. 5, pp. 1492-1504 (2006)
17. L. Deng, A. Acero, M. Plumpe, and X.D. Huang: Large vocabulary speech recognition under adverse acoustic environments. In: Proc. ICSLP, pp. 806-809 (2000)
18. L. Deng, A. Acero, L. Jiang, J. Droppo, and X. Huang: High-performance robust speech recognition using stereo training data. In: Proc. ICASSP, Salt Lake City, Utah (2001)
19. L. Deng, J. Droppo, and A. Acero: Exploiting variances in robust feature extraction based on a parametric model of speech distortion. In: Proc. ICSLP (2002)
20. Li Deng, Jasha Droppo, and Alex Acero: A Bayesian approach to speech feature enhancement using the dynamic cepstral prior. In: Proc. ICASSP, Orlando, Florida (2002)
21. L. Deng, J. Droppo, and A. Acero: Log-domain speech feature enhancement using sequential MAP noise estimation and a phase-sensitive model of the acoustic environment. In: Proc. ICSLP, Denver, Colorado (2002)
22. L. Deng, K. Wang, A. Acero, H. Hon, J. Droppo, C. Boulis, Y. Wang, D. Jacoby, M. Mahajan, C. Chelba, and XD. Huang: Distributed speech processing in MiPad's multimodal user interface. IEEE Trans. on Speech and Audio Processing, vol. 10, no. 8, pp. 605-619 (2002)
23. L. Deng, J. Droppo, and A. Acero: Enhancement of log mel power spectra of speech using a phase-sensitive model of the acoustic environment and sequential estimation of the corrupting noise. IEEE Trans. on Speech and Audio Processing, vol.12, no. 2, pp. 133-143 2004)
24. Li Deng and Xuedong Huang: Challenges in adopting speech recognition. Communications of the ACM, vol. 47, no. 1, pp. 11-13, (2004)
25. Li Deng, Jasha Droppo, and Alex Acero: Recursive estimation of nonstationary noise using iterative stochastic approximation for robust speech recognition. IEEE Trans. on Speech and Audio Processing, vol. 11, no. 6, pp. 568-580 (2003)
26. Li Deng, Jasha Droppo, and Alex Acero: Incremental Bayes Learning with Prior Evolution for Tracking Non-Stationary Noise Statistics from Noisy Speech Data. In: Proc. ICASSP, Hong Kong (2003)
27. Li Deng, Jasha Droppo, and Alex Acero: Estimating cepstrum of speech under the presence of noise using a joint prior of static and dynamic features. IEEE Trans. on Speech and Audio Processing, vol. 12, no. 3, pp. 218-233 (2004)
28. L. Deng, J. Droppo, and A. Acero: Dynamic compensation of HMM variances using the feature enhancement uncertainty computed from a parametric model of speech distortion. IEEE Trans. on Speech and Audio Processing, vol. 12, no. 3, (2005)
29. Li Deng, Mike Seltzer, Dong Yu, Alex Acero, A. Mohamed, and Geoff Hinton: Binary coding of speech spectrograms using a deep auto-encoder. In: Proc. Interspeech (2010)

30. J. Droppo, A. Acero, and L. Deng: Efficient online acoustic environment estimation for FCDCN in a continuous speech recognition system. In: Proc. ICASSP, Salt Lake City, Utah (2001)
31. J. Droppo, A. Acero, and L. Deng: A nonlinear observation model for removing noise from corrupted speech log Mel-spectral energies. In: Proc. ICSLP, Denver, Colorado (2002)
32. J. Droppo, A. Acero, and L. Deng: Uncertainty decoding with SPLICE for noise robust speech recognition. In: Proc. ICASSP, Orlando, Florida (2002)
33. J. Droppo, L. Deng, and A. Acero: Evaluation of SPLICE on the Aurora 2 and 3 Tasks. In: Proc. ICSLP, Denver, Colorado (2002)
34. J. Droppo and A. Acero: Environmental Robustness. In: Handbook of Speech Processing, Springer (2007)
35. Y. Ephraim: A Bayesian estimation approach for speech enhancement using hidden Markov models. IEEE Trans. on Acoustics, Speech, and Signal Processing, 40:725-735 (1992)
36. Y. Ephraim and D. Malah: Speech enhancement using a minimum mean-square error short-time spectral amplitude estimator. IEEE Trans. Acoust., Speech, Signal Process., vol. ASSP-32, no. 6, pp. 1109-1121 (1984)
37. B. Frey, L. Deng, A. Acero, and T.T. Kristjansson: Algonquin: Iterating Laplace's method to remove multiple types of acoustic distortion for robust speech recognition. In: Proc. Eurospeech, Aalborg, Denmark (2001)
38. B. Frey, T. Kristjansson, Li Deng, and Alex Acero: Learning dynamic noise models from noisy speech for robust speech recognition. In: Proc. Advances in Neural Information Processing Systems (NIPS), vol. 14, Vancouver, Canada, 2001, pp. 101-108 (2001)
39. M.J.F. Gales and S.J. Young: Robust speech recognition in additive and convolutional noise using parallel model combination. Computer Speech and Language, 9:289-307 (1995)
40. M. J. F. Gales: Maximum Likelihood Linear Transformations For HMM-Based Speech Recognition. Computer Speech and Language, 12 (January 1998)
41. M.J.F. Gales: Model-based approaches to handling uncertainty. Chapter 5 of this book (2011)
42. G. Hinton, S. Osindero, and Y. Teh: A fast learning algorithm for deep belief nets. Neural Computation, vol. 18, pp. 1527-1554, 2006)
43. R. Haeb-Umbach and V. Ion: Soft features for improved distributed speech recognition over wireless networks. In: Proc. Interspeech (2004)
44. X. He, L. Deng, and W. Chou: Discriminative learning in sequential pattern recognition — A unifying review. IEEE Signal Processing Magazine (2008)
45. J. Hershey, S. Rennie, P. Olsen, and T. Kristjansson: Super-human multi-talker speech recognition: A graphical modeling approach. Computer Speech and Language (June 2010)
46. H. G. Hirsch and D. Pearce: The Aurora experimental framework for the performance evaluation of speech recognition systems under noisy conditions. In: Proc. ISCA ITRW ASR (2000)
47. C. Hsieh and C. Wu: Stochastic vector mapping-based feature enhancement using prior-models and model adaptation for noisy speech recognition. Speech Communication, vol. 50, No. 6, pp. 467-475 (2008)
48. Y. Hu and Q. Huo: Irrelevant variability normalization based HMM training using VTS approximation of an explicit model of environmental distortions. In: Proc. Interspeech (2007)
49. C.-H. Lee and Q. Huo: On adaptive decision rules and decision parameter adaptation for automatic speech recognition. Proc. of the IEEE, vol. 88, No. 8, pp. 1241-1269 (2000)
50. V. Ion and R. Haeb-Umbach: Uncertainty decoding for distributed speech recognition over error-prone networks. Speech Communication, vol. 48, pp. 1435-1446 (2006)
51. V. Ion and R. Haeb-Umbach: A novel uncertainty decoding rule with applications to transmission error robust speech recognition. IEEE Trans. Speech and Audio Processing, vol. 16. No. 5, pp. 1047-1060 (2008)
52. H. Jiang and Li Deng: A Bayesian approach to the verification problem: Applications to speaker verification. IEEE Trans. Speech and Audio Proc., vol. 9, No. 8, pp. 874-884 (2001)
53. H. Jiang and L. Deng: A robust compensation strategy against extraneous acoustic variations in spontaneous speech recognition. IEEE Trans. on Speech and Audio Processing, vol. 10, no. 1, pp. 9-17 (2002)

54. O. Kalinli, M.L. Seltzer, and A. Acero: Noise adaptive training using a vector Taylor series approach for noise robust automatic speech recognition. In: Proc. ICASSP, pages 3825-3828, Taipei, Taiwan (2009)
55. D. Kim and M. Gales: Noisy constrained maximum likelihood linear regression for noise robust speech recognition. IEEE Trans. Audio Speech and Language Processing (2010)
56. D.Y. Kim, C.K. Un, and N.S. Kim: Speech recognition in noisy environments using first-order vector Taylor series. Speech Communication, vol. 24, pp. 39-49 (1998)
57. T.T. Kristjansson and B.J. Frey: Accounting for uncertainty in observations: A new paradigm for robust speech recognition. In: Proc. ICASSP, Orlando, Florida (2002)
58. T.T. Kristjansson, B. Frey, L. Deng, and A. Acero: Towards non-stationary model-based noise adaptation for large vocabulary speech recognition. In: Proc. ICASSP (2001)
59. C.-H. Lee: On stochastic feature and model compensation approaches to robust speech recognition. Speech Communication, vol. 25, pp. 29-47 (1998).
60. V. Leutnant and R. Haeb-Umbach: An analytic derivation of a phase-sensitive observation model for noise robust speech recognition. In: Proc. Interspeech (2009)
61. J. Li, D. Yu, Y. Gong, and Li Deng: Unscented Transform with Online Distortion Estimation for HMM Adaptation. In: Proc. Interspeech (2010)
62. J. Li, D. Yu, L. Deng, Y. Gong, and A. Acero: A unified framework of HMM adaptation with joint compensation of additive and convolutive distortions. Computer Speech and Language, vol. 23, pp. 389-405 (2009)
63. J. Li, L. Deng, D. Yu, Y. Gong, and A. Acero: HMM Adaptation Using a Phase-Sensitive Acoustic Distortion Model for Environment-Robust Speech Recognition. In: Proc. ICASSP, Las Vegas (2008)
64. J. Li, L. Deng, D. Yu, J. Wu, Y. Gong, and A. Acero: Adaptation of compressed HMM parameters for resource-constrained speech recognition. In: Proc. ICASSP, Las Vegas (2008)
65. H. Liao and M. J. F. Gales: Issues with uncertainty decoding for noise robust speech recognition. In: Proc. ICSLP, pp. 1121-1124 (2006)
66. H. Liao and M. J. F. Gales: Adaptive training with joint uncertainty decoding for robust recognition of noisy data. In: Proc. ICASSP, vol. IV, pp. 389-392 (2007)
67. H. Liao and M.J.F. Gales: Joint uncertainty decoding for noise robust speech recognition. In: Proc. Interspeech (2005)
68. Hui Lin, Li Deng, Dong Yu, Yifan Gong, Alex Acero, and Chi-Hui Lee: A study on multilingual acoustic modeling for large vocabulary ASR. In: Proc. ICASSP (2009)
69. R. Lyon: Machine hearing: An emerging field. IEEE Signal Processing Magazine (September 2010)
70. A. Mohamed, D. Yu, and L. Deng: Investigation of full-sequence training of deep belief networks for speech recognition. In: Proc. Interspeech (2010)
71. P. Moreno: Speech Recognition in Noisy Environments. Ph.D. Thesis, Carnegie Mellon University (1996)
72. N. Morgan et al.: Pushing the envelope — Aside. IEEE Signal Processing Magazine, vol. 22, No. 5, pp. 81-88 (2005)
73. R. Munkong and B.-H. Juang: Auditory perception and cognition — Modularization and integration of signal processing from ears to brain. IEEE Signal Processing Magazine, vol. 25, No. 3, pp. 98-117 (2008)
74. C. Rathinavalu and L. Deng: HMM-based speech recognition using state-dependent, discriminatively derived transforms on Mel-warped DFT features. IEEE Trans. on Speech and Audio Processing, pp. 243-256 (1997)
75. S. Rennie, J. Hershey, P. Olsen: Combining variational methods and loopy belief propagation for multi-talker speech recognition. IEEE Signal Processing Magazine, Special issue of Graphical Models for Signal Processing (Eds. M. Jordan et al.), (November 2010)
76. H. Sameti, H. Sheikhzadeh, Li Deng, and R. Brennan: HMM-based strategies for enhancement of speech signals embedded in nonstationary noise. IEEE Trans. on Speech and Audio Processing, vol. 6, no. 5, pp. 445-455 (1998)
77. H. Sameti and Li Deng: Nonstationary-state hidden Markov model representation of speech signals for speech enhancement. Signal Processing, vol. 82, pp. 205-227 (2002)

78. M. Seltzer, K. Kalgaonkar, and A. Acero: Acoustic model adaptation via linear spline interpolation for robust speech recognition. In: Proc. ICASSP (2010)
79. H. Sheikhzadeh and Li Deng: Waveform-based speech recognition using hidden filter models: Parameter selection and sensitivity to power normalization. IEEE Trans. on Speech and Audio Processing, vol. 2, no. 1, pp. 80-91 (1994)
80. G. Shi, Y. Shi, and Q. Huo: A study of irrelevant variability normalizataion based training and unsupervised online adaptation for LVCSR. In: Proc. Interspeech, Makuhari, Japan (2010)
81. V. Stouten,, H. Van hamme, P. Wambacq: Effect of phase-sensitive environment model and higher order VTS on noisy speech feature enhancement. In: Proc. ICASSP, pp. 433-436 (2005)
82. V. Stouten, H. Van hamme, and P. Wambacq: Accounting for the uncertainty of speech estimates in the context of model-based feature enhancement. In: Proc. ICSLP, pp. 105-108, Jeju Island, Korea (2004)
83. D. Yu, Li Deng, Yifan Gong, and Alex Acero: A novel framework and training algorithm for variable-parameter hidden Markov models. IEEE Trans. on Audio, Speech and Language Processing, vol. 17, no. 7, pp. 1348-1360, IEEE (2009)
84. D. Yu and Li Deng: Solving nonlinear estimation problems using Splines. IEEE Signal Processing Magazine, vol. 26, no. 4, pp. 86-90, (2009)
85. D. Yu, Li Deng, J. Droppo, J. Wu, Y. Gong, and A. Acero: Robust speech recognition using cepstral minimum-mean-square-error noise suppressor. IEEE Trans. Audio, Speech, and Language Processing, vol. 16, no. 5 (2008)
86. D. Yu and L. Deng: Deep-Structured Hidden Conditional Random Fields for Phonetic Recognition. In: Proc. Interspeech (2010)
87. D. Zhu and Q. Huo: A maximum likelihood approach to unsupervised online adaptation of stochastic vector mapping function for robust speech recognition. In: Proc. ICASSP (2007)
88. D. Zhu and Q. Huo: Irrelevant variability normalization based HMM training using MAP estimation of feature transforms for robust speech recognition. In: Proc. ICASSP (2008)

Chapter 5
Model-Based Approaches to Handling Uncertainty

M. J. F. Gales

Abstract A powerful approach for handling uncertainty in observations is to modify the statistical model of the data to appropriately reflect this uncertainty. For the task of noise-robust speech recognition, this requires modifying an underlying "clean" acoustic model to be representative of speech in a particular target acoustic environment. This chapter describes the underlying concepts of model-based noise compensation for robust speech recognition and how it can be applied to standard systems. The chapter will then consider important practical issues. These include i) acoustic environment noise parameter estimation; ii) efficient acoustic model compensation and likelihood calculation; and iii) adaptive training to handle multi-style training data. The chapter will conclude by discussing the limitations of the current approaches and research options to address them.

5.1 Introduction

There are many sources of variability in the speech signal, such as inter-speaker variability, intra-speaker variability, background noise conditions, channel distortion and reverberant noise (longer-term channel distortions). A range of approaches has been developed to try and reduce the level of variability: some approaches are based on general linear transformations [18, 38]; others are based on a model of how the variability impacts the acoustic models or features [17, 37]. This chapter will concentrate on one particular form of variability, background noise and convolutional distortion.

Handling background noise is still a fundamental issue in speech recognition. There are often high levels of mismatch between the training conditions of the acoustic models and the test conditions in which they are required to operate. Even

M. J. F. Gales
Cambridge University Engineering Department, Trumpington Street, Cambridge e-mail: mjfg@eng.cam.ac.uk

D. Kolossa and R. Haeb-Umbach (eds.), *Robust Speech Recognition of Uncertain or Missing Data: Theory and Applications*, DOI 10.1007/978-3-642-21317-5_5,
© Springer-Verlag Berlin Heidelberg 2011

with no mismatch, background noise will impact the system performance. As the level of noise increases the speech signal will become masked and the ability of the acoustic models to discriminate between words will decrease. Techniques for handling noise should be able to deal with this increase in uncertainty. This chapter examines approaches that handle background noise and channel distortions by modifying the parameters of the underlying acoustic models, in this case Hidden Markov Models (HMMs) [24, 52]. This class of approaches is often referred to as *model-based noise compensation schemes.*[1]

There is some debate as to whether model-based or feature-based compensation schemes, where the "clean" speech is estimated, are the most appropriate form for noise-robust speech recognition. In practice the best scheme depends heavily on the computational resources available, on whether the scheme needs to act causally, and on the nature of the parametrisation being used. This chapter will briefly mention feature-based schemes, and how uncertainty is included. However, as model-based approaches are more natural for handling additional levels of uncertainty associated with noise-robust speech recognition, this will be the focus of the discussion.

The next section will briefly discuss general forms of model adaptation to speakers or environments with a particular emphasis on how adaptation can be used to handle uncertainty. The impact of noise on speech and the forms of representation that are often used will then be described. This is followed by a brief discussion of feature compensation. Model-based compensation is then described, along with a discussion of computational efficiency and the estimation of all the model parameters. Finally, conclusions are drawn and possible future directions discussed.

5.2 General Acoustic Model Adaptation

Given the range of variability (and related uncertainty) associated with speech, there has been significant research devoted to handling this problem. Currently, one of the most popular approaches is to use linear transformations of the model parameters. This has been applied for rapid adaptation to speaker or environment changes.

Various configurations of linear transforms have been proposed. Note that the notation used in this section is consistent with that in the rest of the chapter. The clean speech parameters (the canonical model) will be represented by an x in the subscript, and y will represent the corrupted speech (target condition) parameters. Thus the corrupted speech mean of component m will be represented as $\boldsymbol{\mu}_{\mathrm{y}}^{(m)}$.

1. **Maximum Likelihood Linear Regression** (MLLR) [18, 23, 38]: one of the earliest and most popular forms of adaptation. Initially, only adaptation of the means was considered [38]. This was extended to adapting the covariance

[1] There has been a large amount of work, and possible variants, for model-based noise compensation schemes. This chapter is not meant as a complete review of all such schemes. The presentation is (naturally) biased towards work performed at Cambridge University. However, it is hoped that all sections are covered with appropriate background references.

matrices as well [18, 23, 57]. Here, for component m

$$\boldsymbol{\mu}_{\mathrm{y}}^{(m)} = \mathbf{A}^{(r_m)}\boldsymbol{\mu}_{\mathrm{x}}^{(m)} + \mathbf{b}^{(r_m)} \tag{5.1}$$

$$\boldsymbol{\Sigma}_{\mathrm{y}}^{(m)} = \mathbf{H}^{(r_m)}\boldsymbol{\Sigma}_{\mathrm{x}}^{(m)}\mathbf{H}^{(r_m)} \tag{5.2}$$

where r_m indicates the regression class to which component m belongs.

2. **Constrained MLLR** (CMLLR) [11, 18]: here the transformations of the means and covariance matrices, $\mathbf{A}^{(r_m)}$ and $\mathbf{H}^{(r_m)}$, are constrained to be the same, hence the name CMLLR. Originally used for diagonal transformation of the means and variances of the acoustic models [11], efficient estimation formulae and full transforms were investigated in [18].

$$\boldsymbol{\mu}_{\mathrm{y}}^{(m)} = \mathbf{H}^{(r_m)}(\boldsymbol{\mu}_{\mathrm{x}}^{(m)} - \mathbf{b}^{(r_m)}) \tag{5.3}$$

$$\boldsymbol{\Sigma}_{\mathrm{y}}^{(m)} = \mathbf{H}^{(r_m)}\boldsymbol{\Sigma}_{\mathrm{x}}^{(m)}\mathbf{H}^{(r_m)} \tag{5.4}$$

Rather than to adapt the model parameters, for full transforms it is more efficient to implement this as a set of transformations of the features [18]. Thus the approach is sometimes referred to as Feature MLLR (FMLLR). Now the likelihood can be expressed as

$$p(\mathbf{y}_t|m) = |\mathbf{A}^{(r_m)}|\mathcal{N}(\mathbf{A}^{(r_m)}\mathbf{y}_t + \mathbf{b}^{(r_m)}; \boldsymbol{\mu}_{\mathrm{x}}^{(m)}, \boldsymbol{\Sigma}_{\mathrm{x}}^{(m)}) \tag{5.5}$$

where $\mathbf{A}^{(r_m)} = \mathbf{H}^{(r_m)\text{-}1}$, $\mathbf{y}_t$ is the corrupted speech observations at time t. This form of adaptation does not require the model parameters to be modified. For large vocabulary systems where there may be hundreds of thousands of components, this is a very important attribute.

3. **Noisy CMLLR** (NCMLLR) [33]: this is an extension to CMLLR that is specifically aimed at handling situations with additional uncertainty. Here,

$$p(\mathbf{y}_t|m) = |\mathbf{A}^{(r_m)}|\mathcal{N}(\mathbf{A}^{(r_m)}\mathbf{y}_t + \mathbf{b}^{(r_m)}; \boldsymbol{\mu}_{\mathrm{x}}^{(m)}, \boldsymbol{\Sigma}_{\mathrm{x}}^{(m)} + \boldsymbol{\Sigma}_{\mathrm{b}}^{(r_m)}). \tag{5.6}$$

Thus NCMLLR may be viewed as a combination of CMLLR with a variance bias transform [57]. This form of transformation has the same structure as various noise model compensation schemes [33].

All of the above approaches involve a transformation of the covariance matrix. Thus in all cases they can model changes in uncertainty in the target conditions by, for example, appropriately scaling the variances.

An interesting extension of these adaptation approaches is adaptive training [4]. Here, the transforms are used during the training process. Rather than training a speaker- (or noise-) independent model-set to be adapted, a "neutral" canonical model is trained that is suitable for adaptation to each of the target conditions. Adaptive training schemes have been derived for all the above transforms [4, 18, 33]. For these adaptive training schemes, changing levels of uncertainty in the training data should be reflected in the contribution of those frames to the canonical model.

Frames with high levels of uncertainty should only make a small contribution to the model updates.

These general adaptation schemes do not rely on explicit models of speaker differences or the impact of noise on the clean speech. Instead, linear transforms, or sets of linear transforms, are estimated given adaptation data. Though advantageous in the sense that these transforms are able to model combinations of differences, they are only linear, or piecewise linear. Furthermore, the number of parameters for each transform can be large, $\mathcal{O}(d^2)$ where d is the size of the feature vector, for full transforms. This makes them impractical for very rapid adaptation, though modifications to improve robustness are possible [7, 20].

To enable very rapid adaptation, some low-dimensional representation of speaker differences or the impact of noise is needed. For speaker adaptation, vocal tract length normalisation [37] is one such scheme. This requires a single parameter, the warping factor, to be estimated. The equivalent for noise robustness is the set of noise models associated with the particular acoustic environment and the *mismatch* function for how the noise alters the clean speech.

5.3 Impact of Noise on Speech

The first stage in any form of feature- or model-based compensation scheme is to specify how the noise alters the clean speech for the parametrisation being used. In this section it is assumed that a "power domain" MFCC-based feature vector is being used.

5.3.1 Static Parameter Mismatch Function

The standard, simplified model of the impact of background additive noise, $\boldsymbol{n}_t$, and convolutional distortion, $\boldsymbol{h}_t$, on the clean $\boldsymbol{x}_t$, is [1]

$$\begin{aligned} \boldsymbol{y}_t &= \mathbf{C}\,\mathbf{log}\left(\mathbf{exp}(\mathbf{C}^{-1}(\boldsymbol{x}_t+\boldsymbol{h}_t))+\mathbf{exp}(\mathbf{C}^{-1}\boldsymbol{n}_t)\right) \\ &= \boldsymbol{f}(\boldsymbol{x}_t,\boldsymbol{h}_t,\boldsymbol{n}_t), \end{aligned} \tag{5.7}$$

where $\boldsymbol{y}_t$ is the corrupted speech observation at time t and $\mathbf{C}$ is the DCT. $\mathbf{exp}()$ and $\mathbf{log}()$ are the element-wise exponential and logarithm functions, respectively. It is simple to see that when the energy level of the noise is far greater than that of the (convolutionally distorted) clean speech, then $\boldsymbol{y}_t \approx \boldsymbol{n}_t$. The clean speech is *masked* by the noise.

Though (5.7) is the most commonly used form, a range of alternative mismatch functions, or interaction functions, have also been proposed [10, 17, 21, 36, 39, 41]. These approaches can be split into two categories, domain-based and phase-based compensation.

1. **Domain-based** [17]: this is the simplest form of modified compensation where the domain of the speech and noise compensation is treated as a tunable parameter. Here,

$$\boldsymbol{y}_t = \frac{1}{\gamma}\mathbf{C}\,\mathbf{log}\left(\mathbf{exp}(\mathbf{C}^{-1}\gamma(\boldsymbol{x}_t+\boldsymbol{h}_t))+\mathbf{exp}(\mathbf{C}^{-1}\gamma\boldsymbol{n}_t)\right). \quad (5.8)$$

γ determines the domain in which the clean speech and noise are combined. $\gamma = 1$ is the power domain, $\gamma = 1/2$ is the magnitude domain. Its value can be empirically tuned for a particular task.

2. **Phase-based** [10]: domain-based approaches are not motivated by the impact of noise on speech; they simply give a degree of flexibility enabling the mismatch function to be optimised. A more precise formulation is derived by taking into account the phase between the clean speech and the noise vectors. Here,

$$\boldsymbol{y}_t = \mathbf{C}\,\mathbf{log}\left(\mathbf{exp}(\mathbf{C}^{-1}(\boldsymbol{x}_t+\boldsymbol{h}_t))+\mathbf{exp}(\mathbf{C}^{-1}\boldsymbol{n}_t)+2\boldsymbol{\alpha}_t\circ\mathbf{exp}(\frac{\mathbf{C}^{-1}}{2}(\boldsymbol{x}_t+\boldsymbol{h}_t+\boldsymbol{n}_t))\right), \quad (5.9)$$

where $\boldsymbol{\alpha}_t$ is the vector of phase factors (the cosine of the angle between the speech and the noise) at time instance t and $\circ$ is element-wise multiplication. There has been a range of approximations within this framework. In [41] a fixed value for all elements of the vector $\boldsymbol{\alpha}$ was empirically determined. This is the closest to the domain-based compensation schemes. Indeed, the two approaches can be shown to be equal for $\gamma = 1$ and $\gamma = 1/2$. The optimal value for the Aurora 2 task yielded similar mismatch functions for the two approaches [21]. More precise forms of compensation treat $\boldsymbol{\alpha}$ as a random variable [10, 39, 65]. In [39] an analytic expression for the moments of $\boldsymbol{\alpha}$ were derived. Rather than to use all the moments of the distribution of $\boldsymbol{\alpha}$, a simpler approach is to assume that it is Gaussian-distributed and use the analysis in [39] to obtain the variance. However, since the phase factor has a physical interpretation, it should lie in the range minus 1 to +1. Thus an extension to this simple Gaussian approximation was used in [65] to compensate acoustic models using sample-based approaches. Here, the variable is treated as

$$p(\alpha_i) \propto \begin{cases} \mathcal{N}(\alpha_i;0,\sigma_{\alpha i}^2) & \alpha\in[-1,+1] \\ 0 & \text{otherwise} \end{cases} \quad (5.10)$$

where $\sigma_{\alpha i}^2$ is the phase factor variance for element i.

The rest of this chapter will focus on the standard form of mismatch function given in (5.7). For some of the alternative mismatch functions, model-based compensation has also been examined [21, 39, 41, 65].

5.3.2 Dynamic Parameter Mismatch Functions

The discussion so far has only considered the static parameters. The feature vector used for decoding usually consists of static and dynamic parameters. The standard form for the dynamic parameters is

$$\boldsymbol{\Delta y}_t = \frac{\sum_{\tau=1}^{w} \tau(\boldsymbol{y}_{t+\tau} - \boldsymbol{y}_{t-\tau})}{2\sum_{\tau=1}^{w} \tau^2} \tag{5.11}$$

where w is the window width used to determine the delta parameters. Similar expressions are used for the delta-delta parameters, $\boldsymbol{\Delta}^2\boldsymbol{y}_t$. The form of (5.11) allows the dynamic parameters to be represented as a linear transform of the static parameters. This is the approach used in [8, 64]. The observation vector for decoding can be expressed as

$$\begin{bmatrix} \boldsymbol{y}_t \\ \boldsymbol{\Delta y}_t \\ \boldsymbol{\Delta}^2\boldsymbol{y}_t \end{bmatrix} = \mathbf{D} \begin{bmatrix} \boldsymbol{y}_{t+w} \\ \vdots \\ \boldsymbol{y}_{t-w} \end{bmatrix} \tag{5.12}$$

$\mathbf{D}$ is the linear transform determined from (5.11). Provided the appropriate correlations in the feature vector are modelled, this allows the mismatch functions in the previous section to be used. Though yielding an accurate form of delta compensation, this form is computationally expensive and requires non-standard clean-speech model statistics to be estimated. A similar style of formulation has been used for simple-difference delta and delta-delta parameters [17].

The above scheme is computationally expensive. Thus the most common form of mismatch function used is the continuous time approximation [25]. Here, the following approximation is used:

$$\boldsymbol{\Delta y}_t \approx \left.\frac{\partial \boldsymbol{y}}{\partial t}\right|_t = \left.\frac{\partial \boldsymbol{y}}{\partial \boldsymbol{x}}\frac{\partial \boldsymbol{x}}{\partial t}\right|_t + \left.\frac{\partial \boldsymbol{y}}{\partial \boldsymbol{n}}\frac{\partial \boldsymbol{n}}{\partial t}\right|_t \approx \frac{\partial \boldsymbol{y}}{\partial \boldsymbol{x}}\boldsymbol{\Delta x}_t + \frac{\partial \boldsymbol{y}}{\partial \boldsymbol{n}}\boldsymbol{\Delta n}_t. \tag{5.13}$$

This is the standard form used in, for example, VTS compensation [2]. For simplicity of presentation, dynamic parameters are not discussed further in this chapter.

5.3.3 Corrupted Speech Distributions

Having derived a representation for the impact of noise on the clean speech, it is useful to examine how it alters the form of the clean speech distribution. Under the mismatch function for the static parameters in (5.7) and the assumption that both the clean speech and the additive noise are Gaussian-distributed, the corrupted speech distribution will be non-Gaussian. This is illustrated for one dimension in Fig. 5.1.

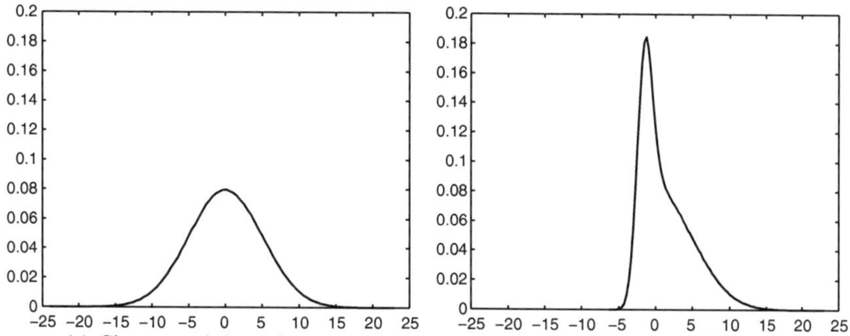

(a) Clean speech (speech mean = 0)

(b) Corrupted speech (noise mean = -2)

Fig. 5.1: Clean speech (a) and corrupted speech (b) distributions in the log-spectral domain

In addition to causing the distribution to be non-Gaussian, noise effects "masking" of low-energy segments of speech. This is seen clearly in Fig. 5.1, where the low-energy speech is completely masked by the noise. This "masking" property has been used for some noise compensation approaches [66] and is also exploited in the missing-feature noise robustness schemes [53, 59].

5.4 Feature Enhancement Approaches

The first forms of noise robustness were based on feature enhancement approaches. Originally, variants on spectral subtraction [6] were popular. These were then replaced by minimum mean squared error estimation (MMSE) schemes [9, 49], either requiring stereo data [3, 9, 48, 49] or using noise model estimates [61]. This section will discuss MMSE style enhancement approaches and how uncertainty has been incorporated into these schemes.

For MMSE-based approaches [9, 49], the estimated clean speech at time t, $\hat{\boldsymbol{x}}_t$, is given by

$$\hat{\boldsymbol{x}}_t = \mathscr{E}\{\boldsymbol{x}|\boldsymbol{y}_t\}. \tag{5.14}$$

The issue is about what form the posterior distribution of the clean speech, given the corrupted speech, should have. For simplicity this is often assumed to be jointly Gaussian. However, given the nonlinear nature of the interaction of speech and noise in (5.7), a mixture of Gaussians is used to improve performance. Thus for *front-end* component n the joint distribution is modelled as

$$\begin{bmatrix}\boldsymbol{y}\\ \boldsymbol{x}\end{bmatrix}\Bigg| n \sim \mathscr{N}\left(\begin{bmatrix}\boldsymbol{\mu}_{\mathrm{y}}^{(n)}\\ \boldsymbol{\mu}_{\mathrm{x}}^{(n)}\end{bmatrix}, \begin{bmatrix}\boldsymbol{\Sigma}_{\mathrm{y}}^{(n)} & \boldsymbol{\Sigma}_{\mathrm{yx}}^{(n)}\\ \boldsymbol{\Sigma}_{\mathrm{xy}}^{(n)} & \boldsymbol{\Sigma}_{\mathrm{x}}^{(n)}\end{bmatrix}\right) \tag{5.15}$$

If the component that generated the distribution at time t is known (here $\hat{n}_t$), then the MMSE estimate of the clean speech will be a linear transform of the corrupted speech [29]:

$$\begin{aligned} \hat{\boldsymbol{x}}_t &= \mathscr{E}\{\boldsymbol{x}|\boldsymbol{y}_t,\hat{n}_t\} && (5.16)\\ &= \boldsymbol{\mu}_{\mathrm{x}}^{(\hat{n}_t)} + \boldsymbol{\Sigma}_{\mathrm{xy}}^{(\hat{n}_t)}\boldsymbol{\Sigma}_{\mathrm{y}}^{(\hat{n}_t)\text{-}1}(\boldsymbol{y}_t - \boldsymbol{\mu}_{\mathrm{y}}^{(\hat{n}_t)}) && (5.17)\\ &= \mathbf{A}^{(\hat{n}_t)}\boldsymbol{y}_t + \mathbf{b}^{(\hat{n}_t)} && (5.18) \end{aligned}$$

In practice, the component is not known, so needs to either be estimated or treated as a latent variable and marginalised over. For the latent variable case,

$$\hat{\boldsymbol{x}}_t = \sum_n P(n|\boldsymbol{y}_t)\mathscr{E}\{\boldsymbol{x}|\boldsymbol{y}_t,n\}. \tag{5.19}$$

There are a number of possible schemes that can be used in terms of both the treatment of the component and the estimation of the compensation parameters $\mathbf{A}^{(n)}$ and $\mathbf{b}^{(n)}$. If the joint distribution (and hence associated marginal distributions) is known, then the posterior can be obtained from

$$P(n|\boldsymbol{y}_t) = \frac{P(n)\mathscr{N}(\boldsymbol{y}_t;\boldsymbol{\mu}_{\mathrm{y}}^{(n)},\boldsymbol{\Sigma}_{\mathrm{y}}^{(n)})}{\sum_m P(m)\mathscr{N}(\boldsymbol{y}_t;\boldsymbol{\mu}_{\mathrm{y}}^{(m)},\boldsymbol{\Sigma}_{\mathrm{y}}^{(m)})}. \tag{5.20}$$

The estimate of a single component can also be found from the best posterior:

$$\hat{n}_t = \underset{n}{\operatorname{argmax}}\{P(n|\boldsymbol{y}_t)\}. \tag{5.21}$$

An interesting alternative is to use an iterative EM-like process [3]. Either the joint distribution or the transform may be estimated from stereo data [3, 49] or using approaches based on model-based compensation [48, 61].

The estimate of the clean speech, $\hat{\boldsymbol{x}}_t$, is then passed to the clean recogniser. Thus the likelihood is approximated for a particular *recognition* component m by

$$p(\boldsymbol{y}_t|m) \approx \mathscr{N}(\hat{\boldsymbol{x}}_t;\boldsymbol{\mu}_{\mathrm{x}}^{(m)},\boldsymbol{\Sigma}_{\mathrm{x}}^{(m)}), \tag{5.22}$$

where $\boldsymbol{\mu}_{\mathrm{x}}^{(m)}$ and $\boldsymbol{\Sigma}_{\mathrm{x}}^{(m)}$ are the mean vector and covariance matrix of the clean speech-trained acoustic model for component m. Thus the underlying assumption behind this model is that the clean speech estimate is "perfect", irrespective of the level of background noise. However, for low-SNR conditions where $\boldsymbol{y}_t \approx \boldsymbol{n}_t$, it is difficult to get an accurate estimate of the clean speech.

One approach to address this problem is to add uncertainty to the estimate of the clean speech [5, 61]. Here, the posterior is assumed to be Gaussian in nature:

$$\boldsymbol{x}|\boldsymbol{y}_t \sim \mathscr{N}(\hat{\boldsymbol{x}}_t,\boldsymbol{\Sigma}_t), \tag{5.23}$$

where $\boldsymbol{\Sigma}_t$ is the "uncertainty" associated with estimate at time instant t. The likelihood is then computed as

$$p(\boldsymbol{y}_t|m) \approx \int p(\boldsymbol{x}|\boldsymbol{y}_t)p(\boldsymbol{x}|m)d\boldsymbol{x} \tag{5.24}$$

$$\approx \int \mathcal{N}(\boldsymbol{x};\hat{\boldsymbol{x}}_t,\boldsymbol{\Sigma}_t)\mathcal{N}(\boldsymbol{x};\boldsymbol{\mu}_{\mathrm{x}}^{(m)},\boldsymbol{\Sigma}_{\mathrm{x}}^{(m)})d\boldsymbol{x}. \tag{5.25}$$

Though intuitively well motivated, from (5.24) it can be seen that the likelihood is not mathematically consistent.

An alternative, more consistent scheme is to propagate the distribution of the corrupted speech given the clean speech [12, 43]. Here,

$$p(\boldsymbol{y}_t|m) \approx \int p(\boldsymbol{y}_t|\boldsymbol{x})p(\boldsymbol{x}|m)d\boldsymbol{x}. \tag{5.26}$$

Again, the acoustic space is represented by a mixture model. Now, the distribution (marginalizing over the components) is

$$p(\boldsymbol{y}_t|\boldsymbol{x}) = \sum_n P(n|\boldsymbol{x})p(\boldsymbol{y}_t|\boldsymbol{x},n). \tag{5.27}$$

Compared to (5.19), this is more complex as the component posterior, $P(n|\boldsymbol{x})$, is conditional on the clean speech latent variable $\boldsymbol{x}$ rather than on the corrupted observation $\boldsymbol{y}_t$. Different approximations for this have been proposed [12, 43].

Though mathematically more consistent, this form of approach has an issue when using the (required) approximations for $P(n|\boldsymbol{x})$. This component posterior term should vary continuously as the "unseen" clean speech $\boldsymbol{x}$ changes. As this is highly computationally expensive to deal with, the approximations used produce a form of "average" component posterior term to use for enhancement. The posterior distribution $p(\boldsymbol{y}_t|\boldsymbol{x})$ is then the same for all "recognition" components m. At very low SNRs the averaged form of component posterior can often result in $p(\boldsymbol{y}_t|\boldsymbol{x}) = p(\boldsymbol{n}_t)$ as the vast majority, but not necessarily all, of clean speech and associated components will be completely masked. As the posterior is now independent of $\boldsymbol{x}$, all recognition components m have the same distribution, so the frame is ignored in terms of acoustic discrimination. The only form of discrimination will be that associated with the language model. The information from any non-masked speech (and components) has been lost. Depending on the task, this can have a large impact on recognition performance. This issue is discussed in detail in [46]. Given that the underlying attribute of feature-based approaches is that enhancement (with or without uncertainty) is decoupled from recognition components, this problem cannot be addressed within an enhancement framework.[2] As soon as there is a coupling

[2] Theoretically, the exact value of $P(n|\boldsymbol{x})$ could be used. However, as $\boldsymbol{x}$ is a function of the recognition component this effectively becomes model-based compensation. Interestingly if no uncertainty is used, as in SPLICE [9], this problem does not occur as only the means, not the variances, of the distributions can be altered.

between the “enhancement” and the recognition component, the scheme becomes a model-based approach, as discussed in the next section.

5.5 Model-Based Noise Compensation

The aim of model-based compensation schemes is to modify the acoustic model parameters so that they are representative of the HMM output distributions in the target domain. The advantage of model-based compensation schemes is that the additional uncertainty that results from the background noise is directly modelled. There is no need to estimate masks or additional uncertainty.

From Fig. 5.1 it is clear that even if the clean speech and noise are Gaussian distributed, the resulting corrupted speech distribution is non-Gaussian. In practice, when considering all elements in the feature vector, the corrupted speech distribution may be highly complicated, with a large number of modes. Some approaches attempt to model this complexity using, for example, GMMs [17]. Alternatively, Gaussian approximations for the likelihood at the observation $\boldsymbol{y}_t$ rather than for the whole distribution of $\boldsymbol{y}$ have been proposed [36]. Finally, non-parametric schemes for the distribution of $\boldsymbol{y}_t$ have been used [65]. A common attribute of all these schemes is that they are computationally very expensive.

In contrast to estimating the “true” distribution, a simple approximation stems from assuming that the distribution of the corrupted speech is Gaussian in nature [2, 17]. Thus

$$p(\boldsymbol{y}_t|m) \approx \mathcal{N}(\boldsymbol{y}_t; \boldsymbol{\mu}_{\mathrm{y}}^{(m)}, \boldsymbol{\Sigma}_{\mathrm{y}}^{(m)}), \tag{5.28}$$

where $\boldsymbol{\mu}_{\mathrm{y}}^{(m)}$ and $\boldsymbol{\Sigma}_{\mathrm{y}}^{(m)}$ are the estimated mean vector and covariance matrix of the corrupted speech for the target environment. The task is now to obtain appropriate estimates of these corrupted model parameters. Using standard ML estimation, these can be obtained using [17]

$$\boldsymbol{\mu}_{\mathrm{y}}^{(m)} = \mathscr{E}\{\boldsymbol{y}|m\}, \quad \boldsymbol{\Sigma}_{\mathrm{y}}^{(m)} = \mathscr{E}\left\{\boldsymbol{y}\boldsymbol{y}^{\mathsf{T}}|m\right\} - \boldsymbol{\mu}_{\mathrm{y}}^{(m)}\boldsymbol{\mu}_{\mathrm{y}}^{(m)\mathsf{T}}. \tag{5.29}$$

There are a number of approximations for these expectations that sit within this *Parallel Model Combination* (PMC) framework. This chapter will only consider two such forms. The first, Vector Taylor Series (VTS) compensation [2, 48], is one of the most popular approaches. The second, based on sampling schemes, aims at improving the approximations underlying VTS. Other forms are possible, for example, the log-normal approximation [17], spline interpolation [58], and Jacobian compensation [56]. However, for all these schemes it is worth emphasising that however accurate the compensation scheme is, the final distribution is approximated by a single Gaussian.

For all these schemes the noise parameters are usually modelled using [17, 48]

$$\boldsymbol{n} \sim \mathcal{N}(\boldsymbol{\mu}_{\mathrm{n}}, \boldsymbol{\Sigma}_{\mathrm{n}}), \quad \boldsymbol{h} = \boldsymbol{\mu}_{\mathrm{h}}. \tag{5.30}$$

Thus the convolutional noise is assumed to be constant. Additionally, the delta and delta-delta noise means are often assumed to be zero [44]. These parameters may be estimated [40], but the motivation for these estimates is not clear.[3] The estimation of these parameters will be described in Section 5.7.

5.5.1 Vector Taylor Series Compensation

A currently popular form of model-based compensation is VTS. Here a first-order Taylor series approximation to the nonlinearity of (5.7) is used. Thus for component m, the random variable for the corrupted speech $\boldsymbol{y}$ is related to the clean speech $\boldsymbol{x}$ and the noise $\boldsymbol{n}$ random variables by [48]:

$$\boldsymbol{y}|m \approx \boldsymbol{f}(\boldsymbol{\mu}_{\mathrm{x}}^{(m)}, \boldsymbol{\mu}_{\mathrm{h}}, \boldsymbol{\mu}_{\mathrm{n}}) + \mathbf{J}^{(m)}\left((\boldsymbol{x} - \boldsymbol{\mu}_{\mathrm{x}}^{(m)}) + (\boldsymbol{h} - \boldsymbol{\mu}_{\mathrm{h}})\right) + (\mathbf{I} - \mathbf{J}^{(m)})(\boldsymbol{n} - \boldsymbol{\mu}_{\mathrm{n}}), \tag{5.31}$$

where the Jacobian $\mathbf{J}^{(m)}$ is defined as

$$\mathbf{J}^{(m)} = \left.\frac{\partial \boldsymbol{y}}{\partial \boldsymbol{x}}\right|_{\boldsymbol{\mu}_{\mathrm{x}}^{(m)}, \boldsymbol{\mu}_{\mathrm{h}}, \boldsymbol{\mu}_{\mathrm{n}}}. \tag{5.32}$$

Using this approximation yields the following estimates for the corrupted speech distribution:

$$\boldsymbol{\mu}_{\mathrm{y}}^{(m)} = \boldsymbol{f}(\boldsymbol{\mu}_{\mathrm{x}}^{(m)}, \boldsymbol{\mu}_{\mathrm{h}}, \boldsymbol{\mu}_{\mathrm{n}}) \tag{5.33}$$

$$\boldsymbol{\Sigma}_{\mathrm{y}}^{(m)} = \mathbf{J}^{(m)}\boldsymbol{\Sigma}_{\mathrm{x}}^{(m)}\mathbf{J}^{(m)\mathsf{T}} + (\mathbf{I} - \mathbf{J}^{(m)})\boldsymbol{\Sigma}_{\mathrm{n}}(\mathbf{I} - \mathbf{J}^{(m)})^{\mathsf{T}}. \tag{5.34}$$

As the Jacobian will be full, this results in a full covariance matrix for the corrupted speech distribution $\boldsymbol{\Sigma}_{\mathrm{y}}^{(m)}$. It is common to diagonalise this covariance matrix to maintain efficient likelihood calculation and control the number of model parameters. Thus in practice the likelihood is computed as

$$p(\boldsymbol{y}_t|m) = \mathcal{N}\left(\boldsymbol{y}_t; \boldsymbol{\mu}_{\mathrm{y}}^{(m)}, \mathrm{diag}(\boldsymbol{\Sigma}_{\mathrm{y}}^{(m)})\right). \tag{5.35}$$

For a discussion of the impact of this approximation see [64].

A nice aspect of VTS, and one of the reasons for its popularity, is that the linearisation simplifies the estimation of the noise and clean speech model parameters [48]. This is discussed in Section 5.7. However, this linearisation may be expected to impact performance; thus alternative schemes are of interest.

[3] These may be interpreted as a general mismatch function rather as motivated by the physical impact of noise on speech.

5.5.2 Sampling-Based Approximations

VTS relies on a first-order (or possibly higher) Taylor series expansion. To improve this form of approximation it is possible to use sampling style approaches. This section briefly describes two of these schemes. Both aim at directly estimating the integrals of the following form, taking the mean of component m as an example:

$$\boldsymbol{\mu}_{\mathrm{y}}^{(m)} = \int \int \boldsymbol{f}(\boldsymbol{x}, \boldsymbol{\mu}_{\mathrm{h}}, \boldsymbol{n}) p(\boldsymbol{x}|m) p(\boldsymbol{n}) d\boldsymbol{n} d\boldsymbol{x}, \tag{5.36}$$

where $\boldsymbol{f}(.)$ is given in (5.7).

The simplest approximation is based on Monte Carlo sampling. As both the clean speech and the noise are Gaussian-distributed there are no problems generating samples from them. This approximation, Data-driven PMC (DPMC) [17], then uses the following update formula for the mean:

$$\boldsymbol{\mu}_{\mathrm{y}}^{(m)} = \frac{1}{K} \sum_{k=1}^{K} \boldsymbol{f}(\boldsymbol{x}^{(k)}, \boldsymbol{\mu}_{\mathrm{h}}, \boldsymbol{n}^{(k)}), \tag{5.37}$$

where $\boldsymbol{x}^{(k)}$ is a sample drawn from $\boldsymbol{x}|m \sim \mathcal{N}(\boldsymbol{\mu}_{\mathrm{x}}^{(m)}, \boldsymbol{\Sigma}_{\mathrm{x}}^{(m)})$ and $\boldsymbol{n}^{(k)}$ is a sample drawn from $\boldsymbol{n} \sim \mathcal{N}(\boldsymbol{\mu}_{\mathrm{n}}, \boldsymbol{\Sigma}_{\mathrm{n}})$. Note that in this case the noise and speech samples are drawn independently.

The advantage of this form of compensation is that in the limit as $K \to \infty$ the compensation will be "exact" (given the assumption that the corrupted speech distribution is Gaussian in nature). However, a major disadvantage of this straightforward scheme is that as the number of dimensions being sampled from increases, the number of samples needs to be increased in order to get robust estimates.

One approach to address these limitations is to use unscented transforms [31]. Rather than drawing independent samples from the speech and noise, a set of samples are jointly drawn given the means and variances of the clean speech and noise. Here, the approximation, again for the mean, has the form

$$\boldsymbol{\mu}_{\mathrm{y}}^{(m)} = \frac{1}{\sum_{k=0}^{K} w^{(k)}} \sum_{k=0}^{K} w^{(k)} \boldsymbol{f}(\boldsymbol{x}^{(k)}, \boldsymbol{\mu}_{\mathrm{h}}, \boldsymbol{n}^{(k)}). \tag{5.38}$$

The samples are drawn in a deterministic fashion. If the overall dimension of the combined vector $\boldsymbol{z}^{(k)}$ has dimensionality $2d$ (the feature vector is d-dimensional),

$$\boldsymbol{z}^{(k)} = \begin{bmatrix} \boldsymbol{x}^{(k)} \\ \boldsymbol{n}^{(k)} \end{bmatrix}. \tag{5.39}$$

$2d+1$ samples are then drawn in a symmetric fashion based on (noting the dependence of the combined vector on the clean speech component m)

$$\mathbf{z}^{(0)} = \boldsymbol{\mu}_{\mathrm{z}}^{(m)}; \; w^{(0)} = \frac{\kappa}{2d+\kappa} \tag{5.40}$$

$$\mathbf{z}^{(k)} = \boldsymbol{\mu}_{\mathrm{z}}^{(m)} + \left[\sqrt{(2d+\kappa)\boldsymbol{\Sigma}_{\mathrm{z}}^{(m)}}\right]_k^{\mathsf{T}}; \; w^{(k)} = \frac{1}{2(2d+\kappa)} \tag{5.41}$$

$$\mathbf{z}^{(k+2d)} = \boldsymbol{\mu}_{\mathrm{z}}^{(m)} - \left[\sqrt{(2d+\kappa)\boldsymbol{\Sigma}_{\mathrm{z}}^{(m)}}\right]_k^{\mathsf{T}}; \; w^{(k+2d)} = \frac{1}{2(2d+\kappa)} \tag{5.42}$$

where $\left[\sqrt{\mathbf{A}}\right]_k^{\mathsf{T}}$ indicates the transpose of the k-th row of the Choleski factorisation of $\mathbf{A}$ and κ is a tunable parameter. The number of samples increases linearly as the number of dimensions increases.

Unscented transform compensation has been applied, with gains over VTS and simpler forms of PMC, for both model compensation and feature-based enhancements [28, 60].

5.6 Efficient Model Compensation and Likelihood Calculation

One of the issues with model-based compensation schemes is that they are computationally expensive. Applying schemes such as VTS to large vocabulary speech recognition systems is currently impractical for real-time compensation. The costs associated with model-based compensation schemes can be split into three parts: i) estimation of the noise parameters; ii) estimation of the compensation parameters; iii) application of the compensation parameters to the acoustic models. The estimation of the noise parameters is not discussed here, but in the next section. This section will briefly describe approaches for reducing the computational load of the remaining two stages.

One approach to addressing the problem of computational cost is to express model-based compensation in a factored form [16]. To improve the efficiency, this can be rewritten in the following approximate form:

$$p(\boldsymbol{y}_t|m) = \int p(\boldsymbol{y}_t|\boldsymbol{x},m)p(\boldsymbol{x}|m)d\boldsymbol{x} \tag{5.43}$$

$$\approx \int p(\boldsymbol{y}_t|\boldsymbol{x},r_m)p(\boldsymbol{x}|m)d\boldsymbol{x}, \tag{5.44}$$

where r_m indicates the regression class that component m belongs to. The distribution of the clean speech is known; it is given by the clean speech HMM. Thus the problem is to find the conditional distribution, $p(\boldsymbol{y}_t|\boldsymbol{x},r_m)$. It is interesting to compare this form with the enhancement schemes in Section 5.4. Here, the posterior is dependent on either the component or the regression class, whereas for feature enhancement it is not. This means that the approximate averaging over the complete acoustic space discussed in [46] and Section 5.4 will not occur for model-based compensation (unless very few regression classes are used).

5.6.1 Compensation Parameter Estimation

For schemes such as VTS the compensation parameters required are the Jacobians associated with each component m, $\mathbf{J}^{(m)}$. This is needed to compensate the covariance matrices. This form of Jacobian can be computed as [2]

$$\mathbf{J}^{(m)} = \mathbf{C}\mathbf{F}^{(m)}\mathbf{C}^{-1}, \tag{5.45}$$

where $\mathbf{C}$ is the DCT matrix and $\mathbf{F}^{(m)}$ is a diagonal covariance matrix where the elements on the leading diagonal are given by

$$f_{ii}^{(m)} = \frac{1}{1+\exp([\mathbf{C}^{-1}]_i(\boldsymbol{\mu}_{\mathrm{n}} - \boldsymbol{\mu}_{\mathrm{x}} - \boldsymbol{\mu}_{\mathrm{h}}))} \tag{5.46}$$

and $[\mathbf{C}^{-1}]_i$ is the ith *row* of $\mathbf{C}^{-1}$. This calculation is dominated by a matrix-matrix multiplication (in the dimensionality of the static parameters) per recognition Gaussian component. For large vocabulary speech recognition this rapidly becomes impractical.

Rather than VTS, the approximation in (5.44) can be used. The aim is to obtain an efficient form for the regression class-specific conditional distribution, $p(\boldsymbol{y}_t|\boldsymbol{x},r)$. One approach is Joint Uncertainty Decoding (JUD) [42]. Here the joint distribution is assumed to be Gaussian at the regression class level. Thus

$$\left.\begin{bmatrix}\boldsymbol{y}\\ \boldsymbol{x}\end{bmatrix}\right| r \sim \mathcal{N}\left(\begin{bmatrix}\boldsymbol{\mu}_{\mathrm{y}}^{(r)}\\ \boldsymbol{\mu}_{\mathrm{x}}^{(r)}\end{bmatrix}, \begin{bmatrix}\boldsymbol{\Sigma}_{\mathrm{y}}^{(r)} & \boldsymbol{\Sigma}_{\mathrm{yx}}^{(r)}\\ \boldsymbol{\Sigma}_{\mathrm{xy}}^{(r)} & \boldsymbol{\Sigma}_{\mathrm{x}}^{(r)}\end{bmatrix}\right). \tag{5.47}$$

The conditional distribution is also Gaussian, where

$$\boldsymbol{\mu}_{\mathrm{y|x}}^{(r)} = \boldsymbol{\mu}_{\mathrm{y}}^{(r)} + \boldsymbol{\Sigma}_{\mathrm{yx}}^{(r)}\boldsymbol{\Sigma}_{\mathrm{x}}^{(r)-1}(\boldsymbol{x} - \boldsymbol{\mu}_{\mathrm{x}}^{(r)}) \tag{5.48}$$

$$\boldsymbol{\Sigma}_{\mathrm{y|x}}^{(r)} = \boldsymbol{\Sigma}_{\mathrm{y}}^{(r)} - \boldsymbol{\Sigma}_{\mathrm{yx}}^{(r)}\boldsymbol{\Sigma}_{\mathrm{x}}^{(r)-1}\boldsymbol{\Sigma}_{\mathrm{xy}}^{(r)}. \tag{5.49}$$

As all distributions are Gaussian, the marginal will also be Gaussian. The likelihood in the joint framework can be computed as

$$p(\boldsymbol{y}_t|m) = \mathcal{N}(\boldsymbol{y}_t; \mathbf{H}^{(r_m)}(\boldsymbol{\mu}_{\mathrm{x}}^{(m)} - \mathbf{b}^{(r_m)}), \mathbf{H}^{(r_m)}(\boldsymbol{\Sigma}_{\mathrm{x}}^{(m)} + \boldsymbol{\Sigma}_{\mathrm{b}}^{(r_m)})\mathbf{H}^{(r_m)\mathsf{T}}), \tag{5.50}$$

where the compensation transform parameters are obtained using

$$\begin{aligned}\mathbf{H}^{(r)} &= \boldsymbol{\Sigma}_{\mathrm{yx}}^{(r)}\boldsymbol{\Sigma}_{\mathrm{x}}^{(r)-1},\\ \mathbf{b}^{(r)} &= \boldsymbol{\mu}_{\mathrm{x}}^{(r)} - \mathbf{H}^{(r)-1}\boldsymbol{\mu}_{\mathrm{y}}^{(r)}\\ \boldsymbol{\Sigma}_{\mathrm{b}}^{(r)} &= \mathbf{H}^{(r)-1}\boldsymbol{\Sigma}_{\mathrm{y}}^{(r)}\mathbf{A}^{(r)-\mathsf{T}} - \boldsymbol{\Sigma}_{\mathrm{x}}^{(r)}.\end{aligned} \tag{5.51}$$

These compensation parameters only need to be computed for each of the R regression classes, rather than for all recognition components. All parameters of the joint distribution, other than the cross term $\boldsymbol{\Sigma}_{\mathrm{xy}}^{(r)}$, can be obtained using, for example, VTS, or from the clean speech training data. For VTS the cross term can be found using [43, 62]

$$\boldsymbol{\Sigma}_{\mathrm{xy}}^{(r)} = \boldsymbol{\Sigma}_{\mathrm{x}}^{(r)} \mathbf{J}^{(r)\mathsf{T}}. \tag{5.52}$$

Computing the compensation parameters per regression class is more expensive than computing them for a single component, but the number of regression classes can be made orders of magnitude smaller than the number of components. It is also flexible as the number of regression classes can be controlled depending on the available computational resources.

5.6.2 Compensating the Model Parameters

After deriving the compensation parameters, the model parameters must then be modified. In a similar fashion to (5.35), whatever form of compensation is used, it should require only diagonal covariance matrix likelihood calculations. Directly applying the VTS compensation parameters requires calculating the means and covariance matrices for every component. For large systems this rapidly becomes impractical. Three alternative options for model compensation are described below.

1. **VTS-JUD** [67]: this form is the most closely related to VTS. Equation (5.50) is used with diagonal covariance matrices. Thus the likelihood is computed as

$$p(\boldsymbol{y}_t|m) = \mathcal{N}\left(\boldsymbol{y}_t; \mathbf{H}^{(r_m)}(\boldsymbol{\mu}_{\mathrm{x}}^{(m)} - \mathbf{b}^{(r_m)}), \mathrm{diag}\left(\mathbf{H}^{(r_m)}(\boldsymbol{\Sigma}_{\mathrm{x}}^{(m)} + \boldsymbol{\Sigma}_{\mathrm{b}}^{(r_m)})\mathbf{H}^{(r_m)\mathsf{T}}\right)\right). \tag{5.53}$$

   This scheme requires all recognition parameters to be transformed. Thus the cost of applying the compensation parameters is comparable to that of standard VTS. However, there is the advantage of only computing the compensation parameters at the regression class level.
2. **JUD** [43]: here the likelihood is computed as

$$p(\boldsymbol{y}_t|m) = |\mathbf{A}^{(r_m)}|\mathcal{N}(\mathbf{A}^{(r_m)}\boldsymbol{y}_t + \mathbf{b}^{(r_m)}; \boldsymbol{\mu}_{\mathrm{x}}^{(m)}, \boldsymbol{\Sigma}_{\mathrm{x}}^{(m)} + \boldsymbol{\Sigma}_{\mathrm{b}}^{(r_m)}), \tag{5.54}$$

   where the compensation transform parameters are obtained using $\mathbf{A}^{(r)} = \mathbf{H}^{(r)\text{-}1}$. This form of compensation only requires a bias to be applied to the clean covariance matrix. However, to limit the computational cost this covariance bias term, $\boldsymbol{\Sigma}_{\mathrm{b}}^{(r)}$, needs to be diagonal. Using a full joint distribution and diagonalising the covariance matrix in (5.54) yields poor performance [43]. To address this problem, the form of the joint distribution can be modified. Here,

$$\begin{bmatrix} \boldsymbol{y} \\ \boldsymbol{x} \end{bmatrix} \Bigg| r \sim \mathcal{N}\left(\begin{bmatrix} \boldsymbol{\mu}_{\mathrm{y}}^{(r)} \\ \boldsymbol{\mu}_{\mathrm{x}}^{(r)} \end{bmatrix}, \begin{bmatrix} \mathrm{diag}(\boldsymbol{\Sigma}_{\mathrm{y}}^{(r)}) & \mathrm{diag}(\boldsymbol{\Sigma}_{\mathrm{yx}}^{(r)}) \\ \mathrm{diag}(\boldsymbol{\Sigma}_{\mathrm{xy}}^{(r)}) & \mathrm{diag}(\boldsymbol{\Sigma}_{\mathrm{x}}^{(r)}) \end{bmatrix}\right). \tag{5.55}$$

This yields diagonal forms for the compensation parameters in (5.51). Note that it will also be more efficient to compute the compensation parameters. This form only requires compensation parameters at the regression class level, and only a variance bias to be applied at the recognition component level.
This form of compensation has exactly the same form as does NCMLLR (5.6) but is derived from a noise compensation perspective. For a discussion of the attributes and comparison of the two approaches see [34].

3. **Predictive CMLLR** (PCMLLR) [22]: this uses the same form of transformation as CMLLR [18]. However, rather than estimating the transform parameters from adaptation data, it estimates them from the model-based corrupted speech distributions. The form of the likelihood calculation is

$$p(\mathbf{y}_t|m) = |\mathbf{A}^{(r_m)}|\mathscr{N}(\mathbf{A}^{(r_m)}\mathbf{y}_t + \mathbf{b}^{(r_m)}; \boldsymbol{\mu}_{\mathrm{x}}^{(m)}, \boldsymbol{\Sigma}_{\mathrm{x}}^{(m)}). \tag{5.56}$$

Here, the model parameters are not altered, but there are additional costs in estimating $\mathbf{A}^{(r)}$ and $\mathbf{b}^{(r)}$ from the compensation form. For a discussion of the computational costs of this see [67]. Though PCMLLR is an approximation to the corrupted distribution, it has additional flexibility. By using full or block-diagonal transformations, correlation changes in the feature vector can be efficiently modelled. This is not possible with standard VTS, where diagonal covariance matrices are used. This flexibility has been found to yield improved performance [67]. Another advantage of this approach is that adaptive training is very simple, as the standard CMLLR adaptive training approach can be used [67].
Interestingly, PCMLLR has exactly the same form as the MMSE estimate in (5.18). However there are two important differences. First, PCMLLR is dependent on the regression class. Second, the compensation parameters are derived from minimising the KL divergence to the estimate of the corrupted speech distribution rather than from an MMSE perspective [63]. As the KL divergence looks at complete distributions (rather than just the first-order moments in MMSE), changes in the uncertainty can be modelled with PCMLLR.

It is simple to show that when the number of regression classes is the same as the number of components, VTS-JUD, JUD and PCMLLR become identical to the standard model compensation scheme being used to derive the joint distribution.

5.7 Adaptive Training and Noise Estimation

So far the discussion has assumed that all the model parameters required for compensation are known. In practice this is rarely the case. Originally, the background noise was simply estimated from periods of “silence” in the test conditions. This required the use of a voice activity detection scheme, and removed any link between the clean model parameters and the estimates of the noise. Furthermore, there is no way to estimate the convolutional noise. For the clean speech parameters it was assumed that clean (high-SNR) training data was always available to estimate the clean models. However, this did not allow application domain, or found, data to be used in the training process. Thus recently there has been growing interest in training both the acoustic models [30, 32, 45] and the noise model [35, 40, 44, 48] in a full ML framework. This research area has parallels with developments in speaker adaptation, where the speaker transform parameters are often estimated in an ML fashion [38] and the canonical model parameters are estimated using adaptive training [4].

The standard approach to estimating the parameters is to maximise the likelihood of the data. Thus the aim is to find the model parameters, $\hat{\mathscr{M}}$, that maximise

$$\mathscr{F}(\hat{\mathscr{M}}) = \sum_{\theta} P(\boldsymbol{\theta}) \prod_{t} \sum_{m \in \theta_t} \mathscr{N}\left(\boldsymbol{y}_t; \hat{\boldsymbol{\mu}}_{\mathrm{y}}^{(m)}, \mathrm{diag}(\hat{\boldsymbol{\Sigma}}_{\mathrm{y}}^{(m)})\right), \tag{5.57}$$

where the summation over $\boldsymbol{\theta}$ includes all possible state sequences for the observation sequence. As with standard HMM parameter training, EM is used. Thus the following auxiliary function is maximised (ignoring all terms independent of the model to be estimated):

$$\mathscr{Q}(\hat{\mathscr{M}}; \mathscr{M}) = \sum_{m,t} \gamma_t^{(m)} \log\left(\mathscr{N}\left(\boldsymbol{y}_t; \hat{\boldsymbol{\mu}}_{\mathrm{y}}^{(m)}, \mathrm{diag}(\hat{\boldsymbol{\Sigma}}_{\mathrm{y}}^{(m)})\right)\right), \tag{5.58}$$

where the posterior of the observation at time t being generated by component m, $\gamma_t^{(m)}$, is determined using the "current" model parameters, $\mathscr{M}$. The task is now to estimate the clean speech model parameters for each of the components, $\hat{\boldsymbol{\mu}}_{\mathrm{x}}^{(m)}$ and $\hat{\boldsymbol{\Sigma}}_{\mathrm{x}}^{(m)}$, and noise model parameters, $\hat{\boldsymbol{\mu}}_{\mathrm{n}}$, $\hat{\boldsymbol{\mu}}_{\mathrm{h}}$ and $\hat{\boldsymbol{\Sigma}}_{\mathrm{n}}$, that maximise (5.58).

Two approaches have been described in the literature. The first is to introduce a second level of EM, where the clean speech, or noise, at time t are considered as continuous latent variables. This will be referred to as the EM approach. The second is a direct approach, based on second-order optimisation schemes. This section gives a summary of some of the attributes of these schemes. Neither of the forms used is exact; a series of approximations is made in each case. The best scheme needs to be determined empirically for the task (and approximations) of interest. For a more detailed analysis and contrast of the two approaches see [15].

5.7.1 EM-Based Approaches

From the VTS approximation (5.31) it can be seen that the corrupted observation can be written in the form of a generative model where

$$\boldsymbol{y}|m \approx \mathbf{J}^{(m)}\boldsymbol{x}^{(m)} + (\mathbf{I} - \mathbf{J}^{(m)})\boldsymbol{n} + \mathbf{J}^{(m)}\hat{\boldsymbol{\mu}}_{\mathrm{h}} + \mathbf{g}^{(m)} \tag{5.59}$$

and

$$\boldsymbol{x}^{(m)} \sim \mathscr{N}(\hat{\boldsymbol{\mu}}_{\mathrm{x}}^{(m)}, \hat{\boldsymbol{\Sigma}}_{\mathrm{x}}^{(m)}) \tag{5.60}$$

$$\boldsymbol{n} \sim \mathscr{N}(\hat{\boldsymbol{\mu}}_{\mathrm{n}}, \hat{\boldsymbol{\Sigma}}_{\mathrm{n}}) \tag{5.61}$$

$$\mathbf{g}^{(m)} = \boldsymbol{f}(\boldsymbol{\mu}_{\mathrm{x}}^{(m)}, \boldsymbol{\mu}_{\mathrm{h}}, \boldsymbol{\mu}_{\mathrm{n}}) - \mathbf{J}^{(m)}(\boldsymbol{\mu}_{\mathrm{x}}^{(m)} + \boldsymbol{\mu}_{\mathrm{h}}) - (\mathbf{I} - \mathbf{J}^{(m)})\boldsymbol{\mu}_{\mathrm{n}}. \tag{5.62}$$

This now has the form of a general factor analysis style model, for which EM-based update formulae can be applied [26, 30, 33, 35, 55]. This allows the clean speech parameters and the noise parameters to be found in an iterative fashion. Note the convolutional noise bias is not estimated within an EM-style framework (as it has no variance) but is estimated in an EM-style approach and is related to the bias transform estimation [57] (and also to the estimation scheme in [48]).

The estimates of the clean speech means and covariances can then be expressed as[4]

$$\hat{\boldsymbol{\mu}}_{\mathrm{x}}^{(m)} = \frac{\sum_t \gamma_t^{(m)} \mathscr{E}\{\boldsymbol{x}|\boldsymbol{y}_t, m\}}{\sum_t \gamma_t^{(m)}} \tag{5.63}$$

$$\hat{\boldsymbol{\Sigma}}_{\mathrm{x}}^{(m)} = \mathrm{diag}\left(\frac{\sum_t \gamma_t^{(m)} \mathscr{E}\{\boldsymbol{x}\boldsymbol{x}^{\mathsf{T}}|\boldsymbol{y}_t, m\}}{\sum_t \gamma_t^{(m)}} - \hat{\boldsymbol{\mu}}_{\mathrm{x}}^{(m)}\hat{\boldsymbol{\mu}}_{\mathrm{x}}^{(m)\mathsf{T}}\right) \tag{5.64}$$

and the noise parameters as

$$\hat{\boldsymbol{\mu}}_{\mathrm{n}} = \frac{\sum_{m,t} \gamma_t^{(m)} \mathscr{E}\{\boldsymbol{n}|\boldsymbol{y}_t, m\}}{\sum_{m,t} \gamma_t^{(m)}} \tag{5.65}$$

$$\hat{\boldsymbol{\Sigma}}_{\mathrm{n}} = \mathrm{diag}\left(\frac{\sum_{m,t} \gamma_t^{(m)} \mathscr{E}\{\boldsymbol{n}\boldsymbol{n}^{\mathsf{T}}|\boldsymbol{y}_t, m\}}{\sum_{m,t} \gamma_t^{(m)}} - \hat{\boldsymbol{\mu}}_{\mathrm{n}}\hat{\boldsymbol{\mu}}_{\mathrm{n}}^{\mathsf{T}}\right) \tag{5.66}$$

where the expectations are over the distribution determined by the current model parameters.

However, compared to the standard general FA-style EM estimation approaches, which are guaranteed not to decrease the likelihood at each iteration, there are two important additional approximations being made:

1. Fixed 'loading matrix' and bias. For this form of FA-style estimation, $\mathbf{J}^{(m)}$ and $\mathbf{g}^{(m)}$ are assumed not to be functions of the clean speech and noise parameters to be estimated. This is not the case as the Jacobian and bias will change as the model parameters change.
2. Diagonal covariance matrices. The use of the form of generative model in (5.59) means that the corrupted speech distribution will be a full covariance matrix (the loading matrix $\mathbf{J}^{(m)}$ is full). However, this covariance matrix is diagonalised for efficient decoding (5.35). The joint distribution between the clean speech and corrupted speech (this is the basis for the FA-style estimation) thus has the form

[4] For simplicity of notation, the multiple noise conditions that would normally be present for adaptive training have been ignored.

$$\begin{bmatrix} \boldsymbol{y} \\ \boldsymbol{x} \end{bmatrix} \bigg| m \sim \mathcal{N}\left(\begin{bmatrix} \boldsymbol{\mu}_{\mathrm{y}}^{(m)} \\ \boldsymbol{\mu}_{\mathrm{x}}^{(m)} \end{bmatrix}, \begin{bmatrix} \mathrm{diag}(\boldsymbol{\Sigma}_{\mathrm{y}}^{(m)}) & \mathbf{J}^{(m)}\boldsymbol{\Sigma}_{\mathrm{x}}^{(m)} \\ \boldsymbol{\Sigma}_{\mathrm{x}}^{(m)}\mathbf{J}^{(m)\mathsf{T}} & \boldsymbol{\Sigma}_{\mathrm{x}}^{(m)} \end{bmatrix}\right). \tag{5.67}$$

However, from the generative model in (5.59), the corrupted speech covariance matrix can be expressed as

$$\hat{\boldsymbol{\Sigma}}_{\mathrm{y}}^{(m)} = \mathbf{J}^{(m)}\hat{\boldsymbol{\Sigma}}_{\mathrm{x}}^{(m)}\mathbf{J}^{(m)\mathsf{T}} + (\mathbf{I}-\mathbf{J}^{(m)})\hat{\boldsymbol{\Sigma}}_{\mathrm{n}}(\mathbf{I}-\mathbf{J}^{(m)})^{\mathsf{T}}. \tag{5.68}$$

From (5.67), this should be diagonal. For these two expressions to be consistent, the off-diagonal terms that result from $\mathbf{J}^{(m)}$ being full and $\hat{\boldsymbol{\Sigma}}_{\mathrm{x}}^{(m)}$ being diagonal must be cancelled out by elements from the noise terms. This is not possible for all components as $\mathbf{J}^{(m)}$ is component-specific whereas the noise is common for all components. Hence, the generative model is not consistent with the joint distribution,[5] so the EM-style approach is not guaranteed to increase the auxiliary function. Similar issues arise for the noise estimation case. An alternative, though approximate solution is to diagonalise the Jacobian. This is the approach adopted in [13]. However, this introduces additional approximations in the form of the generative model.

Both of these approximations mean that the update is not guaranteed to increase the auxiliary function. To overcome this problem it is possible to "back-off" estimates by explicitly evaluating the auxiliary function. This becomes important if multiple iterations are performed. Thus, it is important to be aware of the approximations being used with this approach, though it is mathematically elegant.

One of the advantages of these FA-style training approaches is that it is simple to incorporate discriminative training criteria such as MPE [50] into the adaptive training framework [13].

5.7.2 Second-Order Approaches

Rather than to use an EM-style approach, it is possible to use standard gradient descent-style schemes to directly maximise (5.58) [32, 45]. For a general second-order approach the update has the form

$$\begin{bmatrix} \hat{\boldsymbol{\mu}}_{\mathrm{x}}^{(m)} \\ \hat{\boldsymbol{\sigma}}_{\mathrm{x}}^{(m)2} \end{bmatrix} = \begin{bmatrix} \boldsymbol{\mu}_{\mathrm{x}}^{(m)} \\ \boldsymbol{\sigma}_{\mathrm{x}}^{(m)2} \end{bmatrix} + \zeta \begin{bmatrix} \frac{\partial^2 \mathcal{Q}()}{\partial \hat{\boldsymbol{\mu}}_{\mathrm{x}}^{(m)2}} & \frac{\partial^2 \mathcal{Q}()}{\partial \hat{\boldsymbol{\mu}}_{\mathrm{x}}^{(m)}\partial \hat{\boldsymbol{\sigma}}_{\mathrm{x}}^{(m)2}} \\ \frac{\partial^2 \mathcal{Q}()}{\partial \hat{\boldsymbol{\sigma}}_{\mathrm{x}}^{(m)2}\partial \hat{\boldsymbol{\mu}}_{\mathrm{x}}^{(m)}} & \frac{\partial^2 \mathcal{Q}()}{\partial \left(\hat{\boldsymbol{\sigma}}_{\mathrm{x}}^{(m)2}\right)^2} \end{bmatrix}^{-1} \begin{bmatrix} \frac{\partial \mathcal{Q}()}{\partial \hat{\boldsymbol{\mu}}_{\mathrm{x}}^{(m)}} \\ \frac{\partial \mathcal{Q}()}{\partial \hat{\boldsymbol{\sigma}}_{\mathrm{x}}^{(m)2}} \end{bmatrix}, \tag{5.69}$$

where $\mathcal{Q}(\hat{\mathcal{M}};\mathcal{M})$ is written as $\mathcal{Q}()$ to save space, $\boldsymbol{\sigma}_{\mathrm{x}}^{(m)2}$ is the vector of leading diagonal elements of $\boldsymbol{\Sigma}_{\mathrm{x}}^{(m)}$, and ζ is the learning rate. Considering the estimation of

[5] It is not clear whether the joint covariance matrix in (5.67) is related to any generative model of the form given in (5.59), where the noise model is shared over multiple components.

the clean speech mean, $\boldsymbol{\mu}_{\mathsf{x}}^{(m)}$, the derivative can be written as (the fixed variables are explicitly expressed to make the form of the derivative clear)

$$\left.\frac{\partial \mathscr{Q}()}{\partial \hat{\boldsymbol{\mu}}_{\mathsf{x}}^{(m)}}\right|_{\hat{\boldsymbol{\sigma}}_{\mathsf{x}}^{(m)}} = \left.\frac{\partial \hat{\boldsymbol{\mu}}_{\mathsf{y}}^{(m)}}{\partial \hat{\boldsymbol{\mu}}_{\mathsf{x}}^{(m)}}\right|_{\hat{\boldsymbol{\sigma}}_{\mathsf{x}}^{(m)}} \left.\frac{\partial \mathscr{Q}()}{\partial \hat{\boldsymbol{\mu}}_{\mathsf{y}}^{(m)}}\right|_{\hat{\boldsymbol{\sigma}}_{\mathsf{y}}^{(m)}} + \left.\frac{\partial \hat{\boldsymbol{\sigma}}_{\mathsf{y}}^{(m)2}}{\partial \hat{\boldsymbol{\mu}}_{\mathsf{x}}^{(m)}}\right|_{\hat{\boldsymbol{\sigma}}_{\mathsf{x}}^{(m)}} \left.\frac{\partial \mathscr{Q}()}{\partial \hat{\boldsymbol{\sigma}}_{\mathsf{y}}^{(m)2}}\right|_{\hat{\boldsymbol{\mu}}_{\mathsf{y}}^{(m)}} . \tag{5.70}$$

As with the FA-style approaches, these second-order approaches make a number of approximations:

1. Second-order approximation. In common with all second-order schemes there is the assumption that the "error-surface" is quadratic in nature. In practice, this is not the case. Additionally, the form of the Hessian is often modified, for example, diagonalised, and approximated to simplify optimisation.
2. Approximate derivatives. The mean derivative given in (5.70) is often not used. For example, in [32] the second term in (5.70) is assumed to be zero. Thus the gradient is approximated by

$$\left.\frac{\partial \mathscr{Q}()}{\partial \hat{\boldsymbol{\mu}}_{\mathsf{x}}^{(m)}}\right|_{\hat{\boldsymbol{\sigma}}_{\mathsf{x}}^{(m)}} \approx \mathbf{J}^{(m)} \left.\frac{\partial \mathscr{Q}()}{\partial \hat{\boldsymbol{\mu}}_{\mathsf{y}}^{(m)}}\right|_{\hat{\boldsymbol{\sigma}}_{\mathsf{y}}^{(m)}} . \tag{5.71}$$

   Though simplifying the derivative, this shifts the stationary points of the function.

As there is no guarantee of increasing the likelihood, for noise estimation, backing-off approaches can be used [44]. For the model parameter estimation, additional smoothing can also be added [44].

5.8 Conclusions and Future Research Directions

This chapter has reviewed a number of schemes associated with model-based approaches for handling uncertainty. The discussion concentrates on techniques for handling high levels of background noise and channel distortion as this is one of the most important forms of varying uncertainty in the speech signal. A number of approaches, as well as issues, are highlighted. These include the model compensation process itself; the computational costs associated with this process; and how the parameters of all elements of the process can be estimated from data. Though no performance figures have been given in this chapter, the references given allow a comparison of a number of approaches to be made. In particular, the Aurora 2 test set [27] has been used to evaluate a number of systems within a consistent framework.

One of the interesting aspects of model-based compensation research is that techniques originally developed for general linear transform adaptation schemes (whether to speaker or environment) are being increasingly used. Thus schemes based on ML estimation of the model parameters [38] and adaptive training schemes [4] are becoming popular. Additionally, discriminative training is also being used [13].

Though there have been improvements in noise robustness for speech recognition, there are still a number of issues that need to be addressed. The author feels these will become increasingly important as the complexity of the task and the range of conditions under which ASR systems are required to operate increase.

1. **Impact of noise on speech**: it is not possible to derive representations for the impact of noise on the speech for all forms of parametrisation. This chapter has assumed that MFCC parameters are being used. Even the introduction of basic front-end schemes such as CMN means that the mismatch function cannot be derived, though approaches geared to handling this have been derived [47]. Due to this, and the added problem of delta and delta-delta parameters, feature enhancement schemes based on stereo training [3] are used to combine noise robustness with state-of-the-art front-end processing such as semi-tied transforms [19] and fMPE [51]. Generalising model-based compensation techniques to handle state-of-the-art front-ends will be an important research area.
2. **Handling changes in correlation**: though the Jacobians associated with schemes such as VTS are block diagonal in structure, the resulting covariance matrices are diagonalised for speed of decoding. This is known to degrade recognition performance [43, 64, 67]. Predictive linear transform schemes are one framework for addressing this [22]. However, to date research in addressing this problem has been limited. As performance requirements for robust ASR in low-SNR conditions increase, this topic will become increasingly important.
3. **Improved distribution modelling**: the majority of model-based compensation schemes assume that each speech and noise component pairing will yield a Gaussian-distributed corrupted speech component. As previously discussed, this is not true. Obtaining more efficient non-Gaussian schemes than the current versions may yield improved performance over the Gaussian approximations.
4. **Speed of compensation/parameter estimation**: one of the main objections to model-based approaches is that they are slow. For large vocabulary systems there may be hundreds of thousands of Gaussian components. Improving the speed of all aspects of model compensation is essential for it to be broadly applied. For example, using incremental forms of noise estimation/compensation [14] is one approach to handling this.
5. **Reverberant noise**: this requires extending the range of environments for which model compensation schemes can be used. For example, to handle long-term reverberant noise as well as additive noise, a model-based approach is described in [54].
6. **Improved acoustic modelling**: as the level of background noise increases, and the associated uncertainty of the speech increases, it may become increasingly important to improve the form of the acoustic models being used for the clean speech, noise and corrupted speech. One approach in this direction is to use HMM generative models to obtain scores for use in a discriminative classifier [21].

In summary, model-based compensation schemes are a very natural way of handling uncertainty in speech recognition. However, there is still significant research required to enable these techniques to achieve the levels of performance, in terms of both speed and accuracy, to allow their general deployment in a wide-range of speech applications.

References

1. A. Acero. *Acoustical and Environmental Robustness in Automatic Speech Recognition.* Ph.D. thesis, Carnegie Mellon University, 1990.
2. A. Acero, L. Deng, T. T. Kristjansson, and J. Zhang. HMM adaptation using vector Taylor series for noisy speech recognition. In *Proc. ICSLP*, pages 869–872, Beijing, China, October 2000.
3. M. Afify, X. Cui, and Y. Gao. Stereo-based stochastic mapping for robust speech recognition. In *Proc. ICASSP*, 2007.
4. T. Anastasakos, J. McDonough, R. Schwartz, and J. Makhoul. A compact model for speaker-adaptive training. In *Proc. ICSLP*, 1996.
5. J. A. Arrowood and M. A. Clements. Using observation uncertainty in HMM decoding. In *Proc. ICSLP*, Denver, Colorado, September 2002.
6. S. F. Boll. Suppression of acoustic noise in speech using spectral subtraction. *IEEE Transactions Audio Speech and Signal Processing*, 27:113–120, 1979.
7. W. Chou. Maximum a posterior linear regression with elliptically symmetric matrix variate priors. In *Proc. Eurospeech*, 1999.
8. A. de la Torre, D. Fohr, and J.-P. Haton. Statistical adaptation of acoustic models to noise conditions for robust speech recognition. In *Proc. ICSLP*, pages 1437–1440, 2002.
9. L. Deng, A. Acero, M. Plumpe, and X. D. Huang. Large vocabulary speech recognition under adverse acoustic environments. In *Proc. ICSLP*, pages 806–809, Beijing, China, October 2000.
10. L. Deng, J. Droppo, and A. Acero. Enhancement of log mel power spectra of speech using a phase-sensitive model of the acoustic environment and sequential estimation of the corrupting noise. *IEEE Transactions on Speech and Audio Processing*, 12:133–143, 2004.
11. V. V. Digalakis, D. Rtischev, and L. G. Neumeyer. Speaker adaptation using constrained estimation of Gaussian mixtures. *IEEE Transactions Speech and Audio Processing*, 3:357–366, 1995.
12. J. Droppo, A. Acero, and L. Deng. Uncertainty decoding with SPLICE for noise robust speech recognition. In *Proc. ICASSP*, Orlando, Florida, May 2002.
13. F. Flego and M. J. F. Gales. Discriminative adaptive training with VTS and JUD. In *Proc. ASRU*, 2009.
14. F. Flego and M. J. F. Gales. Incremental predictive and adaptive noise compensation. In *Proc. ICASSP*, Taipei, Taiwan, 2009.
15. F. Flego and M. J. F. Gales. Adaptive Training and Noise Estimation for Model-Based Noise Compensation for ASR. Technical Report CUED/F-INFENG/TR653, University of Cambridge, 2010.
16. B. Frey, L. Deng, A. Acero, and T. T. Kristjansson. ALGONQUIN: Iterating Laplace's method to remove multiple types of acoustic distortion for robust speech recognition. In *Proc. Eurospeech*, Aalbork, Denmark, September 2001.
17. M. J. F. Gales. *Model-Based Techniques for Noise Robust Speech Recognition.* Ph.D. thesis, Cambridge University, 1995.
18. M. J. F. Gales. Maximum likelihood linear transformations for HMM-based speech recognition. *Computer Speech and Language*, 12, January 1998.
19. M. J. F. Gales. Semi-tied covariance matrices for hidden Markov models. *IEEE Transactions on Speech and Audio Processing*, 7:272–281, 1999.
20. M. J. F. Gales. Cluster adaptive training of hidden Markov models. *IEEE Transactions Speech and Audio Processing*, 8:417–428, 2000.
21. M. J. F. Gales and F. Flego. Discriminative classifiers with adaptive kernels for noise robust speech recognition. *Computer Speech and Language*, 2010.
22. M. J. F. Gales and R. C. van Dalen. Predictive linear transforms for noise robust speech recognition. In *Proc. ASRU*, pages 59–64, 2007.
23. M. J. F. Gales and P. C. Woodland. Mean and variance adaptation within the MLLR framework. *Computer Speech and Language*, 10:249–264, 1996.

24. M. J. F. Gales and S. J. Young. The application of hidden Markov models in speech recognition. *Foundation and Trends in Signal Processing*, 1(3):195–304, 2008.
25. R. A. Gopinath, M. J. F. Gales, P. S. Gopalakrishnan, S. Balakrishnan-Aiyer, and M. A. Picheny. Robust speech recognition in noise — performance of the IBM continuous speech recognizer on the ARPA noise spoke task. In *Proc. ARPA Workshop on Spoken Language System Technology*, pages 127–130, Austin, Texas, 1995.
26. R. A. Gopinath, B. Ramabhadran, and S. Dharanipragada. Factor analysis invariant to linear transformations of data. In *Proc. ICSLP*, pages 397–400, 1998.
27. H.-G. Hirsch and D. Pearce. The AURORA experimental framework for the evaluation of speech recognition systems under noisy conditions. In *Proc. ASR*, pages 181–188, September 2000.
28. Y. Hu and Q. Huo. *Chinese Spoken Language Processing*, chapter in An HMM Compensation Approach Using Unscented Transformation for Noisy Speech Recognition. Springer Berlin/Heidelberg, 2006.
29. X. D. Huang, A. Acero, and H. W. Hon. *Spoken Language Processing*. Prentice Hall, 2001.
30. Q. Huo and Y. Hu. Irrelevant variability normalization based HMM training using VTS approximation of an explicit model of environmental distortions. In *Proc. Interspeech*, pages 1042–1045, Antwerp, Belgium, 2007.
31. S. J. Julier and J. K. Uhlmann. Unscented filtering and nonlinear estimation. *Proceedings of the IEEE*, 92(3):401–422, 2004.
32. O. Kalinli, M.L. Seltzer, and A. Acero. Noise adaptive training using a vector taylor series approach for noise robust automatic speech recognition. In *Proc. ICASSP*, pages 3825–3828, Taipei, Taiwan, April 2009.
33. D. Kim and M. J. F. Gales. Adaptive training with noisy constrained maximum likelihood linear regression for noise robust speech recognition. In *Proc. Interspeech*, Brighton, UK, 2009.
34. D. Kim and M. J. F. Gales. Noisy constrained maximum likelihood linear regression for noise robust speech recognition. *IEEE Transactions Audio Speech and Language Processing*, 2010.
35. D. Y. Kim, C. K. Un, and N. S. Kim. Speech recognition in noisy environments using first-order vector Taylor series. *Speech Communication*, 24(1):39–49, June 1998.
36. T. T. Kristjansson. *Speech Recognition in Adverse Environments: A Probabilistic Approach.* Ph.D. thesis, Waterloo University, Waterloo, Canada, 2002.
37. L. Lee and R. C. Rose. Speaker normalisation using efficient frequency warping procedures. In *ICASSP'96*, Atlanta, 1996.
38. C. Leggetter and P. C. Woodland. Maximum likelihood linear regression for speaker adaptation of continuous density HMMs. *Computer Speech and Language*, 9, 1995.
39. V. Leutnant and R. Haeb-Umbach. An analytic derivation of a phase-sensitive observation model for noise robust speech recognition. In *Proc. Interspeech*, pages 2395–2398, 2009.
40. J. Li, L. Deng, D. Yu, Y. Gong, and A. Acero. High-performance HMM adaptation with joint compensation of additive and convolutive distortions via vector Taylor series. In *Proc. ASRU*, pages 65–70, Kyoto, Japan, December 2007.
41. J. Li, L. Deng, D. Yu, Y. Gong, and A. Acero. HMM adaptation using a phase-sensitive acoustic distortion model for environment-robust speech recognition. In *Proc. ICASSP*, pages 4069–4072, April 2008.
42. H. Liao. *Uncertainty Decoding for Noise Robust Speech Recognition*. Ph.D. thesis, Cambridge University, Cambridge, UK, sep 2007.
43. H. Liao and M. J. F. Gales. Joint uncertainty decoding for noise robust speech recognition. In *Proc. Interspeech*, 2005.
44. H. Liao and M. J. F. Gales. Joint uncertainty decoding for robust large vocabulary speech recognition. Technical Report CUED/F-INFENG/TR552, University of Cambridge, 2006. Available from `mi.eng.cam.ac.uk/~mjfg`.
45. H. Liao and M. J. F. Gales. Adaptive training with joint uncertainty decoding for robust recognition of noisy data. In *Proc. ICASSP*, volume 4, pages 389–392, Honolulu, USA, April 2007.

46. H. Liao and M. J. F. Gales. Issues with uncertainty decoding for noise robust speech recognition. *Speech Communication*, 2008.
47. Y. Minami and S. Furui. A maximum likelihood procedure for a universal adaptation method based on HMM composition. In *Proc. ICASSP*, pages 129–132, 1995.
48. P. Moreno. *Speech Recognition in Noisy Environments*. Ph.D. thesis, Carnegie Mellon University, 1996.
49. L. Neumeyer and M. Weintraub. Probabilistic optimum filtering for robust speech recognition. In *Proc. ICASSP*, volume 1, pages 417–420, 1994.
50. D. Povey. *Discriminative Training for Large Vocabulary Speech Recognition*. Ph.D. thesis, Cambridge University, 2003.
51. D. Povey, B. Kingsbury, L. Mangu, G. Saon, H. Soltau, and G. Zweig. fMPE: Discriminatively trained features for speech recognition. In *Proc. ICASSP*, Philadelphia, 2005.
52. L. R. Rabiner. A tutorial on hidden Markov models and selected applications in speech recognition. *Proceedings of the IEEE*, 77(2):257–286, February 1989.
53. B. Raj and R. Stern. Missing feature approaches in speech recognition. *IEEE Signal Processing Magazine*, 22(5):101–116, 2005.
54. C. K. Raut, T. Nishimoto, and S. Sagayama. Maximum likelihood based HMM state filtering approach to model adaptation for long reverberation. In *Proc. ASRU*, 2005.
55. D. Rubin and D. Thayer. EM algorithms for ML factor analysis. *Psychometrika*, 47(1):69–76, March 1982.
56. S. Sagayama, Y. Yamaguchi, S. Takahashi, and J. Takahashi. Jacobian approach to fast acoustic model adaptation. In *Proc. ICASSP*, 1997.
57. A. Sankar and C.-H. Lee. A maximum-likelihood approach to stochastic matching for robust speech recognition. *IEEE Transactions on Speech and Audio Processing*, 4:190–202, May 1996.
58. M. Seltzer, K. Kalgaonkar, and A. Acero. Acoustic model adaptation via linear spline interpolation for robust speech recognition. In *Proc. ICASSP*, 2010.
59. M. Seltzer, B. Raj, and R. Stern. A Bayesian framework for spectrographic mask estimation for missing feature speech recognition. *Speech Communication*, 43(4):379–393, 2004.
60. Y. Shinohara and M. Akamine. Bayesian feature enhancement using a mixture of unscented transformations for uncertainty decoding of noisy speech. In *Proc. ICASSP*, pages 4569–4572, 2009.
61. V. Stouten, H. van Hamme, and P. Wambacq. Accounting for the uncertainty of speech estimates in the context of model-based feature enhancement. In *Proc. ICSLP*, volume I, pages 105–108, Jeju Island, Korea, October 2004.
62. V. Stouten, H. van Hamme, and P. Wambacq. Effect of phase-sensitive environment model and higher order VTS on noisy speech feature enhancement. In *Proc. ICASSP*, volume I, pages 433–436, Philadelphia, USA, March 2005.
63. R. C. van Dalen, F. Flego, and M. J. F. Gales. Transforming features to compensate speech recogniser models for noise. In *Proc. Interspeech*, 2009.
64. R. C. van Dalen and M. J. F. Gales. Extended VTS for noise-robust speech recognition. In *Proc. ICASSP*, Taipei, Taiwan, 2009.
65. R. C. van Dalen and M. J. F. Gales. Asymptotically exact noise-corrupted speech likelihoods. In *Proc. Interspeech*, 2010.
66. A. P. Varga, R. K. Moore, J. Bridle, K. Ponting, and M. Russel. Noise compensation algorithms for use with hidden Markov model based speech recognition. In *Proc. ICASSP*, 1988.
67. H. Xu, M. J. F. Gales, and K. K. Chin. Improving joint uncertainty decoding performance by predictive methods for noise robust speech recognition. In *Proc. ASRU*, 2009.

Chapter 6
Reconstructing Noise-Corrupted Spectrographic Components for Robust Speech Recognition

Bhiksha Raj and Rita Singh

Abstract An effective solution for missing-feature problems is the imputation of the missing components, based on the reliable components and prior knowledge about the distribution of the data. In this chapter we will describe various imputation methods, including those that consider correlation across time and those that do not, and present experimental evaluation of the techniques. We will demonstrate how imputation of missing spectrographic components prior to cepstral feature computation can in fact be superior to techniques that attempt to perform computation directly in the domain with the incomplete data, due to the superior performance obtained with cepstral features.

6.1 Introduction

The performance of automatic speech recognition (ASR) systems degrades when the speech to be recognized is corrupted by noise, particularly when the system has been trained on clean speech. Several algorithms have therefore been proposed in the literature to compensate for the effects of noise on ASR performance. Most of these algorithms attempt to characterize the noise and model its effects on the speech signal explicitly (e.g., [10, 19, 27, 35]) in order to compensate for it. Their performance is usually critically dependent on the ability to measure the noise characteristics accurately, and they fail to be effective when such measurement is difficult, such as when the noise is non-stationary [24].

The missing-feature approach, proposed by researchers at the University of Sheffield in the mid-1990s, is an alternative approach based on exploitation of the inherent redundancy in the speech signal, rather than on explicit characterization of corrupting noise [6]. Speech signals have a large degree of redundancy. For instance, a speech signal that has been either high-pass filtered or low-pass filtered with a

Carnegie Mellon University, 5000 Forbes Avenue, Pittsburgh, PA 15213, USA
e-mail: bhiksha@cs.cmu.edu

D. Kolossa and R. Haeb-Umbach (eds.), *Robust Speech Recognition of Uncertain or Missing Data: Theory and Applications*, DOI 10.1007/978-3-642-21317-5_6,
© Springer-Verlag Berlin Heidelberg 2011

cutoff frequency of 1.800 Hz remains perfectly intelligible [9]. Similarly, speech from which spectral bands have been excised [34] or from which short temporal regions have been deleted [17] remains intelligible. Given the redundancy, one can hope to recognize speech effectively even if only a fraction of the spectro-temporal information in the signal is available.

When the speech is corrupted by noise, some time-frequency components of the signal are more corrupted than others. Because of the redundancy in the speech signal, the high-SNR components in the signal are often sufficient to recognize what was spoken. Missing-feature methods exploit this fact by characterizing low-SNR time-frequency components as being unreliable and effectively *missing*. The remaining reliable components now form an incomplete spectrographic characterization of the signal.

In the two original algorithms proposed by the Sheffield group, recognition was performed by HMM-based recognizers directly with the incomplete spectrographic characterization [7]. Since conventional HMM-based recognizers cannot perform recognition with incomplete representations, their algorithms modified the manner in which state output probabilities were computed within the recognizer. In the first algorithm, referred to as state-based imputation, unreliable time-frequency components were assigned values using state-specific maximum a posteriori (MAP) or minimum mean squared error (MMSE) estimators prior to computing state likelihoods. In the second algorithm, referred to as marginalization, the unreliable components were integrated out of the state output distributions. This latter approach is equivalent to that of the optimal classifier or recognizer, given the incomplete spectrographic data. Later improvements to the algorithms incorporated the assumption that the observed value of unreliable time-frequency components represent an upper bound on their true value, i.e., the value the components would have had in the absence of corrupting noise. This places an upper bound on the estimates of the unreliable components for state-based imputation [14]. For marginalization, this places an upper limit on the domain over which unreliable components must be marginalized out of class distributions [5]. Since these methods modify the recognizer itself, we refer to them as classifier compensation methods.

While classifier compensation methods have been shown to be extremely effective in compensating for noise, they suffer from some important drawbacks:

- In order to use them, the state output distributions of the recognizer must model the distributions of *spectral* vectors derived from the signal, where the reliable and unreliable components are identified. However, speech recognition performance obtained using lower-dimensional *cepstral* vectors derived from spectral vectors has been found to be significantly superior to that obtained with the spectral vectors themselves [8]. Classifier compensation missing-feature methods cannot be employed with cepstra, since cepstra combine the information from both reliable and unreliable components.
- State likelihood computing components of the recognizer must be modified to implement these algorithms. As a result, they can only be used in situations where one has access to the internals of the recognizer.

- Utterance-level preprocessing steps such as mean and variance normalization, which are known to improve recognition performance, cannot be performed with incomplete spectrographic data. The use of difference and double difference features, also common in speech recognizers, becomes difficult and less effective.

These problems arise because classifier compensation methods attempt to perform recognition directly with incomplete time-frequency characterizations of incoming audio, modifying the recognizer to account for the missing components.

In this chapter we present missing-feature algorithms that take an alternative approach: they reconstruct complete time-frequency representations from the incomplete ones prior to recognition [25]. To achieve this, the true values of the unreliable time-frequency components are estimated based on their known statistical relationship to reliable components. Cepstral vectors can now be derived from the complete time-frequency characterizations that result. Normalization procedures such as cepstral mean normalization can also be applied. Recognition performance with the cepstral vectors so obtained is frequently much better than that obtained with any classifier compensation technique. Equally importantly, the recognizer itself need not be modified in any manner. This permits the usage of any form of recognizer, including off-the-shelf recognizers that can take cepstral vectors as input. Since these algorithms work only on incoming feature vectors, we refer to them as *feature compensation* methods.

We present three feature compensation methods in this chapter. *Correlation-based reconstruction* is based on a simple statistical model that represents the sequence of spectral vectors in the spectrogram as the output of a stationary Gaussian random process. A bounded version of the MAP estimation procedure is used to estimate unreliable components, based on the statistical parameters of this process. *Cluster-based reconstruction* is based on the more conventional Gaussian mixture representations of the distributions of clean speech. The reconstruction uses the bounded MAP estimation procedure to obtain Gaussian-specific estimates of unreliable components, which are then combined into a final estimate.

A crucial component of missing-feature methods is the identification of unreliable time-frequency components in the spectrograms. We will refer to the tags which identify the unreliable and reliable components as *spectrographic masks*. We do not actually discuss the issue of spectrographic mask estimation in any detail in this chapter, only briefly mentioning it in Section 6.4 and referring the reader instead to the large volume of excellent literature on the topic. Nevertheless, it is worth mentioning here that most spectrographic mask estimation methods attempt to estimate a *hard* spectrographic mask that unambiguously tags time-frequency components as reliable or not. However, mask estimation is an extremely difficult task and uncertainty is an inherent aspect of the determination of the reliability of spectral components. To account for this, it is found to be advantageous to compute a *soft* spectrographic mask instead that assigns a probability of being unreliable to each time-frequency component.

In [3] Barker et al. present classifier compensation methods that can utilize soft masks, which do not require unambiguous identification of reliable components.

Similarly, feature compensation methods too can work from soft masks. The third spectrogram reconstruction method we present in this chapter [26], which we will creatively call "soft-mask-based reconstruction" composes entire spectrograms from noisy spectrograms and an estimated soft mask.

The rest of this chapter is arranged as follows. Application of missing-feature methods requires an appropriate time-frequency representation of the speech signal. In Section 6.2 we give a brief description of the spectrographic representation we will use. In Section 6.3 we define some notations used in the rest of the chapter. In Section 6.4 we briefly discuss approaches to estimating spectrographic masks. Section 6.5 outlines classifier compensation methods. In Section 6.6 we present the main topic of this chapter: feature compensation missing-feature methods. In Section 6.7 we describe various experiments evaluating missing-feature methods. Finally in Section 6.8 we discuss our findings.

6.2 Spectrographic Representation

In order to apply missing-feature methods, a time-frequency characterization of the signal must first be obtained. Conventionally, this is a mel frequency spectrographic representation [31], which we will refer to as the "mel spectrogram", or simply as a "spectrogram". This consists of a sequence of mel frequency log-spectral vectors (which we will refer to simply as "spectral vectors"), each of which represents the frequency warped log spectrum of a short frame of speech, typically 20 ms wide. Each analysis frame is typically shifted by 10 ms with respect to the previous one. Figure 6.1a shows the mel spectrogram of a typical clean speech signal.

Additive noise affects different regions of the mel spectrogram differently. Figure 6.1b shows the mel spectrogram of the signal in Figure 6.1a when it has been corrupted to 10 dB by white noise. We see that while some regions of the spectrogram are relatively unaffected by the noise, others are badly corrupted. The degree of corruption of any time-frequency component of the spectrogram is dependent on the SNR of that component. Missing-feature methods assume that all time-frequency components that have low SNRs are unreliable. Here a "low-SNR" component is one that has an SNR below a pre-specified threshold. These unreliable components are not entirely uninformative, however. Their observed values are assumed to be the upper bound on their true values, i.e., the values that they would have had in the absence of corrupting noise. This is based on the assumption that the noise is additive and uncorrelated to the speech. All time-frequency components whose SNR lies above the threshold are assumed to be reliable and can therefore be used for recognition. The optimal value of the SNR threshold used to identify reliable components is different for different missing-feature methods. In general, time-frequency components in which the noise energy is comparable to the energy in the speech are deemed to be unreliable.

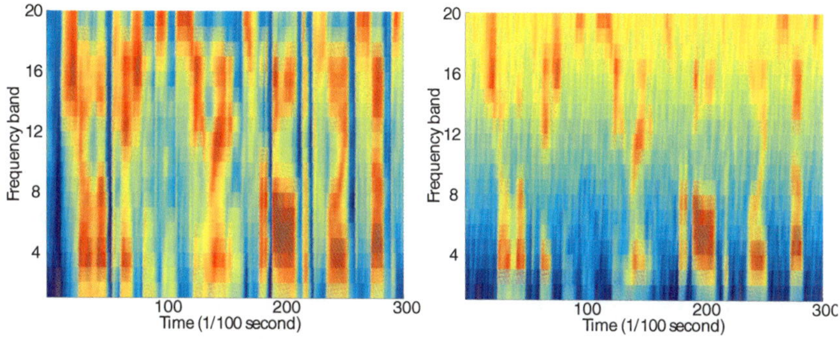

Fig. 6.1: a) Mel spectrogram of a clean speech signal. The utterance is "show locations and C-ratings of all deployed subs". b) The same signal when it has been corrupted to 10 dB by white noise

6.3 Notation

Before proceeding, we will establish some of the notation and terminology used in the rest of this chapter. Every frame of noisy incoming speech has underlying clean speech that has been corrupted by noise. Corresponding to the t-th frame of noisy speech, there is a measured noisy spectral vector $Y(t)$ which includes both reliable and unreliable components. We will arrange the reliable components of $Y(t)$ into a vector $Y_r(t)$, which we will refer to as the *reliable component vector* of $Y(t)$. The unreliable components are arranged in the vector $Y_u(t)$, which we will refer to as the *unreliable component vector* of $Y(t)$. We can express the relation between $Y(t)$, $Y_r(t)$ and $Y_u(t)$ as

$$Y_r(t) = R(t)Y(t), \tag{6.1}$$

$$Y_u(t) = U(t)Y(t), \tag{6.2}$$

$$Y(t) = A(t)[Y_r(t)^\top Y_u(t)^\top]^\top, \tag{6.3}$$

where $R(t)$ and $U(t)$ are permutation matrices that select the reliable and unreliable components, respectively, of $Y(t)$ and arrange them into $Y_r(t)$ and $Y_u(t)$. $[Y_r(t)^\top Y_u(t)^\top]^\top$ is a vector constructed by concatenating the transposes of $Y_r(t)$ and $Y_u(t)$, and $A(t)$ is the permutation matrix that rearranges the components of $[Y_r(t)^\top Y_u(t)^\top]^\top$ to give $Y(t)$.

Corresponding to the noisy spectrogram, i.e., the spectrogram of the noisy speech, is a *true* spectrogram which is the spectrogram that would have been computed had the signal not been corrupted by noise. Corresponding to each noisy spectral vector $Y(t)$ from the noisy spectrogram, there is a true spectral vector $X(t)$ from the true spectrogram. The components of $X(t)$ that correspond to the reliable and unreliable components of $Y(t)$ can also be arranged into vectors $X_r(t)$ and $X_u(t)$. $X_r(t)$ and $X_u(t)$ are related to $Y_r(t)$ and $Y_u(t)$ as follows:

$$X_r(t) = Y_r(t),$$
$$X_u(t) \leq Y_u(t). \tag{6.4}$$

We refer to the components of $X_u(t)$ as the unreliably known components of $X(t)$, since their value is not known, and to $X_u(t)$ as the unreliably known component vector of $X(t)$. Similarly, we refer to the components of $X_r(t)$ as the reliably known components of $X(t)$, and to $X_r(t)$ as the reliably known component vector of $X(t)$.

6.4 A Note on Estimating Spectrographic Masks

The most important aspect of missing-feature methods is the estimation of spectrographic masks – the collection of time-frequency tags which identify reliable and unreliable components. We note at the outset of this section that the main topic discussed in this chapter is the reconstruction of unreliable spectral components as identified by an available spectrographic mask. The actual estimation of the masks is mostly beyond the scope of the chapter. Thus the following description is brief; the reader is referred to the rather large literature on the estimation of masks for further information.

There are several approaches to the problem of estimating spectrographic masks. Approaches based on computational auditory scene analysis [33] attempt to identify time-frequency components that jointly represent speech-related phenomena that are hypothesized to be perceptually detectable. These phenomena include onsets or offsets of sound, common amplitude or frequency modulation, and harmonic structure. Time-frequency components that clearly group together to represent such phenomena are assumed to provide positive evidence towards the identity of the underlying sounds and are hence deemed reliable.

Other approaches attempt to identify unreliable components through the acoustic or statistical properties of the signal, without explicit reference to perceptual phenomena. SNR-based techniques attempt to explicitly estimate the SNR of time-frequency components by estimating the spectrum of the noise at each time [2, 12, 23, 28]. Time-frequency components whose estimated SNR lies below a threshold (typically 0 dB) are assumed to be unreliable.

Classification-based methods [28, 30] utilize a binary classifier to identify reliable components. Here, each time-frequency component is represented by a vector of "features". A classifier is trained from a training corpus whose time-frequency components have been tagged as reliable or unreliable. Time-frequency components of test data are then labelled as reliable or not by the classifier. Figure 6.2a shows the mel spectrogram of a noisy signal corrupted to 10 dB by traffic noise. Figure 6.2b shows a binary spectrographic mask estimated for it by a classifier.

In the case of *Bayesian* classifiers (e.g., [30]), separate distributions are learned for the feature vectors of reliable and unreliable time-frequency components from the training corpus. On test data these distributions are used to compute the a posteriori probability of reliability for each time-frequency component. Any time-

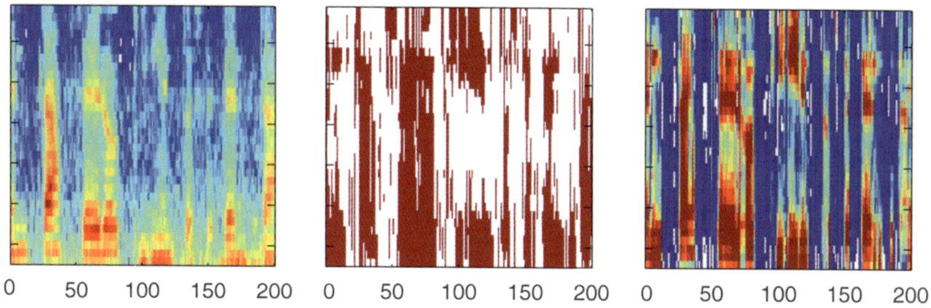

Fig. 6.2: a) Mel frequency spectrogram of a noisy signal b) Estimated binary mask. c) Bayesian soft mask. Higher levels of red indicate more reliable components

frequency component for which the probability exceeds a threshold is deemed to be reliable.

Spectral mask estimation techniques are not perfect. They frequently erroneously dismiss reliable spectral components as being unreliable, or select noisy components as being reliable. Both kinds of errors can affect the performance of missing-feature methods adversely [23]. Feature compensation methods in particular are highly sensitive to such errors.

Bayesian mask estimation also provides additional information. For each time-frequency component the classifier computes an a posteriori probability that that component is reliable. Together, the a posteriori probabilities represent a *soft mask* that assigns a measure of reliability for every time-frequency component of the signal. Figure 6.2c shows the soft mask obtained from the a posteriori probabilities of a Bayesian classifier for the signal in Figure 6.2a. Soft-mask-based missing-feature methods use these soft masks and the associated uncertainty in the reliability of time-frequency components, rather than definitive reliable/unreliable tags.

6.5 Classifier Compensation Methods

We briefly outline the main missing-feature classifier compensation formalisms, namely state-based imputation and marginalization, before proceeding to the main thesis of this chapter, namely feature compensation. Classifier compensation algorithms have been well documented in various papers and we only recapitulate the salient points here for reference. For more detailed information the reader is referred to the several papers on the subject (e.g., [5, 13, 16, 21]).

6.5.1 State-Based Imputation

In most HMM-based speech recognition systems state emission probabilities are modelled as mixtures of Gaussians. For any vector $X(t)$ with reliably known component vector $X_r(t)$ and unreliably known component vector $X_u(t)$, the state output probability of a state s, $P(X(t)|s)$, can be expressed as

$$P(X(t)|s) = \sum_j c_{j,s}\mathcal{N}(X(t);\mu_{j,s},\Theta_{j,s}) = \sum_j c_{j,s}\mathcal{N}(X_r(t),X_u(t);\mu_{j,s},\Theta_{j,s}), \quad (6.5)$$

where $\mathcal{N}(X_r(t),X_u(t);\mu_{j,s},\Theta_{j,s})$ represents the j-th Gaussian in the mixture Gaussian density for s with mean vector $\mu_{j,s}$ and covariance matrix $\Theta_{j,s}$, and $c_{j,s}$ is the mixture weight of the j-th Gaussian. For any noisy spectral vector $Y(t)$ that relates to the underlying clean speech spectrum through Equation (6.4), state-based imputation computes state output probabilities as

$$P(X(t)|s) = \sum_j c_{j,s}G(Y_r(t),\hat{X}_u^s(t);\mu_{j,s},\Theta_{j,s}), \quad (6.6)$$

where $\hat{X}_u^s(t)$ is a *state-specific* MMSE estimate of $X_u(t)$ obtained from $Y_r(t)$, computed as [14]

$$\hat{X}_u^s(t) = \sum_j \gamma_{j,s}(Y(t))U(t)\mu_{j,s}. \quad (6.7)$$

$U(t)$ is the permutation matrix that selects unreliable components from $Y(t)$ to form $Y_u(t)$ and

$$\gamma_{j,s}(Y(t)) = \frac{c_{j,s}\int_{-\infty}^{Y_u(t)}\mathcal{N}(Y_r(t),X_u;\mu_{j,s},\Theta_{j,s})dX_u}{\sum_k c_{k,s}\int_{-\infty}^{Y_u(t)}\mathcal{N}(Y_r(t),X_u;\mu_{k,s},\Theta_{k,s})dX_u}. \quad (6.8)$$

Other forms of the estimate for $\hat{X}_u^s(t)$ have also been proposed [28], but the basic principle behind the implementation of the algorithm remains the same.

6.5.2 Marginalization

In marginalization, unreliable components of spectral vectors are marginalized out of state output distributions, retaining only the bounds implicit in them. State emission probabilities are computed as $P(Y_r(t),X_u(t) \leq Y_u(t)|s)$. When emission probabilities are modelled by mixtures of Gaussians, the probability (or, more accurately, the probability density) for state s is computed as

$$P(Y_r(t),X_u(t) \leq Y_u(t)|s) = \sum_k c_{k,s}\int_{-\infty}^{Y_u(t)}\mathcal{N}(Y_r(t),X_u;\mu_{k,s},\Theta_{k,s})dX_u. \quad (6.9)$$

6.5.3 Marginalization with Soft Masks

Soft-mask-based marginalization [3] is a variant of marginalization that can work from soft masks. Instead of categorically describing components of spectral vectors as reliable or unreliable, we now specify a soft mask value $\gamma(t,k)$, which is the probability that the k-th component of $Y(t)$ is reliable, for every t and k. $\gamma(t,k)$ takes real values between 0 and 1. In order to accommodate soft masks, the Gaussians in the state output densities of the HMMs are constrained to have strictly diagonal covariance matrices. Thus, the j-th Gaussian in the state output density of state s can be represented as

$$\mathscr{N}(X(t);\mu_{j,s},\Theta_{j,s}) = \prod_k P(X(t,k)|s,j), \tag{6.10}$$

$$P(X(t,k)|s,j) = \mathscr{N}(X(t,k);\mu_{j,s}(k),\theta_{j,s}(k)), \tag{6.11}$$

where $X(t,k)$ represents the kth component of $X(t)$, and $\mu_{j,s}(k)$ and $\theta_{j,s}(k)$ represent the k-th component of $\mu_{j,s}$ and the kth diagonal component of $\Theta_{j,s}$ respectively. Soft-mask-based marginalization incorporates the soft mask by modifying the computation of the individual Gaussian terms to

$$\begin{aligned} P(X(t,k)|s,j,\gamma(t,k)) &= \gamma(t,k)P(X(t,k)|s,j) + (1-\gamma(t,k))\frac{\int_{L_k}^{X(t,k)} P(X|s,j)dX}{X(t,k)-L_k}, \\ \mathscr{N}(X(t);\mu_{j,s},\Theta_{j,s}) &\approx \prod_k P(X(t,k)|s,j,\gamma(t,k)), \end{aligned} \tag{6.12}$$

where L_k is a lower bound on acceptable values of $X(t,k)$. Equation (6.12) can be derived from a model assumption that is also used by the MMSE feature compensation algorithm (Section 6.6.3).

6.6 Feature Compensation Methods

Feature compensation methods treat unreliable spectral components as having been erased from the spectrogram and attempt to reconstruct them. The simplest method of estimating these missing values is by simple interpolation between the closest reliable neighboring components. However, as reported by [23], simple interpolation does not result in useful estimates.

The more rigorous approach utilizes the known statistics of speech spectrograms. Below, we present three formalisms for this approach. *Correlation-based* reconstruction models the spectrogram as the output of a Gaussian process. *Cluster-based* reconstruction models spectral vectors as independent draws from a Gaussian mixture density. Both methods assume *hard* spectrographic masks. *Soft-mask-based MMSE estimation* follows the same approach as cluster-based reconstruction, but works from soft masks.

6.6.1 Correlation-Based Reconstruction

In correlation-based reconstruction [25] the sequence of spectral vectors in the spectrogram of a clean speech signal is considered to be the output of a wide-sense stationary Gaussian random process [22]. All clean speech spectrograms are assumed to be individual observations of the same process. The assumption of wide-sense stationarity implies that the means of the spectral vectors and the covariances between components of the spectrogram are independent of their position in the spectrogram. Let $X(t,k)$ represent the kth frequency component of the tth spectral vector of an utterance. If we represent the expected value of $X(t,k)$ as $\mu(t,k)$, and the covariance between $X(t,k_1)$ and $X(t+\tau,k_2)$ as $c(t,t+\tau,k_1,k_2)$, wide-sense stationarity implies that

$$\mu(t,k) = \mu(t+\tau,k) = \mu(k),$$
$$c(t,t+\tau,k_1,k_2) = c(\tau,k_1,k_2).$$

Similarly, the relative covariance $r(t,t+\tau,k_1,k_2)$ between any two components $X(t,k_1)$ and $X(t+\tau,k_2)$ is also dependent only on τ and is given by

$$\begin{aligned} r(t,t+\tau,k_1,k_2) &= r(\tau,k_1,k_2) \\ &= \frac{c(\tau,k_1,k_2)}{\sqrt{c(0,k_1,k_1)c(0,k_2,k_2)}}. \end{aligned} \tag{6.13}$$

The means $\mu(k)$ and the various covariance parameters $c(\tau,k1,k2)$ can be learned from the spectrograms of a training corpus of clean speech. Let $X_j(t,k)$ represent the kth component of the tth spectral vector from the jth training signal. The various mean and covariance values can be estimated as

$$\mu(k) = \frac{1}{\sum_j N_j} \sum_j \sum_t X_j(t,k), \tag{6.14}$$

$$c(\tau,k_1,k_2) = \frac{1}{\sum_j N_j} \sum_j \sum_t (X_j(t,k_1) - \mu(k_1))(X_j(t+\tau,k_2) - \mu(k_2)). \tag{6.15}$$

Relative covariance values can be computed from the covariance values using Equation (6.13). The implication of the assumption of a Gaussian process is that the joint distribution of the components of all the spectral vectors in a sequence of vectors is Gaussian. Consequently, the distribution of any subset of these components is also Gaussian [22]. The estimated mean and covariance values characterize the process completely and no other statistical parameters are required.

The parameters of the process are employed to reconstruct every clean spectral vector $X(t)$. Specifically, this is obtained by setting $X_r(t) = Y_r(t)$. $X_u(t)$ is estimated from $Y(t)$ and the parameters of the Gaussian process. To do so, we construct a neighborhood vector $Y_n(t)$ from all reliable components of the spectrogram that have a relative covariance greater than a threshold value with at least one of the components of $X_u(t)$. Let $X_n(t)$ be the set of components corresponding to the entries of $Y_n(t)$ from the clean speech spectrogram. Since all the components of $Y_n(t)$ are reliable, $X_n(t) = Y_n(t)$ by assumption. The joint distribution of $X_u(t)$ and $X_n(t)$ is Gaussian. The parameters of this distribution are the expected value of $X_u(t)$, $\mu_u(t)$, the expected value of $X_n(t)$, $\mu_n(t)$, the autocorrelation of $X_u(t)$, $C_{uu}(t)$, the autocorrelation of $X_n(t)$, $C_{nn}(t)$, and the cross correlation between $X_u(t)$ and $X_n(t)$, $C_{un}(t)$. These parameters can all be constructed from the mean and covariance terms learned from the training corpus. Figure 6.3 demonstrates the construction of $Y_u(t)$ and $Y_n(t)$ with an example.

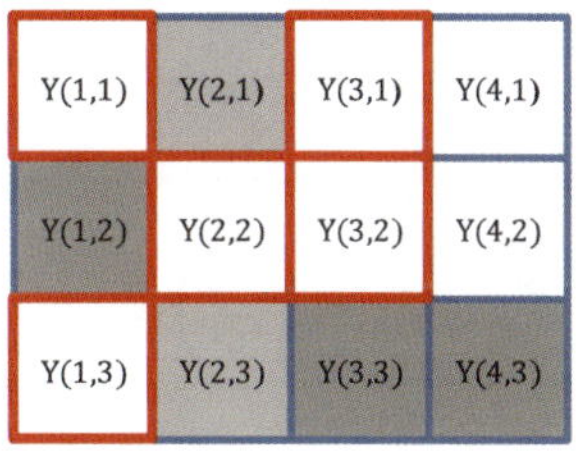

Fig. 6.3: A small spectrogram with four spectral vectors, each with four components. The grey components are missing. We wish to estimate unreliably known components in the second spectral vector. We construct the vector to be estimated as $X_u(2) = [X(2,1)X(2,3)]^\top$. The components outlined in thick red lines represent neighboring reliable components that have a relative covariance greater than 0.5 with at least one of the terms in $X_u(2)$. We therefore compose the neighborhood vector as $Y_n(2) = [Y(1,1)Y(1,3)Y(2,2)Y(3,1)Y(3,2)]^\top$

$X_u(t)$ is now estimated as

$$\hat{X}_u(t) = \underset{X_u}{\operatorname{argmax}} P(X_u | X_n = Y_n(t), X_u \le Y_u(t)). \tag{6.16}$$

Denoting "$X_n = Y_n(t)$" as $Y_n(t)$ for simplicity, and using Bayes' rule, this can be rewritten as

$$\hat{X}_u(t) = \underset{X_u}{\operatorname{argmax}} P(X_u, X_u \le Y_u(t) | Y_n(t)). \tag{6.17}$$

We refer to the estimate given by Equation (6.17) as a bounded MAP estimate. It can be shown that $P(X_u(t)|Y_n(t))$, the distribution of $X_u(t)$ conditioned on $X_n(t)$ being equal to $Y_n(t)$, is a Gaussian with mean $\mu_u(t) + C_{un}(t)C_{nn}^{-1}(t)(Y_n(t) - \mu_n(t))$. As shown in [25] this can be obtained by the following iterative procedure.

Let $X_u(t,k)$ and $Y_u(t,k)$ be the kth components of $X_u(t)$ and $Y_u(t)$ respectively. Let the current estimate of $X_u(t,k)$ be $\bar{X}_u(t,k)$. The estimation procedure can now be stated as follows:

1. Initialize $\bar{X}_u(t,k) = Y_u(t,k), 1 \le k \le K$, where K is the total number of components in $X_u(t)$.
2. For each of the K components,
   a. compute the MAP estimate

$$\tilde{X}_u(t,k) = \underset{X_u(t,k)}{\operatorname{argmax}} P(X_u(t,k)|Y_n(t),\bar{X}_u(t,j)\forall j \neq k); \tag{6.18}$$

this is simply the mean of the Gaussian distribution of $X_u(t,k)$, conditioned on the reliable values $Y_n(t)$ and on all other components of $X_u(t)$ being equal to their current estimates;

b. compute the bounded MAP estimate from the MAP estimate as

$$\bar{X}_u(t,k) = \min(\tilde{X}_u(t,k), Y_u(t,k)). \tag{6.19}$$

3. If all $\bar{X}_u(t,k)$ estimates have converged, set $\hat{X}(t,k) = \bar{X}_u(t,k)\forall k$ to obtain $X_u(t)$, else go back to Step 2.

$X_u(t)$ is estimated as described above for each spectral vector in the spectrogram. This, combined with the reliable components, reconstructs the entire spectrogram.

6.6.2 Cluster-Based Reconstruction

In cluster-based reconstruction [25] all spectral vectors in the spectrogram are modelled as the output of an independent, identically distributed (IID) random process. The probability distribution of the vectors is assumed to be a Gaussian mixture. The unreliable components of spectral vectors are reconstructed from their statistical relationships to the reliable components of the same vector, as represented by the Gaussian mixture.

The nomenclature arises from the fact that the Gaussian mixture model (GMM) is equivalent to assuming that vectors are segregated into clusters, each of which has a Gaussian distribution. Viewing the GMM as a collection of clusters provides useful intuition. The “clusters” localize spectral vectors, enabling estimation of missing or unreliable components, as illustrated by Figure 6.4. If the actual cluster that a spectral vector belongs to were known, its unreliably known components could be well estimated, based on the distribution of the cluster. More realistically, the cluster cannot be known *a priori*. Hence, the actual estimate that we obtain is a *weighted* combination of estimates obtained from each cluster, where the weights are the *a posteriori* probabilities of the cluster.

According to the model, the distribution of the g-th cluster is given by

$$P(X|g) = \frac{exp(-\frac{1}{2}(X-\mu_g)^\top \Theta_g^{-1}(X-\mu_g))}{\sqrt{(2\pi)^d|\Theta_g|}}. \tag{6.20}$$

where X represents an arbitrary vector, d represents the dimensionality of X, and μ_g and Θ_g represent the mean vector and covariance matrix of the cluster, respectively. The overall distribution of spectral vectors, including all clusters, is a Gaussian mixture given by

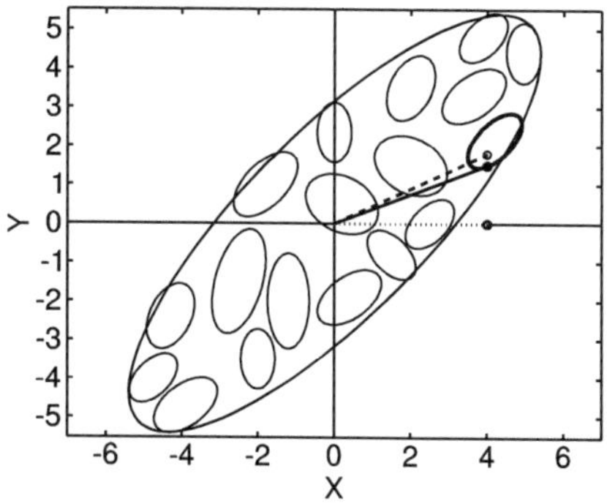

Fig. 6.4: In cluster-based reconstruction, spectral vectors of clean speech are assumed to belong to one of many clusters. The identity of a cluster that a vector belongs to localizes (and provides an estimate for) missing or unreliably known components of the vector. Here, for instance, the Y component of the vector shown by the solid line is assumed unknown and only the X component (shown by the dot on the X axis) is known. Knowing the cluster permits us to estimate the Y, to obtain the estimate given by the dotted line. If the cluster localizes the Y sufficiently, the error can be minimal

$$\begin{aligned} P(X) &= \sum_{g=1}^{G} c_g P(X|g) \\ &= \sum_{g=1}^{G} \frac{c_g}{\sqrt{(2\pi)^d |\Theta_g|}} exp(-\frac{1}{2}(X-\mu_g)^\top \Theta_g^{-1}(X-\mu_g)), \end{aligned} \tag{6.21}$$

where c_g is the *a priori* probability of the gth cluster. The *a priori* probabilities, means, and covariances of the clusters must all be learnt from a training corpus. These can be learned using the Expectation Maximization (EM) algorithm [1].

As before, for any noisy spectral vector $Y(t)$, the following relationships between its components and the components of the true underlying vector are known: $X_u(t) \leq Y_u(t)$ and $X_r(t) = Y_r(t)$. $X_u(t)$ must be estimated. If it were known that the vector was drawn from the gth cluster, we would estimate $X_u(t)$ to be its most probable value, conditioned on the identity of the cluster and on the information provided by $Y(t)$. This estimate $\hat{X}_u^g(t)$ is given by

$$\hat{X}_u^g(t) = \operatorname*{argmax}_{X_u} P(X_u|g, X_u \leq Y_u(t), X_r(t) = Y_r(t)), \tag{6.22}$$

where $P(X_u|g, X_u \leq Y_u(t), X_r(t) = Y_r(t))$ is the distribution of $X_u(t)$, conditioned on $X(t)$ belonging to the gth cluster, $X_u(t)$ being no greater than $Y_u(t)$, and $X_r(t)$ being equal to $Y_r(t)$. Using Bayes' rule and representing the condition "$X_r(t) = Y_r(t)$" simply as $Y_r(t)$ for brevity, this can be written as

$$\hat{X}_u^g(t) = \operatorname*{argmax}_{X_u} P(X_u, X_u \leq Y_u(t)|g, Y_r(t)). \tag{6.23}$$

The operation in Equation (6.23) represents the bounded MAP estimation procedure described in Section 6.6.1. Since all cluster distributions are Gaussian, $P(X_u(t)|g, Y_r(t))$ is also Gaussian. The mean of the gth cluster, μ_g, can be partitioned into the two vectors $\mu_{g,u}(t)$, which represents the expected value of $X_u(t)$, and $\mu_{g,r}(t)$, which represents the means of the components of $X_r(t)$. The components of the covariance matrix of the gth cluster, Θ_g, that represent the components in $X_u(t)$ and $X_r(t)$ can be separated into $\Theta_{g,uu}(t)$ and $\Theta_{g,rr}(t)$ respectively. The cross-correlation between $X_u(t)$ and $X_r(t)$, $\Theta_{g,ur}(t)$, can also be derived from Θ_g. From these terms, the bounded MAP estimate of $X_u(t)$, conditioned on the vector belonging to the g-th cluster can be obtained using the procedure described in Section 6.6.1.

The actual cluster that any vector belongs to is not known. To account for this, the overall estimate of $X_u(t)$ is obtained as a weighted average of estimates obtained from all clusters and is given by

$$\hat{X}_u(t) = \sum_{g=1}^{G} P(g|Y_r(t), X_u(t) \le Y_u(t))\hat{X}_u^g(t). \tag{6.24}$$

$P(g|Y_r(t), X_u(t) \le Y_u(t))$ is the *a posteriori* probability of the gth cluster, and is computed as

$$P(g|Y_r(t), X_u(t) \le Y_u(t)) = \frac{c_g P(Y_r(t), X_u(t) \le Y_u(t)|g)}{\sum_{j=1}^{G} c_j P(Y_r(t), X_u(t) \le Y_u(t)|j)}. \tag{6.25}$$

In order to compute this term, $P(Y_r(t), X_u(t) \le Y_u(t)|g)$ must be stated explicitly in terms of the reliably known and unreliably known component vectors of $X(t)$. Since $P(X(t)|g) = P(X_r(t), X_u(t)|g)$, this gives us

$$P(Y_r(t), X_u(t) \le Y_u(t)|g) = \int_{-\infty}^{Y_u(t)} P(Y_r(t), X_u|g)dX_u. \tag{6.26}$$

The exact form of the above equation is difficult to compute when the covariance matrix of $P(X(t)|g)$ has non-zero off-diagonal elements. We therefore approximate it by considering only the diagonal components of the covariance matrices when computing the *a posteriori* probabilities of clusters, assuming all other components to be 0. Under this assumption, expressing the kth components of $X(t)$, $Y(t)$ and μ_g as $X(t,k)$, $Y(t,k)$ and $\mu_g(k)$ respectively, and the kth diagonal element of Θ_g as $\theta_g(k)$, we obtain the following value for $P(Y_r(t), X_u(t) \le Y_u(t)|g)$:

$$\begin{aligned} P(Y_r(t), X_u(t) \le Y_u(t)|g) = & \prod_{k|X(t,k)\in X_r(t)} \frac{exp(-\frac{(Y(t,k)-\mu_g(k))^2}{2\theta_g(k)})}{\sqrt{2\pi\theta_g(k)}} \\ & \cdot \prod_{k|X(t,k)\in X_u(t)} \int_{-\infty}^{Y(t,k)} \frac{exp(-\frac{(x-\mu_g(k))^2}{2\theta_g(k)})}{\sqrt{2\pi\theta_g(k)}} dx. \end{aligned} \tag{6.27}$$

This can be used in Equation (6.25) to compute $P(g|Y_r(t), X_u(t) \leq Y_u(t))$, the *a posteriori* cluster probabilities. Thereafter, $\hat{X}_u(t)$ can be estimated using Equation (6.24).

As a final note, in order to best accommodate the assumption of diagonal covariance matrices used in the estimation *a posteriori* probabilities, the GMMs are initially learned assuming diagonal covariances for all Gaussians. Full covariance matrices are computed for all clusters in a final pass of the EM algorithm.

6.6.3 Minimum Mean Squared Error Estimation of Spectral Components from Soft Masks

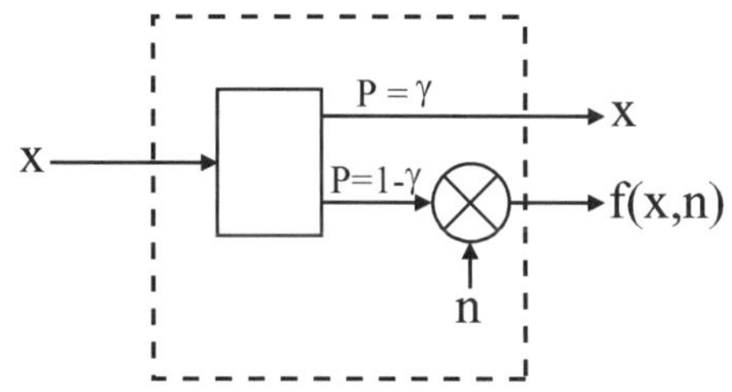

Fig. 6.5: Noisy channel model for soft masks

Correlation-based and cluster-based reconstruction assume binary spectrographic masks that definitively identify time-frequency components as reliable or unreliable. However, soft masks, which only assign a probability of reliability to time-frequency components, being more non-committal, are possibly more realistic characterizations of our knowledge of the state of a spectrogram.

In order to account for the fact that we cannot be certain if time-frequency components are reliable but can only assign a probability to it, we must modify the model we had of unequivocally reliable or unreliable components. Instead, we now model the time-frequency components of noisy speech as the output of a noisy channel [26]. The input to the channel are individual time-frequency components of clean speech. With some probability the channel lets the components through unchanged. Otherwise, it adds an unknown quantity of noise to them. Figure 6.5 illustrates the model. Since we no longer identify components as distinctly reliable or unreliable – rather, every component now has a probability of being unreliable – we must now estimate *every* time-frequency component.

As before, we assume that the log-spectral vectors $X(t)$ are distributed according to a Gaussian mixture; however, unlike in Section 6.6.2 we assume all Gaussians to have diagonal covariances. To clarify the presentation we will use a modified notation for probability distributions that distinguishes clearly between the random variable and the value it takes. We will denote the probability that a random variable

X takes a value Y as $P_X(Y)$. Using this notation, the probability distribution of log-spectral vectors of clean speech is assumed to be

$$\begin{aligned} P_{X(t)}(X(t)) &= \sum_{g=1}^{G} \frac{c_g}{\sqrt{(2\pi)^d|\Theta_g|}} exp(-\frac{1}{2}(X(t)-\mu_g)^\top \Theta_g^{-1}(X(t)-\mu_g)) \\ &= \sum_{g=1}^{G} c_g \prod_k \frac{exp(-\frac{(X(t,k)-\mu_g(k))^2}{2\theta_g(k)})}{\sqrt{2\pi\theta_g(k)}}. \end{aligned} \tag{6.28}$$

Note that the subscript $X(t)$ in $P_{X(t)}(X(t))$ represents the random variable $X(t)$ and the argument $X(t)$ is the value it takes. While the notation might be confusing here, the distinction between the random variable and its value is important for the following description. $\mu_g(k)$ is the kth component of μ_g, the mean of the gth Gaussian, and $\theta_g(k)$ is the kth diagonal element of its covariance matrix Θ_g.

According to the model, each component $X(t,k)$ of the spectral vector $X(t)$ is input to the noisy channel. The channel transmits the input unchanged to the output with a component-specific probability $\gamma(t,k)$. With probability $1-\gamma(t,k)$ it corrupts the input during the transmission. To corrupt the input it randomly draws a noise sample $N(t,k)$ from a distribution $P_{N(t,k)}(n)$, and combines it with the input $X(t,k)$ through a function $f()$ such that $f(X(t,k),N(t,k)) \geq X(t,k)$. If the component goes through the channel uncorrupted, the output $Y(t,k) = X(t,k)$. If not, the noise-corrupted output of the channel $Y(t,k) = f(X(t,k),N(t,k))$. The parameter $\gamma(t,k)$ is the probability of reliability, computed in the spectral mask estimation step. The distribution of the noise $P_{N(t,k)}(n)$ is assumed to be specific to the time-frequency component (t,k).

The *conditional* distribution of the output of the channel $Y(t,k)$ given that the spectral vector $X(t)$ has been drawn from the g-th Gaussian is given by:

$$P_{Y(t,k)}(y|g) = \gamma(t,k)P_{X(t,k)}(y|g) + (1-\gamma(t,k))\int_{-\infty}^{\infty} P_{X(t,k)}(z|g)P_{N(t,k)}(f^{-1}(y,z))dz, \tag{6.29}$$

where $f^{-1}(y,z)$ is the inverse function of $f()$ that computes the set of all scalar noise values n such that $f(z,n) = y$. Since all spectral vectors $X(t)$ are identically distributed according to Equation (6.28), the conditional distribution of the kth component of any vector $X(t)$ is simply

$$P_{X(t,k)}(x|g) = \frac{exp(-\frac{(x-\mu_g(k))^2}{2\theta_g(k)})}{\sqrt{2\pi\theta_g(k)}}. \tag{6.30}$$

We assume that $P_{N(t,k)}(n)$ is a uniform probability distribution which extends between $f^{-1}(Y(t,k),z)$ and $f^{-1}(L_k,z)$, where L_k is some known constant (typically set to the lowest possible value of $X(t,k)$). Using these values, we now get

$$P_{Y(t,k)}(y|g) = \gamma(t,k)P_{X(t,k)}(y|g) + \frac{(1-\gamma(t,k))}{Y(t,k)-L_k}\int_{L_k}^{Y(t,k)} P_{X(t,k)}(z|g)dz \tag{6.31}$$

The overall probability distribution of $Y(t)$ is given by

$$P_{Y(t)}(Y) = \sum_g c_g \prod_k P_{Y(t,k)}(Y_k|g) \tag{6.32}$$

(where Y_k is the k-th component of vector Y).

The a posteriori probability, given $Y(t)$, of the g-th Gaussian is

$$P(g|Y(t)) = \frac{c_g \prod_k P_{Y(t,k)}(Y(t,k)|g)}{\sum_j c_j \prod_k P_{Y(t,k)}(Y(t,k)|j)}. \tag{6.33}$$

It can now be shown that the *a posteriori* probability of $X(t,k)$ given $Y(t)$ and Gaussian index g is given by

$$P_{X(t,k)}(x|Y(t),g) = \begin{cases} \gamma(t,k)\delta_{X(t,k)}(Y(t,k)) + (1-\gamma(t,k))\frac{P_{X(t,k)}(x|g)}{C_{X(t,k)}(Y(t,k)|g) - C_{X(t,k)}(L_k|g)} \\ \qquad \text{if } L_k \le x \le Y(t,k) \\ 0 \qquad \text{else} \end{cases}, \tag{6.34}$$

where $\delta_{X(t,k)}(Y(t,k))$ is a Kronecker delta centered at $Y(t,k)$ and $C_{X(t,k)}(Y(t,k)|g)$ represents the cumulative probability at $Y(t,k)$ of the k-th dimension of the g-th Gaussian. The overall a posteriori probability of $X(t,k)$ is given by

$$P_{X(t,k)}(x|Y(t)) = \sum_g P(g|Y(t)) P_{X(t,k)}(x|Y(t),g). \tag{6.35}$$

The minimum mean-squared error estimate of $X(t,k)$ is simply the expected value of $X(t,k)$ given the observed vector $Y(t)$. To obtain the MMSE estimate of $X(t,k)$ we draw upon the following identity:

$$\int_{-\infty}^{a} x\mathcal{N}(x;\mu,\sigma)dx = \mu \int_{-\infty}^{a} \mathcal{N}(x;\mu,\sigma)dx - \sigma\mathcal{N}(a;\mu,\sigma), \tag{6.36}$$

where $\mathcal{N}(x;\mu,\sigma)$ represents a Gaussian distribution over x, with mean μ and variance σ.

Combining Equations (6.34), (6.35) and (6.36) we get the following MMSE estimate for $X(t,k)$:

$$\hat{X}(t,k) = \gamma(t,k)Y(t,k) + \\ (1-\gamma(t,k)) \sum_g P(g|Y(t,k)) \left(\mu_{g,k} - \theta_{g,k} \frac{P_{X(t,k)}(Y(t,k)|g) - P_{X(t,k)}(L_k|g)}{C_{X(t,k)}(Y(t,k)|g) - C_{X(t,k)}(L_k|g)} \right). \tag{6.37}$$

The MMSE estimates $\hat{X}(t,k)$ are arranged into a vector $\hat{X}(t)$ that is used to compute cepstra that are used for recognition.

6.7 Experimental Evaluation

In this section we describe a series of experiments conducted to evaluate the recognition accuracy obtained using the proposed feature compensation methods, and to contrast this with the accuracy obtained using state-based imputation and marginalization. Initially, in Sections 6.7.1 through 6.7.6 we report results using hard masks. Later, Section 6.7.7 reports experiments on soft masks.

For the hard masks, experiments were conducted on speech corrupted by white noise and segments of music. These noise types represent two extremes of spectral and temporal distortions – white noise has a flat spectrum and is stationary, while music has a very detailed spectral structure and is highly non-stationary.

We initially describe experiments with "oracle" (or perfect) knowledge of the local SNR of time-frequency components in the spectrogram. Within these experiments we evaluate the effect of preprocessing and recognition with cepstra. These experiments establish an upper bound on the recognition performance obtainable with our experimental setup. We then describe results obtained from experiments employing a more realistic scenario where the locations of unreliable components must be estimated.

6.7.1 Experimental Setup

For experiments with hard masks, the DARPA Resource Management database (RM1) [20] was used. The automatic speech recognition system employed was the CMU SPHINX-III HMM-based speech recognition system. Context-dependent HMMs with 2,000 tied states were trained using both the log spectra and cepstra of clean speech. For all experiments except those reported in Figure 6.14 state output distributions were modelled by a Gaussian. For the results in Figure 6.14 state output distributions were modelled as mixtures of Gaussians. In all cases, the Gaussians in the state output distributions were assumed to have diagonal covariance matrices. A simple bigram language model was used. The language weight was kept low in order to emphasize the effect of the noisy acoustics on recognition accuracy. A 20-dimensional mel frequency spectrographic representation was used for the experiments.

Test utterances were corrupted by white noise and randomly chosen samples of music from the Marketplace news program. In all cases both the additive noise and the clean speech samples were available separately, making it possible to evaluate the true SNR of any component in the spectrogram of the noisy utterances.

6.7.2 Recognition Performance with Knowledge of True SNR

Missing-feature methods depend critically on being able to correctly identify unreliable time-frequency components. In the experiments described in this section, we assume that this information is available and accurate. The recognition performance obtained with the various missing-feature methods in this scenario represents an upper bound on the performance that can be obtained within the current experimental setup.

Unreliable components of spectrograms were identified based on the true value of the SNR of time-frequency components, the computation of which was permitted by the experimental setup, as explained in Section 6.7.1. All components whose SNR values lay below a threshold were deemed to be unreliable. A threshold value of 0 dB was found to be optimal or close to optimal at all SNRs for marginalization. For state-based imputation and feature compensation methods, the best threshold across all noise levels was found to be 5 dB. The experiments reported in this section used these threshold values to identify unreliable components.

For state-based imputation and marginalization, recognition was performed with the resulting incomplete spectrograms. For the feature compensation methods, complete spectrograms were reconstructed. Figures 6.6a and 6.6b show example spectrograms obtained by reconstructing unreliable components that have been estimated from their known SNR values, using correlation-based and cluster-based reconstruction. Recognition was performed using either the log-spectral vectors from the reconstructed spectrogram, or 13-dimensional cepstral coefficients derived from the log-spectral vectors.

6.7.3 Recognition with Log Spectra

Figure 6.7 shows recognition accuracies obtained by applying the various missing-feature methods to speech corrupted by white noise and music to various SNRs. Recognition has been performed using log-spectral vectors in all cases. For marginalization, no mean normalization was performed on the features. For all other methods mean normalization was performed. In all cases, HMM state output distributions were modelled as Gaussian. We observe from these plots that marginalization is capable of resulting in remarkable robustness to corruption by noise. In fact, the recognition accuracy at 0 dB is only a relative 20% worse than that obtained at 25 dB. All other methods provide significant improvements over baseline recognition performance (with noisy vectors), but are much worse than marginalization. This is to be expected when recognition is performed with log spectra, since marginalization performs optimal classification with the unreliable data, whereas the other methods do not. Feature compensation methods do, however, perform comparably to or better than state-based imputation.

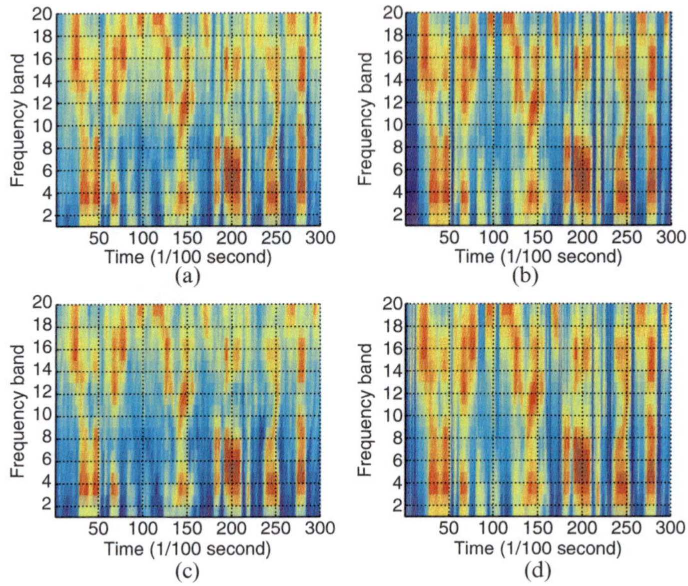

Fig. 6.6: Reconstruction of the mel spectrogram in Figure 6.1b: a) Using correlation-based reconstruction with oracle spectrographic masks obtained from knowledge of the true SNR of time-frequency components. b) With cluster-based reconstruction using oracle masks. c) With correlation-based reconstruction using estimated spectrographic masks. d) With cluster-based reconstruction using estimated masks

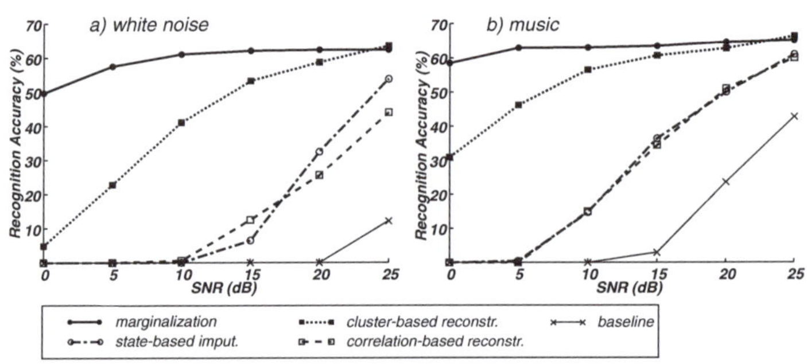

Fig. 6.7: Recognition performance of various missing-feature methods on noisy speech using oracle spectrographic masks. a) Speech corrupted by white noise. b) Speech corrupted by music. The baseline recognition performance with the uncompensated noisy speech is also shown

6.7.4 Effect of Preprocessing

The effect of preprocessing the signal is different on different missing-feature methods. One form of preprocessing commonly used is mean normalization. In this procedure the mean value of the feature vectors is subtracted from all the vectors. This is known to result in significant improvement in recognition performance. When missing-feature methods are applied, however, it is not clear whether this procedure is useful. Figure 6.8 shows the effect of mean normalization on the recognition accuracy obtained with various missing-feature methods on speech corrupted to 10 dB by white noise. Both reliable and unreliable components were used in computing the mean value of the vectors in all cases. We observe that mean normalization is useful in all cases where estimation of unreliable components is performed, i.e., for the feature compensation methods and state-based imputation. For marginalization, however, mean normalization actually results in a degradation of performance.

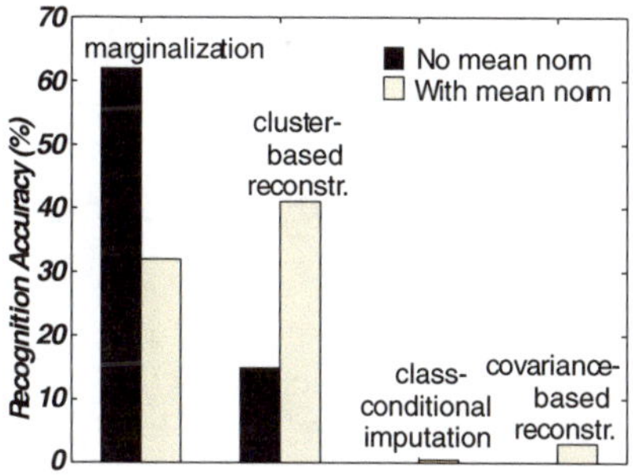

Fig. 6.8: Effect of mean normalization

6.7.5 Recognition with Cepstra

One of the primary arguments for spectrogram reconstruction methods is that the reconstructed spectrograms can now be used to derive cepstral features, and recognition can be performed with cepstra to obtain superior recognition performance. Figure 6.9 shows the recognition results obtained with such a setup. Recognition with cepstra is greatly superior to that with log spectra. As a comparator, recognition performance obtained using spectral subtraction [4], a conventional and highly successful noise compensation method, is also shown. Comparison with Figure 6.7 also shows that, although marginalization greatly outperforms other methods when recognition is performed with log spectra, the recognition performance obtained with cepstra derived from the reconstructed spectrograms results in much better recognition than obtainable with marginalization.

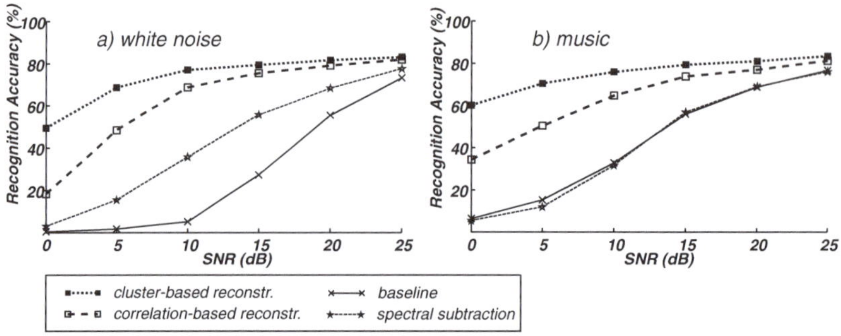

Fig. 6.9: Recognition performance obtained with cepstra derived from spectrograms reconstructed with oracle spectrographic masks. a) Recognition on speech corrupted by white noise to various SNRs. b) Recognition on speech corrupted by music to various SNRs. In both cases the baseline performance with uncompensated noisy speech and the performance with spectral subtraction are shown for contrast

6.7.6 Effect of Errors in Identifying Unreliable Components

The experiments in the previous section only served to establish the upper bound performance obtainable for the various methods when location of unreliable components in the spectrogram is known *a priori*. In reality, however, the location of unreliable components must be estimated. The estimation of these locations can be very errorful, and different missing-feature methods have different sensitivities to errors in identifying unreliable components.

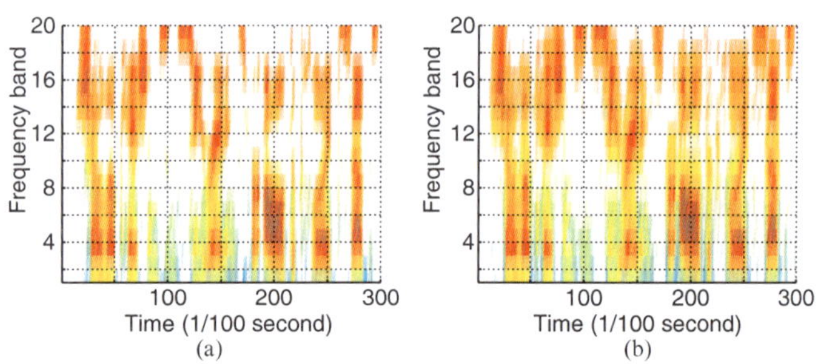

Fig. 6.10: a) Oracle mask for the signal in Figure 6.1b. b) Estimated spectrographic mask

For the experiments reported below we use the Bayesian approach of [30] to estimate spectrographic masks. Figure 6.10 shows an example of a spectrographic mask estimated by this technique (and compares it to the corresponding oracle mask). Figures 6.6c and 6.6d show the reconstructed spectrograms obtained for the spectrogram in Figure 6.1b, when the spectrographic mask was estimated. Comparison to the reconstructions of Figures 6.6a and 6.6b obtained with oracle masks illustrates the kinds of errors introduced by mask estimation.

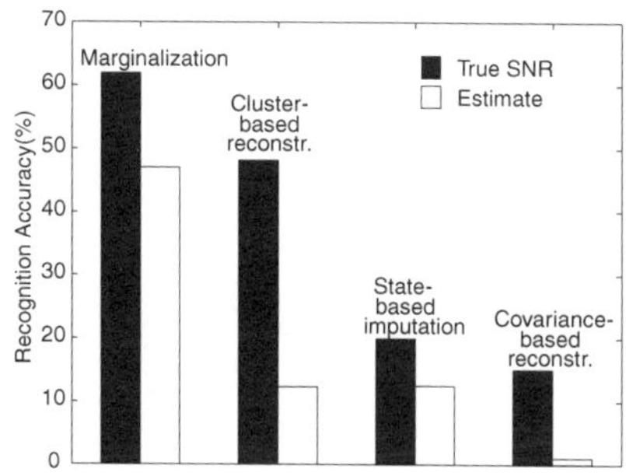

Fig. 6.11: Comparison of the performance obtained with oracle and estimated spectrographic masks

Figure 6.11 shows recognition accuracies obtained for several missing-feature methods applied to speech corrupted by white noise to 10 dB. Recognition has been performed using log spectra in all cases. We compare recognition accuracy obtained using oracle spectrographic masks with that obtained using estimated masks. Marginalization shows the greatest robustness to errors in identification of unreliable components. In general, the classifier compensation methods are much more robust to errors than feature compensation methods.

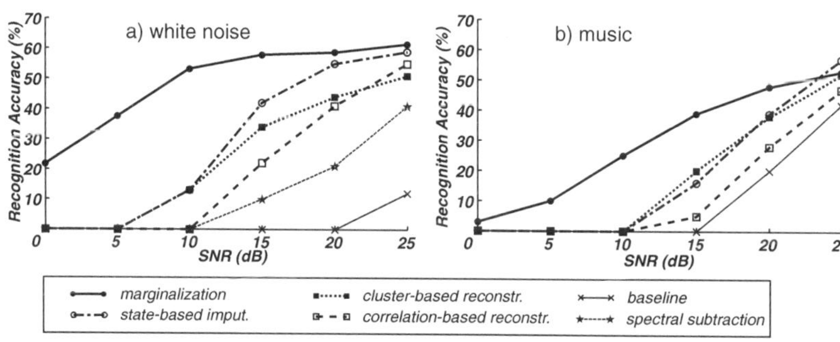

Fig. 6.12: Recognition performance of various missing-feature methods on noisy speech when the spectral mask is estimated. a) Speech corrupted by white noise. b) Speech corrupted by music

More detailed results are shown in Figure 6.12, which shows recognition accuracy obtained using various missing-feature methods with estimated masks as a function of SNR on speech corrupted by white noise and music. In all cases, the HMM state output distributions were modelled by single Gaussians. Mean normalization was performed in the case of the feature compensation methods and state-based imputation, but not for marginalization.

Both classifier compensation methods, marginalization and state-based imputation, are seen to outperform the feature compensation methods. In particular, marginalization is significantly superior to all other methods. The difference between marginalization and the other methods is further enhanced by its greater robustness to errors in identifying unreliable components.

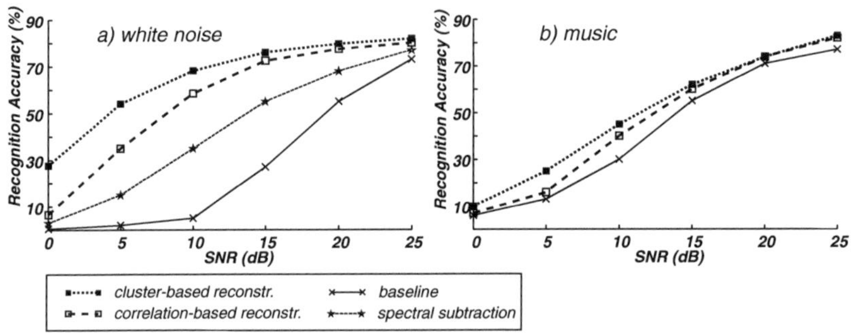

Fig. 6.13: Recognition with cepstra derived from reconstructed spectrograms, when spectrographic masks are estimated. a) Speech corrupted by white noise. b) Speech corrupted by music. As a contrast, baseline performance with the cepstra of noisy speech and, in the case of the white noise, performance with spectral subtraction, are also shown

Once again, however, reconstructed spectrograms can be used to derive cepstra for recognition. Figure 6.13 shows the recognition performance obtained on speech corrupted by white noise and music, with cepstra derived from spectrograms reconstructed by the proposed feature compensation methods. Comparison with Figure 6.12 reveals that even when spectrographic masks are estimated, the recognition accuracy obtained with cepstra derived from reconstructed spectrograms is greater than that obtained with marginalization and log-spectra-based recognition.

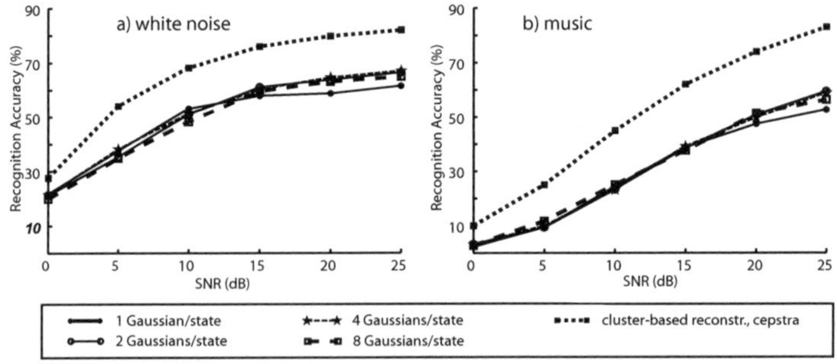

Fig. 6.14: The lower curves show recognition performance of marginalization on HMMs with state output distributions modelled by mixtures of one, two, four and eight Gaussians. The upper curve shows performance obtained with cepstra derived from spectrograms reconstructed by cluster-based reconstruction

In all experiments reported so far state output distributions have been modelled by single Gaussians with diagonal covariances. It may be expected that the recognition performance of the classifier compensation methods can be improved by modelling state output distributions by mixtures of Gaussians instead, thereby better capturing the correlations between spectral components. Figure 6.14 tests this hypothesis. It shows the recognition performance obtained with marginalization when state output distributions are modelled by mixtures of one, two, four and eight Gaussians, for speech corrupted by white noise and music, using estimated spectrographic masks. The figure also shows the performance obtained from cepstra derived from spectrograms reconstructed by cluster-based reconstruction. Although increasing the number of Gaussians results in slightly better performance at higher SNRs, recognition performance obtained with the cepstra derived from reconstructed spectrograms remains better. While the small improvements in recognition performance resulting from increasing the number of Gaussians in the state output densities is explained by the small size of the training corpus for the RM1 database, and it may be expected that larger improvements would be obtained had the training corpus been larger, the overall trends in performance generally do not change: Recognition with cepstra using a recognizer that employs a similar number of parameters remains superior. The ability to perform cepstra-based recognition easily outweighs the advantages due to the optimal classification and those due to the greater robustness to errors in identifying unreliable components that are characteristic of marginalization. The advantage, however, diminishes as the SNR reduces to 0 dB or so.

6.7.7 Experiments with MMSE Estimation from Soft Masks

All previous results assumed hard spectrographic masks. We now report results using soft masks. For this experiment, a Spanish telephone speech database provided by Telefonica Investigacion y Desarrollo (TID) was used. Experiments were performed using the CMU Sphinx-3 speech recognition system as before. Continuous density eight Gaussian/state HMMs with 500 tied states were trained from 3,500 utterances of clean telephone recordings. The test data consisted of telephone recordings corrupted to various SNRs by traffic noise, music, babble recorded in a bar, and noise recordings from a subway. A total of 1,700 test utterances were used in each case.

A Bayesian classifier was used to estimate the spectrographic masks. Details of the classifier can be found in [26]. For purposes of comparison, the classifier was employed to find both a binary and soft spectrographic masks.

Two separate recognition experiments were conducted. In the first, acoustic models were trained with the log-spectral vectors of clean speech. No difference or double difference features were employed. No mean normalization of the training data was performed. Recognition was performed using marginalization with the binary masks and soft-mask-based marginalization with the soft masks.

In the second experiment the recognizer was trained with cepstral vectors from clean speech. Mean normalization was performed. Difference and double difference features were also employed. For the noisy test data complete log spectral vectors were constructed using cluster-based reconstruction employing estimated binary masks and the MMSE algorithm of Section 6.6.3. Since the former utilizes binary spectrographic masks while the latter employs soft masks, the difference in performance between the two shows the improvements to be obtained from the use of soft masks.

The four panels in Figure 6.15 show the recognition performance obtained on speech corrupted by each of the four varieties of noise. In each panel, baseline recognition with cepstra, recognition obtained by marginalization of log spectra using hard masks, the performance obtained by soft-mask-based marginalization of log spectra, and that obtained with cepstra derived from spectra reconstructed using cluster-based reconstruction (using hard masks) and the soft-mask-based MMSE technique are all shown.

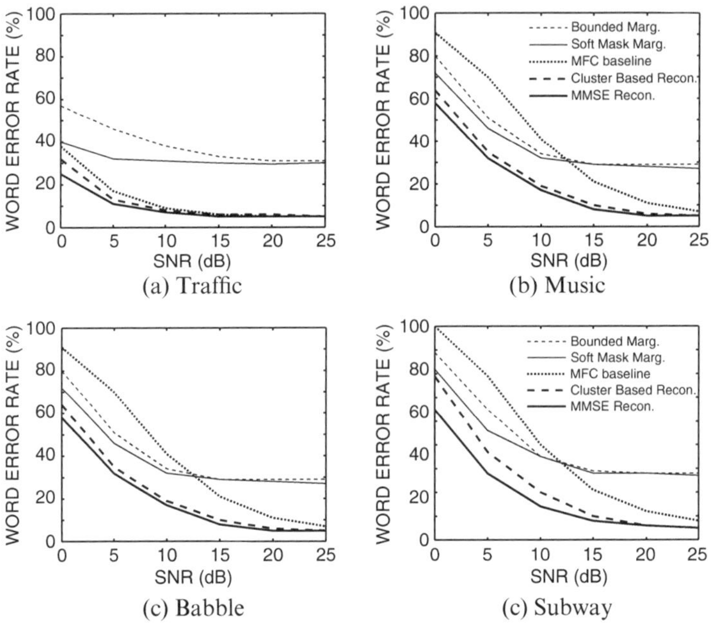

(a) Traffic (b) Music

(c) Babble (c) Subway

Fig. 6.15: Recognition error vs. SNR on speech corrupted by a) traffic noise, b) music, c) babble, and d) subway noise

We observe that soft-mask feature reconstruction combined with cepstrum-based recognition clearly outperforms all other methods. Soft-mask-based marginalization is significantly more robust than marginalization with binary masks. On the other hand the improvement from the use of soft masks is not that great for feature compensation – soft masks contribute significantly less to the robustness of feature compensation methods.

6.8 Conclusion and Discussion

The experiments in Section 6.7 demonstrate that missing-feature methods have great potential – they can generally be very effective when spectrographic masks are correctly known. Even when the masks are estimated and errorful they can be very effective.

When recognition (or classification) is preformed in the same feature domain (log spectra in our case) as the one in which features are found to be unreliable, by far the best techniques to employ are the optimal classification approaches based on marginalization. Not only do they perform optimal classification, they are also far more robust to mask estimation errors, and are particularly amenable to the use of soft masks. Furthermore, [13] presents a variation of marginalization called "fragment decoding" in which mask estimation becomes part of the recognition process itself, and results in better performance still. Other methods have also been proposed [18] that integrate mask estimation into marginalization-based recognition.

However, if recognition (or classification) is to be performed using features that are *derived* from, but not the same as those in which unreliable components are identified, feature compensation methods are clearly the method of choice. In particular, for speech recognition the advantages to be obtained with working in the derived domain (cepstra) far outweigh the advantage of optimal classification provided by marginalization. Moreover, when a fully reconstructed spectrogram is the final goal desired, there is no alternative to feature compensation methods.

The reconstruction methods of Section 6.6 represent only a small, but key subset of approaches one may take to reconstruct complete spectrograms from unreliable data. In [32] latent-variable decompositions of spectrograms are employed to fill in "holes" in spectrograms. Reyes et al. [29] use a modified form of belief propagation to reconstruct damaged regions of spectrograms. In [15] a technique based on non-negative matrix factorization is presented. Several other similar techniques have been proposed in the literature; however, unlike the feature compensation methods in this chapter, most of them are intended for *reconstruction* of spectrograms without specifically intending the reconstructed spectrograms to be useful for recognition. However, we do not expect this to be an impediment in their use as feature compensation techniques. Most methods also work in the log-spectral or magnitude-spectral domains. An exception is the recent work Gemmeke et al. [11], in which principles drawn from compressive sensing have been applied to perform reconstruction directly in a transform domain (which is analogous to cepstra), rather than on log spectra. Notably, this latter approach is actually performed within the recognizer itself, similarly to class-based imputation.

Eventually, the bottleneck in the performance of missing-feature methods remains accurate estimation of spectral masks. Unfortunately for feature compensation approaches, marginalization and its variants remain much more robust to mask estimation errors. Further, as mentioned above, they can in fact integrate mask estimation into the recognition process itself, which makes them more robust still. Currently no equivalent to fragment decoding exists for feature compensation methods, a topic that needs investigation.

Eventually, it is up to the user to evaluate the various tradeoffs that must be considered when choosing a missing-feature technique. We hope this chapter investigates enough of the issues involved to help the reader make an informed choice.

References

1. A. P. Dempster, N.M.L., Rubin, D.B.: Maximum likelihood from incomplete data via the EM algorithm. Journal of the Royal Statistical Society, Series B **39**, 1–38 (1977)
2. A. Vizhinho P. Green, M.C., Josifovski, L.: Missing data theory, spectral subtraction and signal-to-noise estimation for robust ASR: An integrated study. In: Proc. Eurospeech, pp. 2407–2410. Budapest, Hungary (1999)
3. Barker, J., Josifovski, L., Cooke, M.P., Greene, P.D.: Soft decisions in missing data techniques for robust automatic speech recognition. In: Proc. Intl Conf. on Speech and Language Processing. Beijing, China (2000)
4. Boll, S.F.: Suppression of acoustic noise in speech using spectral subtraction. IEEE Transactions on Acoustics, Speech and Signal Processing **27**, 113–120 (1979)
5. Cooke, M., Green, P., Josifovski, L., Vizinho, A.: Robust automatic speech recognition with missing and uncertain acoustic data. Speech Communication **34**, 267–285 (2001)
6. Cooke, M.P., Green, P.G., Crawford, M.D.: Handling missing data in speech recognition. In: Proc. Intl. Conference on Speech and Language Processing, pp. 1555–1558. Yokohama, Japan (1994)
7. Cooke, M.P., Morris, A., Green, P.D.: Missing data techniques for robust speech recognition. In: Proc. IEEE Conf. on Acoustics, Speech and Signal Processing. Munich, Germany (1997)
8. Davis, S., Mermelstein, P.: Comparison of parametric representation for monosyllable word recognition in continuously spoken sentences. IEEE Trans. on Acoustics, Speech, and Signal Processing **28**, 357–366 (1980)
9. Fletcher, H.: Speech and Hearing in Communication. Van Nostrand, New York (1953)
10. Gales, M.J.F., Young, S.J.: Robust continuous speech recognition using parallel model combination. IEEE Tansactions on Speech and Audio Processing **4**, 352–359 (1996)
11. Gemmeke, J.F., Van hamme, H., Cranen, B., Boves, L.: Compressive sensing for missing data imputation in noise robust speech recognition. IEEE Journal of Selected Topics in Signal Processing **4**(2), 272–287 (2010)
12. Gemmeke, J.F., Virtanen, T.: Noise robust exemplar based robust speech recognition. In: IEEE Conf. on Acoustics, Speech and Signal Processing. Dallas, USA (2010)
13. J. Barker N. Ma, A.C., Cooke, M.: Speech fragment decoding techniques for simultaneous speaker identification and speech recognition. Computer Speech and Language **24**, 94–111 (2010)
14. Josifovski, L., Cooke, M., Green, P., Vizinho, A.: State based imputation of missing data for robust speech recognition and speech enhancement. In: Proc. Eurospeech. Budapest, Hungary (1999)
15. LeRoux, J., de Chevigne, A.: Computational auditory induction by missing-data non-negative matrix factorization. In: ISCA tutorial and research workshop on statistical and perceptual audition (SAPA). Brisbane, Australia (2008)
16. Lippmann, R., Carlson, B.: Using missing feature theory to actively select features for robust speech recognition with interruptions, filtering and noise. In: Proc. Eurospeech, pp. 37–40. Rhodes, Greece (1997)
17. Miller, G.A., Licklider, J.C.R.: The intelligibility of interrupted speech. Journal of the Acoustic Society of America **22**, 167–173 (1950)
18. Ming, J., Lin, J., Smith, F.J.: A posterior union model with applications to robust speech and speaker recognition. EURASIP Journal on Applied Signal Processing pp. 1–12 (2006)
19. Moreno, P.: Speech recognition in Noisy Environments. Ph.D. Thesis, Carnegie Mellon University (1996)
20. P. Price W. M. Fisher, J.B., Pallet, D.S.: The DARPA 1000 word resource management database for continuous speech recognition. In: Proc. IEEE Conf. on Acoustics Speech and Signal Processing, pp. 651–654. Seattle, Wa. (1998)
21. Palomaki, K.J., Brown, G.J., Barker, J.: Techniques for handling convolutional distortion with missing data automatic speech recognition. Speech Communication **43**, 123–142 (2004)

22. Papoulis, A.: Probability, Random Variables, and Stochastic Processes. McGraw Hill Inc., New York (1991)
23. Raj, B.: Reconstruction of incomplete spectrograms for robust speech recognition. Ph.D. thesis, Carnegie Mellon University (2000)
24. Raj, B., Parikh, V., Stern, R.M.: The effects of background music on speech recognition accuracy. In: Proc. IEEE Conf. on Acoustics, Speech and Signal Processing. Munich, Germany (1997)
25. Raj, B., Seltzer, M.L., Stern, R.M.: Reconstruction of missing features for robust speech recognition. Speech Communication **43**, 275–296 (2004)
26. Raj, B., Singh, R.: Reconstructing spectral vectors with uncertain spectrographic masks for robust speech recognition. In: Automatic Speech Recognition and Understanding Workshop. Puerto Rico (2006)
27. Raj, B., Virtanen, T., Chaudhuri, S., Singh, R.: Non-negative matrix factorization based compensation of music for automatic speech recognition. In: Proceedings of Interspeech. Makuhari Japan (2010)
28. Renevey, P.: Speech in noisy conditions using missing feature approach. Ph.D. Thesis EPFL No. 2303, Swiss Federal Institute of Technology (2000)
29. Reyes-Gomez, M.J., Jojic, N., Ellis, D.P.W.: Towards single-channel unsupervised source separation of speech mixtures: The layered harmonics/formants separation/tracking model. In: ISCA Tutorial and Research Workshop on Statistical and Perceptual Audition (SAPA). Jeju, Korea (2004)
30. Seltzer, M.L., Raj, B., Stern, R.M.: A bayesian framework for spectrographic mask estimation for missing feature speech recognition. Speech Communication **43**, 379–393 (2004)
31. Shaugnessey, D.O.: Speech Communication – Human and Machine. Addison Wesley (1987)
32. Smaragdis, P., Raj, B., Shashanka, M.: Missing data imputation for spectral audio signals. In: IEEE Intl. Workshop on Machine Learning for Signal Processing. Grenoble, France (2009)
33. Wang, D., Brown, G. (eds.): Computational Auditory Scene Analysis: Principles, Algorithms, and Applications. Wiley-IEEE Press (2006)
34. Warren, R.M., Reiner, K.R., Bashford, J.A., Brubaker, B.S.: Spectral redundancy: Intelligibility of sentences heard through narrow spectral slits. Perception and Pscychophysics **57**, 175–182 (1995)
35. Yu, D., Deng, L., Droppo, J., Wu, J., Gong, Y., Acero, A.: Robust speech recognition using cepstral minimum-mean-square-error noise suppressor. IEEE Transactions on Acoustics, Speech and Language Processing **16**(5) (2008)

Chapter 7
Automatic Speech Recognition Using Missing Data Techniques: Handling of Real-World Data

Jort F. Gemmeke, Maarten Van Segbroeck, Yujun Wang,
Bert Cranen, Hugo Van hamme

Abstract In this chapter, we investigate the performance of a missing data recognizer on real-world speech from the SPEECON and SpeechDat-Car databases. In previous work we hypothesized that in real-world speech, which is corrupted not only by environmental noise, but also by speaker, reverberation and channel effects, the 'reliable' features do not match an acoustic model trained on clean speech. In a series of experiments, we investigate the validity of this hypothesis and explore to what extent performance can be improved by combining MDT with three conventional techniques, viz. multi-condition training, dereverberation and feature enhancement. Our results confirm our hypothesis and show that the mismatch can be reduced by multi-condition training of the acoustic models and feature enhancement, and that these effects combine to some degree. Our experiments with dereverberation reveal that reverberation can have a major impact on recognition performance, but that MDT with a suitable missing data mask is capable of compensating both the environmental noise as well as the reverberation at once.

7.1 Introduction

Automatic speech recognition (ASR) performance drops rapidly when speech is corrupted with increasing levels of unfamiliar background noise (i.e., noise not seen during training) since the observed acoustic features no longer match the acoustic models. One of the most effective approaches to improving the noise robustness of a speech recognizer is to perform multi-condition training [15]: Rather than acoustic

Jort F. Gemmeke, Bert Cranen
Dept. of Linguistics, Radboud University Nijmegen, The Netherlands, e-mail: j.gemmeke,b.cranen@let.ru.nl

Maarten Van Segbroeck, Yujun Wang, Hugo Van hamme
ESAT Department, Katholieke Universiteit Leuven, Belgium, e-mail: yujun.wang, maarten.vansegbroeck, hugo.vanhamme@esat.kuleuven.be

D. Kolossa and R. Haeb-Umbach (eds.), *Robust Speech Recognition of Uncertain or Missing Data: Theory and Applications*, DOI 10.1007/978-3-642-21317-5_7,
© Springer-Verlag Berlin Heidelberg 2011

models being trained on speech from a quiet environment only, they are trained directly on noisy speech signals. By carefully selecting the training speech to reflect the multiple acoustic conditions under which the system must operate, it is possible to minimize the mismatch between training and test/usage conditions. While often effective, recognition accuracies obtained with multi-condition training quickly deteriorate when the noisy environment deviates from the one that was used for training. Another disadvantage of multi-condition training is that the performance for truly clean speech tends to degrade.

Missing Data Techniques (MDTs) [25] are a very different approach to improving noise robustness that ideally overcomes the problems of multi-condition training. MDTs, first proposed in [6], build on two assumptions: The first assumption is that it is possible to estimate, prior to decoding, which spectro-temporal elements in the acoustic representation of noisy speech are reliable (i.e., dominated by speech energy) and which are unreliable (i.e., dominated by background noise). These reliability estimates are referred to as *missing data masks*. The second assumption is that the statistics of the features which are considered as dominated by speech energy match with the statistics of clean speech training data. This assumption implies that the acoustic models of MDT recognizers can be trained using clean speech.

In the unreliable elements, the speech information is considered *missing*, and the challenge is then to do speech recognition with partially observed data. In this work, we focus on the so-called *imputation* approach [24], which handles the missing elements by replacing them with clean speech estimates. Classic imputation methods include correlation and cluster-based reconstruction [23, 25], state-dependent imputation [17], which combines front-end imputation and classifier modification, and the Gaussian-dependent method [32], which additionally allows for reconstruction in the cepstral and PROSPECT domains. The latter method is employed in this chapter.

While imputation has proven effective for increasing noise robustness in the presence of both stationary and non-stationary noise, most of the existing knowledge about the effectiveness of MDT has been acquired using databases with noisy speech that has been constructed by artificially adding noise of various types and intensities to clean speech (see, e.g., [7, 23]). Using artificially corrupted data is attractive as it allows creating a missing data mask based on exact knowledge of the speech and noise power in each time-frequency cell. This facilitates comparison of different MDT approaches and allows for analysis of the influence of errors in reliability estimation.

Real-world recordings, however, are generally not only corrupted by background noise, but can also affected by room acoustics. Moreover, real-world recordings are more likely to introduce a mismatch between the observed speech and the speech on which the recognizer is trained, due to microphone characteristics and speaker-specific behavior such as lip noises and the Lombard effect. Very few reports exist that describe the effectiveness of single-channel MDT recognition on real-world recordings (notable exceptions are [13, 19, 27]). In previous research we have used the SPEECON[16] and SpeechDat-Car[30] databases for that purpose. The SPEECON and SpeechDat-Car databases are recorded in realistic environments such as in an entertainment room, an office, a public hall and a car. The databases contain simultaneous recordings from four microphones placed at different distances from the speaker, one of them being a close-talk microphone. Thus, SPEECON and SpeechDat-Car make it possible to investigate the impact of different degrees of natural distortions (background noise and reverberation) on the performance of ASR systems. Specifically, since the close-talk microphone could be considered as an approximation of 'clean speech', these corpora make it possible to investigate the performance of MDT on real-world speech using an approach similar to what has proven so effective with artificially corrupted databases.

We have found that an MDT recognizer that is trained with speech from the close-talk microphone is not very robust against the distortions that are present in the speech recorded with the three other (far-talk) microphones [13]. Moreover, even when using information from all available channels to estimate a 'cheating' missing data mask, the so-called *semi-oracle mask*, we obtained much lower accuracies than previously obtained on similar recognition tasks (such as Aurora 4 [22]) with artificially corrupted speech [35]. We hypothesize that this is due to a violation of the second assumption underlying MDT, namely that the statistics of the features that are not dominated by background noise match with the statistics of the features from the close-talk microphone. Experiments with artificially added noise all but guarantee that the second assumption holds true: If in some spectro-temporal element the speech energy is higher than the noise energy, the observed signal will fit the distribution of the clean training data. With real-world recordings, however, the speech in the other recording channels is not only affected by additive noise, but also by microphone characteristics and reverberation. This has the effect that the 'reliable' features, while dominated by speech energy, still mismatch the trained speech features. As a result, imputation and recognition accuracy are bound to suffer.

In this chapter, we test this hypothesis and explore whether recognition accuracy can be improved by combining MDT with three conventional techniques, multi-condition training, dereverberation and spectral subtraction. First, we extend the MDT approach, in which the recognizer is trained on close-talk channel 'clean' speech, by using acoustic models that are trained on multi-condition training material from all recording channels. In doing so, we assume that the proven techniques for estimating missing data masks and for imputing missing data can also be applied to real-world speech. Second, the availability of four parallel channels in the SPEECON and SpeechDat-Car databases makes it possible to detect strong reverberation and to create a new kind of 'cheating' missing data mask that labels time-frequency cells dominated by reverberation as 'unreliable'. This missing data mask, reminiscent of the dereverberation technique used in [8, 21], allows us to investigate the impact of reverberation on the recognition accuracy and to explore whether a combination of dereverberation and MDT will improve performance. Third, we investigate the performance obtainable with feature enhancement techniques on real-world recordings and whether feature enhancement can be combined with MDT, either to improve missing data mask estimation or to replace multi-condition training as a means for diminishing the hypothesized mismatch between training and test/use conditions.

The rest of the chapter is organized as follows. In Section 7.2 we introduce MDT and the imputation method used in the chapter. In Section 7.3 we describe the isolated word recognition task used in our experiments. In Section 7.4 we describe the missing data mask estimation techniques used for MDT and the decoding architecture used in later sections. In Section 7.5 we present the experiments with clean and multi-condition acoustic models. In Section 7.6 we investigate the combination of MDT and dereverberation and in Section 7.7 we investigate use of feature enhancement in combination with MDT. Finally, we have a general discussion and present our conclusions in Section 7.8.

7.2 MDT ASR

7.2.1 Missing Data Techniques

In this section, we briefly review the MDT framework [6, 25]. In ASR, the basic representation of speech is a spectro-temporal distribution of acoustic power, a *spectrogram*. In noise-free conditions, the value of each time-frequency cell in this two-dimensional matrix is determined only by the speech signal. In noisy conditions, the value in each cell represents a combination of speech and background noise power.

To mimic human hearing, often a mel frequency scale and logarithmic compression of the power scale are employed. We denote the (mel frequency) log-power spectrograms of noisy speech as **Y**, of clean speech as **S**, and of the background noise as **N**. Elements of **Y** that predominantly contain speech or noise energy are distinguished by introducing a missing data mask **M**. The elements of a mask **M** are either 1, meaning that the corresponding element of **Y** is dominated by speech ('reliable'), or 0, meaning that it is dominated by noise ('unreliable' c.q. 'missing'). Thus, we write

$$M(k,t) = \begin{cases} 1 \stackrel{def}{=} \text{reliable} & \text{if } S(k,t) - N(k,t) > \theta \\ 0 \stackrel{def}{=} \text{unreliable} & \text{otherwise} \end{cases} \tag{7.1}$$

with **M**, **Y**, **S**, and **N** two-dimensional matrices of size $K \times T$, with frequency band index k, $1 \leq k \leq K$ and time frame index t, $1 \leq t \leq T$. θ denotes a constant SNR threshold.

Assuming that only additive noise corrupted the clean speech, the power spectrogram of noisy speech can be approximately described as the sum of the individual power spectrograms of clean speech and noise. As a consequence, in the logarithmic domain, the reliable noisy speech features remain approximately uncorrupted [25] and can be used directly as estimates of the clean speech features.

In the real-world speech recorded by multiple microphones, considered in this chapter, it is questionable whether features that are labeled reliable with such a procedure remain approximately uncorrupted. Most speaker effects (such as the Lombard effect) will show up equally in all recording channels. Environmental noise, channel effects and reverberation, however, are likely to affect the different channel recordings differently. A fundamental problem is thus the definition of 'clean' speech and 'noise' underlying (7.1). Even if a close-talk microphone signal is used for training the acoustic models as an approximation of 'clean' speech, as was done in previous work [13], the 'noise' in the far-talk channels actually constitutes not only the environmental noise, but also extra feature variation due to the way in which channel characteristics and reverberation have affected the speech energy. Conventional mask estimation techniques, however, make the distinction between features dominated by speech or background noise by searching for spectro-temporal elements that have the characteristics of speech. As a result, the resulting 'reliable' features retain all channel and reverberation effects. As a consequence, 'reliable' features that are determined in the conventional way are likely to mismatch the statistics of the features in the close-talk channel used for training.

In Sections 7.5 and 7.7 we explore the impact of reducing this mismatch between the acoustic model and the 'reliable' features. In Section 7.6 we take the opposite approach and see if we can reduce a part of the mismatch by considering the reverberated speech features as 'noise' and modifying the missing data mask accordingly.

7.2.2 Gaussian-Dependent Imputation

Originally, MDT was formulated in the log spectral domain [6]. Here, speech is represented by the log-energy outputs of a filter bank and modeled by a Gaussian Mixture Model (GMM) with diagonal covariance. In the imputation approach to MDT, the GMM is then used to reconstruct clean speech estimates for the unreliable features. When doing *bounded imputation*, the unreliable features are not discarded but used as an upper bound on the log power of the clean speech estimate [7].

Later, it was found the method could be improved by using state- [17] or even Gaussian-dependent [31] clean speech estimates. In these approaches, the unreliable features are imputed during decoding and effectively depend on the hypothesized state identity. However, filter bank outputs are highly correlated and poorly modeled with a GMM with a diagonal covariance. This is the reason why conventional (non-MDT) speech recognizers employ cepstral features, obtained by applying a de-correlating Discrete Cosine Transformation (DCT) on the spectral features.

In [31] a technique was proposed to do Gaussian-dependent (bounded) imputation in the cepstral domain. The drawback of that technique was the increased computational cost, because the imputation of the clean speech was done by solving a Non-Negative Least SQuare (NNLSQ) problem. The Gaussian-dependent imputation approach used in this chapter [32] refines that approach by replacing the DCT used in the generation of cepstra with another data-independent linear transformation that results in computational gains while solving the NNLSQ problem. The resulting PROSPECT features are, just like cepstral coefficients, largely uncorrelated, and therefore allow us to retain the high accuracy at high SNRs as well as the good performance at lower SNRs obtained with Gaussian-dependent imputation.

7.3 Real-World Data: The SPEECON and SpeechDat-Car Databases

In the research reported in this chapter we used the SPEECON [16] and the SpeechDat-Car [30] databases. These databases contain speech recorded in realistic environments with multiple microphones. There are four recording environments: office, public hall, entertainment room and car. The office, public hall and entertainment room material stems from the SPEECON database and contains multichannel data with each channel corresponding to a different microphone position: channel #1 is a headset microphone, #2 is a lavalier microphone and #3 and #4 are medium and far distance microphones placed at 0.5 to 3 meters from the speaker. The car environment contains material from both the SPEECON and the SpeechDat-Car databases, with again channel #1 a headset microphone and #2 a lavalier microphone, while the channel #3 microphone is placed behind the rear-view mirror and the channel #4 microphone is placed near the rear window (SpeechDat-Car) or near the rear-view mirror (SPEECON). The speech material is recorded with a 16 kHz sampling rate.

The use of these databases is a middle ground between the artificially corrupted speech as found in the Aurora databases [14, 22] on the one hand, and the complex real-world conditions on the other.

7.3.1 *Isolated Word Test Set*

For our recognition experiments, we used a subset of the isolated word data in the Flemish part of the SPEECON and the SpeechDat-Car databases. This isolated word data contains command words, nouns and verbs. We constructed a test set containing a balanced mixture of SNR conditions. Using the SNR estimates obtained in [13] we created six SNR subsets, each with a 5 dB bin width, spanning a 0 dB to 30 dB range. The SNR subsets were filled by randomly selecting 700 utterances per SNR subset, ensuring a uniform word occurrence. The SNR bins do not contain equal numbers of utterances from the four channels: Generally speaking, the highest SNR bins mostly contain utterances from channel #1, while the lowest SNR bins mostly contain channel #4 speech.

The resulting test set contains $16,535$ utterances,[1] with 565 unique words, 54 minutes of speech embedded in 13 hours of audio signal. The test set is spoken by 232 speakers, 115 male and 117 female.

[1] The observant reader will have noticed that the total number of words does not add up to $4 \times 6 \times 700 = 16,800$. This is because one subset (the entertainment room environment in the $[0-5]$ dB SNR bin) only contains 435 utterances rather than 700 due to data scarcity.

7.3.2 *Training Sets*

The clean training set contains 40 hours of speech embedded in 63.5 hours of signal. Among the utterances used for training are command words and read sentences. All the 61,940 utterances in this set are from channel #1 data, with an estimated SNR range of 15 to 50 dB. The clean training set was spoken by 191 speakers, 82 of them male and 109 female. There is no overlap between speakers in the test and training sets.

The multi-condition training set contains 127 hours of speech embedded in 205 hours of signal, 231,849 utterances in total. Beside all channel #1 data included in the clean training set, the multi-condition set contains all utterances from channels #2, #3 and #4 which have an estimated SNR of 10 dB and higher. The 10 dB cut-off is necessary to prevent frame/state alignment issues during training and to ensure the acoustic models trained on this data remain sufficiently discriminative. The multi-condition training set thus contains an additional 55 hours of channel #2 data (54,381 utterances), 54 hours of channel #3 data (53,248 utterances) and 32 hours of channel #4 data (31,975 utterances). While the training sets differ in size, they do not differ in terms of speech-related observations since the data stems from multi-channel recordings.

The speech in the training set is taken from three noise environments: office, public hall, and car. The channel #1 speech used in the clean training set, 63.5 hours of signal in total, is composed of 47 hours of signal from the office environment, 2.2 hours from the public hall environment and 14.3 hours from the car environment. For the multi-condition model, the 205 hours of signal originate from 168 hours of signal recorded in the office environment, six hours from the public hall environment and 31 hours from the car environment.

7.4 Experimental Setup

7.4.1 *Mask Estimation*

7.4.1.1 Semi-Oracle Masks

With the recordings used in this paper, the underlying clean speech and noise are not known exactly. As a consequence, oracle masks useful for obtaining an estimate of the upper bound on recognition performance with MDT cannot be computed. We can, however, use the multi-channel data to estimate the so-called *semi-oracle mask*. In order to calculate this mask, we use the channel #1 data, which is obtained from a headset microphone, as an estimate for the underlying clean speech in the other channels. In order to compensate for the delay and microphone differences between channel #1 and the other channels, we use an acoustic echo canceler (AEC) to predict the clean speech component.

By minimizing the (energy) difference between a filtered version of the channel #1 signal and a far-talk microphone signal, the AEC estimates a Finite Impulse Response (FIR), which can be considered as the best possible estimate of the transmission path from the close-talk microphone to the far-talk microphone. Thus, the remaining differences between the filtered far-talk channel and the unfiltered original can be attributed to the noise in the far-talk channel and can serve as a noise estimate. By thresholding the difference between the speech and noise estimates using (7.1) we obtain the semi-oracle mask.

For the AEC, we used the PEM-AFROW algorithm [38], using second-order pre-whitening filters and a 25 ms FIR filter. Because we cannot guarantee that the distance between the speaker and the microphone is constant for all utterances in a session, the filters are re-estimated for every utterance, and multiple iterations over the same utterance are used to improve the convergence. Since this is a 'cheating' missing data mask, we manually selected the optimal mask threshold after recognition over a large interval of threshold values for each recording environment and each acoustic model.

7.4.1.2 Vector Quantization Masks

As a first approach to estimating spectrographic masks from a single recording channel, we employ the Vector Quantization (VQ) strategy proposed in [37]. Here, the key idea is to estimate masks by making only weak assumptions about the noise, while relying on a strong model for the speech. The speech model is expressed as a set of codewords (a codebook) containing the periodic and aperiodic part of training speech. The periodic part consists of the harmonics at pitch multiples and the remaining spectral energy is considered the aperiodic part. Both parts are obtained using the harmonic decomposition method described in [33].

During decoding, we apply harmonic decomposition to the observed speech. We then use the periodic and aperiodic parts of the observed speech to recover a clean speech estimate from the set of stored codewords by minimizing a cost function that is robust against additive noise corruptions. The aperiodic part of the observation is used to provide a noise estimate by taking its long-term minimum as in [20]. Finally, the spectrographic VQ-based mask is estimated by thresholding the ratio of speech and noise power estimates using (7.1). To compensate for linear channel distortions, the VQ system self-adjusts the codebook to the channel during recognition.

Since the codebook only represents a model for the human voice, decoding of non-speech (or noise) frames will lead to incorrect codebook matching and misclassification of mask elements. Therefore, a Voice Activity Detector (VAD) segments speech from non-speech frames in order to restrict mask estimation to frames containing (noisy) speech. For a frame labeled as non-speech, all mask values are set to zero, indicating that all components are unreliable.

The VQ codebook was trained on features extracted from the close-talk channel SPEECON training database. The number of codebook entries was 500. The VAD was inspired by the integrated bi-spectrum method described in[26]. Recognition tests on the complete test set using a large interval of threshold values revealed that the threshold setting was not very sensitive. The (optimal) results presented in this work were obtained with $\theta = 8$ dB.

7.4.1.3 SVM Masks

A different approach to mask estimation is to use machine learning to classify each feature as either reliable or unreliable. A machine learning algorithm can be used to associate noisy speech features with reliability scores that are obtained from suitable training material. Such training material must necessarily consist of oracle masks and therefore requires the use of artificially corrupted clean speech for training.

In [28] it was proposed to use a Bayesian classification approach for mask estimation. In this work, we use Support Vector Machine (SVM) classifiers, a machine learning algorithm which is known for its excellent performance on binary classification tasks and generalization power when trained on relatively small data sets [4]. From the machine learning perspective, mask estimation is a multi-class classification problem with 2^K classes. Since such high-dimensional multi-class classification is infeasible, we assume that the reliability estimates are independent between frequency bands and train a separate SVM classifier for each of the K mel frequency bands.

Each classifier used the same set of single-frame-based $(7 \times K + 1)$-dimensional features consisting of the K-dimensional noisy speech features themselves, the harmonic and aperiodic parts and the long-term energy noise estimate described in Section 7.4.1.2, the gain factor described in[33], the 'sub-band energy to sub-band noise floor ratio' and 'flatness' features derived from the noisy mel spectral features described in [28], and finally a single VAD feature. The training material was taken from another corpus, Aurora 4 [22], which contains artificially noisified Wall Street Journal (WSJ) utterances.

SVMs were trained using LIBSVM [5] on 75,000 frames (amounting to 12.5 minutes of audio signal) randomly extracted from the Aurora 4 multi-condition training set. Reliability labels used in training were obtained from the oracle mask, derived by using the (available) clean speech and noise sources in (7.1) with $\theta = -3$ dB (cf. [37]). We used an RBF kernel, and hyper-parameters were optimized by doing five-fold cross-validation on the training set.

7.4.2 Recognizer Setup

7.4.2.1 Recognizer

The MDT-based recognizer was built by adding the required MDT modifications to the speaker-independent large vocabulary continuous speech recognition (LVCSR) system that has been developed by the ESAT speech group of K. U. Leuven; cf. [1]

for a detailed description of the system. This recognizer was chosen because of its fast experiment turnaround time and good baseline accuracy. Decoding is done with a time-synchronous beam search algorithm.

The recognition performance will be expressed in terms of the word error rate (WER), which is defined as the number of word errors, i.e., insertions, deletions and substitution errors, divided by the total number of words in the reference transcription. The word startup cost was tuned over all noise environments and channels jointly, but has only minor importance given the nature of the task.

The (word-independent) word insertion penalty was tuned over all noise environments and channels jointly by maximizing recognition accuracy. The word insertion penalty only marginally affects the accuracy via the pruning mechanism because in an isolated word task the same penalty is applied to all hypotheses.

7.4.2.2 Preprocessing

The acoustic feature vectors consisted of mel frequency log power spectra: $K = 22$ frequency bands with center frequencies starting at 200 Hz (the first mel band is not used). The spectra were created by framing the 16 kHz signal with a Hamming window with size 25 ms and a frame shift of 10 ms. The decoder also uses the first and second time derivative of these features, resulting in a 66-dimensional feature vector. During training, mean normalization is applied to the features. During decoding, the features are normalized by a more sophisticated technique which works by updating an initial channel estimate through maximization of the log-likelihood of the best-scoring state sequence of a recognized utterance [36]. In the MDT experiments, as described in Section 7.2.2, the spectra and their derivatives are transformed to the PROSPECT domain [32] during decoding. Missing data masks for the derivative features were created by taking the first and second derivatives of the missing data mask [34]. In the uncompensated baseline experiments, the mel frequency features are transformed using the Mutual Information Discriminant Analysis (MIDA) linear transformation [10]. The MIDA transformation maximizes class separation much as Linear Discriminant Analysis (LDA) does, but is based on a mutual information criterion.

7.4.2.3 Acoustic Model Training

First, for each of the two training sets an acoustic model was trained based on the MIDA feature representation. After training, the clean speech model contains $2,534$ tied states using $28,917$ Gaussians. Because the multi-condition training data is larger in size and richer in variation (the clean data plus its noisy variants), more tied states $(4,476)$ and slightly more Gaussians $(32,747)$ are retained by the decision tree inference algorithm. We have chosen to allow the multi-condition model to exploit the augmented data set to maximize its accuracy and hence not to constrain its size. For the MDT experiments, we have then created two new acoustic models in the PROSPECT domain by single-pass retraining. This retraining procedure has consisted of replacing the means and variances of the MIDA acoustic model with their PROSPECT counterparts.

For each of the two training sets an acoustic model was trained based on the MIDA feature representation. This is achieved in several steps. First, a set of 46 context-independent phone models plus four filler models and a silence model are trained using Viterbi re-estimation. Each HMM state has a set of up to 256 unshared Gaussians. Subsequently, a phonetic decision tree (cf. [11]) defines the $2,534$ tied states (for the clean training data) in the cross-word context-dependent models. The final acoustic models are obtained by allowing sharing across all Gaussians and subsequently retaining only those with maximum occupancy [12], resulting in an average of 96.4 Gaussians per state for the clean training data.

Because the multi-condition training data is larger in size and richer in variation (the clean data plus its noisy variants), more tied states $(4,476)$ and slightly more Gaussians $(32,747)$ are retained by the decision tree inference algorithm. We have chosen to allow the multi-condition model to exploit the augmented data set to maximize its accuracy and hence not to constrain its size. On average, 90.2 Gaussians per state are retained in this model. For the MDT experiments, we then create two new acoustic models in the PROSPECT domain by single-pass retraining. This retraining procedure consists of replacing the means and variances of the MIDA acoustic model with their PROSPECT counterparts.

7.5 MDT and Multi-Condition Training

In this section, we investigate the effectiveness of a classical MDT recognizer on speech recorded in real-world environments in combination with a multi-condition-trained acoustic model. To that end we determine the recognition accuracy using a number of different mask estimation methods: the semi-oracle mask (cf. Section 7.4.1.1), the VQ mask (cf. Section 7.4.1.2) and the SVM mask (cf. Section 7.4.1.3). Each mask estimation method is tested using two different acoustic models: a model trained on the clean speech training set and a model trained on the multi-condition training set. In order to give a baseline recognition result, we also discuss recognition experiments without using any additional noise-robust preprocessing other than what is inherent in the acoustic model and channel compensation. In Section 7.5.1 we describe the results of our experiments and in Section 7.5.2 we discuss the results.

7.5.1 Recognition Results

The speech recognition results from our experiments, depicted as word error rate (WER) as a function of SNR, are displayed in Fig. 7.1. The left pane corresponds

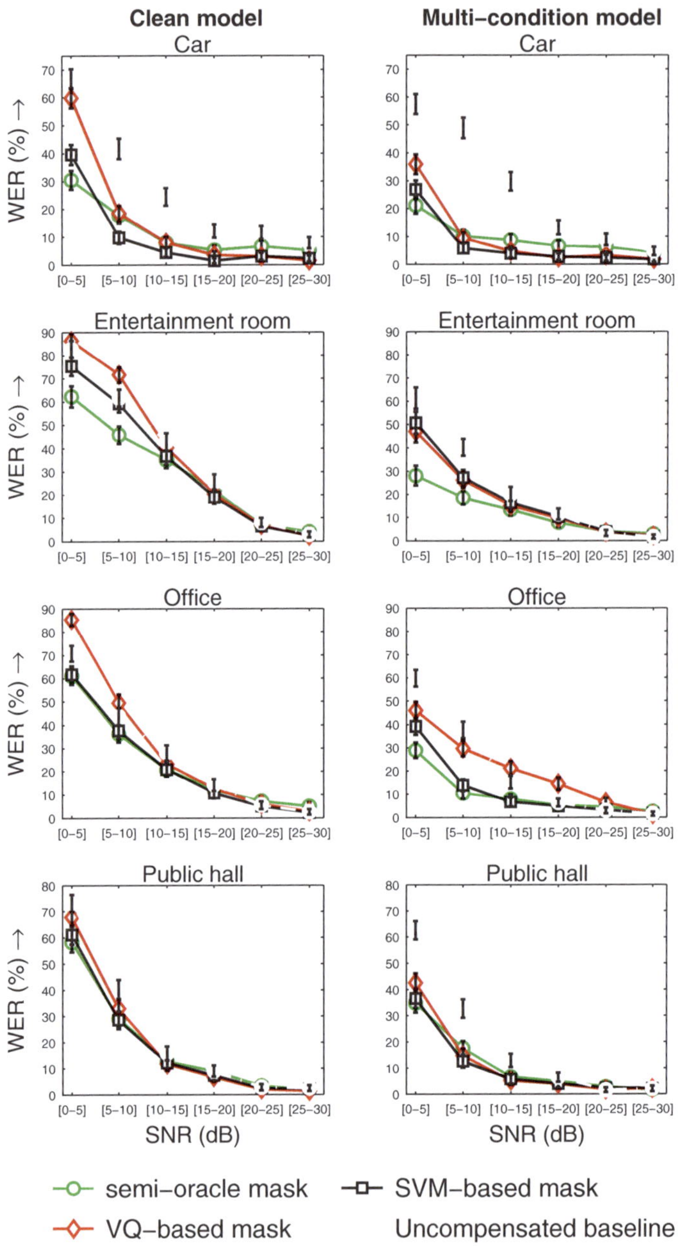

Fig. 7.1: Word error rate (WER) as a function of SNR displayed for clean speech (left) and multi-condition trained models (right). From top to bottom the rows represent different noisy environments, viz. car, entertainment room, office, and public hall. In each panel the results are shown for the uncompensated baseline, semi-oracle mask, VQ mask and SVM mask. Vertical bars around the data points indicate 95% confidence intervals

to recognition using the clean acoustic model while the right pane corresponds to recognition with the multi-condition acoustic model. From top to bottom, the respective rows represent different noise environments, viz. car, entertainment room, office and public hall, respectively.

Within each plot we display the results of the following methods:

- The uncompensated baseline, with no further noise robustness processing beyond what is inherent in the channel compensation or acoustic model training.
- MDT with the 'cheating' semi-oracle mask described in Section 7.4.1.1
- MDT with the VQ-based missing data mask described in Section 7.4.1.2
- MDT with the SVM-based missing data masks described in Section 7.4.1.3

In the top left plot of Fig. 7.1, corresponding to recognition in the car environment using a clean speech acoustic model, we can observe large differences between the approaches. It is apparent that all MDT approaches improve substantially over the baseline. The 'cheating' semi-oracle mask (SO) performs better than the other missing data masks in the $0-5$ dB range; at higher SNRs it is outperformed by the estimated masks. In the highest SNR bins the VQ mask and the SVM mask perform comparably, while at lower SNRs the SVM mask performs better than the VQ mask.

When using a multi-condition acoustic model (right pane, top row), the WERs at SNRs below 10 dB are much lower. Especially, the VQ mask benefits, achieving up to 24% lower WERs (absolute) in the $0-5$ dB range. The baseline, on the other hand, has performance gains of $\approx 10\%$ at lower SNRs. The ranking of the missing data methods is roughly the same as when using the clean acoustic model, although the differences between the methods are much smaller. Importantly, there is no significant difference in speech performance at high SNRs when using a multi-condition acoustic model: 1.5% WER (multi-condition model) vs. 1.8% WER (clean model) for the VQ mask.

Compared to the car environment, the entertainment room (second row of Fig. 7.1) is a more challenging environment: even the semi-oracle, which does best in this environment, has a 62.3% WER in the $0-5$ dB range when using a clean acoustic model. While SVM performs up to 8% better (absolute) than the baseline, the VQ mask performs worse than the baseline in the $0-10$ dB range. As before, there is no significant difference between the methods at high SNRs. When using a multi-condition acoustic model, there is again a substantial overall drop in WER. As in the car environment, the VQ mask benefits especially and now performs comparably to the SVM mask.

In the office and public hall environments (third and fourth rows in Fig. 7.1), we can observe many of the same trends described for the entertainment room environment. The SVM and semi-oracle masks perform comparably, and both methods perform up to $\approx 12\%$ better (absolute WER) than the baseline. The VQ mask does worse in the office environment but comparably in the public hall environment. When using a multi-condition acoustic model, overall WERs are much lower and the gap between the MDT and the baseline is much bigger.

7.5.2 Discussion

7.5.2.1 Effect of Multi-Condition Training

Comparing the uncompensated baseline scores in Fig. 7.1 with those obtained with the MDT, it is clear that the MDT manages to substantially improve upon the clean model baseline, reaching recognition accuracies comparable to that of the multi-condition model baseline. Moreover, comparing the left- and right-hand panes of Fig. 7.1 we can see that the use of a multi-condition acoustic model improves MDT recognition accuracy substantially, especially at lower SNRs. In fact, the performance increase when using a multi-condition model with MDT is much larger than that for the uncompensated baseline. From these results we conclude that our hypothesis — that 'reliable' features do not match the statistics of the clean acoustic model — is correct.

The mismatch of the reliable features with the acoustic model has two causes. First, mask estimation techniques make errors and sometimes unjustly label features dominated by noise as 'reliable', *false reliables*. On real-world data, conventional mask estimation techniques do not take into account the fact that the speech signal can be corrupted by channel and reverberation effects as well as by environmental noise. As a result, speech features that should have been masked because they are dominated by any of these effects are also unjustly labeled 'reliable'. The multi-condition model (partly) corrects false reliables because the acoustic model matches a much larger variance of the speech features. Second, even if all features which are not too heavily affected by additive noise or reverberation would correctly be labeled as 'reliable', the resulting features can, in contrast to artificially noisified data, still mismatch the speech distributions trained on close-talk channel data due to remaining microphone characteristics and reverberated speech energy. The multi-condition acoustic models will also compensate for this effect.

7.5.2.2 Mask Estimation Accuracy

The semi-oracle mask, a 'cheating' mask created with knowledge of all channels, in general hardly performs better than the estimated masks except in the lowest SNR bins. And even there, the differences are small, unlike the performance differences between estimated masks and 'true' oracle masks in experiments with artificially noisified data [14, 22]. While it cannot be established what the performance of a 'true' oracle mask would be, especially given the test/training mismatch issues discussed above, we can point out two shortcomings of the semi-oracle mask. First, the semi-oracle is derived under the assumption that the close-talk signal can be considered as 'clean' speech, resulting in all-reliable masks for close-talk channel speech. Even the features of close-talk channel speech, however, may be occasionally corrupted and should have been labeled 'unreliable'. Second, the AEC captures not only the transmission path between the close-talk microphone and a far-talk microphone, but also reverberation. The semi-oracle thus does not label speech dominated by reverberation of channel effects as unreliable. We will explore this effect in more detail in Section 7.6.

Compared to the other mask estimation methods, the VQ mask has a lower performance and in various conditions does even worse than the baseline. When using a mask derived with multi-condition acoustic models, however, the VQ mask performs much better. As with the semi-oracle mask, this is probably because the codebook is created using channel #1 speech, under the assumption that it contains clean speech. Because in the other channels the observed speech results in harmonic decomposition components that will often be poorly described by the codebook, many reliable features are likely to get mislabeled as unreliable or vice versa.

The SVM mask generally performs very well, often performing comparably to the semi-oracle mask. Its performance is a testament to the generalizability of SVM-based machine learning; after all, the mask estimation was trained using the Aurora 4 corpus. The use of this corpus, which contains noisified Wall Street Journal (WSJ) speech, means there is a mismatch in noise type, language and content. Still, it generally performs better than the VQ mask, even though they share many features such as the harmonic decomposition. Moreover, the SVM mask does not require the tuning of a threshold parameter. A downside of the SVM method, not discussed in this chapter, is its high computational cost.

7.6 MDT and Dereverberation

In order to investigate whether MDT can be combined with dereverberation and to what extent reverberation can affect the recognition performance of our MDT recognizer, we performed two experiments that will be described in more detail in this section.

In the first experiment, we determined to what extent MDT can be used to improve recognition accuracy of artificially reverberated speech. In [8, 21] it was shown that treating features which are dominated by reverberation as unreliable can be quite effective when using clean speech models. Here, we investigate to what extent this approach can be combined with the use of multi-condition models to provide robustness for the artificial reverberation. Using artificially reverberated speech constructed by filtering clean speech with a known room impulse response filter, we created an oracle mask by considering the difference between the reverberated and the non-reverberated versions of the speech as 'noise'. Using the conventional mask definition of (7.1) and the previously trained clean speech models and multi-condition models (cf. Section 7.4.2.3), we investigated to what extent the recognition accuracy of our MDT recognizer can be improved using this oracle mask, which labels features unreliable when they are dominated by reverberation.

In the second experiment we applied the insights gained from the first experiment to real-world data. Under the assumption that the channel #1 data has negligible reverberation and that the reverberation effects become more pronounced in microphone channels #2 – #4, we use the estimated room impulse response filter of the

AEC to estimate the underlying non-reverberated speech signal. Thus, we can construct, similarly to the artificially reverberated speech, an improved (semi-)oracle mask that not only labels features unreliable when they are affected too severely by environmental noise, but also when they are dominated by reverberation. By comparing improvements in recognition accuracy, both for acoustic models that were trained on clean speech and for multi-condition models, we try to estimate an upper and lower bound on the impact reverberation can have in our MDT recognition experiments on real data.

7.6.1 Experimental Setup

7.6.1.1 MDT for Dereverberation of Artificially Reverberated Data

We created artificially reverberated speech as follows. First, we measured two room impulse responses (RIRs) using a microphone at 261 cm from the speaker. The room of 36 m^3 has curtains on all walls. In the first RIR, the curtains were closed ($T_{60} = 140$ ms), while in the second RIR they were open ($T_{60} = 250$ ms). The RIRs were measured with Gaussian white noise excitation using a least-squares estimation approach. The resulting FIR filter had a length of 125 ms (2001 coefficients at 16 kHz). Next, these two FIR filters were applied to the channel #1 utterances from all four environments in the $20 - 25$ dB bin of our SPEECON and SpeechDat-Car data. This results in two new artificially reverberated test sets, each containing 1236 utterances.

Subsequently, we created a delayed version of the original signal by filtering it with the same FIR, but with the tail of the filter coefficients (representing the echos from the non-direct path) zeroed out. This was done by manually setting all FIR filter coefficients of the AEC filter beyond 3 ms after the first peak to zero. Then, we calculated the residual between the delayed and the reverberated channel #1 data. Using the residual as the 'noise' and with the delayed channel #1 data taking the place of the clean speech, we applied (7.1) to obtain our oracle mask. This oracle mask was then used in the MDT recognizer to decode the artificially reverberated signal.

In order to obtain the optimal oracle mask, we performed experiments with a large number of SNR thresholds. The results that will be presented pertain to the oracle masks obtained with the threshold that resulted in the best accuracy. Tuning the threshold on the test data is justified by the fact that with these artificial test data we are only interested in an estimate of the upper bound of the improvement that can be achieved.

As before, all experiments are done both with the acoustic models trained on clean speech and multi-condition speech. No new models were trained for the experiments described here.

7.6.1.2 MDT Using a Reverb-Masking Semi-Oracle Mask on Real-World Data

In this experiment, we try to estimate which spectro-temporal features in the SPEECON and SpeechDat-Car data are associated with speech energy that reaches the microphone via a direct path rather than being dominated by first and higher order reflections or additive noise. To that end, we estimate a dereverberated version of the clean speech signal by modifying the room impulse response filter in the AEC described in Section 7.4.1.1. Analogously to the experiments on artificially reverberated data, the dereverberated clean speech estimate is obtained by manually setting all FIR filter coefficients of the AEC filter to zero beyond 3 ms after the first peak. Thus, the resulting FIR filter ideally should only represent the transfer function of the direct path between the two microphones, discarding any reflections caused by the room acoustics.

Next, we consider the residual of the features of the observed signal and the non-reverberated clean speech estimate as noise. We use (7.1) to label only those speech features reliable which can be assumed not to be excessively affected by additive noise or reverberation. This improved semi-oracle mask, which we will denote by the term reverb-masking semi-oracle (RMSO) mask, is then used to decode the original noisy speech features. This allows us to test whether or not in our framework masking reverberated speech can improve recognition accuracy as it did in [8, 21].

As before, only the thresholds with the best accuracy are shown and all experiments are done both with the acoustic models trained on clean speech and with multi-condition models.

7.6.2 Results and Discussion

7.6.2.1 MDT for Dereverberation of Artificially Reverberated Data

In Fig. 7.2 we can observe a clear trend that on artificially reverberated speech, masking the reverberation improves performance. This holds both for the clean speech model (left panel) and for the multi-condition model (right panel). Moreover, comparing the overall performance of the multi-condition model and the clean model in the various reverb conditions clearly shows that the multi-condition model is the most robust against the artificial reverberation. Surprisingly, the performance for the no-reverb condition shows a similar trend: performance is better with the multi-condition model than with the clean model. Although we do not have a solid explanation for why the apparent mismatch between the current test set and the clean training data would be greater than that with the multi-condition training data, we take this observation as yet another illustration that recognition with a clean speech model is sensitive to even the slightest training/test mismatch and that such a mismatch can often be compensated for by using a multi-condition model.

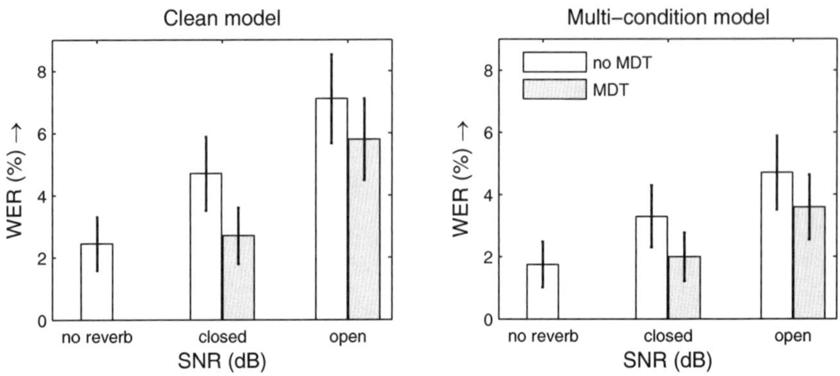

Fig. 7.2: Word error rate (WER) for recognition with clean (left) and multi-condition models (right). Vertical bars around the maxima indicate 95% confidence intervals. In each bar graph results are shown for the non-reverberated 'clean' test set, the reverberated test set with closed curtains and the reverberated test set with open curtains

In the case of the reverberation caused by an RIR of a room with closed curtains, the oracle missing data mask is able to completely compensate for the performance loss due to reverberation, yielding a performance which is indistinguishable from the no-reverb condition. This holds true both for the clean speech model and the multi-condition model.

In summary, these results suggest that loss in recognition accuracy due to reverberation can be alleviated by a multi-condition acoustic model as well as by a suitable missing data mask which labels the features affected by reverberation unreliable. In fact, it seems that these two approaches are to some extent complementary and can be combined to combat reverberation.

7.6.2.2 MDT Using a Reverb-Masking Semi-Oracle Mask on Real-World Data

In Fig. 7.3 we observe that when using the clean acoustic model, masking the reverberation generally increases performance substantially. Especially in the entertainment room and public hall environments, performance differences can be up to 23% (absolute WER). From this we conclude that we were successful in estimating which features were excessively affected by reverberation and therefore should be labeled unreliable, and were thus better able to approximate the 'true' oracle mask (cf. Section 7.5.2.2). Moreover, these results show that MDT can be used to compensate for both noise and reverberation at once by using a suitably chosen missing data mask.

In none of the environments, however, does the reverb-masking semi-oracle mask (RMSO) in combination with a clean acoustic model perform better than the original semi-oracle mask (SO) in combination with a multi-condition acoustic model.

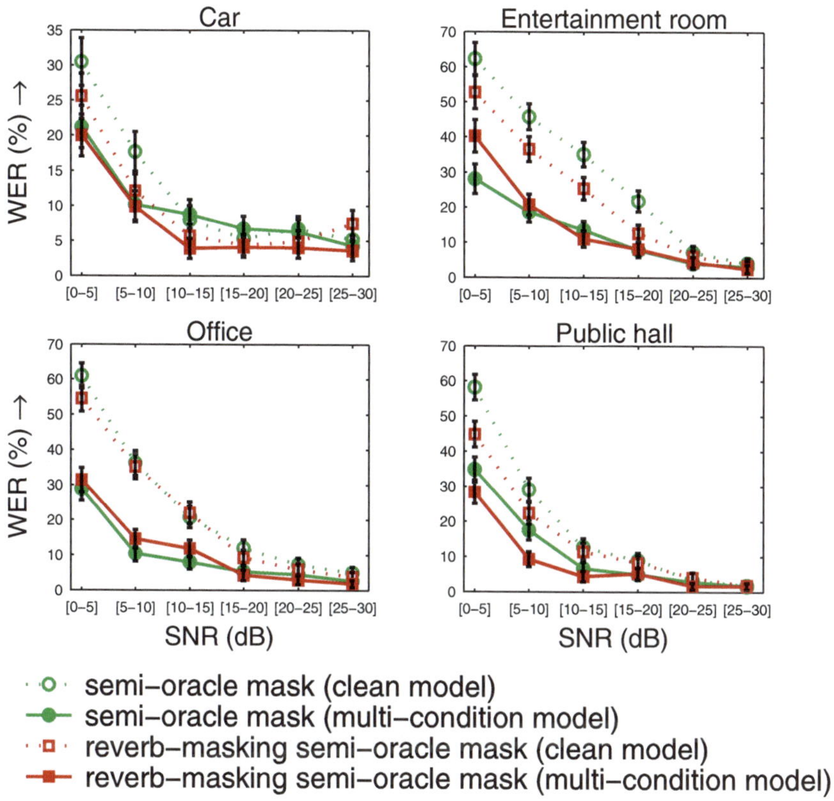

Fig. 7.3: Word error rate (WER) for four different noise environments. In each panel the results are shown for recognition with the semi-oracle masks in combination with the clean acoustic model, semi-oracle mask with the multi-condition acoustic model, and reverb-masking semi-oracle mask with the clean or multi-condition acoustic model. Vertical bars around the data-points indicate 95% confidence intervals

This means that masking the reverberation effects (using the modified RIR filter approach) does not account for the entire mismatch between the noisy speech features and the clean acoustic model. When using the RMSO mask in combination with a multi-condition model, the results vary between environments. In the entertainment room and office environments, the performance is worse than when using the SO mask. This implies that in these environments, the reverberation is already fully compensated for by the multi-condition model, and masking more features only results in masking features that are useful for imputation or recognition.

In the car environment, the RMSO and SO masks perform comparably at high and low SNRs, with the RMSO mask performing better in the $5-25$ dB SNR range. In the public hall environment, which is the most reverberant environment, masking the reverberation lowers the WER by $\approx 7\%$ (absolute) at SNRs < 15 dB. It seems that in the public hall environment, the impact of reverberation is substantial in the channels that contribute most to the lower SNR bins (i.e., channels #3 and #4).

From these results we can roughly estimate an upper bound on the impact of reverberation on our recognition results by taking the difference between the WER obtained with the worst performing method (SO mask with a clean acoustic model) and that obtained with the best performing method (usually RMSO mask when using the multi-condition acoustic model). This would imply that at most 10% to 35% of the errors in the lower SNR bins, depending on the environment, can be attributed to reverberation.

For some environments, we can also try to establish a lower bound by taking the difference in performance between the RMSO mask and the original SO mask in combination with the multi-condition acoustic model. The rationale behind this is that the multi-condition acoustic model already accounts for various sources of variation, to some extent including reverberation. So, if explicitly masking reverberation helps, it must be due to the reverberation that multi-condition models did *not* capture. Following this approach, we might conclude that in the public hall environment, at least about 7% WER loss (absolute) at SNRs under 15 dB is due to reverberation.

7.7 MDT and Feature Enhancement

In the sections above we successfully combined MDT with conventional noise robustness techniques, such as multi-condition training and dereverberation. In Section 7.5 it was shown that replacing a clean speech model by a multi-condition acoustic model dramatically improves results, confirming the hypothesis that the 'reliable' features no longer match the clean speech distributions. We argued that the improvement from using a multi-condition model is partly caused by a greater robustness against mask estimation errors. Multi-condition training material, however, is costly to acquire and the computational effort of training multi-condition acoustic models is substantial. In this section, we explore the combination of MDTs with conventional feature enhancement techniques. Our aim is to reduce the mismatch between reliable features and the clean acoustic model, and to improve mask estimation on real-world recorded speech.

First, as an alternative to multi-condition training, we try to reduce the reliable feature mismatch by applying feature enhancement to the noisy features prior to missing data imputation. In doing so, we keep the estimation procedures for the missing data masks unaltered. For unreliable features, feature enhancement may also be beneficial since the unreliable features are used as an upper bound during imputation. Feature enhancement yields tighter imputation bounds [9]. Here, care must be taken since the uncertainty on the enhanced speech energy increases as the underlying speech signal becomes weaker and the imputation bound may become inaccurate. We therefore opt for spectral subtraction (SS) [18] as a feature enhancement method since the amount of noise suppression is easily controlled.

Secondly, we explore whether mask estimation on real-world data, which was argued to be more difficult than in artificially corrupted databases (cf. Section 7.2.1), can be improved by applying feature enhancement on the speech features used in mask estimation while not preprocessing (NP) the features used for imputation and recognition. The ratio behind this is that on real-world data, the features used for mask estimation mismatch the clean speech features used to train or tune the mask estimation technique, just as the close-talk channel acoustic models mismatch the observed features in recognition. In our experiments, we used the SS feature enhancement technique with the VQ-based mask estimation technique. The VQ mask was chosen because recreating the codebook was less computationally demanding than retraining the SVM mask estimators used for the SVM mask.

Finally, we combine the two approaches described above, leading to four MDT scenarios: whether SS is applied to the features used in mask estimation (mSS vs. mNP), and whether SS is applied to the features used in recognition (fSS vs. fNP). These approaches are summarized below:

mNPfNP neither the VQ-based mask estimation nor the recognizer features are preprocessed with spectral subtraction. This is the VQ-mask result in Section 7.5.1.

mSSfNP only the VQ-based mask estimation is preprocessed with spectral subtraction; the recognizer features are not preprocessed.

mNPfSS features for mask estimates are not preprocessed; only the recognizer features are preprocessed with spectral subtraction.

mSSfSS both the VQ-based mask estimation and the recognizer features are preprocessed with spectral subtraction.

SS SS feature enhancement is used without MDT.

AFE the ETSI AFE front-end is used without MDT.

The last two configurations serve as a baseline. The SS scenario allows us to evaluate the quality of the feature enhancement method. The advanced front-end feature extraction (AFE) baseline is included in order to let us compare our approach to what is currently regarded as a very good feature enhancement method (though it cannot be tuned to control the amount of noise suppression, needed for combination with MDT).

7.7.1 Experimental Setup

7.7.1.1 Spectral Subtraction

The basic principle of spectral subtraction (SS) is to provide an estimate of clean speech features (feature enhancement) by subtracting a direct estimate of the magnitude spectrum of noise from the noisy speech. In our approach, spectral subtraction was done using the multi-band spectral subtraction approach described in [18]. In summary, spectral subtraction is performed independently in each frequency band. The first 20 non-speech frames (as decided by the VAD in Section 7.4.1.2) are used to provide a noise estimate, which is then assumed constant throughout the utterance. Negative values in the enhanced features are floored to the noisy spectrum using a flooring parameter β set to 0.1 (cf. [18]). Other parameter settings are the same as in [18].

For experiments with SS, new acoustic models are generated through retraining with aligned MIDA and cepstral feature streams. This retraining procedure consists of replacing the means and variances of the MIDA acoustic model with their cepstral counterparts.

7.7.1.2 AFE

The AFE algorithm proposed by ETSI [2] is based on a two-stage Wiener filtering noise reduction. Since the parameters of the two Wiener filters are updated on a frame-by-frame basis the ETSI AFE can deal with dynamically changing noise. After estimating the linear spectrum of each frame, the power spectral density is smoothed along the time axis. A voice activity detector (VAD) determines whether a frame contains speech or background noise; the estimated spectrum of both speech and noise are used to calculate the frequency domain Wiener filter coefficients. Frames labeled as non-speech by the VAD are dropped. The AFE produces cepstral features which are directly used for recognition.

As for the experiments with SS, new acoustic models are generated for AFE through retraining with aligned MIDA and AFE feature streams. When retraining, we do not drop the non-speech frames in order to properly align the MIDA and AFE features. The resulting AFE model is then updated using one pass of Viterbi training. Finally, the AFE model is updated by another pass of Viterbi retraining on AFE features with frame dropping enabled.

7.7.1.3 VQ Mask Estimation After SS

The VQ codebook for the mSSfNP and mSSfSS experiments was learned from the same close-talk channel data used in Section 7.4.1.2 to which SS was applied. The VAD and the harmonic decomposition method were applied to the spectral-subtracted noisy test utterances. The mask threshold was again optimized over the complete test set and set to $\theta = 8$ dB.

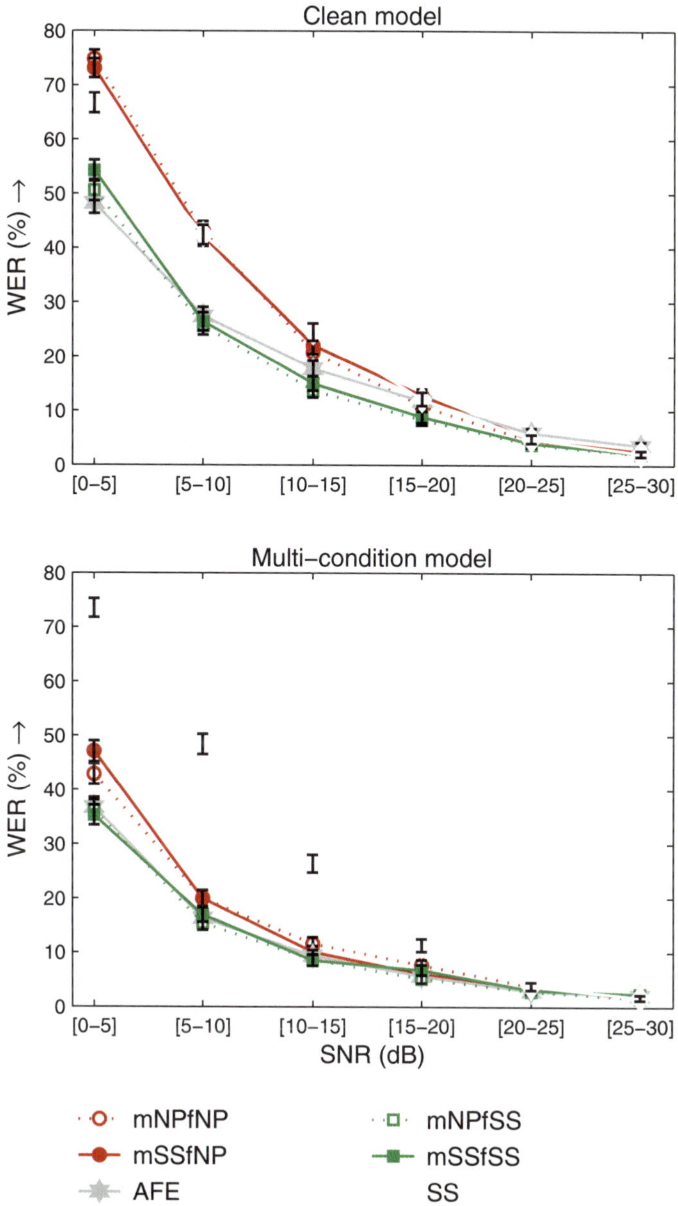

Fig. 7.4: Word error rate (WER), averaged over the four noise environments, is displayed as a function of SNR for the clean (top) and multi-condition (bottom) acoustic models. Vertical bars around the data points indicate 95% confidence intervals. In each figure we display the results of spectral subtraction (SS), AFE, and the four combinations of applying SS to the VQ mask features (m) or to the features used in recognition (f): mNPfNP, mSSfNP, mNPfSS, mSSfSS, with NP indicating the use of the original noisy features

7.7.2 Results and Discussion

7.7.2.1 MDT Versus Feature Enhancement

Comparing the results of the SS and AFE baseline feature enhancement scores with the MDT recognition scores, it becomes apparent there is a vast difference between the methods. SS, on the one hand, performs the worst of all methods, often worse than the uncompensated baseline in Fig. 7.1 in Section 7.5.1. The AFE, on the other hand, has a performance that is among the best of all methods. The low performance of the SS method is misleading, however, since it was set to be conservative in its noise suppression. This was necessary to prevent the MDT method, in combination with which it is used, from having upper imputation bounds that are too tight.

The competitive AFE performance underlines that MDT on real-world data is difficult, since in previous work on artificially corrupted data our MDT framework was superior to AFE. While never compared directly, this can be seen for Aurora 4 by comparing the AFE recognition accuracies in [29] with the MDT results in [35], and for Aurora 2 by comparing the AFE scores in [33] with the MDT results in [35].

7.7.2.2 Spectral Subtraction to Improve Mask Estimation

First, we compare recognition performance when doing recognition on the original noisy features in combination with the unmodified VQ mask (mNPfNP) and the mask in which SS is applied to the noisy features used in mask estimation (mSSfNP). We observe in Fig. 7.4 that applying spectral subtraction to improve mask estimation does not lead to significant improvements. When comparing the two MDT approaches in which spectral subtraction is also applied to the features used in recognition, i.e., mNPfSS and mSSfSS, we again observe no significant improvement when applying spectral subtraction to the features used in mask estimation. If anything, there is a trend towards higher WERs, especially at lower SNRs and when using the multi-condition acoustic model.

The reason for this failure to improve mask estimation may stem from the fact SS mostly compensates for stationary noise, something which is already covered to some degree in the VQ mask estimation because a noise tracker is employed that uses the long-term energy minimum (cf. Section 7.4.1.2). In other words, SS may simply not be powerful enough a technique to reduce a substantial part of the mask estimation errors which are likely due to non-stationary noises. Another possibility is that the harmonic decomposition underlying the VQ-based method fails after speech has been processed by SS because the speech now contains musical noise. Although the VQ codebook is trained on speech processed with SS, it is conceivable that not all such errors are covered ('learned') by the codebook.

7.7.2.3 Spectral Subtraction to Improve MDT Recognition

Comparing the results of the MDT approaches in which spectral subtraction is applied to the features used in recognition (mNPfSS and mSSfSS) with those in which the recognizer uses the original noisy speech features (mNPfNP and mSSfNP) in Fig. 7.4, we observe a substantial decrease in WER when using SS. With a clean acoustic model, the use of SS improves the results significantly at SNRs < 15 dB, with differences as large as 18% (absolute WER) in the $0-5$ dB SNR range. When using a multi-condition acoustic model, SS improves results at SNRs < 10 dB, with differences up to 9% (absolute WER).

The results show that combining SS and MDT can be beneficial if SS is used to modify the features used in recognition. Moreover, our results show that multi-condition training and SS are complementary, making it advantageous to combine the two approaches. With the multi-condition model the impact of applying SS on recognition performance is smaller (both in absolute and relative terms).

As discussed in Section 7.5.2, a mismatch exists between the reliable features (or false reliables) and the clean acoustic model. It is likely that the improvement in recognition performance is at least partly due to SS reducing this mismatch. The smaller impact of SS when using a multi-condition model could be explained by assuming the multi-condition model already compensates for part of the test/training mismatch. As mentioned above, however, the application of SS also results in tighter imputation bounds when applied to the unreliable features, which may also improve recognition accuracy. Our experiments do not allow us to investigate the relative contribution of these two factors when using SS; further research is needed for that. One way to do that would be to only apply SS to either reliable or unreliable features.

The improvements found when applying SS raise the question about the extent to which recognition with other types of estimated masks, such as the SVM mask described in Section 7.4.1.3, can benefit from feature enhancement. The features used for mask estimation in the SVM mask largely overlap with those used in VQ-based mask estimation. Therefore, given the results in Section 7.7.2.2, it is doubtful whether SVM mask estimation improves after applying SS. Given the success of applying SS to the features used in recognition, however, it seems likely that SVM-based mask estimation will also benefit, and without the additional cost of retraining the SVM and re-estimating the missing data masks.

The AFE front-end, used as a baseline in this chapter, was not used in combination with MDT due to its inflexibility. Its competitive performance, however, merits the question about whether a feature enhancement technique based on, or similar to, the Wiener filtering used in AFE can be used for combination with MDT. A very similar combination of techniques is described in [3].

7.8 General Discussion and Conclusions

In this chapter, we have investigated the performance of a missing data recognizer on speech recorded in real-world environments. We hypothesized that on real-world speech, which is corrupted not only by noise, but also by speaker, reverberation and channel effects, the 'reliable' features no longer match an acoustic model trained on clean speech. We investigated the validity of this hypothesis and explored to what extent performance can be improved by combining MDT with three conventional techniques, viz. multi-condition training, dereverberation and feature enhancement. Using a multi-condition-trained acoustic model in combination with MDT, we confirmed the hypothesis and showed that recognition accuracy improves substantially in all noise environments and at all SNR-levels. When comparing the performance of conventional mask estimation techniques, we found that even a 'cheating' semi-oracle missing data mask did not perform better than VQ- or SVM-based estimated masks. We argued this was at least partly due to the semi-oracle missing data mask not being designed to label speech features dominated by reverberation as unreliable.

In a second experiment (cf. Section 7.6), we combined MDT with dereverberation by doing recognition with the reverberated part of speech labeled 'unreliable', both on real-world recordings and on artificially reverberated speech. The experiment with artificially reverberated speech confirmed previous findings that masking reverberation improves recognition accuracy, but also revealed that the multi-condition trained acoustic model is intrinsically more robust against reverberation. To some degree, these two methods can work together for an even better performance. The experiment on real-world recordings showed that the semi-oracle mask also improved when the reverberant part of speech was labeled unreliable, and thus that with a suitable missing data mask, MDT can compensate for noise and reverberation at once. Finally, the experiments showed that reverberation has a major impact on the recognition performance in far-talk channels.

Third, we did an experiment (cf. Section 7.7) in which we combined MDT with feature enhancement techniques. We investigated whether spectral subtraction (SS) could reduce the mismatch of reliable features to such an extent that it might serve as an alternative to multi-condition training. We also investigated whether spectral subtraction could improve the performance of VQ-based missing data mask estimation, which was found to be unexpectedly poor when using clean acoustic models. The application of spectral subtraction to the features used in VQ mask estimation did not improve results, but the application of SS to the features used in recognition proved to be quite successful: WERs decreased when using a clean as well as a multi-condition model. We argued that this is either due to a reduction of the test/training mismatch in the reliable features, or due to tighter imputation bounds on the unreliable features. Finally, even though in previous work MDT was shown to be superior to the ETSI advanced front-end (AFE) on artificially corrupted speech, we could show only a small advantage of MDT in the case where multi-condition training is not an option, while MDT performs comparably with the AFE under a multi-condition scenario.

From our findings we conclude that two issues make applying MDT on real-world speech difficult. The first issue is that one of the assumptions underlying MDT, viz. that reliable features remain uncorrupted, can be violated. The second issue is that conventional mask estimation techniques are not able to deal with the the fact that real-world speech can be affected not only by environmental noise, but also by effects such as reverberation. In this chapter we showed that the first issue can be dealt with to some degree with conventional noise reduction techniques such as multi-condition training and feature enhancement. With even 'cheating' missing data performing only marginally better than estimated missing data masks, it is clear that in order to deal with the second issue, (much) more effort is needed to improve mask estimation techniques.

Based on our results, however, it is not at all obvious whether MDT can beat a well-designed feature enhancement technique such as the ETSI advanced front-end, which operates at a fraction of the computational cost. Yet, the fact that our work shows that MDT can be combined with conventional noise robustness techniques and since mask estimation allows us to integrate additional knowledge sources (e.g., harmonicity), or to use classifiers that do not integrate easily in HMMs (e.g., SVM-based classifiers), means there is still potential for improving the results, for example, by combination with the aforementioned feature enhancement technique.

7.9 Acknowledgements

This research was financed by the MIDAS project of the Nederlandse Taalunie under the STEVIN programme. The research of Maarten Van Segbroeck was financed by the Institute for the Promotion of Innovation through Science and Technology in Flanders (IWT-Vlaanderen). We acknowledge Lou Boves for his help with the manuscript and Toon van Waterschoot for providing the PEM-AFROW code and the room impulse response estimates.

References

1. Spraak: Speech processing, recognition and automatic annotation kit. Website (1996). http://www.spraak.org/
2. ETSI standard document: Speech processing, transmission and quality aspects (STQ); distributed speech recognition; advanced front-end feature extraction algorithm; ES 202 050 V1.1.5 (2007)
3. Astudillo, R.F., Kolossa, D., Mandelartz, P., Orglmeister, R.: An uncertainty propagation approach to robust ASR using the ETSI advanced front-end. IEEE Journal of Selected Topics in Signal Processing **4**, 824 833 (2010)
4. C. Cortes, C., Vapnik, V.: Support-vector networks. Machine Learning pp. 273–297 (1995)
5. Chang, C., Lin, C.: LIBSVM: A library for support vector machines (2001)
6. Cooke, M., Green, P., Crawford, M.: Handling missing data in speech recognition. In: ICSLP-1994, pp. 1555–1558 (1994)
7. Cooke, M., Green, P., Josifovksi, L., Vizinho, A.: Robust automatic speech recognition with missing and unreliable acoustic data. Speech Communication **34**, 267–285 (2001)
8. Delcroix, M., Nakatani, T., Watanabe, S.: Static and dynamic variance compensation for recognition of reverberant speech with dereverberation preprocessing. IEEE Transactions on Audio, Speech, and Language Processing **17**(2), 324–334 (2009)
9. Demange, S., Cerisara, C., Haton, J.P.: Accurate marginalization range for missing data recognition. In: Proc. of Interspeech, pp. 27–31 (2007)
10. Demuynck, K., Duchateau, J., Compernolle, D.V.: Optimal feature sub-space selection based on discriminant analysis. In: Proc. of European Conference on Speech Communication and Technology, vol. 3, pp. 1311–1314 (1999)
11. Duchateau, J., Demuynck, K., Compernolle, D.V.: Fast and accurate acoustic modelling with semicontinuous HMMs. Speech Communication **24**(1), 5–17 (1998)
12. Duchateau, J., Demuynck, K., Wambacq, D.V.C.P.: Improved parameter tying for efficient acoustic model evaluation in large vocabulary continuous speech recognition. In: Proc. ICSLP, vol. V, pp. 2215–2218. Sydney, Australia (1998)
13. Gemmeke, J.F., Wang, Y., Van Segbroeck, M., Cranen, B., Van hamme, H.: Application of noise robust MDT speech recognition on the SPEECON and SpeechDat-Car databases. In: Proc. of Interspeech. Brighton, UK (2009)
14. Hirsch, H., Pearce, D.: The aurora experimental framework for the performance evaluation of speech recognition systems under noisy conditions. In: Proc. of ISCA ASR2000 Workshop, pp. 181–188 (2000)
15. Huang, X., Acero, A., Hon, H.W.: Spoken Language Processing: A Guide to Theory, Algorithm and System Development. Prentice Hall (2001)
16. Iskra, D., Grosskopf, B., Marasek, K., van den Heuvel, H., Diehl, F., Kiessling, A.: SPEECON — speech databases for consumer devices: Database specification and validation. In: Proc. of LREC, pp. 329–333 (2002)
17. Josifovski, L., Cooke, M., Green, P., Vizinho, A.: State based imputation of missing data for robust speech recognitionand speech enhancement. In: Proc. of Eurospeech (1999)
18. Kamath, S., Loizou, P.: A multi-band spectral subtraction method for enhancing speech. In: Proc. of ICASSP (2002)
19. Kim, W., Hansen, J.H.L.: Time-frequency correlation-based missing-feature reconstruction for robust speech recognition in band-restricted conditions. IEEE Transactions on Audio, Speech, and Language Processing **17**(7), 1292–1304 (2009)
20. Martin, R.: Noise power spectral density estimation based on optimal smoothing and minimum statistics. IEEE Transactions on Speech and Audio Processing **9**, 504–512 (2001)
21. Palomäki, K.J., Brown, G.J., Barker, J.: Techniques for handling convolutional distortion with "missing data" automatic speech recognition. Speech Communication **43**, 123–142 (2004)
22. Parihar, N., Picone, J.: An analysis of the Aurora large vocabulary evaluation. In: Proc. of Eurospeech, pp. 337–340 (2003)

23. Raj, B., Seltzer, M., Stern, R.: Reconstruction of missing features for robust speech recognition. Speech Communication **43**, 275–296 (2004)
24. Raj, B., Singh, R., Stern, R.: Inference of missing spectrographic features for robust automatic speech recognition. In: Proc. of International Conference on Spoken Language Processing, pp. 1491–1494 (1998)
25. Raj, B., Stern, R.: Missing-feature approaches in speech recognition. Signal Processing Magazine **22**(5), 101–116 (2005)
26. Ramírez, J., Górriz, J., Segura, J., Puntonet, C., Rubio, A.: Speech/non-speech discrimination based on contextual information integrated bispectrum LRT. In: IEEE Signal Processing Letters (2006)
27. Remes, U., Palomäki, K.J., Kurimo, M.: Missing feature reconstruction and acoustic model adaptation combined for large vocabulary continuous speech recognition. In: Proc. of EUSIPCO (2008)
28. Seltzer, M., Raj, B., Stern, R.: A Bayesian classifier for spectrographic mask estimation for missing feature speech recognition. Speech Communication **43**, 379–393 (2004)
29. Stouten, V.: Robust automatic speech recognition in time-varying environments. Ph.D. thesis, K. U. Leuven (2006)
30. van den Heuvel, H., Boudy, J., Comeyne, R., Communications, M.N.: The SpeechDat-Car multilingual speech databases for in-car applications. In: Proc. of the European Conference on Speech Communication and Technology, pp. 2279–2282 (1999)
31. Van hamme, H.: Robust speech recognition using missing feature theory in the cepstral or LDA domain. In: Proc. of European Conference on Speech Communication and Technology, pp. 3089–3092 (2003)
32. Van hamme, H.: Prospect features and their application to missing data techniques for robust speech recognition. In: Proc. of Interspeech, pp. 101–104 (2004)
33. Van hamme, H.: Robust speech recognition using cepstral domain missing data techniques and noisy masks. In: Proc. of ICASSP, vol. 1, pp. 213–216 (2004)
34. Van hamme, H.: Handling time-derivative features in a missing data framework for robust automatic speech recognition. In: Proc. of ICASSP (2006)
35. Van Segbroeck, M.: Robust large vocabulary continuous speech recognition using missing data techniques. Ph.D. thesis, K. U. Leuven (2010)
36. Van Segbroeck, M., Van hamme, H.: Handling convolutional noise in missing data automatic speech recognition. In: Proc. of ICASSP, pp. 2562–2565 (2006)
37. Van Segbroeck, M., Van hamme, H.: Vector-quantization based mask estimation for missing data automatic speech recognition. In: Proc. of ICSLP, pp. 910–913 (2007)
38. van Waterschoot, T., Rombouts, G., Verhoeve, P., Moonen, M.: Double-talk-robust prediction error identification algorithms for acoustic echo cancellation. IEEE Transactions on Signal Processing **55**(3), 846–858 (2007)

Chapter 8
Conditional Bayesian Estimation Employing a Phase-Sensitive Observation Model for Noise Robust Speech Recognition

Volker Leutnant and Reinhold Haeb-Umbach

Abstract In this contribution, conditional Bayesian estimation employing a phase-sensitive observation model for noise robust speech recognition will be studied.

After a review of speech recognition under the presence of corrupted features, termed uncertainty decoding, the estimation of the posterior distribution of the uncorrupted (clean) feature vector will be shown to be a key element of noise robust speech recognition.

The estimation process will be based on three major components: an a priori model of the unobservable data, an observation model relating the unobservable data to the corrupted observation and an inference algorithm, finally allowing for a computationally tractable solution. Special stress will be laid on a detailed derivation of the phase-sensitive observation model and the required moments of the phase factor distribution. Thereby, it will not only be proven analytically that the phase factor distribution is non-Gaussian but also that all central moments can (approximately) be computed solely based on the used mel filter bank, finally rendering the moments independent of noise type and signal-to-noise ratio.

The phase-sensitive observation model will then be incorporated into a model-based feature enhancement scheme and recognition experiments will be carried out on the Aurora 2 and Aurora 4 databases. The importance of incorporating phase factor information into the enhancement scheme is pointed out by all recognition results. Application of the proposed scheme under the derived uncertainty decoding framework further leads to significant improvements in both recognition tasks, eventually reaching the performance achieved with the ETSI advanced front-end.

Volker Leutnant
Department of Communications Engineering, University of Paderborn, Warburger Straße 100, 33098 Paderborn, Germany, e-mail: `leutnant@nt.uni-paderborn.de`

Reinhold Haeb-Umbach
Department of Communications Engineering, University of Paderborn, Warburger Straße 100, 33098 Paderborn, Germany, e-mail: `haeb@nt.uni-paderborn.de`

D. Kolossa and R. Haeb-Umbach (eds.), *Robust Speech Recognition of Uncertain or Missing Data: Theory and Applications*, DOI 10.1007/978-3-642-21317-5_8,
© Springer-Verlag Berlin Heidelberg 2011

8.1 Introduction

Current state-of-the-art automatic speech recognition (ASR) systems are embedded into a sound statistical framework. Though the used models and the search algorithm itself are subject to severe approximations, such systems perform well when training and test data are obtained under the same (matched) conditions. However, these systems are susceptible to a mismatch of training and testing conditions — noticeable through a quick degradation of the recognition performance.

Approaches to compensate for the mismatch and thus to make the recognizer more robust usually fall into one of the three categories [17]: (i) finding features that are inherently robust to changes in the acoustic environment, (ii) compensating for the detrimental effect of the distortion on the features (so-called *front-end* methods) and (iii) modifying the acoustic models used in the recognizer to better match the incoming distorted features (so-called *back-end* methods).

Unifying (ii) and (iii) in a statistical framework (so-called uncertainty decoding) has intensively been investigated [7, 10, 17, 22] and the posterior distribution of the uncorrupted speech feature vector has been found to be the key element to improving the recognition performance under mismatched conditions.

In recent years, the application of *model-based feature enhancement* to this problem has gained considerable interest. Thereby, the uncorrupted (clean) feature vectors are estimated from the corrupted observations based on a priori models of the clean speech feature vectors and the noise-only feature vectors and an observation model relating the two to the corrupted observation. However, besides the estimates of the uncorrupted feature vectors, a measure of their reliability (the estimation error covariances) is also provided by the feature enhancement scheme, enabling the application of uncertainty decoding. While some researchers rely on a Gaussian mixture model (GMM) as an a priori model (e.g., [22, 29]), a switching linear dynamic model (SLDM) is employed in this contribution.

As the relationship between the clean speech and noise-only feature vectors, $\mathbf{x}_t$ and $\mathbf{n}_t$, and those of the noisy speech $\mathbf{y}_t$ is highly nonlinear, several approximations have been proposed to model the observation probability density $p(\mathbf{y}_t|\mathbf{x}_t,\mathbf{n}_t)$. The most prominent and at the same time simplest approach neglects any phase difference between the speech and the noise signal. However, it is well known that a more accurate model is obtained if a phase factor $\boldsymbol{\alpha}_t$, which results from the unknown phase between the complex speech and noise short-term discrete-time Fourier transforms, is taken into account [9, 11, 14, 28].

Since a numerical evaluation of the integrals to obtain the observation probability $p(\mathbf{y}_t|\mathbf{x}_t,\mathbf{n}_t)$ is computationally very demanding, $p(\mathbf{y}_t|\mathbf{x}_t,\mathbf{n}_t)$ is usually approximated by a Gaussian, where the effect of the phase factor is modeled as a contribution to the mean [14], to the variance [11] or to both the mean and the variance [28].

While in most cases the probability density of the phase factor is assumed to be a zero mean Gaussian whose variance is determined experimentally on stereo training data, Faubel et al. [14] determined the density by Monte Carlo simulations and showed experimentally that it is non-Gaussian, approaching a Gaussian density only for higher mel filter bank indices. Subsequently, the observation probability density $p(\mathbf{y}_t|\mathbf{x}_t, \mathbf{n}_t)$ can be determined either by a vector Taylor series approximation up to linear [5] or higher-order terms [28] or by Monte Carlo integration [14]. An analytic expression can be found in the case where the phase factor is assumed to be distributed according to a Gaussian [9].

In this contribution it is shown how the moments of the phase factor can be computed analytically, rendering stereo training data for the considered scenario obsolete. The derivation confirms the experimental observation made by others that the density of the phase factor is of non-Gaussian nature and that it is independent of the noise type and the signal-to-noise ratio. A Taylor series expansion is then carried out to obtain the mean and the variance of the observation probability density $p(\mathbf{y}_t|\mathbf{x}_t, \mathbf{n}_t)$, which is assumed to be Gaussian. The phase factor thereby delivers a contribution to both the mean (a bias term) and the variance of the observation probability, and the best recognition results are obtained if both are accounted for.

This contribution begins with a short introduction to the statistical framework of speech recognition and in particular the application of the aforementioned uncertainty decoding rule. With the posterior distribution of the clean feature vector being a key element thereof, it will be shown how it can be estimated under a Bayesian framework. However, practical realizations are subject to approximations which affect the a priori models for the clean speech and the noise-only feature vector, the observation model relating the two to the corrupted observation and the inference. These three components of the estimation process are therefore discussed next in more detail, and special stress is laid on the derivation and application of the phase-sensitive observation model and the derivation of the moments of the phase factor distribution. Finally, recognition results on the Aurora 2 and Aurora 4 databases are presented, providing experimental evidence of the superiority of the phase-sensitive observation model to its phase-insensitive counterpart and the benefit of propagating the obtained uncertainty information into the back-end.

8.2 Uncertainty Decoding for ASR

Speech recognition in a statistical framework basically reduces to the application of Bayesian decision rule. While in theory optimal, the application of the decision rule underlies certain restrictions. Thus, for instance, the true statistics of the feature vectors (the acoustic model), representative of the spoken words, are not available and have to be inferred from training data. Moreover, several approximations are required to arrive at computationally tractable solutions to the optimization problem. While employment of the decision rule works well for situations where the conditions the recognizer shall be operated in match the conditions the acoustic models (e.g., hidden Markov models, HMMs) have been trained on, a mismatch of the two leads to severe degradation of the recognition performance. Therefore, an elegant way of coping with mismatched/corrupted test data while using the acoustic model trained on uncorrupted (clean) training data neatly fitting into the statistical framework for speech recognition, termed uncertainty decoding (UD), will be presented in the following. A detailed derivation thereof can be found in Chap. 2.

8.2.1 Bayesian Framework of Speech Recognition

Given a sequence of feature vectors $\mathbf{x}_{1:T} = (\mathbf{x}_1, ..., \mathbf{x}_T)$ of length T, the maximum a posteriori speech recognition amounts to finding that sequence of words $\widehat{\mathbf{W}}$ from a given vocabulary which maximizes the posterior $p(\mathbf{x}_{1:T}|\mathbf{W})$ or, equivalently, where

$$\widehat{\mathbf{W}} = \underset{\mathbf{W}}{\operatorname{argmax}}\, p(\mathbf{x}_{1:T}|\mathbf{W})\, P(\mathbf{W}). \tag{8.1}$$

The prior probability of a word sequence $P(\mathbf{W})$, termed *language model*, is used to probabilistically weight the currently hypothesized word sequence $\mathbf{W}$, and the *acoustic model* is concerned with modeling $p(\mathbf{x}_{1:T}|\mathbf{W})$, the probability density of the observed feature vector sequence $\mathbf{x}_{1:T}$, given the hypothesized word sequence $\mathbf{W}$. In HMM-based speech recognition, this is accomplished by introducing a hidden state sequence $q_{1:T}$ underlying the observations $\mathbf{x}_{1:T}$ as

$$p(\mathbf{x}_{1:T}|\mathbf{W}) = \sum_{\{q_{1:T}\}} p(\mathbf{x}_{1:T}|q_{1:T}, \mathbf{W})\, P(q_{1:T}|\mathbf{W}), \tag{8.2}$$

where the summation is carried out over all possible state sequences $\{q_{1:T}\}$ within $\mathbf{W}$.

Assuming conditional independence between consecutive feature vectors while exploiting the first-order Markov property of the state process modeling the transition probability $P(q_t|q_{1:t-1}, \mathbf{W})$ finally gives

$$p(\mathbf{x}_{1:T}|\mathbf{W}) = \sum_{\{q_{1:T}\}} \prod_{t=1}^{T} P(\mathbf{x}_t|q_t, \mathbf{W})\, P(q_t|q_{t-1}, \mathbf{W}), \tag{8.3}$$

for which the Viterbi algorithm can be applied to compute an approximate value.

8.2.2 Corrupted Features

While employment of the decoding rule (8.1) with (8.2) works well for situations where the training and the test conditions match, a mismatch of the two leads to severe degradation of the recognition performance. This can be expressed by the fact that the sequence of test features $\mathbf{x}_{1:T}$, which are representative of the training conditions, and which are denoted as *clean* features in the following, is not observable. Instead, a corrupted version $\mathbf{y}_{1:T}$ is observed, where the corruption is caused, e.g., by acoustic environmental noise.

The recognition task is now stated as that of finding the most probable word sequence given the *noisy* observations $\mathbf{y}_{1:T}$:

$$\widehat{\mathbf{W}} = \underset{\mathbf{W}}{\operatorname{argmax}}\, p(\mathbf{y}_{1:T}|\mathbf{W})P(\mathbf{W}). \tag{8.4}$$

Taking the $\mathbf{y}_{1:T}$ as if they were the clean, uncorrupted data, i.e., interpreting $\mathbf{y}_{1:T}$ as point estimates of $\mathbf{x}_{1:T}$ to be *plugged* into Bayesian decision rule (8.1), while retaining the acoustic model trained on clean data, results in the well-known performance degradation of speech recognition in the presence of a mismatch between training and testing conditions.

The estimation of the uncorrupted feature vectors together with a measure of their reliability in the *front-end* in conjunction with a modification of the acoustic model in the *back-end* has been proposed as a possible solution to this problem [7, 10, 17, 22, 24, 28]. While this approach, termed uncertainty decoding , is theoretically optimal, practical realizations ask for approximations which, in general, render the decoding rule suboptimal.

Strictly following the derivation in Chap. 2, the only difference between the conventional decoding in the classical HMM framework (8.3) and the decoding in the uncertainty decoding framework lies in the computation of the acoustic likelihood, which is now given by

$$p(\mathbf{y}_{1:T}|\mathbf{W}) \propto \sum_{\{q_{1:T}\}} \prod_{t=1}^{T} \int \frac{p(\mathbf{x}_t|\mathbf{y}_{1:T})}{p(\mathbf{x}_t)} p(\mathbf{x}_t|q_t,\mathbf{W})\,\mathrm{d}\mathbf{x}_t\, p(q_t|q_{t-1},\mathbf{W}). \tag{8.5}$$

Equation (8.5) can now be compared to (8.3): Instead of evaluating the likelihood $p(\mathbf{x}_t|q_t)$ for a point estimate $E[\mathbf{x}_t|\mathbf{y}_{1:T}]$, which amounts to setting the posterior distribution to a Dirac-delta distribution centered at $E[\mathbf{x}_t|\mathbf{y}_{1:T}]$, i.e. $p(\mathbf{x}_t|\mathbf{y}_{1:T}) \approx \delta(\mathbf{x}_t - E[\mathbf{x}_t|\mathbf{y}_{1:T}])$, the entire posterior distribution is now taken into account.

An analytical solution to the integral in (8.5) can be obtained if all involved distributions are assumed to be (mixtures of) Gaussians.

Thus, the state-conditioned clean feature vector distribution will be modeled by a mixture of I Gaussians according to

$$p(\mathbf{x}_t|q_t{=}j) = \sum_{i=1}^{I} c_{i,j}\, \mathcal{N}\Big(\mathbf{x}_t;\, \boldsymbol{\mu}_{i,j}, \boldsymbol{\Sigma}_{i,j}\Big), \tag{8.6}$$

where $c_{i,j}$ is the weight, $\boldsymbol{\mu}_{i,j}$ is the mean vector and $\boldsymbol{\Sigma}_{i,j}$ is the covariance matrix of the ith mixture component of state $q_t{=}j$ at time instant t. Moreover, the distribution of the clean speech prior will be modeled by a single Gaussian,

$$p(\mathbf{x}_t) = \mathcal{N}(\mathbf{x}_t;\, \boldsymbol{\mu}_{\mathbf{x}}, \boldsymbol{\Sigma}_{\mathbf{x}}), \tag{8.7}$$

with mean $\boldsymbol{\mu}_{\mathbf{x}}$ and covariance $\boldsymbol{\Sigma}_{\mathbf{x}}$ — an assumption that has been proven to be *quite* valid by examining clean training data of the Aurora 2 [27] and Aurora 4 [15] databases. Finally, the posterior distribution will be modeled by

$$p(\mathbf{x}_t|\mathbf{y}_{1:T}) = \sum_{m=1}^{M'} c_{t,m}\, \mathcal{N}\Big(\mathbf{x}_t;\, \boldsymbol{\mu}_{\mathbf{x}_t|\mathbf{y}_{1:T},m}, \boldsymbol{\Sigma}_{\mathbf{x}_t|\mathbf{y}_{1:T},m}\Big), \tag{8.8}$$

a mixture of M' Gaussians with time-variant weight $c_{t,m} = P(r_t{=}m|\mathbf{y}_{1:T})$, mean vector $\boldsymbol{\mu}_{\mathbf{x}_t|\mathbf{y}_{1:T},m}$ and covariance matrix $\boldsymbol{\Sigma}_{\mathbf{x}_t|\mathbf{y}_{1:T},m}$ of mixture component $r_t{=}m$ at time instant t (an extension of Chap. 2, where the derivation only incorporates a single Gaussian posterior distribution). Under these assumptions, one can obtain

$$\begin{aligned}
&\int \frac{p(\mathbf{x}_t|\mathbf{y}_{1:T})}{p(\mathbf{x}_t)} p(\mathbf{x}_t|q_t{=}j, \mathbf{W})\, \mathrm{d}\mathbf{x}_t \\
&= \sum_{i=1}^{I}\sum_{m=1}^{M'} c_{i,j}\, c_{t,m} \int \frac{\mathcal{N}\Big(\mathbf{x}_t;\, \boldsymbol{\mu}_{\mathbf{x}_t|\mathbf{y}_{1:T},m}, \boldsymbol{\Sigma}_{\mathbf{x}_t|\mathbf{y}_{1:T},m}\Big)}{\mathcal{N}(\mathbf{x}_t;\, \boldsymbol{\mu}_{\mathbf{x}}, \boldsymbol{\Sigma}_{\mathbf{x}})} \mathcal{N}\Big(\mathbf{x}_t;\, \boldsymbol{\mu}_{i,j}, \boldsymbol{\Sigma}_{i,j}\Big)\, \mathrm{d}\mathbf{x}_t \qquad (8.9) \\
&= \sum_{i=1}^{I}\sum_{m=1}^{M'} c_{i,j}\, c_{\mathbf{e}_t,m}\, \mathcal{N}\Big(\boldsymbol{\mu}_{\mathbf{e}_t,m};\, \boldsymbol{\mu}_{i,j}, \boldsymbol{\Sigma}_{i,j} + \boldsymbol{\Sigma}_{\mathbf{e}_t,m}\Big), \qquad (8.10)
\end{aligned}$$

where the time-variant equivalent means $\boldsymbol{\mu}_{\mathbf{e}_t,m}$, equivalent covariances $\boldsymbol{\Sigma}_{\mathbf{e}_t,m}$ and equivalent weights $c_{\mathbf{e}_t,m}$ are formally given by

$$\boldsymbol{\Sigma}_{\mathbf{e}_t,m} = \left(\boldsymbol{\Sigma}^{-1}_{\mathbf{x}_t|\mathbf{y}_{1:T},m} - \boldsymbol{\Sigma}^{-1}_{\mathbf{x}}\right)^{-1}, \tag{8.11}$$

$$\boldsymbol{\mu}_{\mathbf{e}_t,m} = \boldsymbol{\Sigma}_{\mathbf{e}_t,m}\left(\boldsymbol{\Sigma}^{-1}_{\mathbf{x}_t|\mathbf{y}_{1:T},m}\boldsymbol{\mu}_{\mathbf{x}_t|\mathbf{y}_{1:T},m} - \boldsymbol{\Sigma}^{-1}_{\mathbf{x}}\boldsymbol{\mu}_{\mathbf{x}}\right), \tag{8.12}$$

$$c_{\mathbf{e}_t,m} = c_{t,m} \frac{\mathcal{N}\Big(\mathbf{0};\, \boldsymbol{\mu}_{\mathbf{x}_t|\mathbf{y}_{1:T},m}, \boldsymbol{\Sigma}_{\mathbf{x}_t|\mathbf{y}_{1:T},m}\Big)}{\mathcal{N}(\mathbf{0};\, \boldsymbol{\mu}_{\mathbf{x}}, \boldsymbol{\Sigma}_{\mathbf{x}})\; \mathcal{N}\Big(\mathbf{0};\, \boldsymbol{\mu}_{\mathbf{e}_t,m}, \boldsymbol{\Sigma}_{\mathbf{e}_t,m}\Big)}. \tag{8.13}$$

Equation (8.10) states that the originally trained, state-conditioned feature vector distribution has to be modified by *increasing* the mixtures' covariances by the corresponding equivalent covariances $\boldsymbol{\Sigma}_{\mathbf{e}_t,m}$ and evaluating it at the corresponding equivalent means $\boldsymbol{\mu}_{\mathbf{e}_t,m}$. Since the mean(s) and the covariance(s) of the posterior distribution will be obtained by the front-end (feature extraction and enhancement) while their application comes into effect at the back-end, uncertainty decoding reconciles front-end and back-end methods as a measure to increase the noise robustness in automatic speech recognition.

Note that if the Gaussian mixture components in Eq. (8.8) are approximated by Dirac-delta distributions centered at $\boldsymbol{\mu}_{\mathbf{x}_t|\mathbf{y}_{1:T},m}$, i.e., $\mathcal{N}(\mathbf{x}_t;\, \boldsymbol{\mu}_{\mathbf{x}_t|\mathbf{y}_{1:T},m}, \boldsymbol{\Sigma}_{\mathbf{x}_t|\mathbf{y}_{1:T},m}) \approx \delta(\mathbf{x}_t - \boldsymbol{\mu}_{\mathbf{x}_t|\mathbf{y}_{1:T},m})$, the uncertainty decoding rule employing (8.9) reduces to the conventional decoding rule with

$$\int \frac{p(\mathbf{x}_t|\mathbf{y}_{1:T})}{p(\mathbf{x}_t)} p(\mathbf{x}_t|q_t{=}j, \mathbf{W})\, \mathrm{d}\mathbf{x}_t \approx \sum_{i=1}^{I}\sum_{m=1}^{M'} c_{i,j}\tilde{c}_{t,m} \mathcal{N}\Big(\boldsymbol{\mu}_{\mathbf{x}_t|\mathbf{y}_{1:T},m};\, \boldsymbol{\mu}_{i,j}, \boldsymbol{\Sigma}_{i,j}\Big). \tag{8.14}$$

Thus, the estimated means $\boldsymbol{\mu}_{\mathbf{x}_t|\mathbf{y}_{1:T},m}$ are simply *plugged* into the decoding rule as if they were the clean features, however, with modified mixture weights $\tilde{c}_{t,m} =$

$c_{t,m} \cdot \mathcal{N}(\boldsymbol{\mu}_{\mathbf{x}_t|\mathbf{y}_{1:T},m}; \boldsymbol{\mu}_{\mathbf{x}}, \boldsymbol{\Sigma}_{\mathbf{x}})^{-1}$, i.e., inversely proportional to the a priori distribution evaluated at $\boldsymbol{\mu}_{\mathbf{x}_t|\mathbf{y}_{1:T},m}$.

Thus, uncertainty in the estimate of the clean speech feature $\mathbf{x}_t$ is only accounted for by the use of multiple, weighted hypotheses $\boldsymbol{\mu}_{\mathbf{x}_t|\mathbf{y}_{1:T},m}$, $m \in \{1, ..., M'\}$.

Note that while the assumption of a (single) Gaussian a priori distribution is required for the integral in (8.5) to be analytically solvable, the *sifting property* of the Dirac-delta distribution allows the a priori distribution to take any arbitrary *shape* (e.g., a mixture of Gaussians).

The likelihood computation (8.14) can also be applied if reliable estimates of the covariances $\boldsymbol{\Sigma}_{\mathbf{x}_t|\mathbf{y}_{1:T},m}$ are not available or to simply reduce the computational burden coming with the computation and integration of the equivalent means and covariances, as proposed by Stouten et al. [29].

8.3 Conditional Bayesian Estimation of the Feature Posterior

Looking at Eq. (8.5), the feature posterior $p(\mathbf{x}_t|\mathbf{y}_{1:T})$ can be considered to be the key element of the uncertainty decoding rule. Hence, the effectiveness of the uncertainty decoding rule depends on how well this feature posterior can be determined.

However, when targeting noise robust speech recognition, the posterior $p(\mathbf{x}_t|\mathbf{y}_{1:T})$ depends on the noise-only feature vectors. Incorporation of the noise-only feature vector can thereby be done either by modeling $\mathbf{n}_t$ as a deterministic parameter or as a realization of a random process. Since the latter approach is capable of dealing with uncertainty about the noise-only feature vector, the noise-only feature vector will be incorporated into the estimation process as a random variable — instead of solely estimating the posterior distribution $p(\mathbf{x}_t|\mathbf{y}_{1:T})$ of the clean speech feature vector $\mathbf{x}_t$, the joint a posteriori distribution $p(\mathbf{x}_t, \mathbf{n}_t|\mathbf{y}_{1:T})$ of the clean speech feature vector $\mathbf{x}_t$ and the noise-only feature vector $\mathbf{n}_t$ will be estimated. The posterior distribution $p(\mathbf{x}_t|\mathbf{y}_{1:T})$ of the clean speech feature can then be obtained by marginalization with respect to $\mathbf{n}_t$.

After introducing the shorthand notation $\mathbf{z}_\tau = (\mathbf{x}_\tau, \mathbf{n}_\tau)$ for the joint random variable consisting of clean speech feature vector $\mathbf{x}_\tau$ and noise-only feature vector $\mathbf{n}_\tau$ at time instant τ, the joint posterior $p(\mathbf{x}_t, \mathbf{n}_t|\mathbf{y}_{1:t}) = p(\mathbf{z}_t|\mathbf{y}_{1:t})$ can conceptually be estimated recursively by introducing the past value $\mathbf{z}_{t-1}$ of the joint random variables first and marginalizing over them afterwards using[1]

$$p(\mathbf{z}_t|\mathbf{y}_{1:t-1}) = \int p(\mathbf{z}_t|\mathbf{z}_{t-1}, \mathbf{y}_{1:t-1})\, p(\mathbf{z}_{t-1}|\mathbf{y}_{1:t-1})\, \mathrm{d}\mathbf{z}_{t-1} \tag{8.15}$$

and approximating $p(\mathbf{z}_{t-1}|\mathbf{y}_{1:t-1})$ by

$$p(\mathbf{z}_{t-1}|\mathbf{y}_{1:t-1}) \propto p(\mathbf{y}_{t-1}|\mathbf{z}_{t-1})\, p(\mathbf{z}_{t-1}|\mathbf{y}_{1:t-2}). \tag{8.16}$$

[1] Here, restriction to causal processing applies, i.e. rather than computing $p(\mathbf{z}_t|\mathbf{y}_{1:T})$, $p(\mathbf{z}_t|\mathbf{y}_{1:t})$ is computed.

Equations (8.15) and (8.16) reveal the recursive nature of the estimation process — the computation of $p(\mathbf{z}_t|\mathbf{y}_{1:t-1})$ requires $p(\mathbf{z}_{t-1}|\mathbf{y}_{1:t-1})$ and the computation thereof is based on $p(\mathbf{z}_{t-1}|\mathbf{y}_{1:t-2})$.

In practice, the marginalization (8.15) and the underlying recursion (8.16) may turn out to be very complicated and, moreover, computationally intractable. Thereby, two components can be determined to play a mayor role for the inference of the clean feature posterior, namely the *a priori model* $p(\mathbf{z}_t|\mathbf{z}_{t-1})$ for the joint random variable $\mathbf{z}_t$, which will be composed of an a priori model $p(\mathbf{x}_t|\mathbf{x}_{t-1})$ for the clean speech feature vector and an a priori model $p(\mathbf{n}_t|\mathbf{n}_{t-1})$ for the noise-only feature vector, and the *observation model* $p(\mathbf{y}_t|\mathbf{z}_t)$, relating the joint feature vector $\mathbf{z}_t$, consisting of the clean speech feature vector $\mathbf{x}_t$ and the noise-only feature vector $\mathbf{n}_t$, to the noisy observation $\mathbf{y}_t$.

Modeling these probability densities such that they both reflect the true dependencies as accurately as possible while at the same time being computationally tractable is a major challenge to the conditional Bayesian estimation of the feature posterior and is usually ensured by assuming all involved distributions to be (mixtures of) Gaussians.

The resulting scheme for the inference of the clean speech posterior and the final recognition is given in Fig. 8.1.

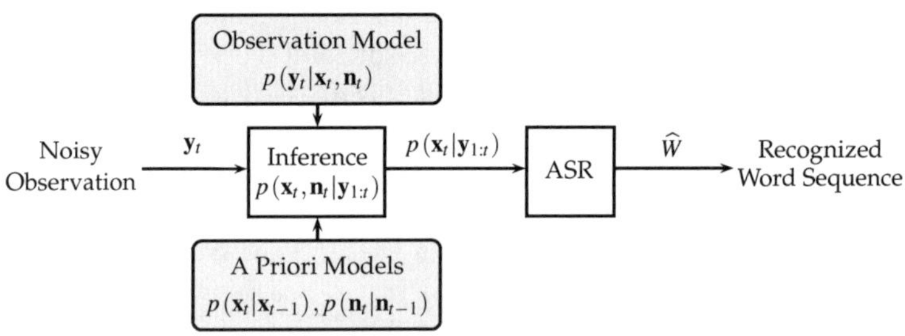

Fig. 8.1: Feature enhancement and recognition scheme

8.3.1 The A Priori Model

Following the explanations at the beginning of Sect. 8.3, the a priori model is composed of an a priori model for clean speech feature vector trajectory and an a priori model for the noise-only feature vector trajectory.

8.3.1.1 A Priori Model for Speech

The clean speech feature vector trajectory will, with respect to Eqs. (8.15) and (8.16), be modeled by a linear, auto-regressive model of order 1, driven by Gaussian noise. Since the speech feature vector trajectory includes segments of different dynamics — more than what a single linear dynamic model (LDM) could capture — a total of M switching linear dynamic models, an SLDM, is employed, with a *regime* variable s_t determining the model active at the given time instant t. The interaction of the M linear dynamic models is modeled as a (homogeneous) Markov process of order one, characterized by the time-invariant transition probabilities $P(s_t = m|s_{t-1} = l)$ and the initial state probabilities $P(s_1 = m)$.

Each of the M models thereby describes a different part of the dynamics of the speech feature vector trajectory as a transition from the clean feature vector $\mathbf{x}_{t-1}$ at time instant $t-1$ to the clean feature vector $\mathbf{x}_t$ at the current time instant t.

The distribution of the clean speech feature under the linear dynamic model m ($m \in \{1,...,M\}$) will be modeled as

$$p(\mathbf{x}_t|\mathbf{x}_{t-1}, s_t = m) = \mathcal{N}(\mathbf{x}_t;\ \mathbf{A}_m\mathbf{x}_{t-1} + \mathbf{b}_m, \mathbf{C}_m), \tag{8.17}$$

with $\mathbf{A}_m$ being the state transition matrix rendering the influence of $\mathbf{x}_{t-1}$ to $\mathbf{x}_t$, $\mathbf{b}_m$ being the state offset and $\mathbf{C}_m$ being the covariance matrix of the zero mean noise driving the state process — all under the regime variable $s_t = m$.

Further, the a priori distribution of the first clean speech feature vector $\mathbf{x}_1$ is modeled as a mixture of Gaussians with

$$p(\mathbf{x}_1|s_1 = m) = \mathcal{N}\big(\mathbf{x}_1;\ \boldsymbol{\mu}_{\mathbf{x}_1,m}, \boldsymbol{\Sigma}_{\mathbf{x}_1,m}\big) \tag{8.18}$$

and the first state probabilities (the mixture weights) $P(s_1 = m)$.

The set of linear dynamic model parameters $\{\mathbf{A}_m, \mathbf{b}_m, \mathbf{C}_m\}_{m=1}^{M}$, the transition probabilities $\{P(s_t = m|s_{t-1} = l)\}_{m,l=1}^{M}$ and the first states' model parameters and probabilities $\{\boldsymbol{\mu}_{\mathbf{x}_1,m}, \boldsymbol{\Sigma}_{\mathbf{x}_1,m}, P(s_1 = m)\}_{m=1}^{M}$ are estimated on clean training data using the expectation maximization (EM) algorithm [4].

A detailed derivation of the iterative estimation of the parameters of an SLDM by application of the EM algorithm is given in [25]. Note that the EM algorithm in general only guarantees the solution to the estimation problem to be locally optimal. Hence, its performance is sensitive to the initial set of parameters, calling for smart initialization routines [20].

8.3.1.2 A Priori Model for Noise

The incorporation of the noise-only feature vector into the estimation process as a random variable rather than an unknown parameter allows the uncertainty over an estimate of the noise-only feature vector and the correlation between the noise-only feature vector and the clean speech feature vector, both accessible through the joint posterior $p(\mathbf{x}_t, \mathbf{n}_t | \mathbf{y}_{1:t})$, to be tracked and considered during the inference.

The a priori model of the noise-only feature vector $\mathbf{n}_t$ will be described by a single linear dynamic model with the corresponding distributions

$$p(\mathbf{n}_t | \mathbf{n}_{t-1}) = \mathcal{N}(\mathbf{n}_t;\ \mathbf{B}\mathbf{n}_{t-1} + \mathbf{c}, \mathbf{D}) \tag{8.19}$$

for time instants $t > 1$ and

$$p(\mathbf{n}_1) = \mathcal{N}\left(\mathbf{n}_1;\ \boldsymbol{\mu}_{\mathbf{n}_1}, \boldsymbol{\Sigma}_{\mathbf{n}_1}\right) \tag{8.20}$$

for time instant $t = 1$. Training of the transition matrix $\mathbf{B}$, the offset $\mathbf{c}$ and the covariance matrix $\mathbf{D}$, as well as the initial mean $\boldsymbol{\mu}_{\mathbf{n}_1}$ and the initial covariance matrix $\boldsymbol{\Sigma}_{\mathbf{n}_1}$ of the noise model, is usually carried out by applying *maximum likelihood* estimation on non-speech frames of the given noisy signal, e.g., on feature vectors of the first and last frames of an utterance [8, 19].

Since the clean speech feature $\mathbf{x}_t$ and the noise-only feature $\mathbf{n}_t$ are a priori independent, the combination of the two a priori models $p(\mathbf{x}_1 | s_1 = m)$ and $p(\mathbf{n}_1)$ as well as $p(\mathbf{x}_t | \mathbf{x}_{t-1}, s_t = m)$ and $p(\mathbf{n}_t | \mathbf{n}_{t-1})$ into joint a priori models $p(\mathbf{z}_1 | s_1 = m)$ and $p(\mathbf{z}_t | \mathbf{z}_{t-1}, s_t = m)$ for speech and noise gives

$$\begin{aligned}
p(\mathbf{z}_1 | s_1 = m) &= p(\mathbf{x}_1, \mathbf{n}_1 | s_1 = m) = p(\mathbf{x}_1 | s_1 = m)\, p(\mathbf{n}_1) && (8.21)\\
&= \mathcal{N}\left(\mathbf{x}_1;\ \boldsymbol{\mu}_{\mathbf{x}_1,m}, \boldsymbol{\Sigma}_{\mathbf{x}_1,m}\right) \mathcal{N}\left(\mathbf{n}_1;\ \boldsymbol{\mu}_{\mathbf{n}_1}, \boldsymbol{\Sigma}_{\mathbf{n}_1}\right) && (8.22)\\
&= \mathcal{N}\left(\begin{bmatrix}\mathbf{x}_1\\ \mathbf{n}_1\end{bmatrix};\ \begin{bmatrix}\boldsymbol{\mu}_{\mathbf{x}_1,m}\\ \boldsymbol{\mu}_{\mathbf{n}_1}\end{bmatrix}, \begin{bmatrix}\boldsymbol{\Sigma}_{\mathbf{x}_1,m} & \mathbf{0}\\ \mathbf{0} & \boldsymbol{\Sigma}_{\mathbf{n}_1}\end{bmatrix}\right) && (8.23)\\
&= \mathcal{N}\left(\mathbf{z}_1;\ \boldsymbol{\mu}_{\mathbf{z}_1,m}, \boldsymbol{\Sigma}_{\mathbf{z}_1,m}\right) && (8.24)
\end{aligned}$$

and

$$\begin{aligned}
p(\mathbf{z}_t | \mathbf{z}_{t-1}, s_t = m) &= p(\mathbf{x}_t, \mathbf{n}_t | \mathbf{x}_{t-1}, \mathbf{n}_{t-1}, s_t = m) && (8.25)\\
&= p(\mathbf{x}_t | \mathbf{x}_{t-1}, s_t = m)\, p(\mathbf{n}_t | \mathbf{n}_{t-1}) && (8.26)\\
&= \mathcal{N}(\mathbf{x}_t;\ \mathbf{A}_m \mathbf{x}_{t-1} + \mathbf{b}_m, \mathbf{C}_m)\, \mathcal{N}(\mathbf{n}_t;\ \mathbf{B}\mathbf{n}_{t-1} + \mathbf{c}, \mathbf{D}) && (8.27)\\
&= \mathcal{N}\left(\begin{bmatrix}\mathbf{x}_t\\ \mathbf{n}_t\end{bmatrix};\ \begin{bmatrix}\mathbf{A}_m & \mathbf{0}\\ \mathbf{0} & \mathbf{B}\end{bmatrix}\begin{bmatrix}\mathbf{x}_{t-1}\\ \mathbf{n}_{t-1}\end{bmatrix} + \begin{bmatrix}\mathbf{b}_m\\ \mathbf{c}\end{bmatrix}, \begin{bmatrix}\mathbf{C}_m & \mathbf{0}\\ \mathbf{0} & \mathbf{D}\end{bmatrix}\right) && (8.28)\\
&= \mathcal{N}(\mathbf{z}_t;\ \mathbf{A}_{\mathbf{z},m}\mathbf{z}_{t-1} + \mathbf{b}_{\mathbf{z},m}, \mathbf{C}_{\mathbf{z},m}), && (8.29)
\end{aligned}$$

with the introduced definitions for $\boldsymbol{\mu}_{\mathbf{z}_1,m}$ and $\boldsymbol{\Sigma}_{\mathbf{z}_1,m}$ as well as $\mathbf{A}_{\mathbf{z},m}$, $\mathbf{b}_{\mathbf{z},m}$ and $\mathbf{C}_{\mathbf{z},m}$ becoming clear by comparing (8.23) to (8.24) and (8.28) to (8.29), respectively.

8.3.2 The Observation Model

The observation model, the second key element to the inference of the clean speech posterior, is derived with respect to the feature extraction process of the ETSI standard front-end [12], which has been modified by using the power spectrum rather than the magnitude spectrum as the input to the mel filter bank. The feature extraction process is depicted in Fig. 8.2. In order to avoid dealing with the discrete cosine transform matrix and its pseudoinverse, the derivation, which in general follows [6], will be carried out in the logarithmic mel domain. However, generalization to the cepstral domain is straightforward.

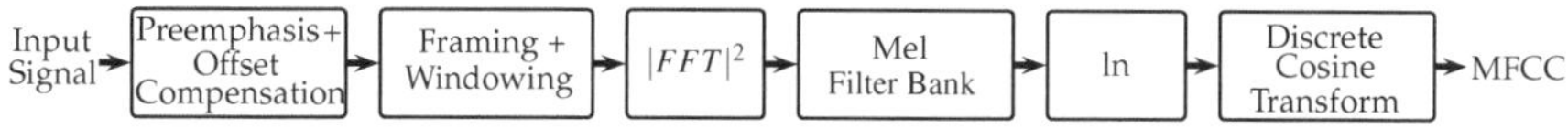

Fig. 8.2: Modified feature extraction process of the ETSI standard front-end

Assuming the speech signal to be corrupted by additive environmental noise and denoting the complex short-term discrete-time Fourier transform coefficients of the framed and windowed clean, noise-only and noisy signal in the kth frequency bin ($k \in \{1,\dots,K\}$) by $X_{t,k}$, $N_{t,k}$ and $Y_{t,k}$, respectively, their relationship in the power spectral domain is given by

$$|Y_{t,k}|^2 = |X_{t,k}|^2 + |N_{t,k}|^2 + 2|X_{t,k}||N_{t,k}|\cos\left(\theta_{t,k}\right), \tag{8.30}$$

with $\theta_{t,k}$ being the relative phase between the complex short-term discrete-time Fourier coefficients of the clean speech and the noise.

The representation in the ith mel frequency bin ($i \in \{1,\dots,I\}$) thus becomes

$$\tilde{Y}_{t,i} = \tilde{X}_{t,i} + \tilde{N}_{t,i} + 2\alpha_{t,i}\sqrt{\tilde{X}_{t,i}\tilde{N}_{t,i}}, \tag{8.31}$$

where

$$\tilde{Y}_{t,i} = \sum_{k=K_l(i)}^{K_u(i)} W_{i,k}|Y_{t,k}|^2,\ \tilde{X}_{t,i} = \sum_{k=K_l(i)}^{K_u(i)} W_{i,k}|X_{t,k}|^2,\ \tilde{N}_{t,i} = \sum_{k=K_l(i)}^{K_u(i)} W_{i,k}|N_{t,k}|^2 \tag{8.32}$$

denote the ith mel frequency coefficient of the noisy signal, the clean speech signal and the noise-only signal, respectively. The ith triangular-shaped mel filter has non-zero elements $W_{i,k}$ only in the range between a lower frequency bin $K_l(i)$ and an upper frequency bin $K_u(i)$. The introduced phase factor of the ith mel filter, denoted by $\alpha_{t,i}$, is given by

$$\alpha_{t,i} = \frac{\sum\limits_{k=K_l(i)}^{K_u(i)} W_{i,k}|X_{t,k}||N_{t,k}|\cos\left(\theta_{t,k}\right)}{\sqrt{\sum\limits_{k=K_l(i)}^{K_u(i)} W_{i,k}|X_{t,k}|^2}\sqrt{\sum\limits_{k=K_l(i)}^{K_u(i)} W_{i,k}|N_{t,k}|^2}} = \sum_{k=K_l(i)}^{K_u(i)} c_{t,i,k}\cos\left(\theta_{t,k}\right), \tag{8.33}$$

a weighted summation of the cosines of the relative phase between clean speech and noise for the ith mel filter.

With the absolute value of the cosine of the relative phase $\theta_{t,k}$ always being lower than or equal to 1, the absolute value of $\alpha_{t,i}$ is upper bounded by

$$|\alpha_{t,i}| \leq \frac{\sum\limits_{k=K_l(i)}^{K_u(i)} \sqrt{W_{i,k}|X_{t,k}|^2}\sqrt{W_{i,k}|N_{t,k}|^2}}{\sqrt{\sum\limits_{k=K_l(i)}^{K_u(i)} W_{i,k}|X_{t,k}|^2}\sqrt{\sum\limits_{k=K_l(i)}^{K_u(i)} W_{i,k}|N_{t,k}|^2}} = \frac{\vec{\mathbf{X}}_{t,i}\vec{\mathbf{N}}_{t,i}}{|\vec{\mathbf{X}}_{t,i}||\vec{\mathbf{N}}_{t,i}|} \leq 1, \tag{8.34}$$

which is interpreted as the normalized inner product of the vectors

$$\vec{\mathbf{X}}_{t,i} = \begin{pmatrix} \sqrt{W_{i,K_l(i)}|X_{t,K_l(i)}|^2} \\ \vdots \\ \sqrt{W_{i,K_u(i)}|X_{t,K_u(i)}|^2} \end{pmatrix}, \; \vec{\mathbf{N}}_{t,i} = \begin{pmatrix} \sqrt{W_{i,K_l(i)}|N_{t,K_l(i)}|^2} \\ \vdots \\ \sqrt{W_{i,K_u(i)}|N_{t,K_u(i)}|^2} \end{pmatrix}, \tag{8.35}$$

whose absolute value by definition is always lower than or equal to 1.

Transforming (8.31) to the logarithmic mel domain results in the phase-sensitive observation model as given in [9]:

$$y_{t,i} = y_i(x_{t,i}, n_{t,i}, \alpha_{t,i}) = \ln\left(e^{x_{t,i}} + e^{n_{t,i}} + 2\alpha_{t,i}e^{\frac{x_{t,i}+n_{t,i}}{2}}\right) \tag{8.36}$$

$$= \ln\left(e^{x_{t,i}} + e^{n_{t,i}}\right) + \underbrace{\ln\left(1 + 2\alpha_{t,i}\frac{e^{\frac{n_{t,i}-x_{t,i}}{2}}}{1+e^{n_{t,i}-x_{t,i}}}\right)}_{\varphi(\alpha_{t,i},x_{t,i},n_{t,i})}. \tag{8.37}$$

Assuming stereo data to be given, i.e., besides the noisy features also clean features and noise-only features, Eq. (8.36) can be used to empirically determine the distribution $p(\boldsymbol{\alpha}_t)$ by assuming the phase factor $\boldsymbol{\alpha}_t$ to be a realization of a white, stationary, ergodic process. Figure 8.3 shows the (empirically found) distribution $p(\boldsymbol{\alpha}_t)$ based on test data of the Aurora 2 database [27].

The components of the phase factor $\boldsymbol{\alpha}_t$ take values between -1 and $+1$, as shown in Eq. (8.34). The distribution approaches a Gaussian for high indices of the mel

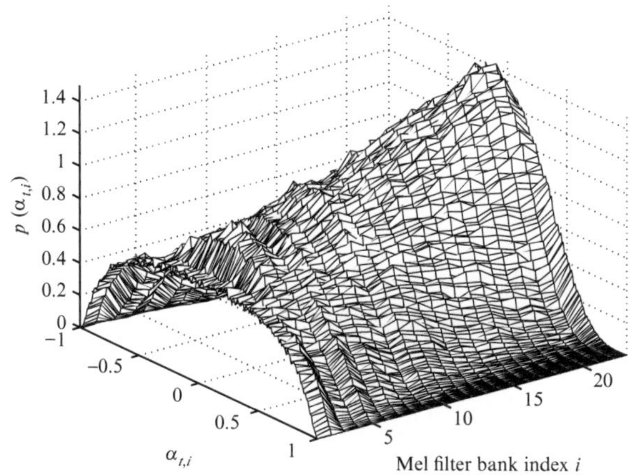

Fig. 8.3: Empirically found distribution $p(\alpha_{t,i})$ of the phase factor $\alpha_{t,i}$ for 23 mel filters based on stereo data of the Aurora 2 database

filter bank, but is clearly non-Gaussian for lower indices (compare [9, 14]).[2] Due to this non-Gaussianity and the fact that stereo data is usually not available, a common approximation of (8.37) is

$$y_{t,i} \approx \ln\left(e^{x_{t,i}} + e^{n_{t,i}}\right) = y_i(x_{t,i}, n_{t,i}, \alpha_{t,i} = 0), \tag{8.38}$$

where the phase factor-dependent term $\varphi\left(\alpha_{t,i}, x_{t,i}, n_{t,i}\right)$ is neglected. While the error due to this approximation is rather small when clean speech and noise mix at different levels, i.e., $|n_{t,i} - x_{t,i}| \gg 0$, it may become large when $n_{t,i} \approx x_{t,i}$.

Since feature enhancement based on a phase-sensitive observation model as given in (8.36) can be assumed to outperform approaches based on the phase-insensitive observation model (8.38), a way to incorporate the information provided by the phase-dependent term $\varphi\left(\alpha_{t,i}, x_{t,i}, n_{t,i}\right)$ into model-based feature enhancement will be derived next.

8.3.2.1 Vector Taylor Series Expansion

The nonlinearity of (8.36) makes the direct application of the phase-sensitive observation model to speech feature enhancement rather impractical. A common way to circumvent dealing with this nonlinearity is to expand the observation model into a vector Taylor series and truncate it to linear terms. This linearization is usually carried out with respect to the hidden clean speech feature vector $\mathbf{x}_t$ and the noise-only

[2] Note that the phase-factor distribution will, due to the limited range of $\boldsymbol{\alpha}_t$, formally never be Gaussian.

feature vector $\mathbf{n}_t$, while terms of order two or higher are disregarded [5]. Aiming at the incorporation of the vector of phase factors $\boldsymbol{\alpha}_t$, the observation model (8.36) is expanded into a vector Taylor series with respect to $\mathbf{x}_t$, $\mathbf{n}_t$ and $\boldsymbol{\alpha}_t$. After truncation to linear terms in $\mathbf{x}_t$ and $\mathbf{n}_t$, the remaining terms up to and including order two are used to model the linearization error.

Denoting the ith component of the expansion vectors of the clean speech, the noise and the phase factor at time instant t by $x_{t,i}^{(0)}$, $n_{t,i}^{(0)}$ and $\alpha_{t,i}^{(0)}$, respectively, a Taylor series expansion of $y_{t,i}$ with respect to $x_{t,i}$, $n_{t,i}$ and $\alpha_{t,i}$ gives

$$\begin{aligned} y_{t,i} &= y_i\left(x_{t,i}^{(0)}, n_{t,i}^{(0)}, \alpha_{t,i}^{(0)}\right) + J_{t,i}^{x}\left(x_{t,i} - x_{t,i}^{(0)}\right) + J_{t,i}^{n}\left(n_{t,i} - n_{t,i}^{(0)}\right) + J_{t,i}^{\alpha}\left(\alpha_{t,i} - \alpha_{t,i}^{(0)}\right) \\ &\quad + \frac{1}{2}H_{t,i}^{xx}\left(x_{t,i} - x_{t,i}^{(0)}\right)^2 + \frac{1}{2}H_{t,i}^{nn}\left(n_{t,i} - n_{t,i}^{(0)}\right)^2 + \frac{1}{2}H_{t,i}^{\alpha\alpha}\left(\alpha_{t,i} - \alpha_{t,i}^{(0)}\right)^2 \\ &\quad + H_{t,i}^{xn}\left(x_{t,i} - x_{t,i}^{(0)}\right)\left(n_{t,i} - n_{t,i}^{(0)}\right) + H_{t,i}^{x\alpha}\left(x_{t,i} - x_{t,i}^{(0)}\right)\left(\alpha_{t,i} - \alpha_{t,i}^{(0)}\right) \\ &\quad + H_{t,i}^{n\alpha}\left(n_{t,i} - n_{t,i}^{(0)}\right)\left(\alpha_{t,i} - \alpha_{t,i}^{(0)}\right) + HOTs \qquad (8.39) \\ &= y_i\left(x_{t,i}^{(0)}, n_{t,i}^{(0)}, \alpha_{t,i}^{(0)}\right) + J_{t,i}^{x}\left(x_{t,i} - x_{t,i}^{(0)}\right) + J_{t,i}^{n}\left(n_{t,i} - n_{t,i}^{(0)}\right) + \varepsilon_{t,i} + HOTs \qquad (8.40) \end{aligned}$$

with

$$J_{t,i}^{x} = \left.\frac{\partial y_i}{\partial x_i}\right|_{x_i^{(0)}, n_i^{(0)}, \alpha_i^{(0)}}, \quad J_{t,i}^{n} = \left.\frac{\partial y_i}{\partial n_i}\right|_{x_i^{(0)}, n_i^{(0)}, \alpha_i^{(0)}}, \quad J_{t,i}^{\alpha} = \left.\frac{\partial y_i}{\partial \alpha_i}\right|_{x_i^{(0)}, n_i^{(0)}, \alpha_i^{(0)}} \qquad (8.41)$$

denoting the diagonal elements of the Jacobian matrices of $\mathbf{y}_t$ with respect to $\mathbf{x}_t$, $\mathbf{n}_t$ and $\boldsymbol{\alpha}_t$, all evaluated at the expansion points $x_{t,i}^{(0)}$, $n_{t,i}^{(0)}$ and $\alpha_{t,i}^{(0)}$, and

$$H_{t,i}^{xx} = \left.\frac{\partial^2 y_{t,i}}{\partial x_{t,i}^2}\right|_{x_{t,i}^{(0)}, n_{t,i}^{(0)}, \alpha_{t,i}^{(0)}}, \quad H_{t,i}^{nn} = \left.\frac{\partial^2 y_{t,i}}{\partial n_{t,i}^2}\right|_{x_{t,i}^{(0)}, n_{t,i}^{(0)}, \alpha_{t,i}^{(0)}}, \quad H_{t,i}^{xn} = \left.\frac{\partial^2 y_{t,i}}{\partial x_{t,i}\partial n_{t,i}}\right|_{x_{t,i}^{(0)}, n_{t,i}^{(0)}, \alpha_{t,i}^{(0)}} \qquad (8.42)$$

$$H_{t,i}^{\alpha\alpha} = \left.\frac{\partial^2 y_{t,i}}{\partial \alpha_i^2}\right|_{x_{t,i}^{(0)}, n_{t,i}^{(0)}, \alpha_{t,i}^{(0)}}, \quad H_{t,i}^{x\alpha} = \left.\frac{\partial^2 y_{t,i}}{\partial x_{t,i}\partial \alpha_{t,i}}\right|_{x_{t,i}^{(0)}, n_{t,i}^{(0)}, \alpha_{t,i}^{(0)}}, \quad H_{t,i}^{n\alpha} = \left.\frac{\partial^2 y_{t,i}}{\partial n_{t,i}\partial \alpha_{t,i}}\right|_{x_{t,i}^{(0)}, n_{t,i}^{(0)}, \alpha_{t,i}^{(0)}} \qquad (8.43)$$

denoting the corresponding diagonal elements of the Hessian matrices.

By further neglecting higher-order terms (*HOTs*) and assuming the linearization error $\varepsilon_{t,i}$ to be Gaussian, the observation probability density $p(y_{t,i}|x_{t,i}, n_{t,i})$ becomes Gaussian, too, with mean and variance given by

$$E\left[y_{t,i}|x_{t,i}, n_{t,i}\right] = y_i\left(x_{t,i}^{(0)}, n_{t,i}^{(0)}, \alpha_{t,i}^{(0)}\right) + J_{t,i}^{x}\left(x_{t,i} - x_{t,i}^{(0)}\right) + J_{t,i}^{n}\left(n_{t,i} - n_{t,i}^{(0)}\right) + E\left[\varepsilon_{t,i}\right], \qquad (8.44)$$

$$E\left[\left(y_{t,i} - E\left[y_{t,i}|x_{t,i}, n_{t,i}\right]\right)^2 \middle| x_{t,i}, n_{t,i}\right] = E\left[\left(\varepsilon_{t,i} - E\left[\varepsilon_{t,i}\right]\right)^2\right]. \qquad (8.45)$$

8.3.2.2 Application to Model-Based Feature Enhancement

The quality of the approximation of Eq. (8.39) by its linearized and truncated version is quite sensitive to the choice of the expansion points $x_{t,i}^{(0)}, n_{t,i}^{(0)}, \alpha_{t,i}^{(0)}$. However, the feature enhancement scheme introduced at the beginning of Sect. 8.3 motivates the usage of a priori knowledge for the estimation problem at time instant t. With respect to the joint estimation of the posterior $p(\mathbf{x}_t, \mathbf{n}_t | \mathbf{y}_{1:t})$ of the clean speech feature and the noise-only feature, this a priori knowledge is given by $p(\mathbf{x}_t, \mathbf{n}_t | \mathbf{y}_{1:t-1})$. The involved means $\boldsymbol{\mu}_{\mathbf{x}_t|\mathbf{y}_{1:t-1}}$ and $\boldsymbol{\mu}_{\mathbf{n}_t|\mathbf{y}_{1:t-1}}$ provide the expansion vectors $\mathbf{x}_t^{(0)}$ and $\mathbf{n}_t^{(0)}$. Prior knowledge about the phase factor $\boldsymbol{\alpha}_t$ is only given in terms of the a priori distribution $p(\boldsymbol{\alpha}_t)$. Its mean $\boldsymbol{\mu}_{\boldsymbol{\alpha}}$ provides the expansion vector $\boldsymbol{\alpha}_t^{(0)}$. Introducing the short-hand notations

$$\tilde{x}_{t,i} = x_{t,i} - x_{t,i}^{(0)}, \quad \tilde{n}_{t,i} = n_{t,i} - n_{t,i}^{(0)}, \quad \tilde{\alpha}_{t,i} = \alpha_{t,i} - \alpha_{t,i}^{(0)}, \tag{8.46}$$

an approximate value for the mean $E[\varepsilon_{t,i}] = \mu_{\varepsilon_t,i}$ and the variance $E[(\varepsilon_{t,i} - E[\varepsilon_{t,i}])^2] = \sigma^2_{\varepsilon_t,i}$ of the linearization error can be obtained from (8.39) as

$$E[\varepsilon_{t,i}] = \frac{1}{2} H_{t,i}^{xx} E\left[\tilde{x}_{t,i}^2\right] + \frac{1}{2} H_{t,i}^{nn} E\left[\tilde{n}_{t,i}^2\right] + \frac{1}{2} H_{t,i}^{\alpha\alpha} E\left[\tilde{\alpha}_{t,i}^2\right] + H_{t,i}^{xn} E\left[\tilde{x}_{t,i}\tilde{n}_{t,i}\right] \tag{8.47}$$

and $E[(\varepsilon_{t,i} - E[\varepsilon_{t,i}])^2] = E[\varepsilon_{t,i}^2] - E[\varepsilon_{t,i}]^2$ with

$$\begin{aligned} E\left[\varepsilon_{t,i}^2\right] = {} & \left(J_{t,i}^{\alpha}\right)^2 E\left[\tilde{\alpha}_{t,i}^2\right] + \frac{1}{4}\left(H_{t,i}^{xx}\right)^2 E\left[\tilde{x}_{t,i}^4\right] + \frac{1}{4}\left(H_{t,i}^{nn}\right)^2 E\left[\tilde{n}_{t,i}^4\right] + \frac{1}{4}\left(H_{t,i}^{\alpha\alpha}\right)^2 E\left[\tilde{\alpha}_{t,i}^4\right] \\ & + \left(\left(H_{t,i}^{x\alpha}\right)^2 + \frac{1}{2} H_{t,i}^{xx} H_{t,i}^{\alpha\alpha}\right) E\left[\tilde{x}_{t,i}^2\right] E\left[\tilde{\alpha}_{t,i}^2\right] + H_{t,i}^{xx} H_{t,i}^{xn} E\left[\tilde{x}_{t,i}^3 \tilde{n}_{t,i}\right] \\ & + \left(\left(H_{t,i}^{n\alpha}\right)^2 + \frac{1}{2} H_{t,i}^{nn} H_{t,i}^{\alpha\alpha}\right) E\left[\tilde{n}_{t,i}^2\right] E\left[\tilde{\alpha}_{t,i}^2\right] + H_{t,i}^{nn} H_{t,i}^{xn} E\left[\tilde{n}_{t,i}^3 \tilde{x}_{t,i}\right] \\ & + \left(\left(H_{t,i}^{xn}\right)^2 + \frac{1}{2} H_{t,i}^{xx} H_{t,i}^{nn}\right) E\left[\tilde{x}_{t,i}^2 \tilde{n}_{t,i}^2\right] \\ & + \left(H_{t,i}^{\alpha\alpha} H_{t,i}^{xn} + H_{t,i}^{x\alpha} H_{t,i}^{n\alpha}\right) E\left[\tilde{\alpha}_{t,i}^2\right] E\left[\tilde{x}_{t,i}\tilde{n}_{t,i}\right]. \end{aligned} \tag{8.48}$$

Thereby, the clean speech feature $\mathbf{x}_t$ and the noise-only feature $\mathbf{n}_t$ are assumed to be jointly Gaussian and uncorrelated with the phase factor $\boldsymbol{\alpha}_t$. Further, all *odd* moments of $\mathbf{x}_t$, $\mathbf{n}_t$ and $\boldsymbol{\alpha}_t$ are assumed to be zero — an assumption that holds for the clean speech feature $\mathbf{x}_t$ and the noise-only feature $\mathbf{n}_t$ once their joint distribution is Gaussian, and also for the phase factor $\boldsymbol{\alpha}_t$, as will be shown in Sect. 8.3.3.

The estimates of the clean speech feature vector and the noise-only feature vector are first, with respect to the expansion points, uncorrelated, too. Thus their cross-covariance $\boldsymbol{\Sigma}_{\mathbf{x}_t,\mathbf{n}_t|\mathbf{y}_{1:t-1}} = \mathbf{0}$. However, in an iterative enhancement scheme, as it will be used later on, the observation model may introduce correlations between the estimates of $\mathbf{x}_t$ and $\mathbf{n}_t$ which then need to be taken into account. Since speech and noise are assumed to be jointly Gaussian, all moments related to $\mathbf{x}_t$ and $\mathbf{n}_t$ can be obtained from the covariance of their joint distribution as [18]:

$$E\left[\tilde{x}_{t,i}\tilde{n}_{t,i}\right] = \sigma_{\mathbf{x}_t,\mathbf{n}_t|\mathbf{y}_{1:t-1},i}, \quad E\left[\tilde{x}_{t,i}^2\right] = \sigma^2_{\mathbf{x}_t|\mathbf{y}_{1:t-1},i}, \quad E\left[\tilde{n}_{t,i}^2\right] = \sigma^2_{\mathbf{n}_t|\mathbf{y}_{1:t-1},i}, \tag{8.49}$$

$$E\left[\tilde{x}_{t,i}^2\tilde{n}_{t,i}^2\right] = \sigma^2_{\mathbf{x}_t|\mathbf{y}_{1:t-1},i}\sigma^2_{\mathbf{n}_t|\mathbf{y}_{1:t-1},i} + 2\sigma^2_{\mathbf{x}_t,\mathbf{n}_t|\mathbf{y}_{1:t-1},i}, \tag{8.50}$$

$$E\left[\tilde{x}_{t,i}^3\tilde{n}_{t,i}\right] = 3\sigma^2_{\mathbf{x}_t|\mathbf{y}_{1:t-1},i}\sigma_{\mathbf{x}_t,\mathbf{n}_t|\mathbf{y}_{1:t-1},i}, \quad E\left[\tilde{n}_{t,i}^3\tilde{x}_{t,i}\right] = 3\sigma^2_{\mathbf{n}_t|\mathbf{y}_{1:t-1},i}\sigma_{\mathbf{n}_t,\mathbf{x}_t|\mathbf{y}_{1:t-1},i}, \tag{8.51}$$

$$E\left[\tilde{x}_{t,i}^4\right] = 3\sigma^4_{\mathbf{x}_t|\mathbf{y}_{1:t-1},i}, \quad E\left[\tilde{n}_{t,i}^4\right] = 3\sigma^4_{\mathbf{n}_t|\mathbf{y}_{1:t-1},i}. \tag{8.52}$$

Thus, only the second central moments (the covariance matrices $\boldsymbol{\Sigma}_{\mathbf{x}_t|\mathbf{y}_{1:t-1}}$, $\boldsymbol{\Sigma}_{\mathbf{n}_t|\mathbf{y}_{1:t-1}}$ and $\boldsymbol{\Sigma}_{\boldsymbol{\alpha}}$ and the cross-covariance matrix $\boldsymbol{\Sigma}_{\mathbf{x}_t,\mathbf{n}_t|\mathbf{y}_{1:t-1}}$) and the fourth central moments are additionally employed to approximate the error $\boldsymbol{\varepsilon}_t$ made during linearization. Its mean (8.47) and its variance (8.48) can therefore be written as

$$\mu_{\boldsymbol{\varepsilon}_t,i} = \frac{1}{2}\left(H^{xx}_{t,i}\sigma^2_{\mathbf{x}_t|\mathbf{y}_{1:t-1},i} + H^{nn}_{t,i}\sigma^2_{\mathbf{n}_t|\mathbf{y}_{1:t-1},i} + H^{\alpha\alpha}_{t,i}\sigma^2_{\boldsymbol{\alpha},i}\right) + H^{xn}_{t,i}\sigma_{\mathbf{x}_t,\mathbf{n}_t|\mathbf{y}_{1:t-1},i} \tag{8.53}$$

and

$$\begin{aligned}
\sigma^2_{\boldsymbol{\varepsilon}_t,i} = {} & \left(J^{\alpha}_{t,i}\right)^2\sigma^2_{\boldsymbol{\alpha},i} + \frac{3}{4}\left(H^{xx}_{t,i}\right)^2\sigma^4_{\mathbf{x}_t|\mathbf{y}_{1:t-1},i} + \frac{3}{4}\left(H^{nn}_{t,i}\right)^2\sigma^2_{\mathbf{n}_t|\mathbf{y}_{1:t-1},i} + \frac{1}{4}\left(H^{\alpha\alpha}_{t,i}\right)^2 E\left[\tilde{\alpha}^4_{t,i}\right] \\
& + \left(\left(H^{x\alpha}_{t,i}\right)^2 + \frac{1}{2}H^{xx}_{t,i}H^{\alpha\alpha}_{t,i}\right)\sigma^2_{\mathbf{x}_t|\mathbf{y}_{1:t-1},i}\sigma^2_{\boldsymbol{\alpha},i} + H^{xx}_{t,i}H^{xn}_{t,i}\,3\,\sigma^2_{\mathbf{x}_t|\mathbf{y}_{1:t-1},i}\sigma_{\mathbf{x}_t,\mathbf{n}_t|\mathbf{y}_{1:t-1},i} \\
& + \left(\left(H^{n\alpha}_{t,i}\right)^2 + \frac{1}{2}H^{nn}_{t,i}H^{\alpha\alpha}_{t,i}\right)\sigma^2_{\mathbf{n}_t|\mathbf{y}_{1:t-1},i}\sigma^2_{\boldsymbol{\alpha},i} + H^{nn}_{t,i}H^{xn}_{t,i}\,3\,\sigma^2_{\mathbf{n}_t|\mathbf{y}_{1:t-1},i}\sigma_{\mathbf{x}_t,\mathbf{n}_t|\mathbf{y}_{1:t-1},i} \\
& + \left(\left(H^{xn}_{t,i}\right)^2 + \frac{1}{2}H^{xx}_{t,i}H^{nn}_{t,i}\right)\left(\sigma^2_{\mathbf{x}_t|\mathbf{y}_{1:t-1},i}\sigma^2_{\mathbf{n}_t|\mathbf{y}_{1:t-1},i} + 2\sigma^2_{\mathbf{x}_t,\mathbf{n}_t|\mathbf{y}_{1:t-1},i}\right) \\
& + \left(H^{\alpha\alpha}_{t,i}H^{xn}_{t,i} + H^{x\alpha}_{t,i}H^{n\alpha}_{t,i}\right)\sigma^2_{\boldsymbol{\alpha},i}\sigma_{\mathbf{x}_t,\mathbf{n}_t|\mathbf{y}_{1:t-1},i} - \mu^2_{\boldsymbol{\varepsilon}_t,i}.
\end{aligned} \tag{8.54}$$

While the Gaussian assumption for the joint posterior of the clean speech and the noise-only may reasonably be justified, it does not hold for the phase factor distribution, which is of non-Gaussian nature as stated at the beginning of Sect. 8.3.2 and illustrated in Fig. 8.3.

A related approach has been proposed by Stouten et al. [28]. However, they only employ the first two terms in (8.53) and only the first term in (8.54).

8.3.3 Moments of the Phase Factor Distribution

The central moments of the posterior distribution $p(\mathbf{x}_t, \mathbf{n}_t|\mathbf{y}_{1:t-1})$ are a by-product of the enhancement scheme. The only moments further required to determine the expansion points and to compute (8.53) and (8.54) are the first, the second and the fourth central moments of the phase factor distribution $p(\boldsymbol{\alpha}_t)$. In contrast to [9], explicit knowledge of the complete phase factor distribution is not required.

Since stereo training data comprising the noisy observations $\mathbf{y}_t$, the clean speech feature vectors $\mathbf{x}_t$ and the noise-only feature vectors $\mathbf{n}_t$ are usually not given, an analytical solution to the required moments of the phase factor distribution is desirable.

Recalling (8.33), the phase factor for the ith mel filter at time instant t is given by

$$\alpha_{t,i} = \frac{\sum\limits_{k=K_l(i)}^{K_u(i)} W_{i,k}|X_{t,k}||N_{t,k}|\cos\left(\theta_{t,k}\right)}{\sqrt{\sum\limits_{k=K_l(i)}^{K_u(i)} W_{i,k}|X_{t,k}|^2}\sqrt{\sum\limits_{k=K_l(i)}^{K_u(i)} W_{i,k}|N_{t,k}|^2}} = \sum_{k=K_l(i)}^{K_u(i)} c_{t,i,k}\cos\left(\theta_{t,k}\right), \tag{8.55}$$

where the summation is carried out over the range of non-zero frequency bins $k \in \{K_l(i), ..., K_u(i)\}$ of the ith mel filter. Thereby, the weight $c_{t,i,k}$ is collating all terms not depending on $\theta_{t,k}$.

The relative phases $\theta_{t,k}$ ($k \in \{1, ..., K\}$) between the complex-valued clean speech and the noise-only short-term discrete-time Fourier transforms are now assumed to be statistically independent, identically distributed random variables, each drawn from a uniform distribution over $-\pi \leq \theta_{t,k} < \pi$.

The distribution $p\left(u_{t,k}\right)$ of the random variable $u_{t,k} = \cos\left(\theta_{t,k}\right)$ can thus be determined to be

$$p\left(u_{t,k}\right) = \begin{cases} \frac{1}{\pi}\frac{1}{\sqrt{1-u_{t,k}^2}}, & \text{for } |u_{t,k}| < 1 \\ 0, & \text{otherwise} \end{cases}, \tag{8.56}$$

with the corresponding characteristic function $\Phi_{u_{t,k}}(\tau)$ given by

$$\Phi_{u_{t,k}}(\tau) = E\left[e^{j\tau u_{t,k}}\right] = \int e^{j\tau u_{t,k}} p\left(u_{t,k}\right) \mathrm{d}u_{t,k} = J_0(\tau), \tag{8.57}$$

which is bounded to $|\Phi_{u_{t,k}}(\tau)| \leq |\Phi_{u_{t,k}}(0)| = 1$ and formally equals (up to a factor of 2π) the inverse Fourier transform of $p\left(u_{t,k}\right)$. $J_0(\tau)$ thereby denotes the 0th-order Bessel function.

The characteristic function $\Phi_{\tilde{u}_{t,i,k}}(\tau)$ of the random variable $\tilde{u}_{t,i,k} = c_{t,i,k}u_{t,k}$ is then obtained by applying standard Fourier transform rules, yielding

$$\Phi_{\tilde{u}_{t,i,k}}(\tau) = \Phi_{u_{t,k}}(c_{t,i,k}\tau) = J_0(c_{t,i,k}\tau). \tag{8.58}$$

Since neighboring short-term discrete-time Fourier transform bins and thus their relative phases are asymptotically independent [23], the probability density $p\left(\alpha_{t,i}\right)$ can be expressed as the convolution of the probability densities $p\left(\tilde{u}_{t,i,k}\right)$ of all terms $k \in \{K_l(i), ..., K_u(i)\}$ under the sum in (8.55). Applying standard Fourier transform rules, the characteristic function of the random variable $\alpha_{t,i}$ can thus be found to be

$$\Phi_{\alpha_{t,i}}(\tau) = \prod_{k=K_l(i)}^{K_u(i)} J_0(c_{t,i,k}\tau), \tag{8.59}$$

whereas the weights $c_{t,i,k}$ are treated as unknown parameters rather than random variables. Finally, the νth moment of $\alpha_{t,i}$ can be obtained via

$$E\left[\alpha_{t,i}^{\nu}\right] = \frac{1}{j^{\nu}} \frac{\mathrm{d}^{\nu}\Phi_{\alpha_{t,i}}(\tau)}{\mathrm{d}\tau^{\nu}}\bigg|_{\tau=0}. \tag{8.60}$$

In particular, one obtains

$$E\left[\alpha_{t,i}\right] = \mu_{\boldsymbol{\alpha},i} = 0 = E\left[\alpha_{t,i}^{2w-1}\right], \text{ for } w \in \mathbb{N}, \tag{8.61}$$

$$E\left[\alpha_{t,i}^{2}\right] = \sigma_{\boldsymbol{\alpha},i}^{2} = \frac{1}{2}\sum_{k=K_l(i)}^{K_u(i)} c_{t,i,k}^{2} = \frac{1}{2}\frac{\sum_{k=K_l(i)}^{K_u(i)} W_{i,k}^{2}|X_{t,k}|^{2}|N_{t,k}|^{2}}{\left(\sum_{k=K_l(i)}^{K_u(i)} W_{i,k}|X_{t,k}|^{2}\right)\left(\sum_{k=K_l(i)}^{K_u(i)} W_{i,k}|N_{t,k}|^{2}\right)}, \tag{8.62}$$

$$E\left[\alpha_{t,i}^{4}\right] = \frac{3}{4}\left(\sum_{k=K_l(i)}^{K_u(i)} c_{t,i,k}^{2}\right)^{2} - \frac{3}{8}\sum_{k=K_l(i)}^{K_u(i)} c_{t,i,k}^{4} \tag{8.63}$$

$$= 3\sigma_{\boldsymbol{\alpha},i}^{4} - \frac{3}{8}\frac{\sum_{k=K_l(i)}^{K_u(i)} W_{i,k}^{4}|X_{t,k}|^{4}|N_{t,k}|^{4}}{\left(\sum_{k=K_l(i)}^{K_u(i)} W_{i,k}|X_{t,k}|^{2}\right)^{2}\left(\sum_{k=K_l(i)}^{K_u(i)} W_{i,k}|N_{t,k}|^{2}\right)^{2}}, \tag{8.64}$$

with (8.61) confirming the zero-mean assumption made in literature. Recalling that the fourth central moment of a Gaussian distribution is equal to three times the square of the second central moment, Eq. (8.63) also imposingly points out the non-Gaussian nature of the phase factor distribution, with $E[\alpha_{t,i}^{4}]$ approaching $3E[\alpha_{t,i}^{2}]^{2}$ with increasing width of the triangular shaped mel filter i.

Assuming the magnitude spectra of the clean speech $|X_{t,k}|$ and the noise $|N_{t,k}|$ to be constant over the width of the ith mel filter renders the central moments (8.62) and (8.63) independent of the clean speech and the noise, and thus independent of both, noise type and signal-to-noise ratio (SNR). Thus, the estimated central moments formally only depend on the particular realization of the mel filter bank. However, they also depend on the shape of the analysis window used during the feature extraction process.

To be more specific, Brillinger [3] found that if h_l ($l \in \{1,..,L\}$) are the taps of a tapered window of length L, e.g., the commonly used Hamming window, the variance of the smoothed periodograms is larger than that obtained with an untapered (rectangular) window by a factor of

$$F_h = L\sum_{l=1}^{L} h_l^{4} \Big/ \left(\sum_{l=1}^{L} h_l^{2}\right)^{2} \geq 1. \tag{8.65}$$

Hence, strictly speaking, Eqs. (8.62) and (8.63) hold for a rectangular analysis window only. With $F_h = 1$ for a rectangular analysis window, the central moments for tapered and untapered analysis windows are generally given by

$$E\left[\alpha_{t,i}^2\right] = F_h \, \frac{1}{2} \sum_{k=K_l(i)}^{K_u(i)} W_{i,k}^2 \Bigg/ \left(\sum_{k=K_l(i)}^{K_u(i)} W_{i,k} \right)^2 , \tag{8.66}$$

$$E\left[\alpha_{t,i}^4\right] = F_h^2 \left(\frac{3}{4} \left(\sum_{k=K_l(i)}^{K_u(i)} W_{i,k}^2 \right)^2 - \frac{3}{8} \sum_{k=K_l(i)}^{K_u(i)} W_{i,k}^4 \right) \Bigg/ \left(\sum_{k=K_l(i)}^{K_u(i)} W_{i,k} \right)^4 . \tag{8.67}$$

Comparing the resulting second and fourth central moments with the ones determined by evaluating Eq. (8.36) on stereo data of the Aurora 2 database for all noise types and all noise conditions with SNR 20 dB to SNR −5 dB validates these findings, as depicted in Fig. 8.4.

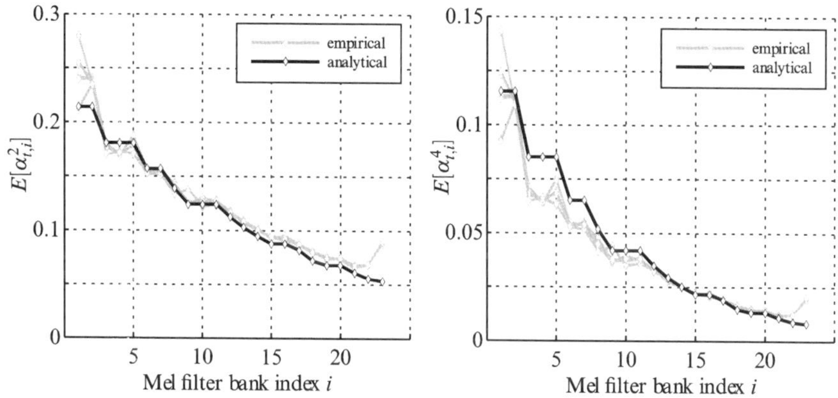

Fig. 8.4: Analytically found and empirically determined second (left) and fourth (right) central moments of the phase factor; each light gray line represents empirically found moments on stereo data of the Aurora 2 database for a different noise type/SNR combination

8.3.4 Inference

The a priori models for speech and noise-only feature vectors given in Sects. 8.3.1.1 and 8.3.1.2 and the observation model given in Sect. 8.3.2 build the fundamental components to the inference of the joint posterior $p(\mathbf{x}_t, \mathbf{n}_t | \mathbf{y}_{1:t})$ and finally the clean speech posterior $p(\mathbf{x}_t | \mathbf{y}_{1:t})$. Although all involved probability densities are (mixtures of) Gaussians, computationally tractable solutions to the posteriors under a multiple model approach can only be obtained by introducing approximations. A detailed discussion on possible approximations is given in [1].

The presented *dynamic multiple model estimators*, namely the generalized pseudo Bayesian estimators of orders one and two (GPB1 and GPB2) , and the interacting multiple model (IMM) estimator differ in the way the exponentially increasing number of possible histories of the regime variable (at time instant t there are M^t possible paths through the trellis spanned by possible values of the regime variable) is approximated and incorporated into the estimation process.

Following the GPB1 approach while again using the shorthand notation $\mathbf{z}_\tau = (\mathbf{x}_\tau, \mathbf{n}_\tau)$ introduced in Sect. 8.3, the joint posterior $p(\mathbf{x}_t, \mathbf{n}_t|\mathbf{y}_{1:t}) = p(\mathbf{z}_t|\mathbf{y}_{1:t})$ is formally modeled by a mixture of Gaussians by employing the total probability theorem with respect to all current models s_t at time instant t as[3]

$$p(\mathbf{z}_t|\mathbf{y}_{1:t}) = \sum_{m=1}^{M} p(\mathbf{z}_t|\mathbf{y}_{1:t}, s_t = m) P(s_t = m|\mathbf{y}_{1:t}), \tag{8.68}$$

with the Gaussian posterior distribution under model $s_t = m$ given by

$$p(\mathbf{z}_t|\mathbf{y}_{1:t}, s_t = m) = \mathcal{N}\left(\begin{bmatrix}\mathbf{x}_t\\ \mathbf{n}_t\end{bmatrix}; \begin{bmatrix}\boldsymbol{\mu}_{\mathbf{x}_t|\mathbf{y}_{1:t},m}\\ \boldsymbol{\mu}_{\mathbf{n}_t|\mathbf{y}_{1:t},m}\end{bmatrix}, \begin{bmatrix}\boldsymbol{\Sigma}_{\mathbf{x}_t|\mathbf{y}_{1:t},m} & \boldsymbol{\Sigma}_{\mathbf{x}_t,\mathbf{n}_t|\mathbf{y}_{1:t},m}\\ \boldsymbol{\Sigma}_{\mathbf{n}_t,\mathbf{x}_t|\mathbf{y}_{1:t},m} & \boldsymbol{\Sigma}_{\mathbf{n}_t|\mathbf{y}_{1:t},m}\end{bmatrix}\right) \tag{8.69}$$

$$= \mathcal{N}\left(\mathbf{z}_t; \boldsymbol{\mu}_{\mathbf{z}_t|\mathbf{y}_{1:t},m}, \boldsymbol{\Sigma}_{\mathbf{z}_t|\mathbf{y}_{1:t},m}\right). \tag{8.70}$$

The model-specific posterior $p(\mathbf{z}_t|\mathbf{y}_{1:t}, s_t = m)$, following (8.16), can be expressed as

$$p(\mathbf{z}_t|\mathbf{y}_{1:t}, s_t = m) \propto p(\mathbf{y}_t|\mathbf{x}_t, \mathbf{n}_t, s_t = m) p(\mathbf{z}_t|\mathbf{y}_{1:t-1}, s_t = m), \tag{8.71}$$

with

$$p(\mathbf{z}_t|\mathbf{y}_{1:t-1}, s_t = m) \approx \int p(\mathbf{z}_t|\mathbf{z}_{t-1}, s_t = m) p(\mathbf{z}_{t-1}|\mathbf{y}_{1:t-1}) \mathrm{d}\mathbf{z}_{t-1}. \tag{8.72}$$

After the model-specific estimates of the joint posterior $p(\mathbf{z}_t|\mathbf{y}_{1:t-1}, s_t = m)$ ($m \in \{1, ..., M\}$) are obtained, the mixture density (8.68) is approximated by a single Gaussian by merging all model-specific posterior estimates as

$$p(\mathbf{z}_t|\mathbf{y}_{1:t}) \approx \mathcal{N}\left(\begin{bmatrix}\mathbf{x}_t\\ \mathbf{n}_t\end{bmatrix}; \begin{bmatrix}\boldsymbol{\mu}_{\mathbf{x}_t|\mathbf{y}_{1:t}}\\ \boldsymbol{\mu}_{\mathbf{x}_t|\mathbf{y}_{1:t}}\end{bmatrix}, \begin{bmatrix}\boldsymbol{\Sigma}_{\mathbf{x}_t|\mathbf{y}_{1:t}} & \boldsymbol{\Sigma}_{\mathbf{x}_t,\mathbf{n}_t|\mathbf{y}_{1:t}}\\ \boldsymbol{\Sigma}_{\mathbf{n}_t,\mathbf{x}_t|\mathbf{y}_{1:t}} & \boldsymbol{\Sigma}_{\mathbf{n}_t|\mathbf{y}_{1:t}}\end{bmatrix}\right) \tag{8.73}$$

$$= \mathcal{N}\left(\mathbf{z}_t; \boldsymbol{\mu}_{\mathbf{z}_t|\mathbf{y}_{1:t}}, \boldsymbol{\Sigma}_{\mathbf{z}_t|\mathbf{y}_{1:t}}\right), \tag{8.74}$$

finally avoiding an exponentially increasing number of model histories to be explicitly accounted for.

Thus, the history of the regime variable is represented by the estimate of the mean and the covariance of this *pseudo* Gaussian, only. These are given by

---

[3] For ease of notation, the inference algorithm is only derived for time instants $t > 1$. However, one has to keep in mind, that the a priori models for time instant $t = 1$ differ from those for time instants $t > 1$.

$$\boldsymbol{\mu}_{\mathbf{z}_t|\mathbf{y}_{1:t}} = \sum_{m=1}^{M} P(s_t = m|\mathbf{y}_{1:t})\, \boldsymbol{\mu}_{\mathbf{z}_t|\mathbf{y}_{1:t},m}, \quad \Delta\boldsymbol{\mu}_{\mathbf{z}_t|\mathbf{y}_{1:t},m} = \boldsymbol{\mu}_{\mathbf{z}_t|\mathbf{y}_{1:t}} - \boldsymbol{\mu}_{\mathbf{z}_t|\mathbf{y}_{1:t},m}, \tag{8.75}$$

$$\boldsymbol{\Sigma}_{\mathbf{z}_t|\mathbf{y}_{1:t}} = \sum_{m=1}^{M} P(s_t = m|\mathbf{y}_{1:t}) \left(\boldsymbol{\Sigma}_{\mathbf{z}_t|\mathbf{y}_{1:t},m} + \Delta\boldsymbol{\mu}_{\mathbf{z}_t|\mathbf{y}_{1:t},m} \Delta\boldsymbol{\mu}^{\dagger}_{\mathbf{z}_t|\mathbf{y}_{1:t},m} \right), \tag{8.76}$$

with † denoting the vector and matrix transpose operator.

Note that although the chosen inference scheme requires the approximation of the Gaussian mixture (8.68) by a single Gaussian according to (8.73), this single Gaussian does not necessarily have to be passed to the recognizer. By passing the Gaussian mixture (8.68) to the recognizer, the multi-modality of the posterior distribution can be taken into account.

The probability $P(s_t = m|\mathbf{y}_{1:t})$ of model $s_t = m$ being active at the time instant t needed to compute (8.75) and (8.76) is obtained recursively by applying the theorem of total probability with respect to the regime variable s_{t-1} as

$$P(s_t = m|\mathbf{y}_{1:t}) \propto p(\mathbf{y}_t|\mathbf{y}_{1:t-1}, s_t = m) \sum_{l=1}^{M} P(s_t = m|s_{t-1} = l)\, P(s_{t-1} = l|\mathbf{y}_{1:t-1}). \tag{8.77}$$

Once the regime variable s_t at time instant t is given, the condition on all past observations $\mathbf{y}_{1:t-1}$ in the observation likelihood $p(\mathbf{y}_t|\mathbf{y}_{1:t-1}, s_t = m)$ is replaced by the condition on the merged statistics of the joint vector of clean speech and noise for time instant $t-1$ given all past observations $\mathbf{y}_{1:t-1}$:

$$p(\mathbf{y}_t|\mathbf{y}_{1:t-1}, s_t = m) \approx p(\mathbf{y}_t|\boldsymbol{\mu}_{\mathbf{z}_{t-1}|\mathbf{y}_{1:t-1}}, \boldsymbol{\Sigma}_{\mathbf{z}_{t-1}|\mathbf{y}_{1:t-1}}, s_t = m). \tag{8.78}$$

The calculation of this likelihood as well as the calculation of the posterior distribution $p(\mathbf{z}_t|\mathbf{y}_{1:t}, s_t = m)$ thus utilize the (collapsed) posterior $p(\mathbf{z}_{t-1}|\mathbf{y}_{1:t-1})$ at time instant $t-1$.

With the nonlinear observation model introduced in Sect. 8.3.2, both entities can be computed by employing the iterated extended Kalman filter (IEKF), which can be shown to be an application of the Gauss-Newton method for approximating a maximum a posteriori estimate of $\boldsymbol{\mu}_{\mathbf{z}_t|\mathbf{y}_{1:t},m}$ [1, 2]. To distinguish the phase-sensitive observation model from its phase-insensitive counterpart, the application of the phase-sensitive observation model, which, with the inherent modeling of the mean and the covariance of the linearization error, represents a modification of the standard IEKF, will be denoted by IEKF$-\boldsymbol{\alpha}$.

With all involved distributions approximated by Gaussians, the estimation of the joint posterior $p(\mathbf{z}_t|\mathbf{y}_{1:t})$ under the IEKF$-\boldsymbol{\alpha}$ in the GPB1 approach reduces to the update of its mean and its covariance matrix based on the joint posterior $p(\mathbf{z}_{t-1}|\mathbf{y}_{1:t-1})$ with its corresponding mean and covariance matrix. Thereby, the GPB1 algorithm requires the posterior probabilities $P(s_{t-1} = l|\mathbf{y}_{1:t-1})$ to recursively compute the posterior probabilities $P(s_t = m|\mathbf{y}_{1:t})$.

An overview of the IEKF$-\boldsymbol{\alpha}$-based feature enhancement under the GPB1 multiple model estimator is given in Alg. 1 with details on the IEKF$-\boldsymbol{\alpha}$, which will be iterated $\Upsilon+1$ times, given in Alg. 2.[4] Note that the IEKF$-\boldsymbol{\alpha}$ reduces to a) the extended Kalman filter with the phase-sensitive observation model (EKF$-\boldsymbol{\alpha}$) if the number of filter iterations is set to $\Upsilon=0$ and b) the standard IEKF/EKF if Step 5 in Alg. 2 is replaced with the assignment of fixed values for the mean $\boldsymbol{\mu}_{\boldsymbol{\varepsilon}_t}^{(\upsilon)}$ and the covariance $\boldsymbol{\Sigma}_{\boldsymbol{\varepsilon}_t}^{(\upsilon)}$ of the linearization error.

Algorithm 1 The GPB1 algorithm for the IEKF$-\boldsymbol{\alpha}$

Require:

- Parameters of the a priori model for the clean speech: $\{\,\mathbf{A}_m,\,\mathbf{b}_m,\,\mathbf{C}_m\,\}_{m=1}^{M}$, $\{\,P(s_t=m|s_{t-1}=l)\,\}_{m,l=1}^{M}$ and $\{\,\boldsymbol{\mu}_{\mathbf{x}_1,m},\,\boldsymbol{\Sigma}_{\mathbf{x}_1,m},\,P(s_1=m)\,\}_{m=1}^{M}$
- Parameters of the a priori model for the noise: $\{\,\mathbf{B},\,\mathbf{c},\,\mathbf{D}\,\}$ and $\{\,\boldsymbol{\mu}_{\mathbf{n}_1},\,\boldsymbol{\Sigma}_{\mathbf{n}_1}\,\}$
- Moments of the phase factor distribution: $\{\,E[\boldsymbol{\alpha}_{t,i}],\,E[\boldsymbol{\alpha}_{t,i}^2],\,E[\boldsymbol{\alpha}_{t,i}^4]\,\}_{i=1}^{I}$

Input: $\mathbf{y}_1,\ldots,\mathbf{y}_T$

Output:

- Means $\{\,\boldsymbol{\mu}_{\mathbf{z}_1|\mathbf{y}_1,m},\,\ldots,\,\boldsymbol{\mu}_{\mathbf{z}_T|\mathbf{y}_{1:T},m}\,\}_{m=1}^{M}$ and covariances $\{\,\boldsymbol{\Sigma}_{\mathbf{z}_1|\mathbf{y}_1,m},\,\ldots,\,\boldsymbol{\Sigma}_{\mathbf{z}_T|\mathbf{y}_{1:T},m}\,\}_{m=1}^{M}$ of the model-specific Gaussian posterior distributions with the corresponding model probabilities $\{\,P(s_1=m|\mathbf{y}_1),\,\ldots,\,P(s_T=m|\mathbf{y}_{1:T})\,\}_{m=1}^{M}$
- Means $\{\,\boldsymbol{\mu}_{\mathbf{z}_1|\mathbf{y}_1},\,\ldots,\,\boldsymbol{\mu}_{\mathbf{z}_T|\mathbf{y}_{1:T}}\,\}$ and covariances $\{\,\boldsymbol{\Sigma}_{\mathbf{z}_1|\mathbf{y}_1},\,\ldots,\,\boldsymbol{\Sigma}_{\mathbf{z}_T|\mathbf{y}_{1:T}}\,\}$ of the *collapsed* Gaussian posterior distributions

for $t=1$ **to** T **do**
  for $m=1$ **to** M **do**
    Perform Υ iterations of the IEKF$-\boldsymbol{\alpha}$ according to Alg. 2 on p. 209:
    Input: $\boldsymbol{\mu}_{\mathbf{z}_{t-1}|\mathbf{y}_{1:t-1}},\,\boldsymbol{\Sigma}_{\mathbf{z}_{t-1}|\mathbf{y}_{1:t-1}},\,m$
    Output: $\boldsymbol{\mu}_{\mathbf{z}_t|\mathbf{y}_{1:t},m},\,\boldsymbol{\Sigma}_{\mathbf{z}_t|\mathbf{y}_{1:t},m}$ **and** $p(\mathbf{y}_t|\mathbf{y}_{1:t-1},s_t=m)$
  end for
  for $m=1$ **to** M **do**
    Obtain the posterior probability $P(s_t=m|\mathbf{y}_{1:t})$:
    if $t=1$ **then**

$$P(s_t=m|\mathbf{y}_{1:t})\propto p(\mathbf{y}_t|\mathbf{y}_{1:t-1},s_t=m)\,P(s_t=m); \tag{8.79}$$

    else

$$P(s_t=m|\mathbf{y}_{1:t})\propto p(\mathbf{y}_t|\mathbf{y}_{1:t-1},s_t=m)\sum_{l=1}^{M}P(s_t=m|s_{t-1}=l)\,P(s_{t-1}=l|\mathbf{y}_{1:t-1}); \tag{8.80}$$

    end if
  end for
  Update the mean $\boldsymbol{\mu}_{\mathbf{z}_t|\mathbf{y}_{1:t}}$ and covariance $\boldsymbol{\Sigma}_{\mathbf{z}_t|\mathbf{y}_{1:t}}$ of the posterior $p(\mathbf{z}_t|\mathbf{y}_{1:t})$:

$$\boldsymbol{\mu}_{\mathbf{z}_t|\mathbf{y}_{1:t}}=\sum_{m=1}^{M}P(s_t=m|\mathbf{y}_{1:t})\,\boldsymbol{\mu}_{\mathbf{z}_t|\mathbf{y}_{1:t},m},\quad \Delta\boldsymbol{\mu}_{\mathbf{z}_t|\mathbf{y}_{1:t},m}=\boldsymbol{\mu}_{\mathbf{z}_t|\mathbf{y}_{1:t}}-\boldsymbol{\mu}_{\mathbf{z}_t|\mathbf{y}_{1:t},m}, \tag{8.81}$$

$$\boldsymbol{\Sigma}_{\mathbf{z}_t|\mathbf{y}_{1:t}}=\sum_{m=1}^{M}P(s_t=m|\mathbf{y}_{1:t})\left(\boldsymbol{\Sigma}_{\mathbf{z}_t|\mathbf{y}_{1:t,m}}+\Delta\boldsymbol{\mu}_{\mathbf{z}_t|\mathbf{y}_{1:t},m}\Delta\boldsymbol{\mu}^{\dagger}_{\mathbf{z}_t|\mathbf{y}_{1:t},m}\right); \tag{8.82}$$

end for

[4] Note that the notation $p(\mathbf{y}_t|\mathbf{y}_{1:t-1},s_t=m)$ is kept, though strictly speaking, $\mathbf{y}_{1:t-1}$ is not well defined for time instant $t=1$ and should therefore be understood as $p(\mathbf{y}_t|s_t=m)$.

Algorithm 2 The IEKF$-\boldsymbol{\alpha}$ with Υ iterations

Input: $\boldsymbol{\mu}_{\mathbf{z}_{t-1}|\mathbf{y}_{1:t-1}}, \boldsymbol{\Sigma}_{\mathbf{z}_{t-1}|\mathbf{y}_{1:t-1}}, m$

Output: $\boldsymbol{\mu}_{\mathbf{z}_t|\mathbf{y}_{1:t},m}, \boldsymbol{\Sigma}_{\mathbf{z}_t|\mathbf{y}_{1:t},m}$ **and** $p(\mathbf{y}_t|\mathbf{y}_{1:t-1}, s_t = m)$

1: Predict the mean $\boldsymbol{\mu}_{\mathbf{z}_t|\mathbf{y}_{1:t-1},m}$ and the covariance $\boldsymbol{\Sigma}_{\mathbf{z}_t|\mathbf{y}_{1:t-1},m}$ of the posterior $p(\mathbf{z}_t|\mathbf{y}_{1:t-1}, m)$:
if $t = 1$ **then**

$$\boldsymbol{\mu}_{\mathbf{z}_t|\mathbf{y}_{1:t-1},m} = \boldsymbol{\mu}_{\mathbf{z}_1,m}, \tag{8.83}$$

$$\boldsymbol{\Sigma}_{\mathbf{z}_t|\mathbf{y}_{1:t-1},m} = \boldsymbol{\Sigma}_{\mathbf{z}_1,m}; \tag{8.84}$$

else

$$\boldsymbol{\mu}_{\mathbf{z}_t|\mathbf{y}_{1:t-1},m} = \mathbf{A}_{\mathbf{z},m}\boldsymbol{\mu}_{\mathbf{z}_{t-1}|\mathbf{y}_{1:t-1}} + \mathbf{b}_{\mathbf{z},m}, \tag{8.85}$$

$$\boldsymbol{\Sigma}_{\mathbf{z}_t|\mathbf{y}_{1:t-1},m} = \mathbf{A}_{\mathbf{z},m}\boldsymbol{\Sigma}_{\mathbf{z}_{t-1}|\mathbf{y}_{1:t-1}}\mathbf{A}^{\dagger}_{\mathbf{z},m} + \mathbf{C}_{\mathbf{z},m}; \tag{8.86}$$

end if

2: Assign the initial Taylor series expansion vector $\boldsymbol{\mu}^{(0)}_{\mathbf{z}_t|\mathbf{y}_{1:t-1},m}$ and covariance $\boldsymbol{\Sigma}^{(0)}_{\mathbf{z}_t|\mathbf{y}_{1:t-1},m}$:

$$\boldsymbol{\mu}^{(0)}_{\mathbf{z}_t|\mathbf{y}_{1:t-1},m} = \boldsymbol{\mu}_{\mathbf{z}_t|\mathbf{y}_{1:t-1},m}, \quad \boldsymbol{\Sigma}^{(0)}_{\mathbf{z}_t|\mathbf{y}_{1:t-1},m} = \boldsymbol{\Sigma}_{\mathbf{z}_t|\mathbf{y}_{1:t-1},m}; \tag{8.87}$$

3: **for** $\upsilon = 0$ **to** Υ **do**

4: Linearize the observation model at $\boldsymbol{\mu}^{(\upsilon)}_{\mathbf{z}_t|\mathbf{y}_{1:t-1},m}$ and $E[\boldsymbol{\alpha}_t] = \mathbf{0}$ to obtain the Jacobian matrix $\mathbf{J}^{(\upsilon)}_{\mathbf{z}_t|\mathbf{y}_{1:t-1},m}$ with respect to $\mathbf{z}_t$:

$$\mathbf{J}^{(\upsilon)}_{\mathbf{z}_t|\mathbf{y}_{1:t-1},m} = \nabla\mathbf{y}(\mathbf{z}_t, \boldsymbol{\alpha}_t)\big|_{\mathbf{z}_t=\boldsymbol{\mu}^{(\upsilon)}_{\mathbf{z}_t|\mathbf{y}_{1:t-1},m}, \boldsymbol{\alpha}_t=\mathbf{0}}; \tag{8.88}$$

5: Calculate the linearization error's mean $\boldsymbol{\mu}^{(\upsilon)}_{\boldsymbol{\varepsilon}_t}$ and covariance matrix $\boldsymbol{\Sigma}^{(\upsilon)}_{\boldsymbol{\varepsilon}_t}$ according to (8.53) and (8.54) employing $\boldsymbol{\Sigma}^{(\upsilon)}_{\mathbf{z}_t|\mathbf{y}_{1:t-1},m}$ and the moments of the phase factor distribution;

6: Compute the mean $\boldsymbol{\mu}^{(\upsilon)}_{\mathbf{y}_t|\mathbf{y}_{1:t-1},m}$ and the covariance matrix $\boldsymbol{\Sigma}^{(\upsilon)}_{\mathbf{y}_t|\mathbf{y}_{1:t-1},m}$ of the probability density $p(\mathbf{y}_t|\mathbf{y}_{1:t-1}, s_t = m)$ at the current filter iteration υ:

$$\boldsymbol{\mu}^{(\upsilon)}_{\mathbf{y}_t|\mathbf{y}_{1:t-1},m} = \mathbf{y}(\mathbf{z}_t = \boldsymbol{\mu}^{(\upsilon)}_{\mathbf{z}_t|\mathbf{y}_{1:t-1},m}, \boldsymbol{\alpha}_t = \mathbf{0}) + \boldsymbol{\mu}^{(\upsilon)}_{\boldsymbol{\varepsilon}_t}, \tag{8.89}$$

$$\boldsymbol{\Sigma}^{(\upsilon)}_{\mathbf{y}_t|\mathbf{y}_{1:t-1},m} = \mathbf{J}^{(\upsilon)}_{\mathbf{z}_t|\mathbf{y}_{1:t-1},m}\boldsymbol{\Sigma}^{(0)}_{\mathbf{z}_t|\mathbf{y}_{1:t-1},m}\mathbf{J}^{(\upsilon)^{\dagger}}_{\mathbf{z}_t|\mathbf{y}_{1:t-1},m} + \boldsymbol{\Sigma}^{(\upsilon)}_{\boldsymbol{\varepsilon}_t}; \tag{8.90}$$

7: Compute the Kalman gain $\mathbf{K}^{(\upsilon)}_m$:

$$\mathbf{K}^{(\upsilon)}_m = \boldsymbol{\Sigma}^{(0)}_{\mathbf{z}_t|\mathbf{y}_{1:t-1},m}\mathbf{J}^{(\upsilon)^{\dagger}}_{\mathbf{z}_t|\mathbf{y}_{1:t-1},m}\left(\boldsymbol{\Sigma}^{(\upsilon)}_{\mathbf{y}_t|\mathbf{y}_{1:t-1},m}\right)^{-1}; \tag{8.91}$$

8: Update the mean and the corresponding covariance matrix:

$$\boldsymbol{\mu}^{(\upsilon+1)}_{\mathbf{z}_t|\mathbf{y}_{1:t-1},m} = \boldsymbol{\mu}^{(0)}_{\mathbf{z}_t|\mathbf{y}_{1:t-1},m} + \mathbf{K}^{(\upsilon)}_m\left(\mathbf{y}_t - \boldsymbol{\mu}^{(\upsilon)}_{\mathbf{y}_t|\mathbf{y}_{1:t-1},m} + \mathbf{J}^{(\upsilon)}_{\mathbf{z}_t|\mathbf{y}_{1:t-1},m}\left(\boldsymbol{\mu}^{(\upsilon)}_{\mathbf{z}_t|\mathbf{y}_{1:t-1},m} - \boldsymbol{\mu}^{(0)}_{\mathbf{z}_t|\mathbf{y}_{1:t-1},m}\right)\right), \tag{8.92}$$

$$\boldsymbol{\Sigma}^{(\upsilon+1)}_{\mathbf{z}_t|\mathbf{y}_{1:t-1},m} = \left(\mathbf{I} - \mathbf{K}^{(\upsilon)}_m\mathbf{J}^{(\upsilon)}_{\mathbf{z}_t|\mathbf{y}_{1:t-1},m}\right)\boldsymbol{\Sigma}^{(0)}_{\mathbf{z}_t|\mathbf{y}_{1:t-1},m}; \tag{8.93}$$

9: **end for**

10: Calculate the observation likelihood $p(\mathbf{y}_t|\mathbf{y}_{1:t-1}, s_t = m)$ for the current observation $\mathbf{y}_t$ based on the initial estimates at iteration $\upsilon = 0$:

$$p(\mathbf{y}_t|\mathbf{y}_{1:t-1}, s_t = m) = \mathcal{N}\left(\mathbf{y}_t; \boldsymbol{\mu}^{(0)}_{\mathbf{y}_t|\mathbf{y}_{1:t-1},m}, \boldsymbol{\Sigma}^{(0)}_{\mathbf{y}_t|\mathbf{y}_{1:t-1},m}\right); \tag{8.94}$$

11: Update the mean $\boldsymbol{\mu}_{\mathbf{z}_t|\mathbf{y}_{1:t},m}$ and the covariance $\boldsymbol{\Sigma}_{\mathbf{z}_t|\mathbf{y}_{1:t},m}$ of the model specific posterior distribution $p(\mathbf{z}_t|\mathbf{y}_{1:t}, m)$:

$$\boldsymbol{\mu}_{\mathbf{z}_t|\mathbf{y}_{1:t},m} = \boldsymbol{\mu}^{(\Upsilon+1)}_{\mathbf{z}_t|\mathbf{y}_{1:t-1},,m}, \quad \boldsymbol{\Sigma}_{\mathbf{z}_t|\mathbf{y}_{1:t},m} = \boldsymbol{\Sigma}^{(\Upsilon+1)}_{\mathbf{z}_t|\mathbf{y}_{1:t-1},,m}; \tag{8.95}$$

8.4 Experiments

The performances of the phase-sensitive observation model (IEKF$-\boldsymbol{\alpha}$) and its phase-insensitive counterpart (IEKF) in a model-based speech feature enhancement scheme are evaluated in recognition experiments conducted on the Aurora 2 and the Aurora 4 databases [15, 27].

8.4.1 Aurora 2 Database

The Aurora 2 database [27] is a subset of the TIDigits database [21], comprising connected digits spoken in American English recorded at 20 kHz, which has been decimated to 8 kHz and to which noise has artificially been added. The database comprises two training sets (*clean* data and *multi-condition* data) and three test sets (A, B and C).

While training of the recognizer's hidden Markov models and the parameters of the switching linear dynamic models is based on the $8,440$ utterances of the clean training data only, the recognition experiments consider clean and noisy data of test sets A and B. Test set C, which has been filtered with a different frequency characteristic, is excluded from all recognition experiments.

Test set A consists of four noise types, namely *suburban train*, *babble*, *car* and *exhibition hall* noise. The noise signals are added to each subset of $1,001$ utterances at SNRs of 20 dB, 15 dB, 10 dB, 5 dB, 0 dB and -5 dB, giving a total of $7,007$ utterances per noise type to be processed (clean speech included). Test set B is created in the same way, however, with four different noise types, namely *restaurant*, *street*, *airport* and *train station* noise. While some of the noises are quite stationary, e.g. car and exhibition hall noise, others contain non-stationary segments, e.g., street and airport noise.

Recognition results are usually given in terms of the word accuracy (Acc), which is defined via the word error rate (E) as

$$\mathrm{Acc} = 100\% - \mathrm{E}, \tag{8.96}$$

$$\mathrm{E} = \frac{\mathrm{I}+\mathrm{S}+\mathrm{D}}{\mathrm{N}} \times 100\%. \tag{8.97}$$

Thus, the sum of the insertion errors (I), substitution errors (S) and deletion errors (D) is divided by the total number (N) of labels in the reference transcriptions to obtain the word error rate (E). When reporting recognition results on the Aurora 2 database, results for SNR -5 dB are usually excluded and the average word accuracies over all noise types and all SNRs from 20 dB to 0 dB are reported for each test set.

8.4.2 Aurora 4 Database

The Aurora 4 database [15] is based on the DARPA Wall Street Journal (WSJ0) Corpus [26] and has been designed for evaluation of the specified 5,000-word closed-vocabulary recognition task.

The training data are taken from the clean data of the SI-84 WSJ0 training set recorded with a Sennheiser microphone at 16 kHz and decimated to 8 kHz, leaving 7,138 utterances for the HMM and SLDM training. The test data consist of the WSJ0 November'92 NIST evaluation test set to which noise at varying SNR levels and of varying types has been added.

The official Aurora 4 selection test set comprises 166 utterances recorded with a Sennheiser microphone at 16 kHz, which have also been decimated to 8 kHz. Beside the clean data of the test set, six versions of the test set with artificially added noises at randomly chosen SNR conditions between 5 dB and 15 dB are examined, namely *car*, *babble*, *restaurant*, *street traffic*, *airport* and *train station* noise. Thus, a total of 1,162 utterances is used for evaluation (including the clean data).

Recognition results are usually given in terms of insertion errors (I), substitution errors (S), deletion errors (D) and word errors (E) for each noise type and averaged over all noise types and the clean data.

8.4.3 Training of the HMMs

Training of HMMs is carried out using the Hidden Markov Model Toolkit (HTK) [30] on the clean training data using either the ETSI advanced front-end [13] or the modified ETSI standard front-end with cepstral mean subtraction (CMS) on the static features. The extracted static cepstral features are appended by dynamic features of the first and second order, giving a feature vector of length 39.

For the Aurora 2 task, each digit is modeled by an HMM with 16 emitting states in a linear (*loop-next*) topology, i.e., whole-word HMMs are used. The number of Gaussians per state has been raised to 20, as opposed to [27], where the baseline system is specified with only three Gaussians per state. With the components of the feature vector assumed to be independent, only diagonal covariances are trained. Besides the whole-word digit HMMs, additional models for *short-pause* and *silence* are trained. The silence model consists of 3 emitting states with 36 Gaussians per state. The short-pause is modeled by a single emitting state tied to the middle state of the silence model.

For the Aurora 4 task, word-internal triphone HMMs with three emitting states in a linear topology and a mixture of ten Gaussians per state are used. The additional silence model also follows a three-state linear topology but is modeled by a mixture of 20 Gaussians per state. The short-pause model is tied to the three states of the silence model; however, the HMM topology for the short-pause is extended by *skip* transitions.

8.4.4 Training of the A Priori Models

For both databases, the SLDM, describing the a priori speech feature vector trajectory, consists of $M = 16$ interacting linear dynamic models for the clean speech trajectory and is trained on the clean training data, coded into 13-dimensional (static) cepstral feature vectors by the modified ETSI standard front-end.

The training is carried out under the *expectation maximization* framework, with the regime variable s_t being the unobservable part of the complete data. Due to the sensitivity of the EM algorithm to its initial set of model parameters, smart initialization routines, e.g., given in [20], are in demand.

The initialization for all following experiments takes its cue from the initialization routine of GMMs [30]. It iteratively *splits* an existing model m into two new models by shifting the means $\mathbf{b}_m$ and $\boldsymbol{\mu}_{\mathbf{x}_1,m}$ in two directions specified by the covariance matrix $\mathbf{C}_m$ and $\boldsymbol{\Sigma}_{\mathbf{x}_1,m}$, respectively. The new covariance and transition matrices are simply copied from model m, and the model's state transition and first state probabilities are divided equally between the two new models. Instead of splitting just one model at a time, all existing models are split at once. Thus, starting from $M = 1$ model, the training proceeds to $M = 2$, $M = 4$, $M = 8$ and finally $M = 16$ models. Each split is followed by 20 iterations of the EM algorithm, subject to the relative change in likelihood, which is required to be larger than 10%.

The single a priori model used for the noise-only feature vector trajectory will be trained on a per-utterance basis. This allows the model to capture noise properties that are specific to the current utterance. Further, the transition matrix $\mathbf{B}$ in the LDM will be set to $\mathbf{B} = \mathbf{0}$, calling for less training data than required by a fully specified LDM. The model's parameters $\mathbf{c} = \boldsymbol{\mu}_{\mathbf{n}_1}$ and $\mathbf{D} = \boldsymbol{\Sigma}_{\mathbf{n}_1}$ (constrained to be diagonal) are trained on the first and last ten and 15 feature vectors of each utterance of the Aurora 2 and the Aurora 4 databases, respectively.

8.4.5 Experimental Setup

Besides the IEKF and IEKF$-\boldsymbol{\alpha}$, the established and standardized ETSI front-ends [12, 13] are also considered, eventually giving the baseline to which to compare the IEKF and the IEKF$-\boldsymbol{\alpha}$. Though the derivation of the phase-sensitive observation model has been carried out in the logarithmic mel domain, feature enhancement and recognition are applied to the cepstral features.

The distribution of the error made by neglecting the phase-dependent term when using the IEKF is modeled by a Gaussian, whose mean $\boldsymbol{\mu}_{\boldsymbol{\varepsilon}}$ and covariance $\boldsymbol{\Sigma}_{\boldsymbol{\varepsilon}}$ are obtained from stereo data. Therefore, the clean speech feature vectors $\mathbf{x}_t$ and the noise-only feature vectors $\mathbf{n}_t$ are *plugged* into Eq. (8.38) and the outcome is subtracted from the corresponding noisy observations $\mathbf{y}_t$. Consequently, Step 5 of Algorithm 2 in Sect. 8.3.4 is replaced with the assignment of these fixed values for the mean $\boldsymbol{\mu}_{\boldsymbol{\varepsilon}_t}^{(\upsilon)} = \boldsymbol{\mu}_{\boldsymbol{\varepsilon}}$ and the covariance $\boldsymbol{\Sigma}_{\boldsymbol{\varepsilon}_t}^{(\upsilon)} = \boldsymbol{\Sigma}_{\boldsymbol{\varepsilon}}$.

Thus, the following front-end and back-end schemes will be subject to experiments on the Aurora 2 and the Aurora 4 speech recognitions tasks:

- ETSI advanced front-end (AFE);
- ETSI standard front-end with cepstral mean subtraction (SFE+CMS);
- Iterated extended Kalman filter with cepstral mean subtraction (IEKF+CMS);
- Iterated extended Kalman filter with the phase-sensitive observation model and cepstral mean subtraction (IEKF$-\boldsymbol{\alpha}$+CMS);
- Iterated extended Kalman filter with the phase-sensitive observation model and cepstral mean subtraction in the uncertainty decoding framework using a mixture of $M' \in \{1,16\}$ diagonal-covariance Gaussian posterior distributions (IEKF$-\boldsymbol{\alpha}$+CMS+UD$-M'$); for $M'=1$, the *collapsed* Gaussian posterior from the enhancement scheme is employed, whereas for $M'=M=16$, model-specific posterior distributions are applied in the back-end;

Note that the HMMs in the back-end will be matched to the front-end feature extraction scheme, i.e., if the AFE is used, the HMM is also trained on clean features extracted by the AFE. The same applies for the SFE with cepstral mean subtraction. However, when using the feature enhancement scheme (e.g., IEKF+CMS or IEKF$-\boldsymbol{\alpha}$+CMS), the HMMs in the back-end are the ones trained on clean features coded by the SFE+CMS. Thus, we do not account for any distortions the feature enhancement scheme may introduce when processing clean feature vectors.

Cepstral mean subtraction (CMS) is applied to the static features, i.e., prior to the computation of the dynamic features. If the uncertainty decoding rule is employed, the variances of the dynamic features are required and calculated according to [16]. Since, in general, severe approximations are required at all stages of the estimation of the clean feature vector posterior — at the a priori models, the observation model and the inference — the quality of the estimate, i.e., the posterior's mean(s) and covariance(s), is quite questionable. However, with the estimate of the posterior's mean(s) being considerably more reliable than the estimate of the posterior's covariance(s), this issue mainly affects the application of the derived uncertainty decoding rule.

Thus, for instance, Droppo et al. [10] and Liao et al. [22] both reported that posing an upper bound on the posterior's variances $\sigma^2_{\mathbf{x}_t|\mathbf{y}_{1:t},d}$ with respect to the clean speech prior's variances $\sigma^2_{\mathbf{x}_t,d}$ is beneficial to the performance of a recognizer under the uncertainty decoding framework. Following these heuristics, the dth component of the posterior variances is upper bounded as

$$\sigma^2_{\mathbf{x}_t|\mathbf{y}_{1:t},d} := \min\left(\sigma^2_{\mathbf{x}_t|\mathbf{y}_{1:t},d}, \gamma\sigma^2_{\mathbf{x}_t,d}\right) \tag{8.98}$$

in all experiments employing the uncertainty decoding rule. The parameter γ will be kept fixed at $\gamma = 0.05$, i.e., no attempt to find optimal values for γ will be made.

8.4.6 Experimental Results and Discussion

Experimental results for the proposed front-end and back-end schemes on the Aurora 2 and the Aurora 4 databases are presented and discussed next.

8.4.6.1 Aurora 2

Detailed recognition results for the Aurora 2 recognition task on test sets A and B can be found in Tables 8.1 to 8.6, with averaged recognition results given in Fig. 8.5.

Clearly, the baseline recognition accuracies of 66.29% and 71.97% set by the ETSI standard front-end with successive cepstral mean subtraction (SFE+CMS, given in Table 8.2) on test set A and B, respectively, are exceeded by any applied enhancement scheme. Comparing the IEKF+CMS and the IEKF$-\boldsymbol{\alpha}$+CMS, one finds the recognition accuracy to be increased by approximately 2.5% absolute on test sets A and B when using the phase-sensitive observation model IEKF$-\boldsymbol{\alpha}$.

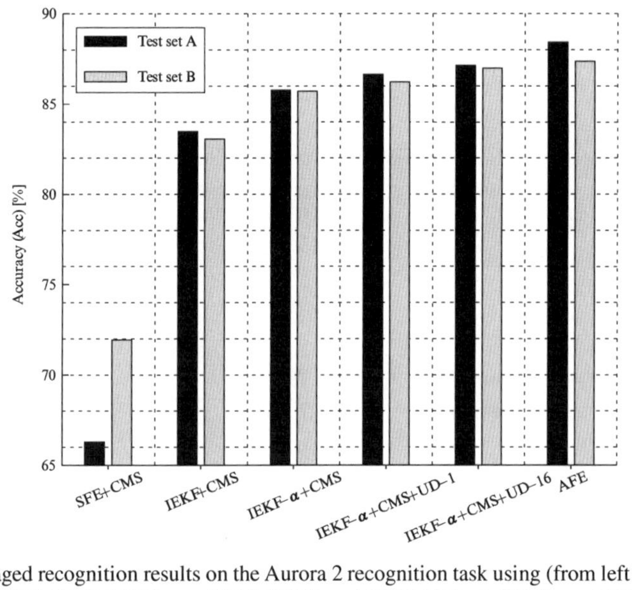

Fig. 8.5: Averaged recognition results on the Aurora 2 recognition task using (from left to right) the SFE+CMS, IEKF+CMS, IEKF$-\boldsymbol{\alpha}$+CMS, IEKF$-\boldsymbol{\alpha}$+CMS+UD$-$1, IEKF$-\boldsymbol{\alpha}$+CMS+UD$-$16 and the AFE on test set A and B

Looking at the results under the IEKF+CMS and the IEKF$-\boldsymbol{\alpha}$+CMS in Tables 8.3 and 8.4 in more detail, an increasing benefit of incorporating phase factor information into the observation model with decreasing SNR values can be observed, confirming the thesis stated in Sect. 8.3.2.

Application of the uncertainty decoding rule using either the *collapsed* Gaussian posterior (i.e., $M' = 1$) or the mixture of $M' = M = 16$ model-specific Gaussian posteriors further increases the recognition accuracies, reaching 87.14% and 86.98% on test set A and B, respectively (see Tables 8.5 and 8.6).

Note that the minor increase of recognition accuracy while moving from single Gaussian posteriors to mixtures of $M = 16$ Gaussian posteriors comes along with an M-fold increased computational load at the likelihood computation in the back-end.

However, the performance of the AFE with 88.42% and 87.37% on test sets A and B, respectively, could not be reached in this recognition task (compare Table 8.1).

Table 8.1: Aurora 2 recognition results using the AFE on test set A (left) and B (right)

SNR	Subway	Babble	Car	Exhibition	AVG
∞	99.69	99.76	99.73	99.78	99.74
20 dB	98.96	99.06	99.25	98.67	98.98
15 dB	97.73	97.76	98.54	97.47	97.88
10 dB	94.57	93.92	96.72	94.23	94.86
5 dB	86.37	82.29	90.43	86.42	86.38
0 dB	66.44	53.69	69.97	65.94	64.01
$\mathrm{AVG}_{0\,\mathrm{dB}}^{20\,\mathrm{dB}}$	88.81	85.34	90.98	88.55	**88.42**

SNR	Restaurant	Street	Airport	Train	AVG
∞	99.69	99.76	99.73	99.78	99.74
20 dB	98.96	98.82	99.25	98.98	99.00
15 dB	96.84	97.82	98.33	97.90	97.72
10 dB	92.02	93.95	95.14	95.50	94.15
5 dB	80.29	85.61	85.86	87.04	84.70
0 dB	52.72	63.69	62.69	65.94	61.26
$\mathrm{AVG}_{0\,\mathrm{dB}}^{20\,\mathrm{dB}}$	84.17	87.98	88.25	89.07	**87.37**

Table 8.2: Aurora 2 recognition results using the SFE+CMS on test set A (left) and B (right)

SNR	Subway	Babble	Car	Exhibition	AVG
∞	99.72	99.67	99.58	99.78	99.69
20 dB	97.39	98.19	97.94	96.95	97.62
15 dB	92.14	94.65	93.68	91.42	92.97
10 dB	77.06	83.80	77.27	75.04	78.29
5 dB	48.20	57.13	42.77	43.41	47.88
0 dB	18.30	21.31	8.17	10.92	14.68
$\mathrm{AVG}_{0\,\mathrm{dB}}^{20\,\mathrm{dB}}$	66.62	71.02	63.97	63.55	**66.29**

SNR	Restaurant	Street	Airport	Train	AVG
∞	99.72	99.67	99.58	99.78	99.69
20 dB	98.56	97.88	98.30	98.09	98.21
15 dB	95.76	94.04	96.33	95.31	95.36
10 dB	86.34	80.83	88.13	83.96	84.81
5 dB	63.09	54.17	63.44	55.17	58.97
0 dB	28.12	17.87	26.51	16.97	22.37
$\mathrm{AVG}_{0\,\mathrm{dB}}^{20\,\mathrm{dB}}$	74.37	68.96	74.54	69.90	**71.94**

Table 8.3: Aurora 2 recognition results using the IEKF+CMS on test set A (left) and B (right)

SNR	Subway	Babble	Car	Exhibition	AVG
∞	99.60	99.64	99.55	99.69	99.62
20 dB	98.74	98.55	99.19	97.93	98.60
15 dB	97.05	96.37	98.66	96.14	97.06
10 dB	92.39	91.17	95.53	91.45	92.64
5 dB	80.20	75.18	85.33	78.22	79.73
0 dB	50.14	39.81	53.53	53.90	49.35
$\mathrm{AVG}_{0\,\mathrm{dB}}^{20\,\mathrm{dB}}$	83.70	80.22	86.45	83.53	**83.47**

SNR	Restaurant	Street	Airport	Train	AVG
∞	99.60	99.64	99.55	99.69	99.62
20 dB	98.00	98.46	98.78	98.89	98.53
15 dB	95.98	97.16	97.67	97.53	97.09
10 dB	90.82	90.66	93.65	94.48	92.40
5 dB	76.08	73.34	80.70	83.37	78.37
0 dB	45.93	42.53	54.10	53.01	48.89
$\mathrm{AVG}_{0\,\mathrm{dB}}^{20\,\mathrm{dB}}$	81.36	80.43	84.98	85.46	**83.06**

Table 8.4: Aurora 2 recognition results using the IEKF$-\alpha$+CMS on test set A (left) and B (right)

SNR	Subway	Babble	Car	Exhibition	AVG
∞	99.63	99.64	99.55	99.69	99.63
20 dB	98.86	98.79	99.31	97.90	98.72
15 dB	97.18	96.52	98.87	96.14	97.18
10 dB	93.55	92.62	96.33	91.89	93.60
5 dB	83.97	80.08	88.13	78.62	82.70
0 dB	59.44	47.67	63.82	55.85	56.70
$\text{AVG}_{0\,\text{dB}}^{20\,\text{dB}}$	86.60	83.14	89.29	84.08	**85.78**

SNR	Restaurant	Street	Airport	Train	AVG
∞	99.63	99.64	99.55	99.69	99.63
20 dB	97.94	98.64	98.84	98.86	98.57
15 dB	96.38	97.58	97.91	97.81	97.42
10 dB	92.08	92.74	93.92	95.06	93.45
5 dB	78.39	80.11	83.66	86.39	82.14
0 dB	51.80	53.66	59.77	62.63	56.96
$\text{AVG}_{0\,\text{dB}}^{20\,\text{dB}}$	83.32	84.55	86.82	88.15	**85.71**

Table 8.5: Aurora 2 recognition results using the IEKF$-\alpha$+CMS+UD-1 on test set A (left) and B (right)

SNR	Subway	Babble	Car	Exhibition	AVG
∞	99.60	99.61	99.55	99.69	99.61
20 dB	98.80	98.49	99.31	97.90	98.62
15 dB	97.48	96.52	98.93	96.39	97.33
10 dB	94.29	92.74	96.90	92.38	94.08
5 dB	85.94	81.08	89.32	80.25	84.15
0 dB	62.17	49.09	66.48	58.32	59.02
$\text{AVG}_{0\,\text{dB}}^{20\,\text{dB}}$	87.74	83.58	90.19	85.05	**86.64**

SNR	Restaurant	Street	Airport	Train	AVG
∞	99.60	99.61	99.55	99.69	99.61
20 dB	97.88	98.73	98.66	98.80	98.52
15 dB	96.01	97.64	97.67	97.62	97.23
10 dB	91.50	93.05	93.71	95.34	93.40
5 dB	79.12	81.68	84.46	87.97	83.31
0 dB	52.99	56.20	61.35	64.09	58.66
$\text{AVG}_{0\,\text{dB}}^{20\,\text{dB}}$	83.50	85.46	87.17	88.76	**86.22**

Table 8.6: Aurora 2 recognition results using the IEKF$-\alpha$+CMS+UD-16 on test set A (left) and B (right)

SNR	Subway	Babble	Car	Exhibition	AVG
∞	99.66	99.61	99.55	99.69	99.63
20dB	98.93	98.88	99.31	97.99	98.78
15dB	97.64	97.07	98.75	96.48	97.48
10dB	94.57	94.29	96.87	92.87	94.65
5dB	86.00	83.34	89.02	82.14	85.12
0dB	62.42	52.18	65.46	58.50	59.64
$\text{AVG}_{0\,\text{dB}}^{20\,\text{dB}}$	87.91	85.15	89.88	85.60	**87.14**

SNR	Restaurant	Street	Airport	Train	AVG
∞	99.66	99.61	99.55	99.69	99.63
20dB	98.25	98.58	98.99	99.01	98.71
15dB	97.18	97.91	98.12	97.96	97.79
10dB	93.49	93.17	94.90	95.46	94.25
5dB	82.56	82.32	86.10	87.75	84.68
0dB	56.06	54.75	63.32	63.68	59.45
$\text{AVG}_{0\,\text{dB}}^{20\,\text{dB}}$	85.51	85.35	88.29	88.77	**86.98**

8.4.6.2 Aurora 4

Detailed experimental results for the Aurora 4 recognition task can be found in Tables 8.7 to 8.12. Figure 8.6 gives an overview of the average number of insertions (I), deletions (D) and substitutions (S), finally composing the average number of errors (E).

Again, the baseline averaged error rate of 38.62% obtained by applying the ETSI standard front-end with successive cepstral mean subtraction (SFE+CMS, given in Table 8.8) is outperformed by any applied enhancement scheme, with major improvements in deletion errors. However, this time the difference between the IEKF+CMS and the IEKF$-\alpha$+CMS, as inferable from Tables 8.9 and 8.10, is only

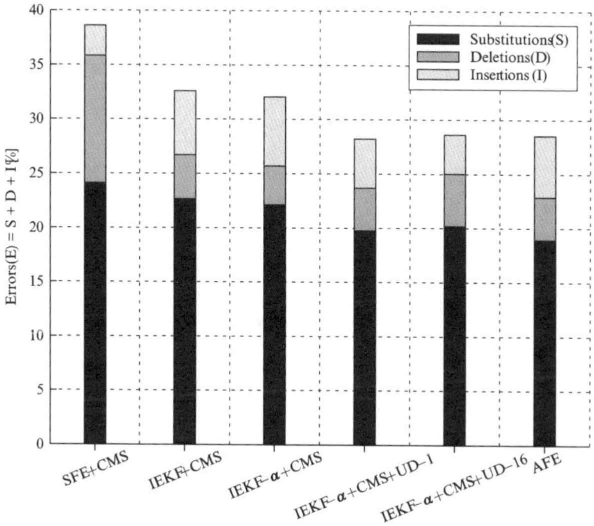

Fig. 8.6: Averaged recognition results on the Aurora 4 recognition task using (from left to right) the SFE+CMS, IEKF+CMS, IEKF$-\boldsymbol{\alpha}$+CMS, IEKF$-\boldsymbol{\alpha}$+CMS+UD-1, IEKF$-\boldsymbol{\alpha}$+CMS+UD-16 and the AFE

marginal. Recalling that the Aurora 4 test set has been created by artificially adding noises at randomly chosen SNRs between 5 dB and 15 dB, this can be attributed to the relatively small portion of test utterances with low SNR and especially to the absence of test utterances with SNR 0 dB, where the Aurora 2 experiments showed the major gain in recognition performance.

Further application of the uncertainty decoding rule using either the *collapsed* Gaussian posterior (i.e., $M' = 1$) or the mixture of $M' = M = 16$ model-specific Gaussian posteriors finally brings the averaged error rate in the vicinity of the error rate obtained by application of the ETSI advanced front-end (compare Tables 8.11 and 8.12 to Table 8.7). In particular, uncertainty decoding with the *collapsed* Gaussian posterior and the mixture of Gaussian posteriors gives error rates of 28.23% and 28.65%, respectively, compared to the 28.46% obtained by the AFE.

Both recognition tasks reveal the superiority of the proposed phase-sensitive observation model (IEKF$-\boldsymbol{\alpha}$) to its phase-insensitive counterpart (IEKF). Experiments on both databases and especially the Aurora 2 task demonstrate the importance of incorporating information about the phase factor into the observation model, and not only at low SNRs.

Application of the uncertainty decoding rule using either the collapsed Gaussian posterior or the mixture of Gaussian posteriors further improves the performance of the speech recognizer, however, subject to the heuristically chosen upper bound of the posterior variances as introduced in Eq. (8.98). In fact, usage of the posterior variances as they are, i.e., without posing an upper bound on them, eventually leads to a degradation of the recognizer's performance compared to using only the mean of the posterior distribution(s) in the standard decoding rule.

Table 8.7: Aurora 4 recognition results using the AFE

Error Statistic	Clean	Car	Babble	Restaurant	Street	Airport	Train	AVG
S [%]	8.69	12.78	21.22	24.49	19.71	22.95	22.03	18.84
D [%]	1.36	2.28	4.75	4.38	4.75	4.60	5.60	3.96
I [%]	2.39	4.53	6.56	7.55	4.49	10.06	4.01	5.66
E [%]	12.45	19.59	32.52	36.43	28.95	37.61	31.64	**28.46**

Table 8.8: Aurora 4 recognition results using the SFE+CMS

Error Statistic	Clean	Car	Babble	Restaurant	Street	Airport	Train	AVG
S [%]	8.36	15.14	27.22	29.91	31.16	27.15	29.91	24.12
D [%]	1.10	4.42	13.78	14.51	15.36	13.92	18.86	11.71
I [%]	1.99	2.73	2.39	3.39	2.95	3.79	2.32	2.79
E [%]	11.45	22.28	43.39	47.81	49.47	44.86	51.09	**38.62**

Table 8.9: Aurora 4 recognition results using the IEKF+CMS

Error Statistic	Clean	Car	Babble	Restaurant	Street	Airport	Train	AVG
S [%]	9.06	11.49	24.86	30.98	26.30	28.55	27.40	22.66
D [%]	1.14	1.73	4.31	5.67	5.34	5.19	4.90	4.04
I [%]	2.36	3.87	6.37	8.40	7.15	8.47	4.90	5.93
E [%]	12.56	17.09	35.54	45.05	38.78	42.21	37.20	**32.63**

Table 8.10: Aurora 4 recognition results using the IEKF$-\alpha$+CMS

Error Statistic	Clean	Car	Babble	Restaurant	Street	Airport	Train	AVG
S [%]	9.13	11.82	23.98	29.80	26.30	27.48	26.45	22.14
D [%]	1.18	1.66	3.79	5.08	4.20	4.90	4.27	3.58
I [%]	2.43	3.90	7.03	9.32	6.81	9.50	5.41	6.34
E [%]	12.74	17.38	34.81	44.20	37.31	41.88	36.13	**32.06**

Table 8.11: Aurora 4 recognition results using the IEKF$-\alpha$+CMS+UD-1

Error Statistic	Clean	Car	Babble	Restaurant	Street	Airport	Train	AVG
S [%]	9.17	11.49	21.73	24.90	23.35	23.98	23.65	19.75
D [%]	1.14	1.80	3.87	5.64	4.83	5.16	5.27	3.96
I [%]	2.43	3.24	4.35	6.19	4.57	7.26	3.61	4.52
E [%]	12.74	16.54	29.94	36.72	32.74	36.39	32.52	**28.23**

8.5 Conclusions

In this contribution, conditional Bayesian estimation employing a phase-sensitive observation model for noise robust speech recognition has been studied.

Table 8.12: Aurora 4 recognition results using the IEKF$-\alpha$+CMS+UD$-$16

Error Statistic	Clean	Car	Babble	Restaurant	Street	Airport	Train	AVG
S [%]	9.13	10.90	21.66	26.59	23.76	23.94	25.41	20.20
D [%]	1.10	1.88	5.49	6.26	6.30	5.60	7.26	4.84
I [%]	2.32	2.69	3.28	4.94	3.79	5.08	3.17	3.61
E [%]	12.56	15.47	30.42	37.79	33.85	34.62	35.84	**28.65**

After a review of speech recognition under the presence of corrupted features, the estimation of the clean feature posterior distribution as the key element of the introduced uncertainty decoding rule has been derived.

Three major components of the estimation process, namely the a priori model, the observation model and the inference algorithm, have further been discussed in more detail. Thereby, most stress has been laid on the derivation of the phase-sensitive observation model and the required moments of the phase factor distribution. While common approaches assume the involved phase factor to be Gaussian-distributed and base the estimation of the moments on available stereo training data, a way to analytically compute all central moments solely based on the used mel filter bank has been derived. Thereby, it has been proven that the distribution is of non-Gaussian nature and independent of noise type and signal-to-noise ratio.

The incorporation of the phase-sensitive observation model into a model-based feature enhancement scheme and its application to the Aurora 2 and Aurora 4 recognition tasks have finally revealed the superiority of the phase-sensitive observation model to its phase-insensitive counterpart and the importance of incorporating phase factor information into the enhancement scheme.

Though subject to many approximations, model-based feature enhancement in combination with uncertainty decoding has eventually reached the performance under the ETSI advanced front-end, however, with a considerably increased computational load.

Acknowledgements This work has been supported by Deutsche Forschungsgemeinschaft (DFG) under contract no. Ha3455/6-1.

References

1. Bar-Shalom, Y., Rong Li, X., Kirubarajan, T.: Estimation with Applications to Tracking and Navigation. John Wiley & Sons, Inc. (2001)
2. Bell, B., Cathey, F.: The iterated Kalman filter update as a Gauss-Newton method. IEEE Transactions on Automatic Control **38**(2), 294–297 (1993)
3. Brillinger, D.R.: Time Series: Data Analysis and Theory. Holt, Rinehart and Winston, Inc. (1975)
4. Dempster, A.P., Laird, N.M., Rubin, D.B.: Maximum likelihood from incomplete data via the EM algorithm. Journal of the Royal Statistical Society: Series B (Statistical Methodology) **39**(1), 1–38 (1977)
5. Deng, J., Bouchard, M., Yeap, T.H.: Noisy speech feature estimation on the Aurora2 database using a switching linear dynamic model. Journal of Multimedia (JMM) **2**(2), 47–52 (2007)
6. Deng, L., Droppo, J., Acero, A.: Enhancement of log mel power spectra of speech using a phase-sensitive model of the acoustic environment and sequential estimation of the corrupting noise. IEEE Transactions on Speech and Audio Processing **12**(2), 133–143 (2004)
7. Deng, L., Droppo, J., Acero, A.: Dynamic compensation of HMM variances using the feature enhancement uncertainty computed from a parametric model of speech distortion. IEEE Transactions on Speech and Audio Processing **13**(3), 412–421 (2005)
8. Droppo, J., Acero, A.: Noise robust speech recognition with a switching linear dynamic model. In: A. Acero (ed.) Proc. of IEEE International Conference on Acoustics, Speech and Signal Processing (ICASSP), vol. 1, pp. I–953–6 vol.1. Montreal, Quebec, Canada (2004)
9. Droppo, J., Acero, A., Deng, L.: A nonlinear observation model for removing noise from corrupted speech log mel-spectral energies. In: Proc. of International Conference on Spoken Language Processing (ICSLP). Denver, Colorado (2002)
10. Droppo, J., Acero, A., Deng, L.: Uncertainty decoding with splice for noise robust speech recognition. In: A. Acero (ed.) Proc. of IEEE International Conference on Acoustics, Speech and Signal Processing (ICASSP), vol. 1, pp. I–57–I–60 vol.1. Orlando, Florida (2002)
11. Droppo, J., Deng, L., Alex, A.: A comparison of three non-linear observation models for noisy speech features. In: Proc. Eurospeech, pp. 681–684. International Speech Communication Association, Geneva, Switzerland (2003)
12. ETSI ES 201 108: Speech processing, transmission and quality aspects; distributed speech recognition; front-end feature extraction algorithm; compression algorithms (2003)
13. ETSI ES 202 050: Speech processing, transmission and quality aspects; distributed speech recognition; advanced front-end feature extraction algorithm; compression algorithms (2007)
14. Faubel, F., McDonough, J., Klakow, D.: A phase-averaged model for the relationship between noisy speech, clean speech and noise in the log-mel domain. In: Proc. of Annual Conference of the International Speech Communication Association (Interspeech). Interspeech, Brisbane, Australia (2008)
15. Hirsch, H.G.: Experimental framework for the performance evaluation of speech recognition front-ends on a large vocabulary task AU/417/02. Tech. rep., STQ AURORA DSR WORKING GROUP (2002)
16. Ion, V., Haeb-Umbach, R.: Uncertainty decoding for distributed speech recognition over error-prone networks. Speech Commununication **48**(11), 1435–1446 (2006)
17. Ion, V., Haeb-Umbach, R.: A novel uncertainty decoding rule with applications to transmission error robust speech recognition. IEEE Transactions on Audio, Speech, and Language Processing **16**(5), 1047–1060 (2008)
18. Isserlis, L.: On a formula for the product-moment coefficient of any order of a normal frequency distribution in any number of variables. Biometrika **12**(1/2), 134–139 (1918)
19. Kim, N.S., Lim, W., Stern, R.: Feature compensation based on switching linear dynamic model. IEEE Signal Processing Letters **12**(6), 473–476 (2005)
20. Krueger, A., Leutnant, V., Haeb-Umbach, R., Ackermann, M., Bloemer, J.: On the initialization of dynamic models for speech features. In: Proc. of ITG Fachtagung Sprachkommunikation. ITG, Bochum, Germany (2010)

21. Leonard, R.: A database for speaker independent digit recognition. In: Proc. of IEEE International Conference on Acoustics, Speech and Signal Processing (ICASSP), vol. 9, pp. 328–331. ICASSP, San Diego, California (1984)
22. Liao, H., Gales, M.: Issues with uncertainty decoding for noise robust automatic speech recognition. Speech Commununication **50**(4), 265–277 (2008)
23. Martin, R., Lotter, T.: Optimal recursive smoothing of non-stationary periodograms. In: Proc. of International Workshop on Acoustic Echo and Noise Control (IWAENC). Darmstadt, Germany (2001)
24. Morris, A., Barker, J., Bourlard, H.: From missing data to maybe useful data: Soft data modelling for noise robust ASR. In: Proc. of International Workshop on Innovation in Speech Processing (WISP), 06. Stratford-upon-Avon, England (2001)
25. Murphy, K.P.: Switching Kalman filters. Tech. rep., U.C. Berkeley (1998)
26. Paul, D.B., Baker, J.M.: The design for the Wall Street Journal-based CSR corpus. In: HLT '91: Proceedings of the workshop on Speech and Natural Language, pp. 357–362. Association for Computational Linguistics, Morristown, NJ, USA (1992)
27. Pearce, D., Hirsch, H.G.: The Aurora experimental framework for the performance evaluation of speech recognition systems under noisy conditions. In: Proc. of International Conference on Spoken Language Processing (ICSLP). Beijing, China (2000)
28. Stouten, V., Van hamme, H., Wambacq, P.: Effect of phase-sensitive environment model and higher order VTS on noisy speech feature enhancement. In: Proc. of IEEE International Conference on Acoustics, Speech and Signal Processing (ICASSP), vol. 1, pp. 433–436. Philadelphia, PA, USA (2005)
29. Stouten, V., Van hamme, H., Wambacq, P.: Model-based feature enhancement with uncertainty decoding for noise robust ASR. Speech Commununication **48**(11), 1502–1514 (2006). Robustness Issues for Conversational Interaction
30. Young, S., Evermann, G., Gales, M., Hain, T., Kershaw, D., Liu, X., Moore, G., Odell, J., Ollason, D., Povey, D., Valtchev, V., Woodland, P.: The HTK Book V3.4. Cambridge University Press, Cambridge, UK (2006)

Part III
Applications: Reverberation Robustness

Chapter 9
Variance Compensation for Recognition of Reverberant Speech with Dereverberation Preprocessing

Marc Delcroix, Shinji Watanabe, Tomohiro Nakatani

Abstract The performance of automatic speech recognition is severely degraded in the presence of noise or reverberation. Much research has been undertaken on noise robustness. In contrast, the problem of the recognition of reverberant speech has received far less attention and remains very challenging. In this chapter, we use a dereverberation method to reduce reverberation prior to recognition. Such a preprocessor may remove most reverberation effects. However, it often introduces distortions, causing a *static* and *dynamic* mismatch between speech features and the acoustic model used for recognition. Acoustic model adaptation could be used to reduce this mismatch. However, conventional model adaptation techniques assume a *static* mismatch and may therefore not cope well with a *dynamic* mismatch arising from dereverberation. This chapter introduces a novel model adaptation scheme that is capable of managing both *static* and *dynamic* mismatches. We introduce a parametric representation of Gaussian variances of the acoustic model that includes *static* and *dynamic* components. Adaptive training is used to optimize the variances in order to realize an appropriate interconnection between dereverberation and a speech recognizer.

Marc Delcroix
NTT Communications Science Laboratories, Kyoto 619-0237, Japan, e-mail: `marc.delcroix@cslab.kecl.ntt.co.jp`

Shinji Watanabe
NTT Communications Science Laboratories, Kyoto 619-0237, Japan, e-mail: `watanabe@cslab.kecl.ntt.co.jp`

Tomohiro Nakatani
NTT Communications Science Laboratories, Kyoto 619-0237, Japan, e-mail: `nak@cslab.kecl.ntt.co.jp`

D. Kolossa and R. Haeb-Umbach (eds.), *Robust Speech Recognition of Uncertain or Missing Data: Theory and Applications*, DOI 10.1007/978-3-642-21317-5_9,
© Springer-Verlag Berlin Heidelberg 2011

9.1 Introduction

Building automatic speech recognition (ASR) systems robust to reverberation is essential if we are to extend their use to hands-free situations, such as in meetings. In such cases, the speech captured by a distant microphone is degraded by noise and reverberation. Although many researchers have focused on building ASR systems robust to noise [13, 39], few have investigated the problem of recognition of reverberant speech. Because reverberation is characterized by a long duration convolutive distortion, most conventional approaches for noise robust ASR may not be efficient when dealing with reverberation. The problem of recognizing reverberant speech therefore remains a challenging issue that we address here.

In this chapter, we introduce a new framework for the recognition of reverberant speech. We use a dereverberation preprocessor to remove reverberation energy from the observed speech signal. Such a preprocessor may reduce most reverberation effects but it usually introduces distortions as a result of imperfect reverberation estimation. These distortions are responsible for the mismatch between the dereverberated speech features and the acoustic model used for recognition. Consequently, using the dereverberated speech features for recognition would not result in satisfactory recognition performance. As the distortion caused by the dereverberation preprocessor may change from one speech analysis time frame to another, the induced mismatch is mainly *dynamic*, but it may as well include a *static* part, which may be dependent only on the states of the acoustic model and be constant over all the analysis time frames.

The *dynamic* nature of the mismatch means that it may not be well compensated for with conventional *static* model adaptation methods [11, 37]. Proposals for handling such *dynamic* mismatches, using for example *dynamic* variance compensation,[1] have recently attracted increased interest. The idea consists of accounting for the mismatch by considering features as random variables, with *dynamic* feature variances. In this case, the ASR decoding is achieved by adding the feature variances to the acoustic model Gaussian variances. This approach leads to a great improvement in ASR performance when the correct *dynamic* feature variances are available. However, *dynamic* feature variances are hard to estimate in practice. Moreover, *dynamic* variance compensation does not consider the existence of a *static* variance mismatch.

To address these issues we propose a novel Gaussian variance parametric model that includes both *static* and *dynamic* components. The *dynamic* component consists of a simple estimation of the *dynamic* feature variance derived from a dereverberation preprocessor. The *static* part is given by a weighted version of the Gaussian variance, as with conventional variance adaptation [37]. The model parameters, which consist of weights for the *static* and *dynamic* components, are optimized using adaptive training. We may thus expect to approach the optimal variance compensation performance. The proposed method can be combined with conventional mean adaptation techniques such as maximum likelihood linear regression (MLLR) in order to

[1] This is also often called uncertainty decoding [10, 27]; see Chapters 2 and 4.

further reduce the *static* mismatch of the Gaussian means.[2] Therefore, the proposed method achieves the combination of *static* and *dynamic* model adaptation.

The organization of this chapter is as follows. In Section 9.2 we describe the relations with existing approaches for reverberant speech recognition, and with variance compensation techniques. In Section 9.3 we describe the proposed system for the recognition of reverberant speech. We briefly describe a dereverberation preprocessor, and recall the principles of *dynamic* variance compensation. In section 9.4 we introduce a parametric model of Gaussian variances, and an algorithm for estimating the variance model parameters. Section 9.5 presents experimental results obtained with the proposed framework for a reverberant digit recognition task. We show that the method may be employed for supervised and unsupervised adaptation. Finally, Section 9.6 presents experimental results for a large vocabulary continuous speech recognition (LVCSR) task. This chapter is based on previously presented papers [6–8], but includes additional theoretical formulations related to the proposed approach and some new experimental results.

9.2 Related Work

Let us briefly review the investigations that have been undertaken on reverberant speech recognition and *dynamic* variance compensation.

The problem of speech recognition under reverberant conditions remains largely unsolved. There are two approaches to solving this problem; model-oriented approaches and feature-oriented approaches. Model-oriented approaches [5, 16, 35, 40, 41, 44] modify a clean speech acoustic model to represent reverberant speech. As model oriented approaches work in the domain of recognition, they can achieve a tight connection with the recognizer. Recently there have been several proposals for modeling reverberant speech by combining a clean speech model with a reverberation model [16, 40]. Promising results were achieved. However, as it is difficult to model the long-term time-domain convolution effect of reverberation in the cepstral feature domain, the performance of such approaches may be limited and decoding will become complex.

The feature-oriented approaches propose transforming the observed reverberant signal to obtain features close to clean speech features [14, 20, 21, 31, 46–48]. Typically a speech dereverberation preprocessor is used to remove reverberation prior to feature extraction. Dereverberation methods consist of estimating the reverberation energy and then removing it from the observed signal to obtain estimated clean speech. Recently many dereverberation methods have been proposed that achieve a significant reduction in reverberation. In particular, methods focusing on late reverberation removal appear particularly promising because they can achieve a large reverberation reduction, are robust to ambient noise and have low computational load [21, 46]. However, because most methods may not perfectly estimate the rever-

[2] The *dynamic* mean mismatch is assumed to be well compensated by the dereverberation preprocessor, and is therefore not considered here.

beration energy, some distortion remains in the processed signal. Such remaining distortions are responsible for a *dynamic* mismatch between the speech features and the acoustic model. Therefore, using a dereverberation preprocessor may not result in great improvements in terms of ASR performance. Note that this problem is also observed when using other speech enhancement preprocessors, such as blind source separation or Wiener filtering for noise reduction [2, 24].

Recently there have been several proposals involving the use of feature uncertainty to account for such a *dynamic* mismatch [2, 4, 9, 10, 24, 25, 34]. One approach is to use *dynamic* variance compensation, or uncertainty decoding, which was proposed in [1, 9, 10, 27, 42]. The idea is to reduce the influence of unreliable features on ASR results by adding a *dynamic* feature variance to the acoustic model variances during decoding. In [9, 10, 27, 42] the feature variance could be derived from the output of model-based speech enhancement methods working in the feature domain.

Many speech enhancement techniques do not work in the feature domain or are not designed to output a feature variance estimate. There have been some trials to estimate the feature variance in such cases, including for blind source separation [22–24], Wiener filtering [2] and dereverberation [25]. For example, [25] introduced a model-based dereverberation method working in the log mel spectrum domain.[3] The method relies on an exponential decline model for the room impulse responses. Dereverberation is achieved using switching Kalman filters to estimate the feature mean and variance. The feature mean and variance are then converted to mel-frequency cepstral coefficients (MFCCs) for recognition.

Many studies have reported a great improvement brought about by *dynamic* variance compensation, especially when Oracle feature variances are used, i.e., feature variances obtained as the difference between clean and enhanced speech features [9, 10]. However, performance is limited when using estimated variances, and in many cases, the feature variance needs to be rescaled to obtain an ASR performance improvement [2, 23, 25, 43] (see also the discussion in Section 5 of Chapter 2). This suggests that the estimated feature variance is not optimal.

The *dynamic* variance adaptation discussed in this chapter proposes modeling the acoustic model Gaussian variances with a parametric model given an estimated feature variance. Therefore, our proposed *dynamic* variance adaptation framework could also be used with the estimated feature variance obtained from other speech enhancements.

The combination of uncertainty decoding and acoustic model adaptation has also been reported in [28]. Joint uncertainty decoding was used with adaptive training in order to mitigate the influence of noise in multi-style training, and therefore obtain a noise-free acoustic model, since features with high uncertainty would contribute less to the model parameter adaptation. In this chapter, adaptation is introduced to solve a different problem. Here, adaptation is employed to optimize the parameters of our proposed parametric model of the Gaussian variance in order to approach the optimal performance of *dynamic* variance compensation.

[3] Details of the method can be found in Chapter 10.

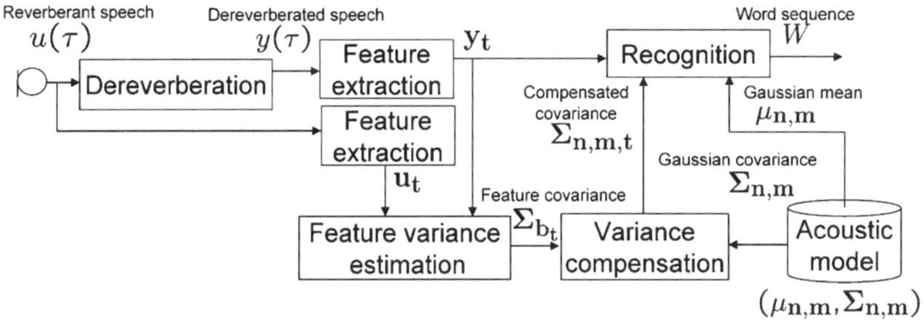

Fig. 9.1: Schematic diagram of the recognition system for reverberant speech

9.3 Overview of Recognition System

Let us describe our proposed framework for the recognition of reverberant speech. Figure 9.1 is a schematic diagram of our proposed ASR system, which consists of the following stages.

1. The observed signal is preprocessed with a dereverberation technique that aims at removing late reverberation.
2. Feature vectors are calculated using standard feature extraction on both dereverberated and observed speech. Here, we use MFCCs as features. Cepstrum Mean Normalization (CMN) is applied to the features to remove short-term reverberation.
3. Recognition is performed using the dereverberated features. *Dynamic* variance compensation is used to mitigate the *dynamic* mismatch between the clean speech acoustic model and the dereverberated features, using a *dynamic* feature variance derived from the observed and dereverberated speech features.

In the following, we describe steps 1 and 3 in more detail.

9.3.1 Dereverberation Based on Late Reverberation Energy Suppression

Let us briefly review the speech dereverberation method used for preprocessing. A reverberant speech signal, $u(\tau)$, can be modeled as the convolution of a clean speech signal $x(\tau)$ with a room impulse response, $h(\tau)$, as

$$u(\tau) = h(\tau) * x(\tau), \tag{9.1}$$

where $*$ is a convolution operator and τ is a discrete time index. The room impulse response, $h(\tau)$, represents the multi-path propagation of speech caused by the reflections of sounds from surfaces in the room. The reverberation time (RT60) of the room impulse response is typically several hundred milliseconds in usual living spaces [26].

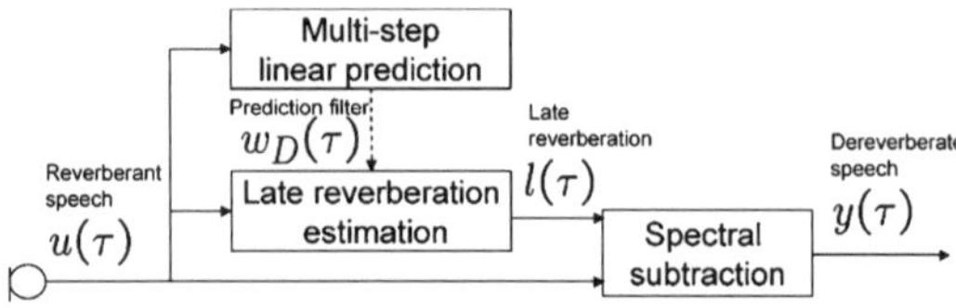

Fig. 9.2: Schematic diagram of the dereverberation preprocessor

We can arbitrarily divide a room impulse response into two parts as

$$h(\tau) \triangleq h_e(\tau) + h_l(\tau), \tag{9.2}$$

where $h_e(\tau)$ represents early reflections, which consist of the direct path and all the reflections arriving at the microphone within around 30 ms of the direct sound, and $h_l(\tau)$ represents late reflections, which consist of all later reflections. Consequently, the microphone signal can be expressed as

$$u(\tau) = \underbrace{h_e(\tau) * x(\tau)}_{\text{early reverberation}} + \underbrace{h_l(\tau) * x(\tau)}_{\text{late reverberation}}. \tag{9.3}$$

In this chapter, we set the duration of the early reflections at the same length as the speech analysis frame used for recognition [21]. Although early reflections generally modify the spectral shape of the speech signal, they can be compensated for to a great degree with current ASR systems using, for example, CMN [33]. In contrast, the later parts of the reverberation, namely late reflections, fall outside the spectral analysis frame, and as a result attenuated versions of past frames are added to the current frame. The induced distortion, i.e., late reverberation, is thus non-stationary and cannot be handled by conventional compensation techniques. Many researchers have reported that late reflections are the main cause of the severe degradation in ASR performance when recognizing reverberant speech [12, 45]. Accordingly, we assume that the main role of the dereverberation preprocessor is to eliminate as far as possible late reverberation.

For this reason, we use a dereverberation method that focuses on late reverberation removal. We adopt the method that was introduced in [21] as it is known to remove much of the late reverberation energy while being relatively robust to ambient noise, and having a low computational complexity [21]. It was shown in [21] that late reverberation could be estimated using multi-step linear prediction, then suppressed by using spectral subtraction. Figure 9.2 is a schematic diagram of the dereverberation preprocessor. A detailed description of the dereverberation method is presented in [21]. Here, we just briefly recall the main steps of the algorithm.

1. First, multi-step linear prediction filter coefficients $w_D(\tau)$ are calculated as

$$\mathbf{w} = (E\{\mathbf{u}(\tau - D)\mathbf{u}(\tau - D)^T\})^+ E\{\mathbf{u}(\tau - D)u(\tau)\}, \tag{9.4}$$

where $\mathbf{w} = [w(0), w(1), \ldots, w(N)]^T$, D is a delay of around 30 ms, $\mathbf{u}(\tau) = [u(\tau), u(\tau-1), \ldots, u(\tau-N)]^T$, N is the filter order, $E\{\}$ is a time averaging operator and $^+$ indicates the Moore-Penrose pseudoinverse. To remove the effect of speech short-term correlation, the filter coefficients are calculated using pre-whitened signals [21].

2. Late reverberation is approximated by a convolution of observed reverberant speech $u(\tau)$ with a multi-step delay linear prediction filter $w_D(\tau)$ as

$$l(\tau) = w_D(\tau) * u(\tau), \tag{9.5}$$

where the prediction filter $w_D(\tau)$ is obtained by adding $D+1$ zeros at the beginning of $\mathbf{w}$, namely $\mathbf{w}_D = [0, 0, ..., 0, w(0), w(1), ..., w(N)]^T$.

3. Finally late reverberation $l(\tau)$ is subtracted from the observed signal $u(\tau)$ by using a spectral subtraction technique [3] in the short time power spectral domain. The dereverberated signal $y(\tau)$ is resynthesized based on the overlap-add synthesis technique by substituting the phase of the observed signal for that of the dereverberated signal.

This dereverberation method may remove late reverberation well. However, due to imperfect estimation of the late reverberation and the use of spectral subtraction, distortions arise that prevent any great improvement in ASR performance.[4] This will be demonstrated experimentally in Section 9.5.2. Since late reverberation $l(\tau)$ is time-varying, the remaining distortions cause a *dynamic* mismatch between the clean speech model and the dereverberated features used for recognition. In the next section, we investigate the use of *dynamic* variance compensation to mitigate the effect of such a mismatch during recognition.

9.3.2 Speech Recognition Based on Dynamic Variance Compensation

Recognition is usually achieved by finding a word sequence, W, that maximizes a likelihood function as

$$\hat{W} = \underset{W}{\operatorname{argmax}}\, p(X|W)p(W), \tag{9.6}$$

where $X = [\mathbf{x}_1, ..., \mathbf{x}_T]$ is a sequence of clean speech feature vectors $\mathbf{x}_t$, t is a frame index, $p(X|W)$ is the acoustic model and $p(W)$ is a language model. The acoustic model is obtained by modeling speech using a Hidden Markov Model (HMM) with the state density given by a Gaussian Mixture (GM):

[4] Note that by using a multi-channel implementation of the dereverberation algorithm, late reverberation estimation accuracy could be improved, leading to somewhat better ASR performance [21]. However, here we consider the more challenging single-channel case.

$$
\begin{aligned}
p(\mathbf{x}_t|n) &= \sum_{m=1}^{M} p(m)p(\mathbf{x}_t|n,m) \\
&= \sum_{m=1}^{M} p(m)N(\mathbf{x}_t;\boldsymbol{\mu}_{n,m},\boldsymbol{\Sigma}_{n,m}),
\end{aligned}
\tag{9.7}
$$

where n is the state index, m is the Gaussian mixture component index, M is the number of Gaussian mixtures, and $\boldsymbol{\mu}_{n,m}$ and $\boldsymbol{\Sigma}_{n,m}$ are mean vector and covariance matrix, respectively. In the following, we consider diagonal covariance matrices and denote the diagonal elements of $\boldsymbol{\Sigma}_{n,m}$ by $\sigma^2_{n,m,i}$, where i is the feature dimension index. The parameters of the acoustic model are trained with clean speech data.

In practice, clean speech features are not available for recognition. A conventional speech recognizer would thus use the dereverberated speech feature vector sequence $Y = [\mathbf{y}_1,...,\mathbf{y}_T]$ as an estimate of the clean features, and perform recognition by replacing X with Y in Eq. (9.6). However, this approach would lead to poor recognition performance since the dereverberated feature vectors $\mathbf{y}_t$ are affected by distortions induced by the dereverberation preprocessor. *Dynamic* variance compensation can be used to mitigate such a mismatch between training and testing conditions [9, 19]. In [9], *dynamic* variance compensation is introduced to reduce the mismatch originating from a noise reduction preprocessor. Here, we follow the same derivation but apply it to speech dereverberation.

We can express the probability of the dereverberated speech feature vector $\mathbf{y}_t$ given state n by marginalizing the joint probability over clean speech:

$$
\begin{aligned}
p(\mathbf{y}_t|n) &= \int_{-\infty}^{+\infty} p(\mathbf{y}_t,\mathbf{x}_t|n)d\mathbf{x}_t, \\
&= \int_{-\infty}^{+\infty} p(\mathbf{y}_t|\mathbf{x}_t,n)p(\mathbf{x}_t|n)d\mathbf{x}_t, \\
&\approx \int_{-\infty}^{+\infty} p(\mathbf{y}_t|\mathbf{x}_t)p(\mathbf{x}_t|n)d\mathbf{x}_t,
\end{aligned}
\tag{9.8}
$$

where $\mathbf{x}_t$ plays the role of a hidden variable, and we assumed that $p(\mathbf{y}_t|\mathbf{x}_t,n) \approx p(\mathbf{y}_t|\mathbf{x}_t)$.

Let us model the mismatch, $\mathbf{b}_t$, between clean speech and dereverberated speech as

$$
\mathbf{y}_t = \mathbf{x}_t + \mathbf{b}_t, \tag{9.9}
$$

where $\mathbf{b}_t$ is modeled as a Gaussian with 0 mean and time-varying covariance matrix $\boldsymbol{\Sigma}_{b_t}$:

$$
p(\mathbf{b}_t) \approx N(\mathbf{b}_t;\mathbf{0},\boldsymbol{\Sigma}_{b_t}). \tag{9.10}
$$

With this model of the mismatch, we assume that the *dynamic* mismatch on the mean could be fully compensated for by the dereverberation preprocessor. Moreover, to simplify the derivation we also consider the case where there is no *static* mismatch for the mean. If the mismatch had a non-zero mean, it could be compensated with conventional model adaptation techniques such as MLLR. We further refer to $\boldsymbol{\Sigma}_{b_t}$

as the feature covariance matrix. Given the model of the mismatch provided by Eq. (9.10), $p(\mathbf{y}_t|\mathbf{x}_t)$ can be expressed as

$$p(\mathbf{y}_t|\mathbf{x}_t) \approx N(\mathbf{y}_t;\mathbf{x}_t,\boldsymbol{\Sigma}_{b_t}). \tag{9.11}$$

Replacing Eqs. (9.7) and (9.11) in Eq. (9.8) and using the probability multiplication rule, the likelihood of reverberant feature $\mathbf{y}_t$ can be rewritten as

$$p(\mathbf{y}_t|n) = \sum_{m=1}^{M} p(m)N(\mathbf{y}_t;\boldsymbol{\mu}_{n,m},\underbrace{\boldsymbol{\Sigma}_{n,m}+\boldsymbol{\Sigma}_{b_t}}_{\triangleq\boldsymbol{\Sigma}_{n,m,t}}). \tag{9.12}$$

Recognition is thus performed by using the dereverberated features as with a conventional recognizer, but the Gaussian covariance matrices are augmented by the feature covariance matrix. Intuitively, this means that unreliable features, which would have a large variance, would lead to a lower likelihood value, and therefore have less effect on the recognition results. $\boldsymbol{\Sigma}_{n,m,t}$ is a time-varying mixture covariance matrix obtained after compensation. It is shown in [9] that *dynamic* variance compensation could greatly improve the recognition performance, especially when the Oracle feature variance is used for $\boldsymbol{\Sigma}_{b_t}$. Here the Oracle feature variance would be obtained as the square of the difference between clean and dereverberated speech features. In practice, the Oracle feature variance is not available, and therefore the compensated covariance matrix $\boldsymbol{\Sigma}_{n,m,t}$ may not be optimal. As a result, the performance is far from the Oracle case. In Section 9.4, we address the issue of estimating the time-varying mixture covariance matrix in order to approach Oracle performance.

9.3.3 Relation with a Conventional Uncertainty Decoding Approach

The principles behind *dynamic* variance compensation, as discussed in Section 9.3.2 are similar to other approaches [1, 9, 10, 27, 42], and mainly differ in terms of the mismatch model used. It is interesting to compare the mismatch model of Eq. (9.11) with the model used by other uncertainty decoding approaches. As an example, we compare our approach with a conventional uncertainty decoding approach based on SPLICE [10]. Let us briefly recall the formalism of SPLICE uncertainty decoding. SPLICE was originally derived for noisy speech recognition. Here, for comparison with our framework, we consider the derivation of SPLICE to compensate for a dereverberation mismatch,[5] and therefore the notations are consistent with the previous sections but have a different meaning from that in [10]. SPLICE formalization relies on a GMM model for the dereverberated speech as

[5] This would be equivalent to applying SPLICE after dereverberation preprocessing.

$$p(\mathbf{y}_t) = \sum_s p(s)p(\mathbf{y}_t|s) = \sum_s p(s)N(\mathbf{y}_t;\boldsymbol{\mu}_s,\boldsymbol{\Sigma}_s), \tag{9.13}$$

where s is the dereverberated speech mixture component index and $\boldsymbol{\mu}_s$, $\boldsymbol{\Sigma}_s$ are the GMM mean vector and covariance matrix, trained from dereverberated speech data. The conditional probability is modeled by a Gaussian model as

$$p(\mathbf{x}_t|\mathbf{y}_t,s) = N(\mathbf{x}_t;\mathbf{y}_t+\mathbf{r}_s,\boldsymbol{\Gamma}_s), \tag{9.14}$$

where $\mathbf{r}_s$ and $\boldsymbol{\Gamma}_s$ represent the mean and covariance mismatch and would need to be trained using stereo data.

The conditional probability of dereverberated speech given clean speech is then expressed as

$$\begin{aligned} p(\mathbf{y}_t|\mathbf{x}_t) &= \frac{\sum_s p(s)p(\mathbf{y}_t|s)p(\mathbf{x}_t|\mathbf{y}_t,s)}{p(\mathbf{x}_t)} \\ &\approx \frac{\sum_s p(s)p(\mathbf{y}_t|s)N(\mathbf{x}_t;\mathbf{y}_t+\mathbf{r}_s,\boldsymbol{\Gamma}_s)}{N(\mathbf{x}_t;\boldsymbol{\mu}_x,\boldsymbol{\Sigma}_x)} \\ &\approx N(\mathbf{x}_t;\boldsymbol{\mu}_{\hat{x}}^{s^*},\boldsymbol{\Sigma}_{\hat{\mathbf{x}}}^{s^*}), \end{aligned} \tag{9.15}$$

where $\boldsymbol{\mu}_x$ and $\boldsymbol{\Sigma}_x$ are the estimated mean and covariance of clean speech obtained by replacing the GMM by a single Gaussian, $s*$ is the mixture component that maximizes $p(s)p(\mathbf{y}_t|s)$ and

$$\begin{aligned} \mu_{\hat{x},i}^{s} &= \frac{\Sigma_{x,i}(y_{t,i}+r_{s,i}) - \Gamma_{s,i}\mu_{x,i}}{\Sigma_{x,i}-\Gamma_{s,i}}, \\ \Sigma_{\hat{x},i}^{s} &= \frac{\Sigma_{x,i}\Gamma_{s,i}}{\Sigma_{x,i}-\Gamma_{s,i}}, \end{aligned} \tag{9.16}$$

where i is the feature index, and we consider diagonal covariance matrices. Details of the derivation can be found in [10].

We can link our feature covariance model with the one used for a SPLICE uncertainty decoding formalism if we consider that the mismatch covariance of our model Σ_{b_t} is somewhat similar to the SPLICE mismatch covariance matrix $\boldsymbol{\Gamma}_s$. With SPLICE, the mismatch covariance is dependent on the dereverberated speech mixture index, and is calculated using stereo training data as the variance of the difference between clean and dereverberated speech given s. In contrast, $\boldsymbol{\Sigma}_{b_t}$ is a time estimate of such a covariance. The SPLICE derivation of the conditional probability $p(\mathbf{y}_t|\mathbf{x}_t)$ therefore becomes equivalent to our model if we can assume that dereverberation almost completely removes the mean mismatch, i.e., $\mathbf{r}_s \approx \mathbf{0}$, and that the mismatch variance is much smaller than the clean speech variance ($\Sigma_{x,i} \gg \Gamma_{s,i}$). These hypotheses may be reasonable here since we are dealing with a mismatch emanating from the preprocessor and not from a noise mismatch as with the original SPLICE formalization.

As discussed in [10], the SPLICE estimate of the feature covariance does not provide the optimal performance improvement. Our proposed *dynamic* variance adaptation approach, described in Section 9.4, could thus be used with SPLICE variance estimates to further improve recognition performance.

9.4 Proposed Method for Variance Compensation

In an effort to approach the performance obtained with *dynamic* variance compensation using Oracle variance, we propose a novel parametric model for compensated covariance matrix $\boldsymbol{\Sigma}_{n,m,t}$, and a procedure for estimating the model parameters using adaptive training implemented with the EM algorithm. By considering a model for $\boldsymbol{\Sigma}_{n,m,t}$ that includes *static* and *dynamic* components, we may compensate for both *static* and *dynamic* mismatches.

9.4.1 *Combination of* Static *and* Dynamic *Variance Adaptation*

Inspired by the compensated mixture covariance matrix of Eq. (9.12), we model the compensated covariance matrix as

$$\boldsymbol{\Sigma}'_{n,m,t} = \boldsymbol{\Sigma}_S + \boldsymbol{\Sigma}_{D,t}, \tag{9.17}$$

where $\boldsymbol{\Sigma}_S$ and $\boldsymbol{\Sigma}_{D,t}$ represent *static* and *dynamic* covariance components, respectively. The $'$ symbol is used here to distinguish the covariance matrix obtained from the parametric model and the one of Eq. (9.12).

We further express $\boldsymbol{\Sigma}_S$ and $\boldsymbol{\Sigma}_{D,t}$ with a parametric representation similar to MLLR. The *static* covariance $\boldsymbol{\Sigma}_S$ can thus be expressed as

$$\boldsymbol{\Sigma}_S(\boldsymbol{\Sigma}_{n,m}, \mathbf{L}) = \mathbf{L}\boldsymbol{\Sigma}_{n,m}\mathbf{L}^T, \tag{9.18}$$

where $\mathbf{L}$ is a matrix of *static* covariance compensation parameters. Equation (9.18) can be simplified if we assume the use of a diagonal covariance matrix, which is widely employed in speech recognition,

$$(\boldsymbol{\Sigma}_S(\boldsymbol{\Sigma}_{n,m}, \boldsymbol{\lambda}))_{i,i} = \lambda_i \sigma^2_{n,m,i}, \tag{9.19}$$

where λ_i can be interpreted as the weight of the variances of the acoustic models [38].

In a similar way, we model the *dynamic* covariance as

$$\boldsymbol{\Sigma}_{D,t}(\boldsymbol{\Sigma}_{\hat{b}_t}, \mathbf{A}) = \mathbf{A}\boldsymbol{\Sigma}_{\hat{b}_t}\mathbf{A}^T, \tag{9.20}$$

where $\mathbf{A}$ is a matrix of *dynamic* covariance compensation parameters, and $\boldsymbol{\Sigma}_{\hat{b}_t}$ is an estimate of the previously defined feature covariance matrix. Ideally, the feature variance should be computed as the squared difference between clean and enhanced speech features [9]. However, this calculation is not possible because the clean speech features are unknown. Here, we consider a diagonal feature covariance and assume that it is proportional to the square of an estimated mismatch, $\hat{\mathbf{b}}_t$, given by the difference between observed reverberant and dereverberated speech features. Intuitively, this means that speech enhancement introduces more distortions when a great amount of reverberation energy is removed. We can then express the *dynamic* variance component $\boldsymbol{\Sigma}_{D,t}$ as,

$$(\boldsymbol{\Sigma}_{D,t}(\hat{\mathbf{b}}_t, \boldsymbol{\alpha}))_{i,i} = \alpha_i \hat{b}_{t,i}^2 \tag{9.21}$$

where the α_i are model parameters.

Given the proposed parametric models for the *static* and *dynamic* variance, we can rewrite the time-varying state variance as,

$$(\boldsymbol{\Sigma}'_{n,m,t})_{i,i} = \alpha_i \hat{b}_{t,i}^2 + \lambda_i \sigma_{n,m,i}^2. \tag{9.22}$$

We further estimate the parameters α_i and λ_i by using adaptive training. Note that if $\alpha_i = 0$ the model is equivalent to that of conventional *static* variance compensation [38] and if α_i is constant and $\lambda_i = 1$ it is equivalent to the conventional *dynamic* variance compensation model [9]. Consequently, the proposed model enables us to combine both *static* and *dynamic* variance compensation within an adaptive training framework. It is important to note that the proposed method can be further combined with mean adaptation techniques such as MLLR [11] to further reduce the gap between model and speech features.

Figure 9.3 is a schematic diagram of the adaptation process used to estimate parameters $(\boldsymbol{\alpha}, \boldsymbol{\lambda})$. The adaptation process consists of the following steps.

1. First dereverberation preprocessing is employed.
2. Feature extraction is performed on both reverberant and dereverberated speech.
3. The estimated mismatch $\hat{\mathbf{b}}_t$ is obtained as

$$\hat{\mathbf{b}}_t = \mathbf{u}_t - \mathbf{y}_t. \tag{9.23}$$

4. Optimal variance parameters $(\boldsymbol{\alpha}, \boldsymbol{\lambda})$ are obtained from the adaptation with the EM algorithm, using the dereverberated feature, the estimated mismatch $\hat{\mathbf{b}}_t$, the acoustic model and reference labels.

In the next subsection, we discuss the adaptation in detail.

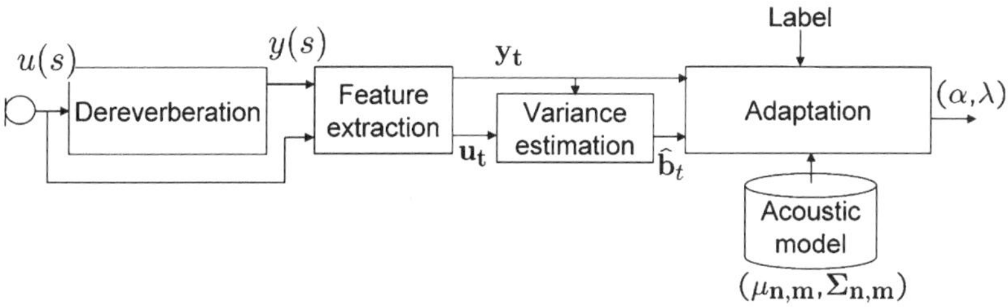

Fig. 9.3: Schematic diagram of adaptation (from [8], ©2009 IEEE)

9.4.2 Adaptation of Variance Model Parameters

The parameters of the variance model, $\boldsymbol{\theta} = (\boldsymbol{\alpha}, \boldsymbol{\lambda})$, can be estimated using adaptive training by maximizing the likelihood as

$$\hat{\boldsymbol{\theta}} = \underset{\boldsymbol{\theta}}{\operatorname{argmax}}\, p(Y|\bar{W}, \boldsymbol{\theta})p(\bar{W}), \tag{9.24}$$

where $\bar{W}$ is the word sequence corresponding to the input speech. To simplify the discussion, we consider supervised adaptation where $\bar{W}$ is known a priori, although the method can also be applied to unsupervised adaptation as discussed in Section 9.5.4.

It is possible to find a solution to Eq. (9.24) using the EM algorithm by defining an auxiliary function $Q(\boldsymbol{\theta}|\tilde{\boldsymbol{\theta}})$ as

$$Q(\boldsymbol{\theta}|\tilde{\boldsymbol{\theta}}) = \sum_S \sum_C \iint_{X+B=Y} p(X,B,S,C|\boldsymbol{\Psi}, \tilde{\boldsymbol{\theta}}) \log(p(X,B,S,C|\boldsymbol{\Psi}, \boldsymbol{\theta}))dXdB, \tag{9.25}$$

where B is a mismatch feature sequence,[6] $B = [\mathbf{b}_1, \ldots, \mathbf{b}_T]$, $S = [s_1, \ldots, s_T]$ is a set of all possible state sequences, $C = [c_1, \ldots, c_T]$ is a set of all mixture components, $\boldsymbol{\Psi}$ represents the acoustic model parameters, and $\tilde{\boldsymbol{\theta}}$ represents an estimate of parameter θ obtained from the previous step of the EM algorithm. If we assume an HMM model for speech with $a_{i,j}$ as the state transition probability, Eq. (9.25) can be rewritten as

$$Q(\boldsymbol{\theta}|\tilde{\boldsymbol{\theta}}) = \sum_S \sum_C \iint_{X+B=Y} p(X,B,S,C|\boldsymbol{\Psi}, \tilde{\boldsymbol{\theta}})$$
$$\log(\prod_{t=1}^{T} a_{s_{t-1},s_t} p(c_t = m|s_t = n)p(\mathbf{x}_t|s_t = n, c_t = m, \boldsymbol{\lambda})p(\mathbf{b}_t|\boldsymbol{\alpha}))dXdB. \tag{9.26}$$

In the following, to simplify the notation, we omit the integration domain, and replace $p(\mathbf{x}_t|s_t = n, c_t = m, \boldsymbol{\lambda})$ with $p(\mathbf{x}_t|n, m, \boldsymbol{\lambda})$.

[6] Note that the mismatch feature sequence B is used here only to simplify the derivation of the EM algorithm. However the solution obtained does not require to explicitly know B.

If we neglect the terms that are independent of $\boldsymbol{\alpha}$ and $\boldsymbol{\lambda}$, and perform a similar derivation as that described in [38], we may express the auxiliary function as

$$\begin{aligned} Q(\boldsymbol{\theta}|\tilde{\boldsymbol{\theta}}) &\propto \sum_{t=1}^{T}\sum_{n=1}^{N}\sum_{m=1}^{M} \iint p(X,B,n,m|\boldsymbol{\Psi},\tilde{\boldsymbol{\theta}}) \log(p(\mathbf{x}_t|n,m,\boldsymbol{\lambda}))dXdB \\ &+\sum_{t=1}^{T}\sum_{n=1}^{N}\sum_{m=1}^{M} \iint p(X,B,n,m|\boldsymbol{\Psi},\tilde{\boldsymbol{\theta}}) \log(p(\mathbf{b}_t|\boldsymbol{\alpha}))dXdB, \\ &= Q_{\boldsymbol{\lambda}}(\boldsymbol{\lambda}|\tilde{\boldsymbol{\theta}}) + Q_{\boldsymbol{\alpha}}(\boldsymbol{\alpha}|\tilde{\boldsymbol{\theta}}), \end{aligned} \tag{9.27}$$

where N is the number of HMM states. The auxiliary function of Eq. (9.27) is similar to that used for stochastic matching [38]. The difference arises from the model of the mismatch given by Eq. (9.21), which includes a *dynamic* part. $\boldsymbol{\theta}$ should be obtained by maximizing Eq. (9.27). We observe that the auxiliary function decomposes into two functions, $Q_{\boldsymbol{\alpha}}(\boldsymbol{\alpha}|\tilde{\boldsymbol{\alpha}},\tilde{\boldsymbol{\lambda}})$ and $Q_{\boldsymbol{\lambda}}(\boldsymbol{\lambda}|\tilde{\boldsymbol{\alpha}},\tilde{\boldsymbol{\lambda}})$. However, there is no closed form solution for the joint estimation of $(\boldsymbol{\alpha},\boldsymbol{\lambda})$. Therefore, we consider the three following cases, $\alpha_i = const.$ (i.e., *Static* Variance Adaptation, SVA), $\lambda_i = const.$ (*Dynamic* Variance Adaptation, DVA) and a combination of the two (*Static* and *Dynamic* Variance Adaptation, SDVA or DSVA).

9.4.2.1 Static Variance Adaptation (SVA, $\alpha_i = const.$)

Let us here consider the maximization of $Q(\boldsymbol{\theta}|\tilde{\boldsymbol{\theta}})$ with respect to $\boldsymbol{\lambda}$ for a constant $\boldsymbol{\alpha}$. By considering the model expressed by Eq. (9.19) and performing a similar calculation to that reported in [38], we can show that a closed form solution for λ_i may be obtained as

$$\lambda_i = \frac{\sum_{t=1}^{T}\sum_{n=1}^{N}\sum_{m=1}^{M} \gamma_t(n,m) \frac{R(x_{t,i},\mathbf{y}_t,n,m,\boldsymbol{\Psi},\tilde{\boldsymbol{\alpha}},\tilde{\boldsymbol{\lambda}})}{\sigma^2_{n,m,i}}}{\sum_{t=1}^{T}\sum_{n=1}^{N}\sum_{m=1}^{M} \gamma_t(n,m)}, \tag{9.28}$$

where $\gamma_t(n,m)$ is the mixture component occupancy probability, which can be obtained using the forward-backward algorithm, and $R(x_{t,i},\mathbf{y}_t,n,m,\boldsymbol{\Psi},\tilde{\boldsymbol{\alpha}},\tilde{\boldsymbol{\lambda}})$ is an estimate of the dereverberated feature variance. The details of the derivation of Eq. (9.28) are given in Appendix 1. Looking at Eq. (9.28), we can interpret λ_i as a weighted average of the ratio between the enhanced feature variance and the model variance.

9.4.2.2 Dynamic Variance Adaptation (DVA, $\lambda_i = const.$)

If we consider $\lambda_i = const.$, we can find a closed form solution to the maximization problem by inserting Eqs. (9.10) and (9.21) in Eq. (9.27) and maximizing with respect to α_i:

$$\alpha_i = \frac{\sum_{t=1}^{T}\sum_{n=1}^{N}\sum_{m=1}^{M}\gamma_t(n,m)\frac{E\{b_{t,i}^2|\mathbf{y}_t,n,m,\boldsymbol{\Psi},\tilde{\boldsymbol{\alpha}},\tilde{\boldsymbol{\lambda}}\}}{\hat{b}_{t,i}^2}}{\sum_{t=1}^{T}\sum_{n=1}^{N}\sum_{m=1}^{M}\gamma_t(n,m)}, \tag{9.29}$$

where $E\{b_{t,i}^2|\mathbf{y}_t,n,m,\boldsymbol{\Psi},\tilde{\boldsymbol{\alpha}},\tilde{\boldsymbol{\lambda}}\}$ is an estimate of the mismatch variance given the enhanced feature and the acoustic model. The derivation of Eq. (9.29) is given in Appendix 2.

Note that it follows from Eq. (9.29) that α_i is simply a weighted average of the ratio between the mismatch variance given the enhanced feature and the acoustic model, and the estimated mismatch variance $\hat{b}_{t,i}^2$.

9.4.2.3 *Static* and *Dynamic* Variance Adaptation (SDVA or DSVA)

It may not be easy to find a closed form solution for the EM algorithm when the maximization relative to $\boldsymbol{\alpha}$ and $\boldsymbol{\lambda}$ is performed at the same time. However, we determined that solutions could be found if we considered the maximization relative to $\boldsymbol{\alpha}$ and $\boldsymbol{\lambda}$ separately. As these two maximization problems involve the same likelihood function, the likelihood would also increase if we performed maximization relative to each parameter in turn, as regards the Expectation Conditional Maximization (ECM) algorithm [29]. This procedure may approach the general solution. With the first case, we start by removing the *static* bias with *static* variance adaptation as described in Section 9.4.2.1 and setting $\alpha_i = 0$. Then, using the previously adapted acoustic model, we perform *dynamic* variance adaptation as shown in Section 9.4.2.2. This is referred to as *Static* and *Dynamic* Variance Adaptation (SDVA). In the experiments described in Section 9.5.5.2 we also investigated the opposite case, where first *dynamic* variance adaptation is performed followed by *static* variance adaptation (i.e., *Dynamic* and *Static* Variance Adaptation, DSVA).

9.5 Digit Recognition Experiments

We carried out experiments to confirm the effectiveness of combining *static* and *dynamic* variance adaptation. We implemented the proposed adaptation scheme and the *dynamic* variance compensation decoding rule of Eq. (9.12) in the SOLON speech recognizer [17]. We first carried out experiments using a small vocabulary task consisting of continuous digit utterances. Experimental results for large vocabulary recognition are provided in Section 9.6.

9.5.1 Experimental Settings

The experimental settings are summarized in Table 9.1. The acoustic model consisted of speaker- (and gender-)independent word-based left-to-right HMMs with 16 states and three Gaussians per state, except for the silent model which consisted of three states. The HMMs were trained using clean speech drawn from the TI-Digits database. The speech signals were downsampled from 20 to 8 kHz. The acoustic feature vector consisted of 39 coefficients: 12 MFCCs, the 0-th cepstral coefficient, the delta and the acceleration coefficients. CMN was applied to the features. The above experimental setting almost corresponds to a clean speech set of Aurora 2 noisy digit recognition tasks [15].

We generated reverberant speech by convolving clean speech with a room impulse response measured in a room with a reverberation time of around 0.5 s. We considered two impulse responses measured for speaker-to-microphone distances of 1.5 m and 2 m.

In the dereverberation described in Section 9.3.1, we set the length of the prediction filter (L) at 4,000 taps (= 0.5 s), the delay D of the multi-step linear prediction at 30 ms, and the frame rate and frame size for spectral subtraction at 3.7 and 30 ms, respectively.

The test set consisted of 561 utterances (6,173 digits) spoken by 104 male and female speakers. The clean speech utterances were obtained from the TI-Digits clean test set. To account for the long distortion caused by reverberation, the test utterances were generated by concatenating two or three utterances from the same speaker without inserting pauses, so that the average duration of the test utterances was around 6 s. We measured the ASR performance using the Word Error Rate (WER).

9.5.2 Baseline Results

Let us first investigate the potential of *dynamic* variance compensation without adaptation. Table 9.2 gives the following baseline recognition results:

- Clean: Recognizing clean speech without preprocessing or variance compensation.
- Reverberant: Recognizing reverberant speech (distance = 1.5 m) without preprocessing or variance compensation.
- Dereverberated: Recognizing dereverberated speech with dereverberation preprocessing [21] and without variance compensation.

Table 9.1: Experimental settings for the digit recognition experiment

Room characteristics	
Reverberation time	0.5 s
Speaker-microphone distance	1.5 m and 2 m
Dereverberation	
Sampling frequency	8 kHz
Prediction filter length (L)	4000 taps
Delay (analysis frame length) (D)	30 ms
Frame rate for spectral subtraction	3.7
Features	
Dimension	39
Type	12 MFCC coefficients + 0th cepstral coefficient + Δ + $\Delta\Delta$
Postprocessing	CMN
Acoustic model	
Type	Word based HMM Speaker independent
Number of HMM states (N)	16 (3 for silent)
Number of Gaussians per state (M)	3
Training data set	
Training data	Clean speech from TI-Digits data Downsampled from 20 to 8 kHz
Testing data set	
Number of utterances	561 (6,173 digits)
Number of speakers	104 male and female
Average duration	6 s

- Variance compensation ($\alpha_i = 1$, $\lambda_i = 1$): Dereverberated + variance compensation with variance given by the square of the estimated mismatch (without adaptation, i.e., $\alpha_i = 1$, $\lambda_i = 1$).
- Variance compensation (Oracle feature variances): Dereverberated + variance compensation with ideal (Oracle) variance given by the square of the mismatch between clean features known a priori and dereverberated speech features.

We observed severe degradation induced by reverberation. Only a small error reduction was achieved when using single-channel dereverberation. Using *dynamic* variance compensation without adaptation can already greatly reduce WER by around 48% compared with dereverberated speech. Note that similar results for *dynamic* variance compensation were obtained in a similar experiment conducted with a different dereverberation algorithm [46]. These results confirmed our intuition that *dynamic* variance compensation is effective in reducing the *dynamic* mismatch caused by a dereverberation preprocessor. However, the performance obtained with variance compensation (WER of 15.9%) is still far from that obtained with Oracle variance (WER of 3.3%), which approaches the WER of clean speech (1.2%). This result suggests that variance compensation performance could be further improved if we could obtain better feature variance estimates. Our objective is to approach the level of performance provided by the Oracle feature variance.

Table 9.2: Baseline ASR results measured by Word Error Rate (WER (%)) (from [8], ©2009 IEEE)

	WER (%)
Clean	1.2
Reverberant	32.7
Dereverberated	31.0
Variance compensation ($\alpha_i = 1$, $\lambda_i = 1$)	15.9
Variance compensation (Oracle feature variances)	3.3

9.5.3 Supervised Adaptation

We first conducted experiments using supervised adaptation. We used speaker-independent adaptation data, i.e., utterances spoken by various speakers, to evaluate the adaptation performance with respect to mitigating only the distortion originating from the preprocessor. The adaptation data consisted of 520 utterances, which were generated by concatenating two or three utterances, similarly to the test set preparation. The utterances were spoken by the same female and male speakers that spoke the test set. To test the influence of the amount of adaptation data, we used subsets of adaptation data containing from two to 512 utterances extracted randomly from the 520 adaptation utterances. The number of iterations of the EM algorithm was set at two for SVA and 30 for DSVA; see the discussion in Section 9.5.5.1.

We investigated the different adaptation schemes for distances between the speaker and the microphones of 1.5 – 2 m. Even if the reverberation time remains the same, using a larger speaker-to-microphone distance reduces the ratio of early to late reverberation energy and therefore affects ASR performance. This can be measured using the Deutlichkeit value (D50) of an impulse response, which is defined as the ratio of the power of the impulse response within the first 50 ms to that of the entire impulse response:

$$\mathrm{D50} = \frac{\sum_{\tau=0}^{F_s*0.05} (h(\tau))^2}{\sum_{\tau=0}^{\infty} (h(\tau))^2}, \tag{9.30}$$

where F_s is the sampling frequency. The D50 values of the two impulse responses with distances of 1.5 – 2 m were 0.694 and 0.629, respectively.

Table 9.3 summarizes the results obtained with variance adaptation for two, 32 and 512 adaptation utterances. The parameters for the variance models were obtained as follows:

- SVA : using SVA with $\alpha_i = 0$, i.e., without *dynamic* feature variance,
- DVA : using DVA with $\lambda_i = 1$, i.e., without *static* variance adaptation,
- SDVA : SVA followed by DVA.

Applying only SVA or DVA achieves approximatively a 50% relative WER reduction compared with dereverberated speech. Furthermore, combining *static* and *dynamic* adaptation (SDVA) achieves up to 56% WER improvement. These results

Table 9.3: WER for distances of 1.5 – 2 m between the speaker and the microphone. The results compare the proposed variance adaptation, variance compensation ($\alpha_i = 1$, $\lambda_i = 1$), and reverberant/dereverberated speech recognition (without any variance adaptation/compensation) (from [8], ©2009 IEEE)

	1.5 m			2 m		
Reverberant	32.7 %			37.1 %		
Dereverberated	31.0 %			36.3 %		
Variance Compensation ($\alpha_i = 1$, $\lambda_i = 1$)	15.9 %			19.5 %		
	2 ut.	32 ut.	512 ut.	2 ut.	32 ut.	512 ut.
SVA	15.1 %	15.1 %	15.2 %	18.4 %	18.4 %	18.5 %
DVA	15.6 %	15.5 %	15.5 %	19.6 %	19.4 %	19.2 %
SDVA	**13.5 %**	13.3 %	13.4 %	**16.6 %**	16.3 %	16.3 %
MLLR(mean)	38.4 %	17.7 %	17.4 %	43.2 %	21.4 %	21.1 %
SVA + MLLR(mean)	25.5 %	11.4 %	11.3 %	34.5 %	12.4 %	12.4 %
DVA + MLLR(mean)	28.3 %	**8.6 %**	8.1 %	31.7 %	**10.2 %**	10.3 %
SDVA + MLLR(mean)	19.8 %	11.2 %	**5.4 %**	25.1 %	12.8 %	**6.5 %**

confirm that variance adaptation can significantly improve ASR performance, and that the best results are obtained using a combination of *static* and *dynamic* variance adaptation. Moreover, we observed that convergence was almost achieved after two utterances; this will be further discussed in Section 9.5.5.1.

We also investigated the combination of variance adaptation with mean adaptation using MLLR. In this experiment, we first performed variance adaptation and then MLLR mean adaptation, because better performance was realized with this processing order. We used global MLLR, i.e., the MLLR mean transforms were shared among all Gaussians of the model. Note that SVA + MLLR (mean) is equivalent to "unconstrained MLLR" where the covariance matrices are transformed independently of the mean vectors [11]. By using MLLR, the performance improves when enough adaptation data are available. In particular, when 512 adaptation utterances are available, we can achieve a WER improvement of up to 82%, and recognition performance close to clean speech.

In Table 9.3 we highlighted the best performance for a given number of adaptation utterances. It is interesting to observe that the best performance is obtained with different adaptation schemes. When only a few adaptation utterances are available, SDVA performs best. When more adaptation data are available, combining variance adaptation with mean adaptation further improves the results. In all cases, the optimal performance is obtained by combining *static* mean or variance adaptation with *dynamic* variance adaptation.

As predicted, recognition performance worsens when using a larger speaker-to-microphone distance. However, we observed the same tendency in the results for distances of 1.5 m and 2 m.

Table 9.4: Comparison of supervised and unsupervised adaptation for SDVA and SDVA + MLLR(mean) (from [8], ©2009 IEEE)

		SDVA		SDVA + MLLR(mean)	
	# of training data	Supervised	Unsupervised	Supervised	Unsupervised
Open	2 utterances	13.5 %	13.9 %	19.8 %	37.5 %
	32 utterances	13.3 %	13.9 %	11.2 %	12.1 %
	512 utterances	13.4 %	13.8 %	5.4 %	11.4 %
Closed	561 utterances	-	13.9 %	-	12.0 %

9.5.4 Unsupervised Adaptation

We have shown the potential of the method in a supervised adaptation experiment. However, as the set of adaptation parameters is globally shared among all the Gaussians of the models, the method has the potential for extension to unsupervised adaptation. Indeed, in this case, the estimation of the global adaptation parameters is far less sensitive to errors in estimated labels than is cluster-based adaptation. Unsupervised adaptation is performed by using HMM state alignments instead of labels. The state alignments are obtained in advance by applying the recognizer to the unlabeled adaptation data. In this experiment we considered open adaptation, i.e., using different data sets for adaptation and recognition, and closed adaptation, i.e., using the same data set for adaptation and recognition. The open adaptation case is equivalent to the previous experiment and is used here for comparison. Table 9.4 gives the WER for unsupervised adaptation in the open and closed cases for SDVA and SDVA + MLLR(mean). The results are given for a distance of 1.5 m between the speaker and the microphone. We observe that using unsupervised SDVA, the performance degrades by only around a WER of 0.4 % compared with the supervised case. Moreover, closed and open adaptation perform similarly. This result shows that adaptation could be performed online or in a semi-batch way, as only two utterances may be sufficient to achieve convergence with unsupervised SDVA. Therefore, we believe that the proposed algorithm could be made robust to changes in the acoustic environment as long as the time scale of the changes is of the order of a couple of seconds.

When using unsupervised SDVA + MLLR(mean), even if the reduction in WER is less than that obtained in the supervised case, we still observe a WER reduction of around 2% compared with unsupervised SDVA when sufficient adaptation data are used. This result confirms that the proposed method may be combined with MLLR even for unsupervised adaptation.

9.5.5 Insights about the Proposed Algorithm

In this section we investigate the convergence of the proposed adaptation scheme and discuss the order of *static* and *dynamic* variance adaptation.

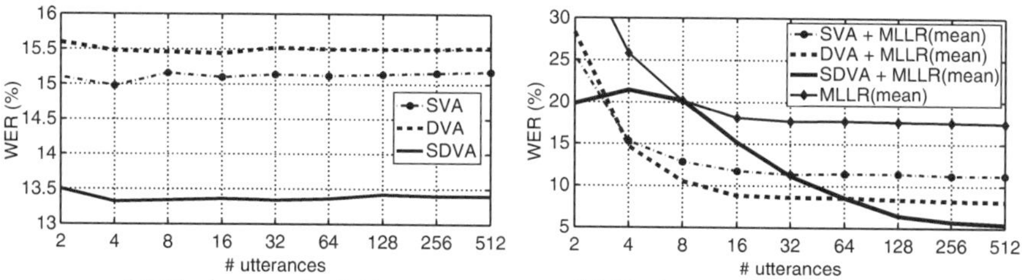

(a) Variance adaptation (b) Variance adaptation + MLLR

Fig. 9.4: WER as a function of the number of adaptation data. (a) plots the results for SVA (dash-dotted line), DVA (dotted line) and SDVA (solid line). (b) plots the results for MLLR (thin solid line), SVA + MLLR (thin dash-dotted line), DVA + MLLR (dotted line) and SDVA + MLLR (thick solid line) (from [7], ©2008 IEEE)

9.5.5.1 Convergence

Here, we discuss the convergence property of the proposed method in terms of the required number of adaptation utterances and the number of steps in the EM algorithm. The results presented here are for supervised adaptation with a distance of 1.5 m between the speaker and the microphone.

Figure 9.4(a) plots the WER as a function of the number of adaptation utterances for SVA ($\alpha_i = 0$), DVA and SDVA. The results are averaged over five randomly generated adaptation data sets. Inter-algorithm WER differences are highly significant, since the two-sided 99% confidence intervals for WER were never wider than $\pm 0.17\%$ for any of the data points in Fig. 9.4(a). We observe that in all cases, convergence is almost achieved after two utterances since the number of adaptation parameters is small, namely 39 for SVA and DVA, and 2×39 for SDVA.

Figure 9.4(b) plots the WER as a function of the number of adaptation data for MLLR(mean), SVA+MLLR(mean), DVA+MLLR(mean) and SDVA+MLLR(mean). The values of the two-sided 99% confidence intervals were up to 3.7% for different samples of adaptation data.We observe that MLLR requires more than eight utterances to achieve convergence. Therefore, SVA + MLLR and DVA + MLLR also needed more than eight utterances to converge. When SDVA + MLLR is used, better performance is achieved at the cost of more adaptation data (here more than 128 utterances). However, when using SDVA + MLLR, we obtain poorer results when too few utterances are used. The problem may arise from instabilities that occur when performing the EM algorithm in turns. One way to solve this problem may be to perform interleaved updates of the mean and variance parameters in the EM algorithm.

Finally, let us briefly discuss the convergence of the proposed adaptation scheme in terms of the number of steps of the EM algorithm. Since the proposed scheme adds a *dynamic* variance term to the acoustic model variances, they will become time-varying and therefore we may expect poor convergence with the EM algorithm. Figure 9.5 plots the distortion (the sum of the negative log-likelihood over all adaptation utterances, averaged over the number of frames) and the WER as a function of the number of iterations of the EM algorithm for SVA, DVA and SDVA.

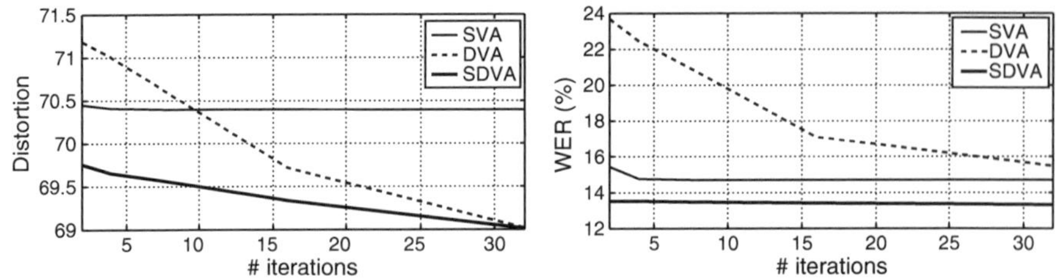

(a) Distortion (b) WER

Fig. 9.5: Distortion and WER as a function of the number of iterations of the EM algorithm for SVA, DVA and SDVA. In this experiment, four utterances were used for adaptation (from [8], ©2009 IEEE)

The EM algorithm achieves convergence after only two iterations for SVA. In contrast, the convergence is much slower with DVA. The poor convergence is due to the difficulty of handling the *dynamic* component, and from the fact that the mismatch may not be well modeled with only a *dynamic* component, as suggested in Section 9.5.3. Such a poor convergence property would pose a problem, especially for online applications. When combining *static* and *dynamic* adaptations, we observe that even though the distortion does not completely converge after 32 iterations, WER converges after two to four iterations. Therefore, a limited number of iterations may be sufficient to attain good performance, suggesting the potential of the method for online uses.

9.5.5.2 Order of Static and Dynamic Adaptation

In our experiments we noticed that performing *dynamic* variance adaptation before *static* adaptation (DSVA) provided slightly worse results than SDVA. For example, in the experiments in Figure 9.4 the DSVA WER was around 0.8% worse than that of SDVA. This result can be explained by looking at the $\boldsymbol{\lambda}$ and $\boldsymbol{\alpha}$ values for SDVA and DSVA.

Figure 9.6 plots the $\boldsymbol{\lambda}$ and $\boldsymbol{\alpha}$ values obtained after adaptation using 16 utterances for SDVA and DSVA. Looking at Figure 9.6(a), when *dynamic* adaptation is performed first, we observe large peaks in $\boldsymbol{\alpha}$ and $\boldsymbol{\lambda}$ for the 13-th component, which corresponds to the 0th cepstral coefficient. It is not surprising that speech enhancement introduces large uncertainty in that coefficient, which is related to feature energy. Looking at Figure 9.6(b), we observe the same large peak in $\boldsymbol{\lambda}$ but the peak has disappeared in $\boldsymbol{\alpha}$. This suggests that the mismatch in the 13th components of the cepstral coefficients is essentially *static*. By using only *dynamic* adaptation, the model may not be well suited for compensating for *static* components and consequently the performance is not optimal. By first removing the *static* mismatch with SVA, we may then focus solely on optimizing the model for the *dynamic* part and consequently obtain better performance. This illustrates the need to include both *static* and *dynamic* variance compensation, and suggests that the order in which the optimization is performed influences the results.

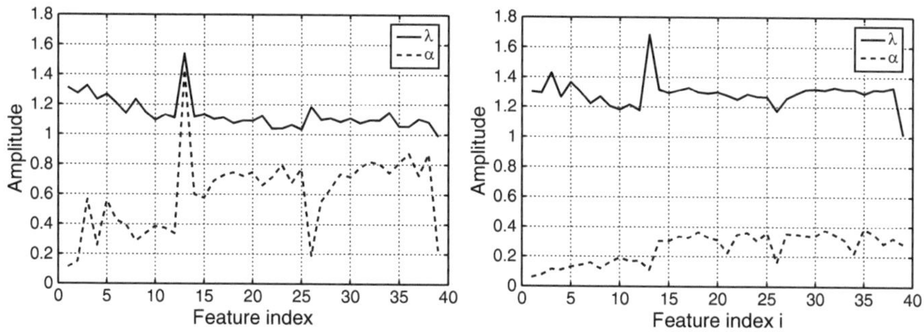

(a) DVA-SVA (b) SVA-DVA

Fig. 9.6: λ_i (solid line) and α_i (dashed line) for DSVA and for SDVA (from [6], Copyright &copy 2007 IEICE)

9.6 Experiment on Large Vocabulary Task

It is important to test the applicability of the proposed method for large vocabulary tasks, as there may be some concerns that variance compensation would increase confusion during decoding, making recognition harder. To show the effectiveness of the proposed method for LVCSR tasks, we performed an experiment with the Wall Street Journal (WSJ) task [32]. This task consists of reading English newspaper sentences. We modified the WSJ evaluation to simulate reverberant speech.

9.6.1 Experimental Settings

The experimental settings are summarized in Table 9.5. We used a speaker- (and gender-)independent acoustic model trained using WSJ training data. The acoustic model consisted of phoneme-based left-to-right HMMs, with three HMM states per phoneme and ten Gaussians per state. As for the previous experiment, we downsampled the speech signals from 16 to 8 kHz. The acoustic model was initially trained with speech sampled at 16 kHz and then retrained with data resampled at 8 kHz. We used about 250 hours of training data. The acoustic feature vector used for recognition consisted of 39 coefficients: 12 MFCCs, the energy, the delta and the acceleration coefficients. We used a standard tri-gram language model with a vocabulary size of 20.000 words. The test data consisted of the si_et_20 test set for the recording with a close microphone. This test set consists of 333 utterances spoken by eight male and female speakers. We generated reverberant speech by convolving the clean speech signals with room impulse responses extracted from the RWCP sound scene database in real acoustic environments [30]. We used impulse responses measured in a Japanese room and a meeting room, which had reverberation times of 0.47 and 0.7 s, respectively. In both cases the distance between the speaker and the microphone was 2 m. The D50 values of the impulse responses were 0.943 and 0.886 for the Japanese room and the meeting room, respectively.

Table 9.5: Settings for the WSJ experiment

Room characteristics	
Reverberation time	Japanese room : 0.48 s Meeting room : 0.78 s
Distance between speaker and microphone	2 m
Dereverberation	
Sampling frequency	8 kHz
Prediction filter length	Japanese room : 4000 taps (0.5 s) Meeting room : 6000 taps (0.75 s)
Delay (analysis frame length) (D)	30 ms
Frame rate for spectral subtraction	3.7
Features	
Dimension	39
Type	12 MFCC coefficients + Energy + Δ + $\Delta\Delta$
Postprocessing	CMN
Acoustic model	
Type	Phoneme based left-to-right HMM Speaker independent Retrained for 8 kHz speech
Number of HMM states (N)	3
Number of Gaussians per state (M)	10
Number of phonemes	44
Number of triphone HMM states	7500
Language model	
Type	tri-gram
Vocabulary size	20 000 words
Training data set	
Training data	Clean speech from WSJ training set
Amount of training data	251 hours
Test data set	
Data set	si_et_20 for close microphone Downsampled to 8kHz
Number of utterances	333
Number of speakers	8 (male and female)
Average sentence duration	7.6 s

9.6.2 Large Vocabulary Results

We tested our method for closed unsupervised speaker independent adaptation, i.e., we used all the 333 test sentences to perform unsupervised adaptation and recognition. For adaptation, we limited the number of iterations for SVA and MLLR to three, and to eight for SDVA.

Table 9.6 shows the WER results we obtained for the Japanese room and the meeting room. The baseline clean speech WER was 11.4%, which is larger than the state-of-the-art WSJ task recognition performance because of the effect of downsampling. The WER seriously detoriated for reverberant speech and fell to 46.5% for the Japanese room and 63.5% for the meeting room. The dereverberation preprocessor achieved an absolute WER reduction of 13.5% and 23.3% for the Japanese room and meeting room, respectively. However, the WER was still very high. Using

Table 9.6: WERs for the Japanese room and meeting room experiments

	Japanese Room	Meeting Room
Clean	11.4%	11.4%
Reverberant	46.5%	63.5%
Dereverberated	33 %	41.2%
Variance compensation (without adaptation)	29 %	36.2%
SVA	29.4%	36.5%
SDVA	28.6%	35 %
MLLR	28 %	32.9%
MLLR + SVA	25.6%	29.6%
MLLR + SDVA	23.8%	28.7%

SDVA can reduce the WER by close to 5%. Furthermore, by combining MLLR and SDVA a relative WER improvement of around 30% could be achieved compared with the recognition of dereverberated speech. As for the previous experiment, here we used global MLLR. Note that in this case we performed MLLR prior to SDVA as a better performance was achieved with this order. As seen in Table 9.6, MLLR alone outperforms SVA and SDVA.[7] This suggests that the *static* mean mismatch is large and that it is best to compensate for it prior to mean adaptation.

Some issues may arise when dealing with large vocabulary tasks. For example, if variance compensation greatly increases the acoustic model variances, there may be a large number of recognition hypotheses, making beam search decoding prone to an increase in search errors. However, this may not be a critical issue with the proposed method, because the acoustic model variances are not excessively increased, since the variance model weights ($\boldsymbol{\lambda}$, $\boldsymbol{\alpha}$) remain small. This is also confirmed by the results of Table 9.6, which clearly show a WER reduction brought by the proposed method without the broadening of the beam width.

Another concern when dealing with large vocabulary tasks is the complexity of the acoustic model. The large number of model parameters often makes adaptation difficult due to overtraining issues. However, the parameters of our model of the compensated Gaussian variance are globally shared among all Gaussians, and therefore the size of the proposed model is not affected by the increased size of the acoustic model.

These experimental results confirm that the proposed method may be applied to LVCSR.

9.7 Conclusion

In this chapter we presented a new framework for improving the recognition performance of reverberant speech preprocessed with dereverberation. We introduced a parametric representation of the feature variance that includes *static* and *dynamic*

[7] These results are the opposite of those obtained for the digit recognition experiment given in Section 9.5.3

components. We showed that the variance model parameters could be optimized using adaptive training with the EM algorithm. Experiments revealed that the proposed method could improve recognition for both supervised and unsupervised adaptation in a digit recognition task and a large vocabulary task. Moreover, combining the method with MLLR provided a WER improvement of up to 80% for digit recognition and 30% for LVCSR.

In this chapter, we focused on the recognition of reverberant speech. However, the proposed *dynamic* variance adaptation is not restricted to a particular speech enhancement preprocessor. Therefore, we believe that the method could be employed with other speech enhancement techniques or combined with other uncertainty decoding approaches to optimize the variance estimate. The interconnection of our proposed method with other speech enhancements constitutes part of our future work.

Appendix 1

λ_i can be obtained by deriving the auxiliary function of Eq. (9.27) with respect to λ_i as

$$\begin{aligned}\frac{\partial Q_{\boldsymbol{\lambda}}(\boldsymbol{\lambda}|\tilde{\boldsymbol{\theta}})}{\partial \lambda_i} &= \sum_{t=1}^{T}\sum_{n=1}^{N}\sum_{m=1}^{M}\iint p(X,B,n,m|\boldsymbol{\Psi},\tilde{\boldsymbol{\theta}})\frac{\partial \log(p(\mathbf{x}_t|n,m,\boldsymbol{\lambda}))}{\partial \lambda_i}dXdB,\\ &= \sum_{t=1}^{T}\sum_{n=1}^{N}\sum_{m=1}^{M}\iint p(X,B,n,m|\boldsymbol{\Psi},\tilde{\boldsymbol{\theta}})\frac{1}{2}(-\frac{1}{\lambda_i}+\frac{(x_{t,i}-\mu_{n,m,i})^2}{\sigma_{n,m,i}^2\lambda_i^2})dXdB.\end{aligned} \tag{9.31}$$

From Eq. (9.31) we can derive the following estimation formula for λ_i:

$$\lambda_i = \frac{\sum_{t=1}^{T}\sum_{n=1}^{N}\sum_{m=1}^{M}\iint p(X,B,n,m|\boldsymbol{\Psi},\tilde{\boldsymbol{\theta}})\frac{(x_{t,i}-\mu_{n,m,i})^2}{\sigma_{n,m,i}^2}dXdB}{\sum_{t=1}^{T}\sum_{n=1}^{N}\sum_{m=1}^{M}\iint p(X,B,n,m|\boldsymbol{\Psi},\tilde{\boldsymbol{\theta}})dXdB}. \tag{9.32}$$

The numerator term can be calculated as

$$
\begin{aligned}
&\iint p(X,B,n,m|\Psi,\tilde{\boldsymbol{\theta}})(x_{t,i}-\mu_{n,m,i})^2 dX dB \\
&= \prod_{\tau}\iint p(\boldsymbol{x}_{\tau},\mathbf{b}_{\tau},n,m|\Psi,\tilde{\boldsymbol{\theta}})(x_{t,i}-\mu_{n,m,i})^2 d\mathbf{x}_{\tau}d\mathbf{b}_{\tau} \\
&= \prod_{\tau\neq t} p(y_{\tau}|\Psi,\tilde{\boldsymbol{\theta}})\iint p(\mathbf{x}_t,\mathbf{b}_t,n,m|\Psi,\tilde{\boldsymbol{\theta}})(x_{t,i}-\mu_{n,m,i})^2 d\mathbf{x}_t d\mathbf{b}_t \\
&= \frac{p(Y|\Psi,\tilde{\boldsymbol{\theta}})}{p(\mathbf{y}_t|\Psi,\tilde{\boldsymbol{\theta}})}\iint p(n,m|\mathbf{x}_t,\mathbf{b}_t,\Psi,\tilde{\boldsymbol{\theta}})p(\mathbf{x}_t,\mathbf{b}_t|\Psi,\tilde{\boldsymbol{\theta}})(x_{t,i}-\mu_{n,m,i})^2 d\mathbf{x}_t d\mathbf{b}_t \\
&= p(Y|\Psi,\tilde{\boldsymbol{\theta}})p(n,m|\mathbf{y}_t,\Psi,\tilde{\boldsymbol{\theta}})\iint p(\mathbf{x}_t|\Psi,\tilde{\boldsymbol{\theta}})p(\mathbf{b}_t|\mathbf{x}_t)(x_{t,i}-\mu_{n,m,i})^2 d\mathbf{x}_t d\mathbf{b}_t \\
&= p(Y|\Psi,\tilde{\boldsymbol{\theta}})\gamma_t(n,m)\underbrace{\left(\frac{\tilde{\alpha}_i\hat{b}_{t,i}^2\tilde{\lambda}_i\sigma_{n,m,i}^2}{\tilde{\alpha}_i\hat{b}_{t,i}^2+\tilde{\lambda}_i\sigma_{n,m,i}^2}+\left(\frac{\tilde{\lambda}_i\sigma_{n,m,i}^2}{\tilde{\alpha}_i\hat{b}_{t,i}^2+\tilde{\lambda}_i\sigma_{n,m,i}^2}(y_{t,i}-\mu_{n,m,i})\right)^2\right)}_{\triangleq R(x_{t,i},\mathbf{y}_t,n,m,\Psi,\tilde{\boldsymbol{\theta}})},
\end{aligned}
\tag{9.33}
$$

where we used the model of speech and distortion given by Eqs. (9.7) and (9.10) and the Gaussian multiplication formula. $p(n,m|\mathbf{y}_t,\Psi,\tilde{\boldsymbol{\theta}}) = \gamma_t(n,m)$ is the state mixture occupancy probability that can be calculated using the forward-backward algorithm [18]. $R(x_{t,i},\mathbf{y}_t,n,m,\Psi,\tilde{\boldsymbol{\theta}})$ can be interpreted as an estimated enhanced feature variance. The derivation of Eq. (9.33) follows from derivations similar to those in [36]. Performing a derivation similar to that in Eq. (9.33), we can show that

$$
\iint_{X+B=Y} p(X,B,n,m|\Psi,\tilde{\boldsymbol{\theta}})dXdB = p(Y|\Psi,\tilde{\boldsymbol{\theta}})\gamma_t(n,m). \tag{9.34}
$$

Replacing Eqs. (9.33) and (9.34) in Eq. (9.32), we can express the update equation for λ_i as

$$
\lambda_i = \frac{\sum_{t=1}^{T}\sum_{n=1}^{N}\sum_{m=1}^{M}\gamma_t(n,m)\frac{R(x_{t,i},\mathbf{y}_t,n,m,\Psi,\tilde{\boldsymbol{\theta}})}{\sigma_{n,m,i}^2}}{\sum_{t=1}^{T}\sum_{n=1}^{N}\sum_{m=1}^{M}\gamma_t(n,m)}. \tag{9.35}
$$

Note that if $\boldsymbol{\alpha} = \mathbf{0}$, the problem is reduced to conventional *static* model variance adaptation as proposed in [38], which is sometimes referred to as variance scaling.

Appendix 2

As for $\boldsymbol{\lambda}$, α_i can be obtained by deriving the auxiliary function given in Eq. (9.27) with respect to α_i. This leads to the following estimation equation:

$$\alpha_i = \frac{\sum_{t=1}^{T}\sum_{n=1}^{N}\sum_{m=1}^{M}\iint p(X,B,n,m|\Psi,\tilde{\boldsymbol{\theta}})\frac{b_{t,i}^2}{\hat{b}_{t,i}^2}dXdB}{\sum_{t=1}^{T}\sum_{n=1}^{N}\sum_{m=1}^{M}\iint p(X,B,n,m|\Psi,\tilde{\boldsymbol{\theta}})dXdB}. \tag{9.36}$$

By performing similar manipulations to those in Appendix 1, we can derive the following expression:

$$\begin{aligned}
&\iint p(X,B,n,m|\Psi,\tilde{\boldsymbol{\theta}})b_{t,i}^2 dXdB \\
&= p(Y|\Psi,\tilde{\boldsymbol{\theta}})\gamma_t(n,m)\iint p(\mathbf{x}_t|\Psi,\tilde{\boldsymbol{\theta}})p(\mathbf{b}_t|\mathbf{x}_t)b_{t,i}^2 d\mathbf{x}_t d\mathbf{b}_t \\
&= p(Y|\Psi,\tilde{\boldsymbol{\theta}})\gamma_t(n,m)\underbrace{\left(\frac{\tilde{\alpha}_i\hat{b}_{t,i}^2\tilde{\lambda}_i\sigma_{n,m,i}^2}{\tilde{\alpha}_i\hat{b}_{t,i}^2+\tilde{\lambda}_i\sigma_{n,m,i}^2}+\left(\frac{\tilde{\alpha}_i\hat{b}_{t,i}^2}{\tilde{\alpha}_i\hat{b}_{t,i}^2+\tilde{\lambda}_i\sigma_{n,m,i}^2}(y_{t,i}-\mu_{n,m,i})\right)^2\right)}_{\triangleq E(b_{t,i}^2|\mathbf{y}_t,n,m,\Psi,\tilde{\boldsymbol{\theta}})},
\end{aligned} \tag{9.37}$$

where $E(b_{t,i}^2|\mathbf{y}_t,n,m,\Psi,\tilde{\boldsymbol{\theta}})$ can be interpreted as an estimate of the mismatch variance. We therefore obtain the following update equation for α_i:

$$\alpha_i = \frac{\sum_{t=1}^{T}\sum_{n=1}^{N}\sum_{m=1}^{M}\gamma_t(n,m)\frac{E(b_{t,i}^2|\mathbf{y}_t,n,m,\Psi,\tilde{\boldsymbol{\theta}})}{\hat{b}_{t,i}^2}}{\sum_{t=1}^{T}\sum_{n=1}^{N}\sum_{m=1}^{M}\gamma_t(n,m)}. \tag{9.38}$$

References

1. Arrowood, J. and Clements, M.: Using observation uncertainty in HMM decoding. In: Proceedings of International Conferences on Spoken Language Processing (ICSLP'02), **3**, 1562–1564 (2002)
2. Astudillo, R. F., Kolossa, D. and Orglmeister, R.: Accounting for the uncertainty of speech estimates in the complex domain for minimum mean square error speech enhancement. In: Proceedings of 10th European Conference on Speech Communication and Technology (Interspeech'09), 2491–2494 (2009)
3. Boll, S. F.: Suppression of acoustic noise in speech using spectral subtraction. IEEE Transactions on Acoustics, Speech and Signal Processing, **27**(2), 113–120 (1979)
4. Cooke, M. P., Green, P. D., Josifovski, L. B. and Vizinho, A.: Robust automatic speech recognition with missing and uncertain acoustic data. Speech Communication, **34**, 267–285 (2001)
5. Couvreur, L. and Couvreur, C.: Blind model selection for automatic speech recognition in reverberant environments. Journal of VLSI Signal Processing Systems, **36**(2–3), 189–203 (2004)
6. Delcroix, M., Nakatani, T. and Watanabe, S.: Dynamic feature variance adaptation for robust speech recognition with a speech enhancement pre-processor. In: IEICE Technical Report, **SP-105**, 55–60 (2007)

7. Delcroix, M., Nakatani, T. and Watanabe, S.: Combined static and dynamic variance adaptation for efficient interconnection of a speech enhancement pre-processor with speech recognizer. In: Proceedings of IEEE International Conference on Acoustics, Speech, and Signal Processing (ICASSP'08), 4073–4076 (2008)
8. Delcroix, M., Nakatani, T. and Watanabe, S.: Static and dynamic variance compensation for recognition of reverberant speech with dereverberation preprocessing. IEEE Transactions on Audio, Speech, and Language Processing, **17**(2), 324–334 (2009)
9. Deng, L., Droppo, J. and Acero, A.: Dynamic compensation of HMM variances using the feature enhancement uncertainty computed from a parametric model of speech distortion. IEEE Transactions on Speech and Audio Processing, **13**(3), 412–421 (2005)
10. Droppo, J., Acero, A. and Deng, L.: Uncertainty decoding with SPLICE for noise robust speech recognition. In: Proceedings of IEEE International Conference on Acoustics, Speech, and Signal Processing (ICASSP'02), **1**, 57–60 (2002)
11. Gales, M. J. F. and Woodland, P. C.: Mean and variance adaptation within the MLLR framework. Computer Speech and Language, **10**, 249–264 (1996)
12. Gillespie, B. W. and Atlas, L. E.: Acoustic diversity for improved speech recognition in reverberant environments. Proceedings of IEEE International Conference on Acoustics, Speech, and Signal Processing (ICASSP'02), **1**, 557–600 (2002)
13. Gong, Y.: Speech recognition in noisy environments: A survey. Speech Communication, **16**, 261–291 (1995)
14. Hikichi, T., Delcroix, M. and Miyoshi, M.: Speech dereverberation algorithm using transfer function estimates with overestimated order. Acoustical Science and Technology, **27**(1), 28–35 (2006)
15. Hirsch, H. G. and Pearce, D.: The AURORA experimental framework for the performance evaluations of speech recognition systems under noisy condition. In: Proceedings of The ISCA Tutorial and Research Workshop on Automatic Speech Recognition: Challenges for the New Millenium (ITRW ASR2000), 18–20 (2000)
16. Hirsch, H. G. and Finster, H.: A new approach for the adaptation of HMMs to reverberation and background noise. Speech Communication, **50**, 244–263 (2008)
17. Hori, T., Hori, C., Minami, Y. and Nakamura, A.: Efficient WFST-based one-pass decoding with on-the-fly hypothesis rescoring in extremely large vocabulary continuous speech recognition. IEEE Transactions on Speech and Audio Processing, **15** (4), 1352–1365 (2007)
18. Huang, X., Acero, A. and Hon, H.W.: Spoken Language Processing: A Guide to Theory, Algorithm and System Development. Prentice Hall, New-Jersey (2001)
19. Ion, V. and Haeb-Umbach, R.: A novel uncertainty decoding rule with applications to transmission error robust speech recognition. IEEE Transactions on Speech and Audio Processing, **16** (5), 1047–1060 (2008)
20. Kameoka, H., Nakatani, T. and Yoshioka, T.: Robust speech dereverberation based on non-negativity and sparse nature of speech spectrograms. In: Proceedings of IEEE International Conference on Acoustics, Speech and Signal Processing (ICASSP'09), 45–48 (2009)
21. Kinoshita, K., Delcroix, M., Nakatani T. and Miyoshi, M.: Suppression of late reverberation effect on speech signal using long-term multiple-step linear prediction. IEEE Transactions on Audio, Speech and Language Processing, **17** (4), 534–545 (2009)
22. Kolossa, D., Sawada, H., Astudillo, R. F., Orglmeister, R. and Makino, S.: Recognition of convolutive speech mixtures by missing feature techniques for ICA. In: Proceedings of The Asilomar Conference on Signals, Systems, and Computers (ACSSC'06), 1397–1401 (2006)
23. Kolossa, D., Araki, S., Delcroix, M., Nakatani, T., Orglmeister, R. and Makino, S.: Missing feature speech recognition in a meeting situation with maximum SNR beamforming. In: Proceedings of The IEEE International Symposium on Circuits and Systems (ISCAS'08), 3218–3221 (2008)
24. Kolossa, D., Klimas A. and Orglmeister, R.: Separation and robust recognition of noisy, convolutive speech mixtures using time-frequency masking and missing data techniques. In: Proceedings of The IEEE Workshop on Applications of Signal Processing to Audio and Acoustics, 82–85 (2005)

25. Krueger, A. and Haeb-Umbach, R.: Model based feature enhancement for automatic speech recognition in reverberant environments. In: Proceedings of 10th European Conference on Speech Communication and Technology (Interspeech'09), 1231–1234 (2009)
26. Kuttruff, H.: Room Acoustics. 3rd ed. (Elsevier Science, London, 1991)
27. Liao, H. and Gales, M. J. F.: Joint uncertainty decoding for noise robust speech recognition. In: Proceedings of 9th European Conference on Speech Communication and Technology (Interspeech'05-Eurospeech), 3129–3132 (2005)
28. Liao, H. and Gales, M. J. F.: Adapative training with joint uncertainty decoding for robust recognition of noisy data. In: Proceedings of IEEE International Conference on Acoustics, Speech, and Signal Processing (ICASSP'07), **4**, 389–392 (2007)
29. Meng, X.-L. and Rubin, D. B.: Maximum likelihood estimation via the ECM algorithm: A general framework. Biometrika, **80**, 267–278 (1993)
30. Nakamura, S. and Nishiura, T.: RWCP sound scene database in real acoustical environments. `http://tosa.mri.co.jp/sounddb/micarray/indexe.htm` Cited 31 May 2010
31. Naylor, P. A. and Gaubitch, N. D.: Speech dereverberation. In: Proceedings of International Workshop on Acoustic Echo and Noise Control (IWAENC'05), iwaenc05.ele.tue.nl/proceedings/papers/pt03.pdf (2005)
32. Paul, D. B. and Baker, J. M. : The design for the Wall Street Journal-based CSR corpus. In: Proceedings of the Workshop on Speech and Natural Language. 357–362 (1992)
33. Quatieri, T. F.: Discrete-Time Speech Signal Processing. (Prentice Hall, New Jersey, 2002)
34. Raj, B. and Stern, R. M.: Missing-feature approaches in speech recognition. IEEE Signal Processing Magazine, **22** (5), 101–116 (2005)
35. Raut, C. K., Nishimoto, T. and Sagayama, S.: Model adaptation by state splitting of HMM for long reverberation. In: Proceedings of 9th European Conference on Speech Communication and Technology (Interspeech'05-Eurospeech), 277–280 (2005)
36. Rose, R. C., Hofstetter, E. M. and Reynolds, D. A.: Integrated models of signal and background with application to speaker identification in noise. IEEE Transactions on Speech and Audio Processing, **2**(2), 245–257 (1994)
37. Sankar, A. and Lee C.-H.: Robust speech recognition based on stochastic matching. In: Proceedings of IEEE International Conference on Acoustics, Speech, and Signal Processing (ICASSP'95), **1**, 121–125 (1995)
38. Sankar, A. and Lee, C.-H.: A maximum-likelihood approach to stochastic matching for robust speech recognition. IEEE Transactions on Speech and Audio Processing, **4**(3), 190–202 (1996)
39. Schuller, B., Wollmer, M., Moosmayr, T. and Rigoll, G.: Recognition of noisy speech: A comparative survey of robust model architecture and feature enhancement. EURASIP Journal on Audio, Speech, and Music Processing **2009**, (2009)
40. Sehr, A. and Kellerman, W.: A new concept for feature-domain dereverberation for robust distant-talking ASR. In: Proceedings of IEEE International Conference on Acoustics, Speech, and Signal Processing (ICASSP'07), **4**, 369–372 (2007)
41. Sehr, A., Maas, R. and Kellerman, W.: Reverberation model-based decoding in the logmelspec domain for robust distant-talking speech recognition. IEEE Transactions on Audio, Speech, and Language Processing, (To appear) (2010)
42. Stouten, V., Van hamme, H. and Wambacq, P.: Model-based feature enhancement with uncertainty decoding for noise robust ASR. Speech Communication, **48**, 1502–1514 (2006)
43. Stouten, V., Van hamme, H. and Wambacq, P.: Accounting for the uncertainty of speech estimates in the context of model-based feature enhancement In: Proceedings of International Conferences on Spoken Language Processing (ICSLP'04), 105108 (2004)
44. Takiguchi, T. and Nishimura, M.: Acoustic model adaptation using first order prediction for reverberant speech. In: Proceedings of IEEE International Conference on Acoustics, Speech, and Signal Processing (ICASSP'04), **1**, 869–972 (2004)
45. Tashev, I. and Allred, D.: Reverberation reduction for improved speech recognition. In: Proceedings of Joint Workshop on Hands-Free Speech Communication and Microphone Arrays (HSCMA'05), (2005)

46. Yoshioka, T.: Speech enhancement in reverberant environments. Ph.D. dissertation, Kyoto University (2010)
47. Wolfel, M.: Enhanced speech features by single-channel joint compensation of noise and reverberation. IEEE Transactions on Audio, Speech, and Language Processing, **17**(2), 312–323 (2009)
48. Wu, M. and Wang, D.: A two-stage algorithm for one-microphone reverberant speech enhancement. IEEE Transactions on Audio, Speech, and Language Processing, **14**, 774–784 (2006)

Chapter 10
A Model-Based Approach to Joint Compensation of Noise and Reverberation for Speech Recognition

Alexander Krueger and Reinhold Haeb-Umbach

Abstract Employing automatic speech recognition systems in hands-free communication applications is accompanied by perfomance degradation due to background noise and, in particular, due to reverberation. These two kinds of distortion alter the shape of the feature vector trajectory extracted from the microphone signal and consequently lead to a discrepancy between training and testing conditions for the recognizer. In this chapter we present a feature enhancement approach aiming at the joint compensation of noise and reverberation to improve the performance by restoring the training conditions. For the enhancement we concentrate on the logarithmic mel power spectral coefficients as features, which are computed at an intermediate stage to obtain the widely used mel frequency cepstral coefficients. The proposed technique is based on a Bayesian framework, to attempt to infer the posterior distribution of the clean features given the observation of all past corrupted features. It exploits information from a priori models describing the dynamics of clean speech and noise-only feature vector trajectories as well as from an observation model relating the reverberant noisy to the clean features. The observation model relies on a simplified stochastic model of the room impulse response (RIR) between the speaker and the microphone, having only two parameters, namely RIR energy and reverberation time, which can be estimated from the captured microphone signal. The performance of the proposed enhancement technique is finally experimentally studied by means of recognition accuracy obtained for a connected digits recognition task under different noise and reverberation conditions using the Aurora 5 database.

Alexander Krueger
Department of Communications Engineering, University of Paderborn, Warburger Straße 100, 33098 Paderborn, Germany, e-mail: krueger@nt.uni-paderborn.de

Reinhold Haeb-Umbach
Department of Communications Engineering, University of Paderborn, Warburger Straße 100, 33098 Paderborn, Germany, e-mail: haeb@nt.uni-paderborn.de

D. Kolossa and R. Haeb-Umbach (eds.), *Robust Speech Recognition of Uncertain or Missing Data: Theory and Applications*, DOI 10.1007/978-3-642-21317-5_10,
© Springer-Verlag Berlin Heidelberg 2011

10.1 Introduction

Automatic speech recognition (ASR) is often considered a key technology for human-machine communication and is today part of numerous applications such as dictation machines, electronic translators and navigation systems. In order to increase the acceptance of such devices in certain applications, it is desirable that the user is able to move freely without wearing a headset to communicate with the system. Consequently, to fulfill such a requirement, which may arise for either convenience or safety reasons, distant-talking microphones are employed instead of close-talking ones. However, the increased speaker-microphone distance results in degraded signal quality at the microphones due to acoustic environmental noise and reverberation. The latter is a convolutive distortion, which is caused by reflections of the speech signals on walls and objects. The source signal is then superposed with its delayed and attenuated versions at the microphone.

While the utterances for the training of ASR systems are typically clean, i.e., non-reverberant and noise-free, this obviously does not hold for those utterances which are recorded in noisy and reverberant environments at testing. The discrepancy between training and testing conditions is one of the major reasons for the degradation of the ASR systems' performance.

Basically, the widely known approaches to deal with the mentioned distortions in ASR can be separated into three categories.

Firstly, there are speech enhancement techniques to be applied to the microphone signal prior to feature extraction, which is the first stage of ASR. A comprehensive overview of these approaches with the main focus on reverberation can be found in [14]. Among these are the blind deconvolution approaches [4, 12, 13], comprising methods which try to estimate the speaker-microphone transfer functions (TFs) from multi-channel microphone signals and reconstruct the clean signal by inverse filtering the reverberant signal. However, in highly reverberant environments such approaches become computationally demanding and have shown high sensitivity to small changes in the TFs. Further, the spectral enhancement techniques [22, 26, 28, 39, 42] estimate the statistical characteristics of the late reverberation and noise to subsequently remove them from the speech signal. Other approaches attempt to enhance the linear prediction residual [40, 41], apply temporal envelope filtering [25, 38] or perform homomorphic deconvolution [20, 37]. A disadvantage of such speech enhancement techniques is, however, that they may introduce artefacts into the speech signal with unpredictable effects on the successive feature extraction, leading to degraded recognition performance.

Secondly, there are methods which try to adapt the model parameters of the recognizer to the effects of reverberation and noise [5, 18]. A special member of this category is the maximum likelihood linear regression (MLLR) approach [10, 11, 27]. It assumes that the distortion on the features can be described by linear transformations. In order to compensate for that distortion in hidden Markov model- (HMM-) based recognizers, linear transformations on means and variances of Gaussian mixture densities of HMMs are applied. A disadvantage of this approach is that long reverberation cannot be described appropriately by linear transformations on static features only and thus few improvements are to be expected in that case. Further, the application of MLLR requires a certain amount of adaptation data and its transcription, which is in practice obtained from a preceding recognition pass. A further very intuitive approach is given by computing a pool of models or transformations for different reverberation [3] or noise [44] conditions in advance at the training stage, and then selecting the appropriate model or transformation for decoding. However, this approach has the drawback that lots of data corresponding to the models have to be stored. To take into account the energy smearing caused by reverberation, Raut et al. [34] have introduced additional HMM states by performing state splitting dependent on the extent of reverberation. Based on a model of reverberation in the feature domain, there have also been attempts to modify the likelihood evaluation within the Viterbi decoding [36].

Finally, the third category consists of feature extraction and normalization methods attempting to compute acoustic features which are insensitive to any kind of channel distortion [15, 21]. A very prominent representative of this category is the cepstral mean normalization (CMN) approach [35], which was developed to compensate for unknown microphone characteristics and environmental influences. It is based on the assumption that the desired source signal is corrupted by the influence of an unknown, linear time-invariant filter. Further, an essential part is the multiplicative transfer function assumption (MTFA) [1], which states that the short time spectrum of the corrupted signal can be expressed as the product of the short time spectrum of the desired source signal and the filter's transfer function. Exploiting the time-invariance of the unknown filter and the property of the logarithmic function, which turns multiplicative relations into additive ones, the influence of the filter can be expressed as an additive constant in the cepstral domain which can be eliminated by subtracting the cepstral means. However, applying MTFA to reverberation is only valid if the duration of the room impulse response is small compared to the duration of the window function. If that condition is not met, which is usually the case, the dereverberation effect achieved by CMN is poor.

In this contribution we will present a feature enhancement approach, which, in the widest sense, can be classified into the third category. Enhancement techniques which operate in the feature domain are desireable since this domain is the one the recognizer is operating in. The idea of the proposed feature enhancement technique is illustrated in Fig. 10.1. It is based on the Bayesian framework of carrying out inference by using information from an a priori model, which describes the dynamics of clean speech and noise only features, and information from an observation model, which links the observeable reverberant noisy features to the nonobserveable clean features, to obtain a posteriori information about the clean features given the reverberant noisy ones. The a posteriori information is given in the form of a posterior density, which allows us to compute a point estimate, e.g., the minimum mean squared error (MMSE) estimate of the clean features given all previously observed reverberant noisy features, as well as an estimate of the quality of the enhancement in the form of an error covariance.

The key component of the enhancement is certainly the observation model, which is derived from the convolutional distortion in the time domain resulting from the

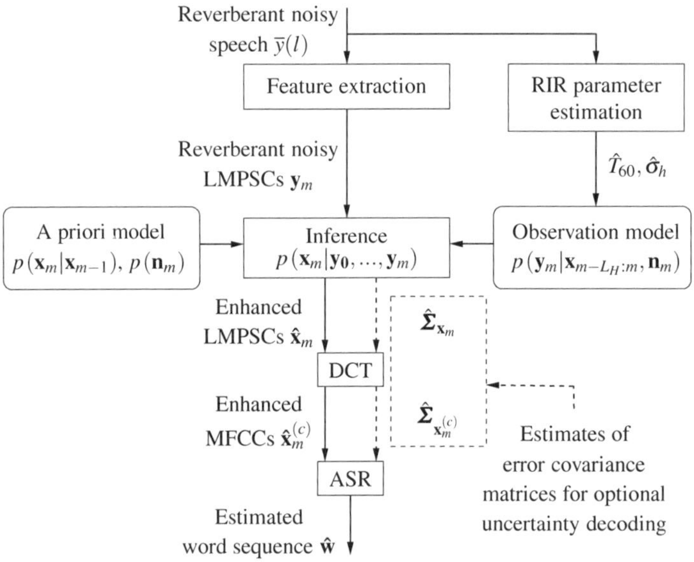

Fig. 10.1: Schematic diagram of proposed feature enhancement algorithm (the notation will be introduced in the later sections)

speaker–microphone room impulse response (RIR). It is shown that the observation model may be formulated using certain parameters extracted from the RIR. Since the RIR is generally time-variant, depending on temperature, humidity, room geometry and speaker location and direction, we propose to use a simplified time-invariant stochastic model, which requires the estimation of only two parameters, RIR energy and reverberation time.

In principle, the Bayesian approach to feature enhancement is feasible for any kind of acoustic features which allow us to formulate an appropriate observation model linking the clean to the reverberant noisy features. The focus of the chapter will, however, be on the so-called mel frequency cepstral coefficients (MFCCs), whose computation is motivated by the human acoustic perception and which are popular and widely used because of their simple computation. This allows us to integrate the enhancement into existing ASR systems without any serious modifications of the front-end, while the back-end is left completely unchanged. In order to avoid any numerical problems arising from the great differences in the ranges of the individual MFCCs, we propose to perform the enhancement on the logarithmic mel power spectral coefficients (LMPSCs), which are computed at an intermediate stage, and subsequently transform the enhanced LMPSCs to the cepstral domain.

The organization of this chapter is as follows. In Sect. 10.2 we introduce the problem of automatic speech recognition in a reverberant noisy environment and recapitulate the feature extraction process to obtain the MFCCs in Sect. 10.3. After explaining the idea of the Bayesian feature enhancement approach in Sect. 10.4, we present the observation model linking the clean to the reverberant LMPSCs. We first derive this functional relation under the assumption of a time-invariant speaker–microphone impulse response. Secondly, we propose a simplified stochastic model for the room impulse response to take into account the time-variant nature of the RIR. In Sect. 10.5, we present the actual enhancement algorithm resulting from the implementation of the Bayesian idea. Finally, in Sect. 10.6 we apply the proposed algorithm to a connected digits recognition task on the Aurora 5 database under different reverberation and noise conditions and evaluate the performance by means of recognition accuracy.

10.2 Problem Formulation

We consider the typical hands-free scenario for automatic speech recognition, which is illustrated in Fig. 10.2. The target speaker is located in a reverberant noisy envi-

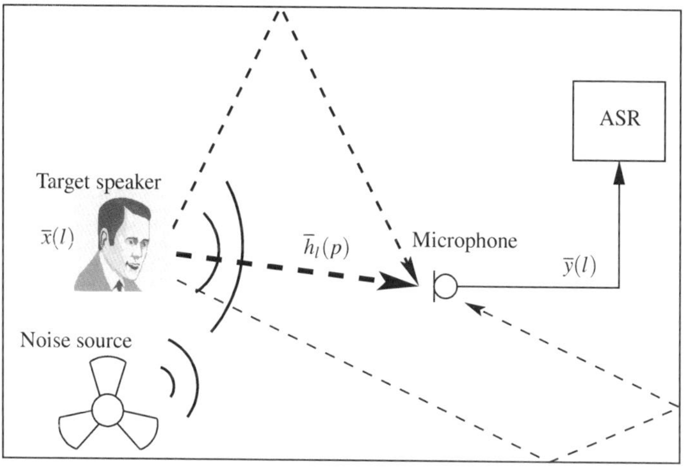

Fig. 10.2: Hands-free automatic speech recognition scenario

ronment at a certain distance from a far-field microphone. The time discrete acoustic signal $\overline{y}(l)$ captured at this microphone, where $l \in \mathbb{N}_0$ denotes the time index, consists of two components, which are given by the reverberant speech signal $\overline{s}(l)$ on the one hand, and the noise signal $\overline{n}(l)$ on the other hand:

$$\overline{y}(l) = \overline{s}(l) + \overline{n}(l). \tag{10.1}$$

The reverberant speech signal results from the convolution of the source speech signal $\overline{x}(l)$ with the time-variant RIR $\overline{h}_l(p)$ from the target speaker to the microphone:

$$\bar{s}(l) = \sum_{p=0}^{\infty} \bar{h}_l(p)\bar{x}(l-p). \tag{10.2}$$

The expression $\bar{h}_l(p)$ can be interpreted as the RIR which is valid at the time index l and where $p \in \mathbb{N}_0$ can be regarded as the corresponding tap index. The noise signal component $\bar{n}(l)$ includes all reverberant noise signals, which originate from noise sources such as fans, located in the environment, as well as inherent microphone noise. The two signal components, $\bar{s}(l)$ and $\bar{n}(l)$, may be modeled as independent stochastic random processes, since they originate from different sources.

The microphone signal $\bar{y}(l)$ is passed to an ASR system which is expected to estimate the word sequence spoken by the target speaker. A typical ASR system consists of two processing units. In the so-called front-end, features are extracted from the incoming microphone signal, which are then used in the back-end for the search for the most probable word sequence. Performing ASR in a noisy reverberant environment leads to performance degradation resulting from the difference of the extracted features from those obtained under clean conditions. Particularly worth mentioning is the dispersive effect of reverberation, which introduces considerable temporal correlations between successive features and thereby causes the conditional independence assumption, which is inherent in HMM-based recognizers (see Chap. 2), to be violated.

The task considered in this contribution is the reconstruction of the clean features from the reverberant noisy ones. The particular challenge of this feature enhancement task consists of assuming a *blind* scenario, meaning that the environment, especially the time-variant speaker-microphone RIR and the noise signal component, is completely unknown to the ASR system. This requires the ASR system to learn certain qualities of the environment from the captured signal to be employed for the feature enhancement.

10.3 Feature Extraction

The way reverberation and noise influence the shape of the features highly depends on the particular feature extraction method. In this contribution we focus on the MFCCs since they are probably the most popular features used for ASR. Their computation is motivated by the human acoustic perception and is standardized according to the ETSI ES 201 108 standard [9]. Figure 10.3 illustrates a slightly modified version of the ETSI standard front-end (SFE). The modifications are the replacement of the magnitude spectrum by the power spectrum and the replacement of the logarithmic frame energy by the zero-th cepstral component.

After offset compensation and preemphasis of the captured speech signal $\bar{y}(l)$, the resulting time signal $\tilde{y}(l)$ is framed and weighted by a Hamming analysis window function $\tilde{w}_a(l)$ of finite length L_w to obtain the frame-dependent windowed signal segments:

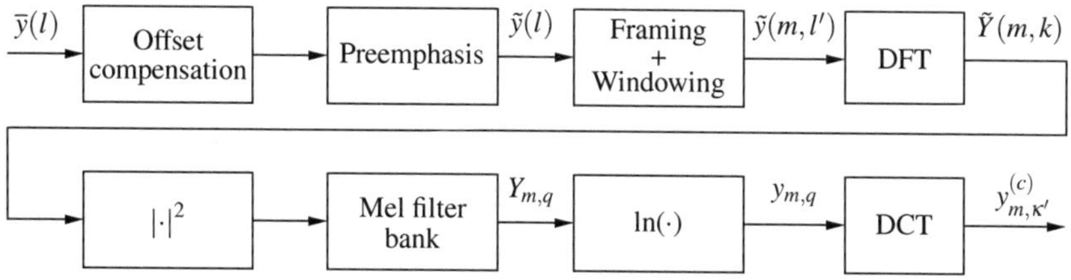

Fig. 10.3: Modified version of the feature extraction according to ETSI ES 201 108 standard [9]. The modification is the replacement of the magnitude spectrum by the power spectrum and the replacement of the logarithmic frame energy by the zeroth cepstral component

$$\tilde{y}(m,l') := \tilde{w}_a(l')\tilde{y}(l' + mB). \tag{10.3}$$

Here $m \in \mathbb{N}_0$ denotes the frame index, $l' \in \{0,...,L_w - 1\}$ denotes the time index within the segment and B denotes the frame shift. The windowed signal segments are subsequently transformed to the frequency domain by applying a discrete Fourier transform (DFT), resulting in the short-time discrete Fourier transform (STDFT) representations

$$\tilde{Y}(m,k) := \sum_{l'=0}^{L_w-1} \tilde{y}(m,l') \cdot e^{-j\frac{2\pi}{K}kl'}, \tag{10.4}$$

where K is the number of frequency bins and $k \in \{0,...,K-1\}$ the frequency bin index. The mel power spectral coefficients $Y_{m,q}$ are then obtained as perceptually weighted sums of the STDFT powers. This is accomplished by applying a bank of Q overlapping, triangular filters Λ_q, $q \in \{0,...,Q-1\}$, which are equally spaced on the mel scale, according to

$$Y_{m,q} := \sum_{k=K_q^{(\mathrm{lo})}}^{K_q^{(\mathrm{up})}} |\tilde{Y}(m,k)|^2 \Lambda_q(k). \tag{10.5}$$

Here $K_q^{(\mathrm{lo})}$ and $K_q^{(\mathrm{up})}$ are the lower and upper bounds of the qth mel band. Hereafter, the LMPSCs are computed by taking the natural logarithm as

$$y_{m,q} := \ln(Y_{m,q}). \tag{10.6}$$

Finally, the LMPSCs are decorrelated by applying a discrete cosine transform (DCT) to obtain the well-known MFCCs

$$y^{(c)}_{m,\kappa'} := \sum_{q=0}^{Q-1} y_{m,q} \cos\left[\frac{\kappa'\pi}{K'}\left(q + \frac{1}{2}\right)\right], \tag{10.7}$$

where κ' denotes the MFCC index and K' the overall number of MFCC components.

10.4 Bayesian Feature Enhancement Approach

The fundamental principle behind the Bayesian feature enhancement approach is to assume the sequences of clean speech, noise and reverberant noisy speech LMPSC vectors, $\{\mathbf{x}_m\}_{m\in\mathbb{N}_0}$, $\{\mathbf{n}_m\}_{m\in\mathbb{N}_0}$ and $\{\mathbf{y}_m\}_{m\in\mathbb{N}_0}$ respectively, to be realizations of individual stochastic vector processes, where each feature vector is composed of all spectral components according to

$$\mathbf{x}_m := \left(x_{m,0},\dots,x_{m,Q-1}\right)^T, \tag{10.8}$$

$$\mathbf{n}_m := \left(n_{m,0},\dots,n_{m,Q-1}\right)^T, \tag{10.9}$$

$$\mathbf{y}_m := \left(y_{m,0},\dots,y_{m,Q-1}\right)^T. \tag{10.10}$$

The key idea is to estimate the posterior distribution $p\left(\mathbf{d}_m|\mathbf{y}_{0:m}\right)$ of the joint feature vector

$$\mathbf{d}_m := \left(\mathbf{x}_m^T, \mathbf{n}_m^T\right)^T \tag{10.11}$$

given all previously observed reverberant noisy speech feature vectors

$$\mathbf{y}_{0:m} := \left(\mathbf{y}_0^T,\dots,\mathbf{y}_m^T\right)^T. \tag{10.12}$$

The knowledge about the posterior distribution allows the computation of different kinds of point estimates for the clean speech feature vector $\mathbf{x}_m$ such as the MMSE estimate

$$\hat{\mathbf{x}}_m^{(\mathrm{MMSE})} := \mathrm{E}\left[\mathbf{x}_m|\mathbf{y}_{0:m}\right]. \tag{10.13}$$

Further, the posterior covariance matrix

$$\hat{\boldsymbol{\Sigma}}_{\mathbf{x}_m}^{(\mathrm{MMSE})} := \mathrm{E}\left[\left(\mathbf{x}_m - \hat{\mathbf{x}}_m^{(\mathrm{MMSE})}\right)\left(\mathbf{x}_m - \hat{\mathbf{x}}_m^{(\mathrm{MMSE})}\right)^T \middle| \mathbf{y}_{0:m}\right] \tag{10.14}$$

can be regarded as a measure of the amount of uncertainty in the point estimate. With the knowledge of the posterior distribution it is also possible to compute the corresponding quantities for the noise components, $\hat{\mathbf{n}}_m^{(\mathrm{MMSE})}$ and $\boldsymbol{\Sigma}_{\mathbf{n}_m}^{(\mathrm{MMSE})}$.

A typical way of computing the posterior distribution is to recursively carry out Bayesian inference, which consists of two steps. In the *prediction step*, a predictive distribution for the joint feature vector $\mathbf{d}_m$ based on the previous reverberant noisy observations $\mathbf{y}_{0:m-1}$ may be computed as[1]

[1] Throughout this contribution we use the expression $p(\cdot)$ to denote any probability density function. The purpose of this actually non-standard notation, where the random variable as subscript of p is omitted, is to keep all occurring expressions concise. However, in each particular case the random variable is supposed to become clear from the arguments of p.

$$p\left(\mathbf{d}_m|\mathbf{y}_{0:m-1}\right) = \int p\left(\mathbf{d}_m|\mathbf{d}_{m-1},\mathbf{y}_{0:m-1}\right) p\left(\mathbf{d}_{m-1}|\mathbf{y}_{0:m-1}\right) d\mathbf{d}_{m-1}, \tag{10.15}$$

requiring a predictive distribution $p\left(\mathbf{d}_m|\mathbf{d}_{m-1},\mathbf{y}_{0:m-1}\right)$ for the clean speech and noise feature vectors. In the *update step*, the desired posterior distribution $p\left(\mathbf{d}_m|\mathbf{y}_{0:m}\right)$ is obtained from

$$p\left(\mathbf{d}_m|\mathbf{y}_{0:m}\right) = \frac{p\left(\mathbf{y}_m|\mathbf{d}_m,\mathbf{y}_{0:m-1}\right) p\left(\mathbf{d}_m|\mathbf{y}_{0:m-1}\right)}{\int p\left(\mathbf{y}_m|\tilde{\mathbf{d}}_m,\mathbf{y}_{0:m-1}\right) p\left(\tilde{\mathbf{d}}_m|\mathbf{y}_{0:m-1}\right) d\tilde{\mathbf{d}}_m}, \tag{10.16}$$

employing the observation $\mathbf{y}_m$ and an observation distribution $p\left(\mathbf{y}_m|\mathbf{d}_m,\mathbf{y}_{0:m-1}\right)$.

For the enhancement of noisy speech feature vectors the dependence of $\mathbf{y}_m$ on the previous observations $\mathbf{y}_{0:m-1}$ is often neglected by the approximation

$$p\left(\mathbf{y}_m|\mathbf{d}_m,\mathbf{y}_{0:m-1}\right) \approx p\left(\mathbf{y}_m|\mathbf{d}_m\right). \tag{10.17}$$

However, it will be shown in the next section that successive reverberant feature vectors are strongly correlated with each other resulting from the dispersive effect of reverberation. For that reason an approximation like (10.17) is not appropriate. Nevertheless, if the extended vector

$$\boldsymbol{\chi}_m := \left(\mathbf{x}_m^T, \mathbf{x}_{m-1}^T, \ldots, \mathbf{x}_{m-L_C+1}^T\right)^T, \tag{10.18}$$

consisting of the current and an appropriate number of $L_C - 1$ previous clean feature vectors, is introduced, and the vector $\mathbf{d}_m$ is replaced with the vector

$$\mathbf{z}_m := \left(\boldsymbol{\chi}_m^T, \mathbf{n}_m^T\right)^T \tag{10.19}$$

in the inference equations (10.15) and (10.17), it is reasonable to neglect the dependence on $\mathbf{y}_{0:m-1}$ to approximate the observation distribution according to

$$p\left(\mathbf{y}_m|\mathbf{z}_m,\mathbf{y}_{0:m-1}\right) \approx p\left(\mathbf{y}_m|\mathbf{z}_m\right). \tag{10.20}$$

Assuming all involved distributions to be Gaussian or mixtures of those, the inference simplifies to recursively computing the mean vector and covariance matrix of the predictive distribution $p\left(\mathbf{z}_m|\mathbf{y}_{0:m-1}\right)$,

$$\mathbf{z}_{m|m-1} := \mathrm{E}\left[\mathbf{z}_m|\mathbf{y}_{0:m-1}\right], \tag{10.21}$$

$$\boldsymbol{\Sigma}_{\mathbf{z}_{m|m-1}} := \mathrm{E}\left[\left(\mathbf{z}_m - \mathbf{z}_{m|m-1}\right)\left(\mathbf{z}_m - \mathbf{z}_{m|m-1}\right)^T \middle| \mathbf{y}_{0:m-1}\right] \tag{10.22}$$

and those of the posterior distribution $p\left(\mathbf{z}_m|\mathbf{y}_{0:m}\right)$,

$$\mathbf{z}_{m|m} := \mathrm{E}\left[\mathbf{z}_m|\mathbf{y}_{0:m}\right], \tag{10.23}$$

$$\boldsymbol{\Sigma}_{\mathbf{z}_{m|m}} := \mathrm{E}\left[\left(\mathbf{z}_m - \mathbf{z}_{m|m}\right)\left(\mathbf{z}_m - \mathbf{z}_{m|m}\right)^T \middle| \mathbf{y}_{0:m}\right]. \tag{10.24}$$

The issue of how to compute these moments in practice is addressed in Section 10.5.

10.4.1 A Priori Model

Since the stochastic vector processes underlying the realizations $\{\mathbf{x}_m\}_{m\in\mathbb{N}_0}$ and $\{\mathbf{n}_m\}_{m\in\mathbb{N}_0}$ have been assumed to be independent, the overall predictive distribution $p(\mathbf{z}_m|\mathbf{z}_{m-1},\mathbf{y}_{0:m-1})$ may be expressed by the product of the individual predictive distributions for the speech and noise components as

$$p(\mathbf{z}_m|\mathbf{z}_{m-1},\mathbf{y}_{0:m-1}) = p\left(\boldsymbol{\chi}_m|\boldsymbol{\chi}_{m-1},\mathbf{y}_{0:m-1}\right) p(\mathbf{n}_m|\mathbf{n}_{m-1},\mathbf{y}_{0:m-1}). \qquad (10.25)$$

In principle, it is possible to use the same kind of models for the speech and noise components. However, in consideration of the fact that the model complexity should match the specific characteristics of speech and noise and be kept as low as possible to reduce the computational effort of the feature enhancement, we propose different kind of models for the individual components.

10.4.1.1 A Priori Speech Model

To explicitly consider the high degree of dynamics in speech, it is appropriate to model the predictive speech distribution as a mixture of I interacting speech prediction models as

$$p\left(\boldsymbol{\chi}_m|\boldsymbol{\chi}_{m-1},\mathbf{y}_{0:m-1}\right) = \sum_{i=1}^{I} p\left(\boldsymbol{\chi}_m|\boldsymbol{\chi}_{m-1},\mathbf{y}_{0:m-1},\gamma_m=i\right) P\left(\gamma_m=i|\boldsymbol{\chi}_{m-1},\mathbf{y}_{0:m-1}\right), \qquad (10.26)$$

where $\gamma_m \in \{1,...,I\}$ is a realization of a hidden state variable indicating the active model at time instant m. The distribution $p\left(\boldsymbol{\chi}_m|\boldsymbol{\chi}_{m-1},\mathbf{y}_{0:m-1},\gamma_m=i\right)$ is completely determined by the distribution $p\left(\mathbf{x}_m|\boldsymbol{\chi}_{m-1},\mathbf{y}_{0:m-1},\gamma_m=i\right)$, which may be approximated by a linear, autoregressive prediction model as

$$p\left(\mathbf{x}_m|\boldsymbol{\chi}_{m-1},\mathbf{y}_{0:m-1},\gamma_m=i\right) \approx p(\mathbf{x}_m|\mathbf{x}_{m-1},\gamma_m=i) \qquad (10.27)$$

$$= \begin{cases} \mathcal{N}\left(\mathbf{x}_0;\boldsymbol{\mu}_{\mathbf{x}}^{(i)},\boldsymbol{\Sigma}_{\mathbf{x}}^{(i)}\right) & \text{for } m=0 \\ \mathcal{N}\left(\mathbf{x}_m;\mathbf{A}^{(i)}\mathbf{x}_{m-1}+\mathbf{b}^{(i)},\mathbf{V}^{(i)}\right) & \text{for } m>0 \end{cases}. \qquad (10.28)$$

According to this model the first clean speech feature vector is modeled by a Gaussian mixture model (GMM) with mean vectors $\boldsymbol{\mu}_{\mathbf{x}}^{(i)}$ and covariance matrices $\boldsymbol{\Sigma}_{\mathbf{x}}^{(i)}$. All successive clean feature vectors are predicted from their predecessors by a linear transformation described by the state transition matrix $\mathbf{A}^{(i)}$ and the bias vector $\mathbf{b}^{(i)}$. The prediction error is assumed to be zero mean, having a Gaussian distribution with the covariance matrix $\mathbf{V}^{(i)}$.

The mixture probabilities may be approximated by

$$P\left(\gamma_m = i|\boldsymbol{\chi}_{m-1}, \mathbf{y}_{0:m-1}\right) = \begin{cases} \alpha^{(i)} & \text{for } m = 0 \\ \sum_{j=1}^{I} a_{ij} P\left(\gamma_{m-1} = j|\mathbf{y}_{0:m-1}\right) & \text{for } m > 0 \end{cases}, \tag{10.29}$$

with time-invariant state transition probabilities

$$a_{ij} := P\left(\gamma_m = i|\gamma_{m-1} = j\right) \tag{10.30}$$

and model probabilities for the first time instant $m = 0$

$$\alpha^{(i)} := P\left(\gamma_0 = i\right). \tag{10.31}$$

This kind of a priori model, known as a switching linear dynamic model (SLDM), explicitly considers correlations between successive speech feature vectors which are due to the speech production process on the one hand and the feature extraction process on the other. It has been successfully employed for noise robust speech recognition [7].

Its parameters are supposed to be learned from training data prior to the feature enhancement by applying the well-known expectation maximization (EM) algorithm [6], which iteratively delivers sets of improved parameter estimates obtained from maximizing the likelihood of the training data based on previous parameter estimates. In general, the algorithm only gives locally optimal solutions. An extensive tutorial on the SLDM parameter estimation is provided by [29] whereas some strategies for the initialization of SLDM parameters may be found in [24].

10.4.1.2 A Priori Noise Model

The variety of noise characteristics is much more extensive than that of speech characteristics, based on the fact that the type of noise sources may differ depending on the environment where the ASR system is employed. Thus, attempting to train a noise model covering the whole range of different noise types would on the one hand require lots of training data, rendering the idea circumstantial. On the other hand, such a model would only deliver a very general description of the noise, resulting in only slight benefits for feature enhancement.

However, often it is reasonable to assume the noise to be stationary at least for short periods of time covering the duration of single speech utterances. Then it is possible to estimate the noise model parameters from either noise-only periods, which might be detected by voice activity detection (VAD), or directly from noisy speech, e.g, by using a minimum statistics approach. In consideration of the fact that mainly the most recent noise or noisy observations should be considered for the estimation of the noise model parameters to be able to track the possibly time-variant noise characteristics, the model complexity should be kept low to allow an online adaptation of the parameters. For that reason, the predictive noise distribution in (10.25) is approximated by a single Gaussian distribution with mean vector $\boldsymbol{\mu}_{\mathbf{n}}$ and covariance matrix $\boldsymbol{\Sigma}_{\mathbf{n}}$:

$$p(\mathbf{n}_m|\mathbf{n}_{m-1},\mathbf{y}_{0:m-1}) \approx p(\mathbf{n}_m) = \mathscr{N}(\mathbf{n}_m;\boldsymbol{\mu}_\mathbf{n},\boldsymbol{\Sigma}_\mathbf{n}) \quad \text{for } m \in \mathscr{I}_\mathrm{u}, \tag{10.32}$$

which are assumed to be constant for the duration of a speech utterance, which is characterized by the index set $\mathscr{I}_\mathrm{u}$.

10.4.2 Observation Model

The observation distribution $p(\mathbf{y}_m|\mathbf{z}_m)$ describes the dependence of the observed reverberant noisy feature vector $\mathbf{y}_m$ on the sequence of clean speech feature vectors $\boldsymbol{\chi}_m$ and the noise feature vector $\mathbf{n}_m$. This dependence is mainly given through the RIR between the speaker and the microphone (see Fig. 10.2). In general, the RIR is highly time-variant depending on temperature, humidity, room geometry, and speaker location and direction, rendering a derivation of an observation model very complicated. For that reason, we assume, for the moment, the RIR to be time-invariant, having a finite length L_h, i.e.,

$$\overline{h}_l(p) = \overline{h}(p) \qquad \forall l \in \mathbb{N}_0, \tag{10.33}$$

$$\overline{h}(p) = 0 \qquad \forall p \geq L_h, \tag{10.34}$$

which can be justified by its physically attributed exponentially decreasing envelope. Based on the assumptions (10.33) and (10.34), we will in the following recapitulate the derivation of an observation model in the logarithmic mel power spectral domain presented in [23], which is extended here to consider additive noise.

10.4.2.1 Observation Model for Given Time-Invariant RIR

Starting from the expressions (10.1) and (10.2) describing the dependence of the reverberant noisy speech signal $\overline{y}(l)$ on the clean speech signal $\overline{x}(l)$, a functional relationship between the corresponding LMPSCs $x_{m,q}$ and $y_{m,q}$ will be derived. Based on the sufficient condition that the STDFT (10.4) is computed at oversampling, which means that the parameters B, L_w and K are chosen to fulfil the conditions $K \geq B$ and $L_w \geq B$, the relationship between the STDFTs of the reverberant noisy microphone signal, the clean speech signal and the noise signal can be approximated as [23]

$$\tilde{Y}(m,k) \approx \sum_{m'=0}^{L_H} \tilde{X}(m-m',k)H_{m',k} + \tilde{N}(m,k), \tag{10.35}$$

where the number of summands in the convolution is determined by

$$L_H = \left\lfloor \frac{L_h + L_w - 2}{B} \right\rfloor. \tag{10.36}$$

The computation of the STDFT-like RIR representation

$$H_{m',k} := \frac{L_w}{B} e^{j\frac{2\pi}{K}k(L_w-1)} \sum_{p=0}^{2L_w-2} \left[\overline{h}(m',p)e^{-j\frac{2\pi}{K}kp}\right] \tag{10.37}$$

essentially comprises the transform of overlapping windowed RIR segments,

$$\overline{h}(m',p) := \overline{h}(m'B+p-L_w+1)\tilde{w}(p-L_w+1), \tag{10.38}$$

to the frequency domain, where the window function

$$\tilde{w}(l') := \sum_{l=0}^{L_w-1} \tilde{w}_a(l)\tilde{w}_s(l-l') \tag{10.39}$$

is computed from the convolution of the analysis window $\tilde{w}_a(l)$ with a corresponding time-reversed synthesis window $\tilde{w}_s(-l)$ of length L_w. Together with the analysis window it has to obey the so called completeness condition [31]:

$$\sum_{m=0}^{\lfloor \frac{L_w}{B} \rfloor} \tilde{w}_\mathrm{s}(l+mB)\tilde{w}_\mathrm{a}(l+mB) = \frac{1}{K}, \quad 0 \le l < B. \tag{10.40}$$

It can be shown that the solution for the synthesis window satisfying (10.40) always exists under the stated assumptions. Although it is not unique in general, a reasonable solution is the one with minimum l^2-norm, which has the property of being well concentrated in the time domain.

The relationship between the corresponding LMPSCs can be expressed as [23]

$$y_{m,q} = \ln\left(\sum_{m'=0}^{L_H} e^{x_{m-m',q}+\overline{h}_{m',q}} + e^{n_{m,q}}\right) + v_{m,q}, \tag{10.41}$$

where the coefficients

$$\overline{h}_{m',q} := \ln\left(\overline{H}^2_{m',q}\right) \tag{10.42}$$

can be interpreted as a logarithmic mel power spectral-like representation of the RIR with

$$\overline{H}^2_{m',q} := \left(\frac{1}{K_q^{(\mathrm{up})} - K_q^{(\mathrm{lo})} + 1}\right) \sum_{k=K_q^{(\mathrm{lo})}}^{K_q^{(\mathrm{up})}} |H_{m',k}|^2 \tag{10.43}$$

denoting the averaged RIR powers per mel band. The error term $v_{m,q}$ is given by

$$v_{m,q} := \ln\left(1 + \frac{\mathscr{E}_{m,q}}{\sum_{m'=0}^{L_H} e^{x_{m-m',q}+\overline{h}_{m',q}} + e^{n_{m,q}}}\right) \tag{10.44}$$

with

$$\mathscr{E}_{m,q} \approx \sum_{m'=0}^{L_H} \sum_{k=K_q^{(\mathrm{lo})}}^{K_q^{(\mathrm{up})}} |\tilde{X}(m-m',k)|^2 \left(|H_{m',k}|^2 - \overline{H}_{m',q}^2\right) \Lambda_q(k)$$
$$+ \sum_{m'=0}^{L_H} \sum_{m''=m'+1}^{L_H} \sum_{k=K_q^{(\mathrm{lo})}}^{K_q^{(\mathrm{up})}} 2\Re\left[\tilde{X}(m-m',k)\tilde{X}^*(m-m'',k)H_{m',k}H_{m'',k}^*\right]\Lambda_q(k)$$
$$+ \sum_{m'=0}^{L_H} \sum_{k=K_q^{(\mathrm{lo})}}^{K_q^{(\mathrm{up})}} 2\Re\left[\tilde{X}(m-m',k)H_{m',k}\tilde{N}^*(m-m',k)\right]\Lambda_q(k), \tag{10.45}$$

where $\Re(\cdot)$ denotes the real part. This error term cannot be expressed by means of the LMPSCs $x_{m-m',q}$, $n_{m,q}$ or the RIR coefficients $\overline{h}_{m',q}$ alone, since it requires additional spectral information contained in $\mathscr{E}_{m,q}$, which gets lost when computing the logarithmic mel power spectrum [23]. In particular, the term $\mathscr{E}_{m,q}$ arises, first, from the approximation (10.35) and, second, from neglecting cross terms in the calculation of the power of (10.35) to obtain the first term in (10.41). This term is highly time-variant and can in principle be deterministically computed with the knowledge of the STDFTs of all involved signals and the STDFT-like RIR representation. When performing feature enhancement in the logarithmic mel power spectral domain, the mentioned STDFTs are obviously unknown. For simplification the error term $v_{m,q}$ may then be approximated to be a realization of a Gaussian, stationary stochastic process. Additionally assuming the stochastic process to be ergodic facilitates computing its moments prior to feature enhancement from some appropriate training data. For the noise-free case it has been shown in [23] that the application of a Gaussian distribution is reasonable.

Defining the mapping

$$f : \mathbb{R}^{Q\times[2(L_H+1)+1]} \to \mathbb{R}^Q,$$
$$f\left(\mathbf{x}_{m-L_H:m}, \overline{\mathbf{h}}_{0:L_H}, \mathbf{n}_m\right) = \ln\left(\sum_{m'=0}^{L_H} e^{\mathbf{x}_{m-m'}+\overline{\mathbf{h}}_{m'}} + e^{\mathbf{n}_m}\right), \tag{10.46}$$

where the mathematical operations are assumed to be applied to each vector component separately, the relationship (10.41) can be reformulated to the following observation mapping:

$$\mathbf{y}_m = f\left(\mathbf{x}_{m-L_H:m}, \overline{\mathbf{h}}_{0:L_H}, \mathbf{n}_m\right) + \mathbf{v}_m. \tag{10.47}$$

This observation mapping mathematically shows the dispersive effect of reverberation on the LMPSCs, since it illustrates the dependence of $\mathbf{y}_m$ on the sequence $\mathbf{x}_{m-L_H:m}$ of previous clean speech feature vectors.

The vector sequence $\{\mathbf{v}_m\}_{m\in\mathbb{N}_0}$ is assumed to be a realization of a white Gaussian stochastic process, where the distribution for each time instant m is given by

$$p(\mathbf{v}_m) = \mathcal{N}(\mathbf{v}_m; \boldsymbol{\mu}_{\mathbf{v}}, \boldsymbol{\Sigma}_{\mathbf{v}}) \tag{10.48}$$

with mean vector $\boldsymbol{\mu}_{\mathbf{v}}$ and covariance matrix $\boldsymbol{\Sigma}_{\mathbf{v}}$, the choice of which is addressed in Section 10.4.2.3.

10.4.2.2 Modeling the Room Impulse Response

Before the observation model (10.47) may be employed for feature enhancement, some aspects concerning the logarithmic mel power spectral-like representation $\overline{\mathbf{h}}_{0:L_H}$ have to be considered. In principle, it may be computed from a given RIR.

An example of an RIR measured in a large office is depicted in Fig. 10.4. In

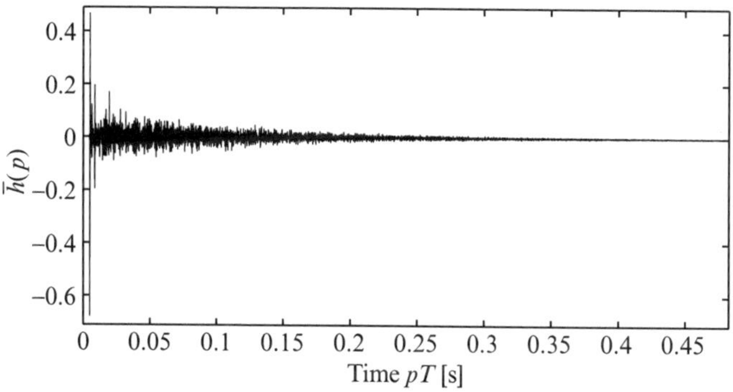

Fig. 10.4: Example of a RIR recorded in an office room having a reverberation time of about 0.75 s

general, an RIR is characterized by a main peak caused by the direct sound, some distinct peaks caused by early reflections and a late reverberation part starting at about 50 ms caused by late reflections. The peaks arising from the late reflections occur more frequently with increasing time. Their magnitude exhibits an exponential decay that can be characterized by the reverberation time T_{60}, which is defined to be the time required for the RIR energy to decay by 60 dB from its initial level.

In the considered "blind" scenario, however, it is assumed that the RIR is not available. Further, the RIR is strongly time-variant, resulting from changes in the position of the speaker and the objects located in the room, the temperature and the humidity. However, the time-variance mainly influences the fine structure of the RIR whereas the overall characteristics such as the power envelope and reverberation time are hardly affected.

Motivated by these considerations, we abstain from estimating the RIR for the computation of its logarithmic mel power spectral-like representation, since an accurate blind estimation is not trivial. Instead, the coarse RIR model [26]

$$\overline{h}(l) = \sigma_h \cdot \overline{u}(l) \cdot \overline{\zeta}(l) \cdot e^{-\frac{l}{\tau}} \tag{10.49}$$

is employed to compute the expected RIR coefficients $\hat{\overline{\mathbf{h}}}_{0:L_H}$ under this model. It is primarily characterized by a realization of a zero mean white Gaussian stochastic process $\overline{\zeta}(l)$ with $\mathrm{E}[\overline{\zeta}^2(l)] = 1$ modeling the random appearance of reflections. The factor $e^{-\frac{l}{\tau}}$ is applied to obtain an exponentially decaying envelope, where the decay parameter τ can be expressed by means of the reverberation time T_{60} and the sampling period T_s as

$$\tau = \frac{T_{60}}{3\ln(10) \cdot T_s}. \tag{10.50}$$

The RIR's support indicator function

$$\overline{u}(l) := \begin{cases} 1 & \text{for } 0 \leq l \leq L_h - 1 \\ 0 & \text{else} \end{cases} \tag{10.51}$$

enforces the RIR to be causal, having a finite length L_h. Further, σ_h is a normalizing scalar which determines the RIR's overall energy.

Based on Monte Carlo simulations, it has been observed that the distributions of the RIR coefficients $\overline{\mathbf{h}}_{0:L_H}$ under the stochastic RIR model (10.49) can be well approximated by Gaussians. Using this assumption, it has been shown in [24] that the means of these distributions may be computed by

$$\hat{\overline{h}}_{m',q} := \mathrm{E}\left[\overline{h}_{m',q}\right] = \frac{1}{2}\ln\left(\frac{\mu^{4(\overline{H}^2)}_{m',q}}{\sigma^{2(\overline{H}^2)}_{m',q} + \mu^{2(\overline{H}^2)}_{m',q}}\right). \tag{10.52}$$

Here, $\mu^{(\overline{H}^2)}_{m',q}$ and $\sigma^{2(\overline{H}^2)}_{m',q}$ denote the means and variances of the averaged RIR powers per mel band (10.43), which can be computed according to [23] as

$$\mu^{(\overline{H}^2)}_{m',q} := \mathrm{E}\left[\overline{H}^2_{m',q}\right] = \sum_{p'=0}^{2L_w-2} \left|\psi_{m',p',0}\right|^2, \tag{10.53}$$

$$\sigma^{2(\overline{H}^2)}_{m',q} := \mathrm{E}\left[\left(\overline{H}^2_{m',q} - \mu^{(\overline{H}^2)}_{m',q}\right)^2\right] \tag{10.54}$$

$$= \frac{2}{\left(K_q^{(\mathrm{up})} - K_q^{(\mathrm{lo})} + 1\right)^2} \sum_{k'=K_q^{(\mathrm{lo})}}^{K_q^{(\mathrm{up})}} \sum_{k''=K_q^{(\mathrm{lo})}}^{K_q^{(\mathrm{up})}} \left[\left|\sum_{p'=0}^{2L_w-2} \psi_{m',p',k'} \Re\left(\psi_{m',p',k''}\right)\right|^2 + \left|\sum_{p'=0}^{2L_w-2} \psi_{m',p',k'} \Im\left(\psi_{m',p',k''}\right)\right|^2\right], \tag{10.55}$$

where $\Im(\cdot)$ denotes the imaginary part and

$$\psi_{m',p,k} := \sigma_h \frac{L_w}{B} \cdot \bar{u}\left(m'B + p - L_w + 1\right) e^{-\frac{m'B+p-L_w+1}{\tau}} \tilde{w}\left(p - L_w + 1\right) e^{-j\frac{2\pi}{N}kp}. \tag{10.56}$$

The great advantage of the RIR model (10.49) is that its application only requires the estimation of two parameters, T_{60} and σ_h, delivering a coarse RIR description, rather than the estimation of detailed RIR characterized by a high number of variables. For the noise-free case there exist maximum likelihood-based approaches to estimate the reverberation time blindly [32, 33], directly from the incoming microphone signal without the requirement of specific excitation sounds. If the incoming microphone signal also contains noise, a denoising may be performed using spectral subtraction or Wiener filtering techniques prior to the application of the reverberation time estimation algorithm.

The blind estimation of the energy parameter σ_h is generally not trivial, resulting from the nonstationary character of speech signals. A suboptimal approach may be derived from the simplified assumption of a stationary source signal $\bar{x}(l)$ and a stationary noise signal $\bar{n}(l)$. Exploiting the independence of the source speech signal from the noise signal and from the RIR and using the assumption that neighboring RIR samples are uncorrelated, we obtain the following relationship between the power of the microphone signal E_y and that of the source signal E_x:

$$\mathrm{E}_{\bar{y}} := \mathrm{E}\left[\bar{y}^2(l)\right] = \mathrm{E}\left[\bar{s}^2(l)\right] + \mathrm{E}\left[\bar{n}^2(l)\right] \tag{10.57}$$

$$= \mathrm{E}\left[\sum_{l'=0}^{L_h-1}\sum_{l''=0}^{L_h-1} \bar{h}(l')\bar{h}(l'')\bar{x}(l-l')\bar{x}(l-l'')\right] + \mathrm{E}_{\bar{n}} \tag{10.58}$$

$$= \sum_{l'=0}^{L_h-1}\sum_{l''=0}^{L_h-1} \mathrm{E}\left[\bar{h}(l')\bar{h}(l'')\right] \mathrm{E}\left[\bar{x}(l-l')\bar{x}(l-l'')\right] + \mathrm{E}_{\bar{n}} \tag{10.59}$$

$$= \sum_{l'=0}^{L_h-1} \mathrm{E}\left[\bar{h}^2(l')\right] \mathrm{E}\left[\bar{x}^2(l-l')\right] + \mathrm{E}_{\bar{n}} \tag{10.60}$$

$$= \mathrm{E}_{\bar{x}}\mathrm{E}\left[\sum_{l'=0}^{L_h-1} \bar{h}^2(l')\right] + \mathrm{E}_{\bar{n}}. \tag{10.61}$$

If we assume that the noisy reverberant signals for the individual utterances are appropriately normalized to ensure $\mathrm{E}_{\bar{x}} = \mathrm{E}_{\bar{s}}$, the RIR must satisfy the constraint $E\left[\sum_{l=0}^{L_h-1} \bar{h}^2(l)\right] = 1$. That constraint allows us to compute the energy parameter σ_h by

$$\sigma_h = \sqrt{\left(e^{-\frac{2}{\tau}} - 1\right) / \left(e^{-\frac{2L_h}{\tau}} - 1\right)}. \tag{10.62}$$

An appropriate normalization of $\bar{y}(l)$ requires an estimate of the noise power, which may be computed from speech pauses detected by a VAD and may be assumed constant for the duration of individual speech utterances.

However, strictly speaking, the RIR model (10.49) is only valid for the diffuse part of the RIR caused by late reverberation. In consequence of its simplicity, it does not consider frequency dependence of reverberation caused by the surface character of the walls of the enclosure. Further, the RIR model does not capture the typical two-sloped decay of RIRs corresponding to microphone positions within the critical distance. These two aspects can be seen from the illustration of the logarithmic mel power spectral- (LMPS-) like representations of the RIR in Fig. 10.5, where in Fig. 5(a) the true RIR coefficients $\overline{h}_{m',q}$ corresponding to the RIR given in Fig. 10.4 have to be compared to those in Fig. 5(b) computed from (10.52) using the simplified RIR model with a reverberation time of $\hat{T}_{60} = 0.75$ s.

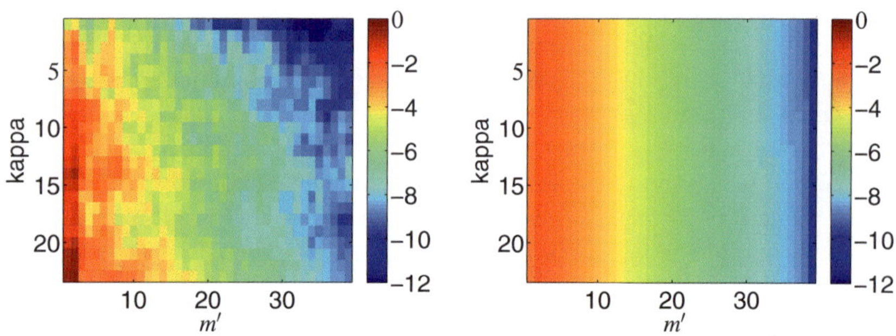

(a) True representation $\overline{h}_{m',q}$ computed from (10.42)

(b) Approximate representation $\hat{\overline{h}}_{m',q}$ computed from (10.52) with $\hat{T}_{60} = 0.75$ s

Fig. 10.5: Logarithmic mel power spectral-like representations of RIR illustrated in Fig. 10.4, where m' denotes the frame index within the RIR and q denotes the mel band index

In principle, it would be possible to alter the model to include frequency-dependent decays. However, this would increase the model complexity and, probably, impair the estimation accuracy for the individual parameters. It is left for further research to investigate the influences of a more detailed RIR model on the performance of the proposed feature enhancement algorithm.

The number $L_H + 1$ of summands in the observation mapping (10.46) depends on the RIR length L_h through (10.36). A reasonable criterion for the choice of L_h should depend on the estimate of the decay parameter τ. One possibility is to introduce an upper bound $0 < \varepsilon_h < 1$ for the relative error ε between the expected energy of the true RIR and that of the RIR truncated to the length of L_h samples, and then to minimize the RIR length L_h under this constraint. Assuming the simplified model (10.49) for the RIR, the error can be expressed as

$$\varepsilon = \varepsilon(L_h) = \frac{E\left[\sum_{l=L_h}^{\infty} \overline{h}^2(l)\right]}{E\left[\sum_{l=0}^{\infty} \overline{h}^2(l)\right]} = \frac{\sum_{l=L_h}^{\infty} e^{-\frac{2l}{\tau}}}{\sum_{l=0}^{\infty} e^{-\frac{2l}{\tau}}} = e^{-\frac{2L_h}{\tau}}. \tag{10.63}$$

The desired estimate $\hat{L}_h$ for the RIR length is then given by

$$\hat{L}_h = \underset{L_h}{\operatorname{argmax}}\, \varepsilon(L_h) \quad \text{s.t.} \quad \varepsilon(L_h) < \varepsilon_h \tag{10.64}$$

$$= \left\lceil -\frac{\tau}{2} \ln(\varepsilon_h) \right\rceil . \tag{10.65}$$

10.4.2.3 Observation Error Model

In Sect. 10.4.2.1 the observation error (10.44) has been derived under the assumption of a given time-invariant RIR. However, this assumption is usually not valid in practice, leading to additional errors introduced on the deterministic part of the observation mapping (10.46). These arise, on the one hand, from the time-variant nature of the RIR and, on the other hand, from using the simplified RIR model (10.49) instead of the true RIR. The latter reason comprises model deficiencies and estimation errors of the model parameters $\hat{T}_{60}$, $\hat{\sigma}_h$ and $\hat{L}_h$. Consequently, the observation error $\mathbf{v}_m$ in the observation model (10.47) has to consider all these effects.

In order to arrive at a manageable description of the observation error, it has been assumed to be a realization of a stationary, ergodic white Gaussian random process (see (10.48)). The unknown probability density function (pdf) parameters $\boldsymbol{\mu}_{\mathbf{v}}$ and $\boldsymbol{\Sigma}_{\mathbf{v}}$ may, in principle, be determined a priori from stereo training data exploiting the assumed ergodicity of the stochastic error process. Using stereo training utterances for a specific reverberant scenario, the errors

$$\hat{\mathbf{v}}_m = \mathbf{y}_m - f\left(\mathbf{x}_{m-\hat{L}_H:m}, \hat{\bar{\mathbf{h}}}_{0:\hat{L}_H}, \mathbf{n}_m\right) \tag{10.66}$$

between the true corrupted LMPSC vectors $\mathbf{y}_m$ and their predictions $f\left(\mathbf{x}_{m-\hat{L}_H:m}, \hat{\bar{\mathbf{h}}}_{0:\hat{L}_H}, \mathbf{n}_m\right)$ from the sequence $\mathbf{x}_{m-\hat{L}_H:m}$ of the corresponding recent clean LMPSC vectors and the noise LMPSC vector $\mathbf{n}_m$ may be computed. The predictions are based on the deterministic part of the observation mapping (10.47) and employ an approximate LMPS-like RIR representation $\hat{\bar{\mathbf{h}}}_{0:\hat{L}_H}$ obtained from (10.52) using estimated values of $\hat{T}_{60}$ and $\hat{L}_H$. The mean vector $\boldsymbol{\mu}_{\mathbf{v}}$ and the covariance matrix $\boldsymbol{\Sigma}_{\mathbf{v}}$ are obtained from (10.66) as empirical estimates of the corresponding first two central moments.

Typically, however, stereo training data for the particular scenario, where the ASR system might be employed, are not available. At least for the noise-free case, a possible way out of this problem is to artificially create stereo training data by convolving the clean utterances with artificial RIRs created using the image method [30]. Thereby, a pool of parameters $\left(\boldsymbol{\mu}_{\mathbf{v}}^{(T_{60})}, \boldsymbol{\Sigma}_{\mathbf{v}}^{(T_{60})}\right)_{T_{60}\in\mathscr{T}}$ for a finite set $\mathscr{T}$ of predefined reverberation times T_{60} may be computed in advance. When carrying out recognition, those parameters may be chosen from this set whose index fits best the estimate $\hat{T}_{60}$ for the particular reverberant hands-free scenario.

In order to consider different recognition setups, we propose using several different artificial RIRs per reverberation time obtained by varying the positions of the speaker and the microphone within the simulated enclosure, which is assumed to be a cuboid of a typical size. Further, to consider estimation errors in the reverberation time T_{60}, which might occur at enhancement, it is possible to vary the reverberation time in a certain interval around the exact value.

The artificial creation of stereo training data for the noisy case is much more demanding, since the parameters of the observation error highly depend on the noise characteristics and the signal-to-noise ratio (SNR). The separate estimation of the observation noise model parameters for each possible combination of noise type and SNR would not only require the corresponding training data to be available, but would also pose the problem of how to decide which of the large number of parameters to choose at enhancement. Alternatively, using all available noise training data jointly would allow us to compute averaged observation noise model parameters, which would avoid the decision problem at the expense of more inaccurate modelling. If no noise training data are available, approximate parameters of the observation noise model may be estimated by considering reverberation only and ignoring the noise, as in Sect. 10.6, to follow.

10.5 Suboptimal Inference

While in Sect. 10.4 the fundamental Bayesian framework for model-based feature enhancement was introduced, this section is concerned with the actual computation of the posterior distribution moments, (10.23) and (10.24).

In general, if all involved distributions were (single) Gaussian and the prediction and observation noise processes were white and mutually independent, the mean and covariance of the posterior distribution would represent a finite-dimensional sufficient statistics that would completely summarize the information given by all past observations [2]. Then, if the observation mapping were linear, these two required moments could be computed using a Kalman filter.

However, in the considered feature enhancement problem, these requirements are violated, and thus only suboptimal filtering algorithms may be employed. First, since the observation mapping is highly nonlinear, an iterated extended Kalman filter (IEKF) may be used for an approximate computation of the posterior moments. Its main idea is to linearize the observation function at a prediction point, which is iteratively recomputed according to an optimization criterion [2]. Alternatively, the IEKF may be replaced by any other filter which is able to propagate the moments of the Gaussian pdf through nonlinear functions, such as the Unscented Kalman Filter (UKF) [19].

Second, employing an SLDM as the a priori speech model requires a Kalman filter to be used for each of the individual speech models. That means that if the a priori distribution is assumed to be Gaussian, the posterior distribution would in general be a GMM, where each of the mixture components would have to be filtered by each individual Kalman filter. Consequently, the number of possible model histories, and thus the number of mixtures in the posterior distribution, would grow exponentially with increasing time index m, rendering the approach not feasible for real-time applications. However, a suboptimal solution to that problem is given by the interacting multiple model (IMM) approach [2], where an appropriate model combination is performed to approximate the posterior and prior distributions by single Gaussians, resulting in a computational effort which is approximately linear in the number of models and constant over time.

Third, from (10.44) it can be seen that the time-variant observation noise is certainly not white. In general, it is not even Gaussian. Hence, applying the IEKF for the inference computation leads, also from that perspective, only to suboptimal solutions.

10.5.1 Feature Enhancement Algorithm

The feature enhancement algorithm presented in Alg. 3 is assumed to be applied to a sequence of noisy reverberant feature vectors $\mathbf{y}_{0:M}$ belonging to a single speech utterance. Basically, the algorithm iteratively computes estimates of the moments (10.23) and (10.24) of the posterior distribution $p(\mathbf{z}_m|\mathbf{y}_{0:m})$ for $0 \leq m \leq M$.

The desired MMSE estimates of the clean speech feature vectors and the corresponding error covariance matrices may then be extracted according to (10.98) and (10.99) further below. It should be noted that an estimate $\hat{\mathbf{z}}_{m|m}$ provides the clean speech feature vector estimates $\hat{\mathbf{x}}_{m-L_C+1|m}$, ..., $\hat{\mathbf{x}}_{m|m}$, which are all conditioned on the observations $\mathbf{y}_{0:m}$. Since the vector $\hat{\mathbf{x}}_{m-L_C+1|m}$ contains the most information regarding the future, it is employed as the final estimate. However, this implicit smoothing introduces a latency of $L_C - 1$ blocks to the algorithm, which is in principle suited for online processing.

Algorithm 3: Feature enhancement

- Initialize the estimates of the mean vectors and error covariance matrices of those clean speech feature vectors, which are assumed to have occurred prior to the time instant $m = 0$ by

$$\hat{\mathbf{x}}_{m|m+L_C-1} = (x_{\min}, \ldots, x_{\min})^T \in \mathbb{R}^Q, \quad -L_H \leq m \leq -1, \tag{10.67}$$

$$\hat{\mathbf{\Sigma}}_{\mathbf{x}_{m|m+L_C-1}} = \sigma^2_{x,\min} \cdot \mathbf{I}_{Q\times Q}, \quad -L_H \leq m \leq -1, \tag{10.68}$$

where $x_{\min} \in \mathbb{R}$, $\sigma^2_{x,\min} \in \mathbb{R}^+$ and $\mathbf{I}_{Q\times Q} \in \mathbb{R}^{Q\times Q}$ denotes the identity matrix. If assuming absence of speech, an appropriate initialization is given by $x_{\min} = -50$ and $\sigma^2_{x,\min} = 10^{-6}$.
- **for** $m = 0..M$ **do**

1. **Perform preprocessing**:
   if $m = 0$ **then**
   a. Set a priori model probabilities $P^{(i)}_{0|-1}$:

$$P^{(i)}_{0|-1} = \alpha^{(i)}, \quad i \in \{1, \ldots I\}. \tag{10.69}$$

   else
   a. Compute the probabilities $\Lambda^{(i,j)}_m := P(\gamma_{m-1} = j|\gamma_m = i, \mathbf{y}_{0:m-1})$ that model j was active at time instant $m-1$ given that model i is active at time instant m conditioned

on the observations $\mathbf{y}_{0:m-1}$:

$$\Lambda_m^{(i,j)} = \frac{1}{c_j} a_{ij} P^{(i)}_{m-1|m-1}, \quad i,j \in \{1,...,I\} \tag{10.70}$$

with

$$c_j = \sum_{i=1}^{I} a_{ij} P^{(i)}_{m-1|m-1}, \quad j \in \{1,\ldots I\}, \tag{10.71}$$

where $P^{(i)}_{m-1|m-1} := P(\gamma_{m-1} = i|\mathbf{y}_{0:m-1})$ is the posterior probability that model i is active at time instant $m-1$ conditioned on the observations $\mathbf{y}_{0:m-1}$.

b. Compute the mean vectors $\hat{\mathbf{z}}^{(0,i)}_{m-1|m-1}$ and covariance matrices $\hat{\boldsymbol{\Sigma}}^{(0,i)}_{\mathbf{z}_{m-1|m-1}}$ to be input to the ith IEKF:

$$\hat{\mathbf{z}}^{(0,i)}_{m-1|m-1} = \sum_{j=1}^{I} \Lambda_m^{(i,j)} \hat{\mathbf{z}}^{(j)}_{m-1|m-1}, \quad i \in \{1,\ldots I\}, \tag{10.72}$$

$$\hat{\boldsymbol{\Sigma}}^{(0,i)}_{\mathbf{z}_{m-1|m-1}} = \sum_{j=1}^{I} \Lambda_m^{(i,j)} \Bigg[\hat{\boldsymbol{\Sigma}}^{(j)}_{\mathbf{z}_{m-1|m-1}} + \left(\hat{\mathbf{z}}^{(0,j)}_{m-1|m-1} - \hat{\mathbf{z}}^{(j)}_{m-1|m-1}\right)\left(\hat{\mathbf{z}}^{(0,j)}_{m-1|m-1} - \hat{\mathbf{z}}^{(j)}_{m-1|m-1}\right)^T \Bigg], \tag{10.73}$$

where $\hat{\mathbf{z}}^{(j)}_{m-1|m-1}$ and $\hat{\boldsymbol{\Sigma}}^{(j)}_{\mathbf{z}_{m-1|m-1}}$ are estimates of the mean vector and the covariance matrix of the posterior distribution $p(\mathbf{z}_{m-1}|\gamma_{m-1} = j, \mathbf{y}_{0:m-1})$.

c. Compute the a priori model probabilities $P^{(i)}_{m|m-1} := P(\gamma_m = i|\mathbf{y}_{0:m-1})$:

$$P^{(i)}_{m|m-1} = \sum_{j=1}^{I} a_{ij} P^{(j)}_{m-1|m-1}, \quad i \in \{1,\ldots I\}. \tag{10.74}$$

end if

2. **Perform iterated extended Kalman filtering based on each model** $i \in \{1,\ldots I\}$**:**
   a. **Perform prediction step:**
   if $m = 0$ **then**
      i. Initialize the estimates of the mean vector and the covariance matrix of the model conditioned predictive distribution $p(\mathbf{z}_0|\gamma_0 = i)$:

$$\hat{\mathbf{z}}^{(i)}_{0|-1} = \left[\left(\boldsymbol{\mu}^{(i)}_{\mathbf{x}}\right)^T \; \mathbf{x}^T_{\text{init}} \; \boldsymbol{\mu}^T_{\mathbf{n}}\right]^T, \tag{10.75}$$

$$\hat{\boldsymbol{\Sigma}}^{(i)}_{\mathbf{z}_{0|-1}} = \begin{bmatrix} \boldsymbol{\Sigma}^{(i)}_{\mathbf{x}} & \mathbf{0}_{Q\times(L_C-1)Q} & \mathbf{0}_{Q\times Q} \\ \mathbf{0}_{(L_C-1)Q\times Q} & \boldsymbol{\Sigma}_{\mathbf{x}_{\text{init}}} & \mathbf{0}_{(L_C-1)Q\times Q} \\ \mathbf{0}_{Q\times Q} & \mathbf{0}_{Q\times(L_C-1)Q} & \boldsymbol{\Sigma}_{\mathbf{n}} \end{bmatrix} \tag{10.76}$$

   with

$$\mathbf{x}_{\text{init}} := \left(\hat{\mathbf{x}}^T_{-1|L_C-2}, \ldots, \hat{\mathbf{x}}^T_{-L_C+1|0}\right)^T \in \mathbb{R}^{(L_C-1)Q}, \tag{10.77}$$

$$\boldsymbol{\Sigma}_{\mathbf{x}_{\text{init}}} := \text{blockdiag}\left(\hat{\boldsymbol{\Sigma}}_{\mathbf{x}_{-1|L_C-2}}, \ldots, \hat{\boldsymbol{\Sigma}}_{\mathbf{x}_{-L_C+1|0}}\right) \in \mathbb{R}^{(L_C-1)Q\times(L_C-1)Q}, \tag{10.78}$$

   where blockdiag$(\cdot)$ denotes the operation of composing a block diagonal matrix from the arguments.
   else
      i. Compute estimates of the mean vector and the covariance matrix of the model conditioned predictive distribution $p(\mathbf{z}_m|\gamma_m = i, \mathbf{y}_{0:m-1})$ based on the input mean vector $\hat{\mathbf{z}}^{(0,i)}_{m-1|m-1}$ and the covariance matrix $\hat{\boldsymbol{\Sigma}}^{(0,i)}_{\mathbf{z}_{m-1|m-1}}$:

$$\hat{\mathbf{z}}^{(i)}_{m|m-1} = \mathscr{A}^{(i)} \hat{\mathbf{z}}^{(0,i)}_{m-1|m-1} + \boldsymbol{\beta}^{(i)}, \tag{10.79}$$

$$\hat{\boldsymbol{\Sigma}}^{(i)}_{\mathbf{z}_{m|m-1}} = \mathscr{A}^{(i)} \hat{\boldsymbol{\Sigma}}^{(0,i)}_{\mathbf{z}_{m-1|m-1}} \left(\mathscr{A}^{(i)}\right)^T + \mathscr{V}^{(i)}, \tag{10.80}$$

with

$$\mathscr{A}^{(i)} := \begin{bmatrix} \mathbf{A}^{(i)} & \mathbf{0}_{Q\times(L_C-1)Q} & \mathbf{0}_{Q\times Q} \\ \mathbf{I}_{(L_C-1)Q\times(L_C-1)Q} & \mathbf{0}_{(L_C-1)Q\times Q} \\ \mathbf{0}_{Q\times L_C Q} & \mathbf{0}_{Q\times Q} \end{bmatrix}, \quad \boldsymbol{\beta}^{(i)} := \begin{bmatrix} \mathbf{b}^{(i)} \\ \mathbf{0}_{(L_C-1)Q\times 1} \\ \boldsymbol{\mu}_{\mathbf{n}} \end{bmatrix}, \tag{10.81}$$

$$\mathscr{V}^{(i)} := \begin{bmatrix} \mathbf{V}^{(i)} & \mathbf{0}_{Q\times(L_C-1)Q} & \mathbf{0}_{Q\times Q} \\ \mathbf{0}_{(L_C-1)Q\times Q} & \mathbf{0}_{(L_C-1)Q\times(L_C-1)Q} & \mathbf{0}_{(L_C-1)Q\times Q} \\ \mathbf{0}_{Q\times Q} & \mathbf{0}_{Q\times(L_C-1)Q} & \boldsymbol{\Sigma}_{\mathbf{n}} \end{bmatrix}, \tag{10.82}$$

where $\mathbf{0}_{s_1\times s_2}\in\mathbb{R}^{s_1\times s_2}$ denotes the zero matrix.

endif

b. **Perform update step:**

i. Set the initial vector at which to linearize the observation mapping to the prediction vector

$$\hat{\mathbf{z}}^{(1,i)}_{m|m} = \hat{\mathbf{z}}^{(i)}_{m|m-1}. \tag{10.83}$$

ii. Iterate the linearization point[2]:

for $r = 1..R$ **do**

A. Compute the predicted observation $\hat{\mathbf{y}}^{(r,i)}_m$ and the corresponding covariance matrix $\hat{\boldsymbol{\Sigma}}^{(r,i)}_{\mathbf{y}_m}$ based on $\hat{\mathbf{z}}^{(r,i)}_{m|m}$:

$$\hat{\mathbf{y}}^{(r,i)}_m = f\left(\hat{\boldsymbol{\chi}}^{(r,i)}_{m|m}, \hat{\mathbf{x}}_{m-L_C|m-1}, \ldots, \hat{\mathbf{x}}_{m-L_H|m-L_H+L_C-1}, \hat{\bar{\mathbf{h}}}_{0:L_H}, \hat{\mathbf{n}}^{(r,i)}_{m|m}\right) + \boldsymbol{\mu}_{\mathbf{v}}, \tag{10.84}$$

$$\hat{\boldsymbol{\Sigma}}^{(r,i)}_{\mathbf{y}_m} = \mathbf{H}^{(i)}_{\mathbf{z}_m}\hat{\boldsymbol{\Sigma}}^{(i)}_{\mathbf{z}_{m|m-1}}\left(\mathbf{H}^{(i)}_{\mathbf{z}_m}\right)^T + \sum_{m'=m-L_C}^{m-L_H} \mathbf{H}^{(i)}_{\mathbf{x}_{m'}}\hat{\boldsymbol{\Sigma}}_{\mathbf{x}_{m'|m'+L_C-1}}\left(\mathbf{H}^{(i)}_{\mathbf{x}_{m'}}\right)^T + \boldsymbol{\Sigma}_{\mathbf{v}}, \tag{10.85}$$

where the Jacobian of f with respect to $\mathbf{z}_m$ is given by $\mathbf{H}^{(r,i)}_{\mathbf{z}_m} = \left[\mathbf{H}^{(r,i)}_{\boldsymbol{\chi}_m}\ \mathbf{H}^{(r,i)}_{\mathbf{n}_m}\right]$ with

$$\mathbf{H}^{(r,i)}_{\boldsymbol{\chi}_m} := \left.\frac{\partial f\left(\boldsymbol{\chi}_m, \hat{\mathbf{x}}_{m-L_C|m-1}, \ldots, \hat{\mathbf{x}}_{m-L_H|m-L_H+L_C-1}, \hat{\bar{\mathbf{h}}}_{0:L_H}, \hat{\mathbf{n}}^{(r,i)}_{m|m}\right)}{\partial \boldsymbol{\chi}_m}\right|_{\boldsymbol{\chi}_m=\hat{\boldsymbol{\chi}}^{(r,i)}_{m|m}}, \tag{10.86}$$

$$\mathbf{H}^{(r,i)}_{\mathbf{n}_m} := \left.\frac{\partial f\left(\hat{\boldsymbol{\chi}}^{(r,i)}_{m|m}, \hat{\mathbf{x}}_{m-L_C|m-1}, \ldots, \hat{\mathbf{x}}_{m-L_H|m-L_H+L_C-1}, \hat{\bar{\mathbf{h}}}_{0:L_H}, \mathbf{n}_m\right)}{\partial \mathbf{n}_m}\right|_{\mathbf{n}_m=\hat{\mathbf{n}}^{(r,i)}_{m|m}}, \tag{10.87}$$

and the Jacobian of f with respect to $\mathbf{x}_{m'}$ is denoted by

$$\mathbf{H}^{(r,i)}_{\mathbf{x}_{m'}} := \left.\frac{\partial f\left(\hat{\boldsymbol{\chi}}^{(r,i)}_{m|m}, \mathbf{x}_{m-L_C}, \ldots, \mathbf{x}_{m-L_H}, \hat{\bar{\mathbf{h}}}_{0:L_H}, \hat{\mathbf{n}}^{(r,i)}_{m|m}\right)}{\partial \mathbf{x}_{m'}}\right|_{\substack{\mathbf{x}_{m-n}=\hat{\mathbf{x}}_{m-n|m-n+L_C-1}\\ \text{for } L_C\le n\le L_H}}. \tag{10.88}$$

---

[2] Note that only the most recent L_C speech feature vectors in $\hat{\boldsymbol{\chi}}^{(r,i)}_{m|m}$ are iterated, while the less recent estimates $\hat{\mathbf{x}}_{m-L_C|m-1}$, ..., $\hat{\mathbf{x}}_{m-L_H|m-L_H+L_C-1}$, which are required for the evaluation of the observation mapping f in (10.84) and its Jacobians in (10.86), (10.87) and (10.88) further below, are regarded as being fixed, but nevertheless containing errors. The value of L_C may thereby be chosen distinctly smaller than that of L_H to reduce the dimension of $\mathbf{z}_m$ and hence to reduce the computational effort.

All Jacobians are evaluated at $\left(\hat{\boldsymbol{\chi}}^{(r,i)}_{m|m}, \hat{\mathbf{x}}_{m-L_C|m-1}, \ldots, \hat{\mathbf{x}}_{m-L_H|m-L_H+L_C-1}, \hat{\bar{\mathbf{h}}}_{0:L_H}, \hat{\mathbf{n}}^{(r,i)}_{m|m}\right)$.

B. Update the linearization vector by

$$\hat{\mathbf{z}}^{(r+1,i)}_{m|m} = \hat{\mathbf{z}}^{(i)}_{m|m-1} + \mathbf{K}^{(r,i)}_m \left[\mathbf{y}_m - \hat{\mathbf{y}}^{(r,i)}_m + \mathbf{H}^{(r,i)}_{\mathbf{z}_m}\left(\hat{\mathbf{z}}^{(r,i)}_{m|m} - \hat{\mathbf{z}}^{(i)}_{m|m-1}\right)\right] \tag{10.89}$$

with the Kalman gain $\mathbf{K}^{(r,i)}_m$ computed by

$$\mathbf{K}^{(r,i)}_m = \hat{\boldsymbol{\Sigma}}^{(i)}_{\mathbf{z}_{m|m-1}} \left(\mathbf{H}^{(r,i)}_{\mathbf{z}_m}\right)^T \left(\hat{\boldsymbol{\Sigma}}^{(r,i)}_{\mathbf{y}_m}\right)^{-1}. \tag{10.90}$$

end for

iii. Compute the estimate of the mean vector and the covariance matrix of the posterior distribution $p(\mathbf{z}_m|\gamma_m = i, \mathbf{y}_{0:m})$ of $\mathbf{z}_m$ by

$$\hat{\mathbf{z}}^{(i)}_{m|m} = \hat{\mathbf{z}}^{(R+1,i)}_{m|m}, \tag{10.91}$$

$$\hat{\boldsymbol{\Sigma}}^{(i)}_{\mathbf{z}_{m|m}} = \left(\mathbf{I} - \mathbf{K}^{(R,i)}_m \mathbf{H}^{(R,i)}_{\mathbf{z}_m}\right) \hat{\boldsymbol{\Sigma}}^{(i)}_{\mathbf{z}_{m|m-1}}. \tag{10.92}$$

3. **Perform postprocessing**

a. Compute the posterior model probabilities by

$$P^{(i)}_{m|m} = \frac{1}{c} p(\mathbf{y}_m|\mathbf{z}^{(i)}_{m|m-1}, \boldsymbol{\Sigma}^{(i)}_{\mathbf{z}_{m|m-1}}) P(\gamma_m = i|\mathbf{y}_{0:m-1}) \tag{10.93}$$

$$= \frac{1}{c} \mathcal{N}\left(\mathbf{y}_m; \hat{\mathbf{y}}^{(1,i)}_m, \hat{\boldsymbol{\Sigma}}^{(1,i)}_{\mathbf{y}_m}\right) P^{(i)}_{m|m-1}, \quad i \in \{1, \ldots, I\}, \tag{10.94}$$

where the normalizing constant c is computed from

$$c = \sum_{i=1}^{I} \mathcal{N}\left(\mathbf{y}_m; \hat{\mathbf{y}}^{(1,i)}_m, \hat{\boldsymbol{\Sigma}}^{(1,i)}_{\mathbf{y}_m}\right) P^{(i)}_{m|m-1}. \tag{10.95}$$

b. Compute mean vector and covariance matrix of posterior distribution $p(\mathbf{z}_m|\mathbf{y}_{0:m})$ of $\mathbf{z}_m$ by model combination:

$$\hat{\mathbf{z}}_{m|m} = \sum_{j=1}^{I} P^{(j)}_{m|m} \hat{\mathbf{z}}^{(j)}_{m|m}, \tag{10.96}$$

$$\hat{\boldsymbol{\Sigma}}_{\mathbf{z}_{m|m}} = \sum_{j=1}^{I} P^{(j)}_{m|m} \left[\hat{\boldsymbol{\Sigma}}^{(j)}_{\mathbf{z}_{m|m}} + \left(\hat{\mathbf{z}}_{m|m} - \hat{\mathbf{z}}^{(j)}_{m|m}\right)\left(\hat{\mathbf{z}}_{m|m} - \hat{\mathbf{z}}^{(j)}_{m|m}\right)^T\right]. \tag{10.97}$$

c. **if** $m \geq L_C - 1$ **then**

- Extract the enhanced LMPSC vector and its estimation error covariance matrix by

$$\hat{\mathbf{x}}_{m-L_C+1|m} = \mathbf{M}_{\text{EXTR}} \cdot \hat{\mathbf{z}}_{m|m}, \tag{10.98}$$

$$\hat{\boldsymbol{\Sigma}}_{\mathbf{x}_{m-L_C+1|m}} = \mathbf{M}_{\text{EXTR}} \cdot \hat{\boldsymbol{\Sigma}}_{\mathbf{z}_{m|m}} \cdot \mathbf{M}^T_{\text{EXTR}}, \tag{10.99}$$

where $\mathbf{M}_{\text{EXTR}} = \Big(\underbrace{\mathbf{0}_{Q\times Q}, \ldots, \mathbf{0}_{Q\times Q}}_{(L_C-1)\times}, \mathbf{I}_{Q\times Q}, \mathbf{0}_{Q\times Q}\Big)$.

- Compute the enhanced MFCC vector and its estimation error covariance matrix by

$$\hat{\mathbf{x}}^{(c)}_{m-L_C+1|m} = \mathbf{M}_{\text{DCT}} \cdot \hat{\mathbf{x}}_{m-L_C+1|m}, \tag{10.100}$$

$$\hat{\boldsymbol{\Sigma}}_{\mathbf{x}^{(c)}_{m-L_C+1|m}} = \mathbf{M}_{\text{DCT}} \cdot \hat{\boldsymbol{\Sigma}}_{\mathbf{x}_{m-L_C+1|m}} \cdot \mathbf{M}^T_{\text{DCT}}, \tag{10.101}$$

where $\mathbf{M}_{\text{DCT}} \in \mathbb{R}^{K' \times Q}$ denotes the DCT matrix.

end if

end for

10.6 Experiments

To evaluate the performance of the presented feature enhancement approach, we carried out experiments, where we considered a connected digits recognition task in noisy reverberant environments. The speech data for the evaluation of the ASR performance was taken from the Aurora 5 database [16], which contains utterances of connected digits sampled at a rate of $1/T_s = 8$ kHz.

The database comprises 8,623 clean nonreverberant training utterances, which were employed for the training of the acoustic model of the speech recognizer as well as for the training of the SLDM. For the SLDM training, a "k-means++"-like algorithm [24] was used to initialize the parameters of $I = 4$ models, which were then refined using four iterations of the EM algorithm similar to [29]. For the recognition experiments, an HMM-based recognizer was utilized, whose acoustic models consisted of speaker-independent word-based HMMs with 16 states per word and four Gaussian mixture components per state. Simple left-to-right models without skips over states were used. The training of the HMM parameters and Viterbi decoding for the recognition was carried out using the hidden Markov model toolkit (HTK) software [43].

The test set consisted of noisy reverberant utterances under different reverberation conditions and different signal-to-noise ratios (SNRs). The considered reverberant environments comprise an office and a living room. The individual reverberant utterances of the database were initially created by convolving the corresponding clean utterances with artificial RIRs [17], where the reverberation time T_{60} was varied in the range between 0.3 s and 0.4 s for the office, and in the range between 0.4 s and 0.5 s for the living room. The noisy reverberant utterances were created by adding interior noise at different SNR levels, where the noise signals partly contained non-stationary segments. For each condition there were 8,700 test utterances.

The parameters for the feature extraction were chosen according to the ETSI ES 201 108 standard defined in [9]. They are summarized in Table 10.1. In all considered cases, the features consisted of 39-dimensional vectors resulting from the MFCCs as well as their corresponding delta and delta-delta features.

Table 10.1: Parameters for feature extraction according to the ETSI 201 108 standard

B	L_w	K	Q	K'	Window type
80	200	256	23	13	Hamming

10.6.1 Baseline Recognition Results

The baseline results were obtained by passing the extracted features directly to the recognizer without applying any enhancement techniques. They are depicted in Table 10.2 for all considered test conditions. For the sake of completeness, the results for the non-reverberant test data are also given. It can be seen that the recognition performance severely degrades if the reverberation increases or the SNR decreases.

Table 10.2: Baseline word accuracies [%] for SFE

SNR [dB] \ Test environment	Non-reverberant	Office	Living room
∞	99.36	93.68	85.06
15	87.96	80.07	64.42
10	66.78	55.25	42.62
5	37.29	28.27	20.99
0	16.74	11.90	10.28

In Table 10.3 the corresponding recognition results are given for the ETSI advanced front-end (AFE) according to the ETSI ES 202 050 standard [8], which is basically an extension of the SFE by a two-stage Wiener filter and a blind equalization to compensate for noise and acoustic mismatch caused by input devices, respectively. It can be observed that the AFE manages to provide a high degree of robustness against noise. However, it fails to enhance noise-free reverberant speech, since it does not consider the correlation between the direct and reverberant components of speech. Correspondingly, the achieved recognition performance using the AFE for dereverberation is similarly as poor as that of the SFE.

Table 10.3: Baseline word accuracies [%] for AFE

SNR [dB] \ Test environment	Non-reverberant	Office	Living room
∞	99.34	93.89	85.47
15	97.10	89.08	78.69
10	93.86	82.74	70.83
5	85.07	69.91	56.94
0	64.52	48.59	37.35

10.6.2 Choice of Model Parameters

Before applying the model-based feature enhancement, we address the choice of the required parameters, which are given in Table 10.4. Although for the Aurora 5 database the reverberation times for the individual test utterances are explicitly given, we used averaged fixed values of $\hat{T}_{60}$ for all utterances belonging to a single environment to simulate estimation errors. Further, the energy parameter was chosen to satisfy (10.62). Since the complete utterances of the Aurora 5 database are already appropriately normalized such that the energy of the reverberant signal part $\bar{s}(l)$ is equal to that of its non-reverberant counterpart, further normalization was not necessary.

Table 10.4: Feature enhancement parameters for the considered test conditions

Test environment	$\hat{T}_{60}$	$\hat{\sigma}_h$	$\hat{L}_h$	$\hat{L}_H$
Office	0.35 s	0.0705	935	14
Living room	0.45 s	0.0622	1202	18

The values for $\hat{L}_h$ and $\hat{L}_H$, defining the approximate length of the RIR and its logarithmic mel power spectral-like representation, are completely determined by ε_h. The choice of ε_h has to be considered in connection with the choice of the parameters of the observation error model, as will be explained in the following.

As already mentioned in Sect. 10.4.2.3 the observation error model parameters may be computed from real or artificial stereo training data. We performed the estimation with both types of data to analyze the effect of artificial data on the recognition performance, where we first restricted the estimation to the noise-free conditions. As real stereo data we used 575 clean utterances and their noise-free reverberant counterparts for each of the two considered scenarios.

The artificially reverberated training data were obtained by convolving 575 clean training utterances with RIRs computed using the image method [30]. For the creation of the RIRs we used a virtual room having the shape of a cuboid of fixed size of 6 m × 5 m × 3 m, which is displayed in Fig. 10.6. The speaker and mi-

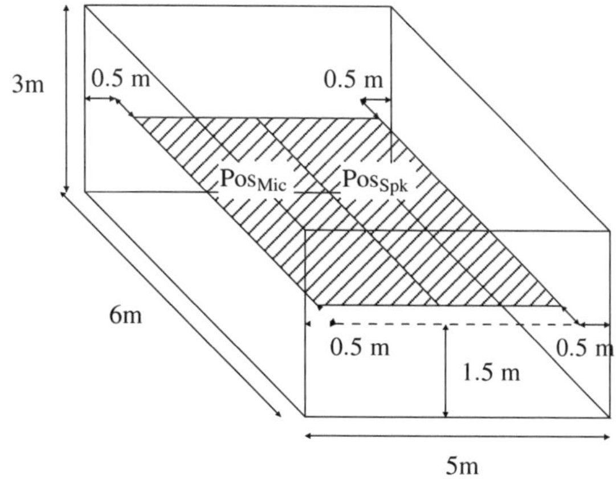

Fig. 10.6 The cuboid employed as enclosure for the image method to create the artificial RIRs. The positions of the speaker and the microphone were uniformly varied within the planes denoted by Pos_{Spk} and Pos_{Mic}, respectively

crophone positions within the enclosure were assumed to be uniformly distributed on the planes denoted by Pos_{Spk} and Pos_{Mic}, respectively, and were independently drawn for each utterance. The reverberation times were uniformly varied within the intervals [0.3 s, 0.4 s] and [0.4 s, 0.5 s] for the office and the living room, respectively.

For the determination of an appropriate value of ε_h we performed a preliminary experiment, where we estimated the variances $\sigma^2_{\mathbf{v},q}$ of the individual observation error components for different values of ε_h. Here, we assume $\operatorname{diag}(\boldsymbol{\Sigma}_\mathbf{v}) = \left(\sigma^2_{\mathbf{v},0}, \ldots, \sigma^2_{\mathbf{v},Q-1}\right)^T$, where $\operatorname{diag}(\cdot)$ denotes the operation of extracting the diagonal of a matrix to a vector. From the results for the real and artificial training data, which are depicted in Figure 10.7, it can be observed that the decrease of ε_h from 10^{-1} to 10^{-2} considerably decreases the observation error variances, while a further decrease has only minor effects. This effect comes as no surprise, since decreasing the

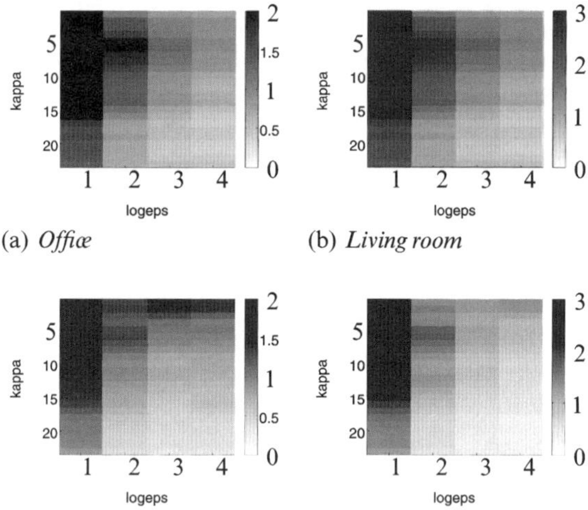

(a) *Office* (b) *Living room*

(c) *Office* (d) *Living room*

Fig. 10.7: Variances $\sigma^2_{\mathbf{v},q}$ of the observation error estimated from real 7(a)–(b) and artificial 7(c)–(d) noise-free stereo training utterances for the two considered environments

value of ε_h only makes sense as long as the RIR model is accurate enough. Further, it can be seen that the observation error variances increase with growing reverberation time, while those obtained by the artificial training data have similar values as those estimated from real data, however, on average being slightly lower. As a trade-off between model complexity and accuracy, we employed a value of 10^{-2} for ε_h for the following experiments, resulting in the estimates $\hat{L}_h$ and $\hat{L}_H$ given in Table 10.4.

Exemplarily, the corresponding histograms of the observation error components $\hat{v}_{m,0}$ and $\hat{v}_{m,22}$ for the two different environments are given in Figure 10.8, where those computed from real stereo training data are denoted by "Real" and those computed from artificial stereo training data are denoted by "Model". The empirical pdfs for the two different kinds of stereo training data look quite similar, encouraging the usage of artificial stereo training data.

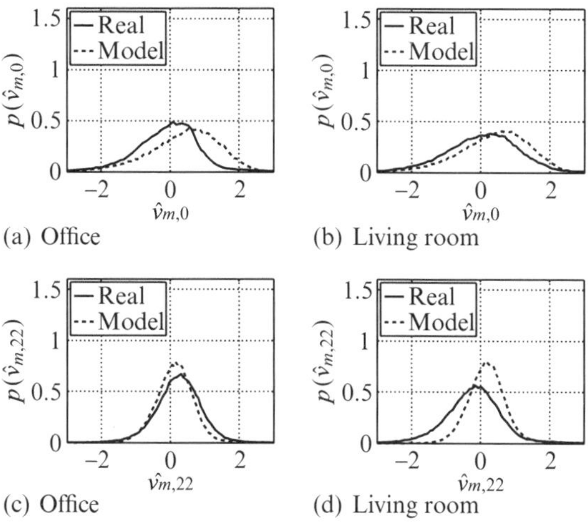

Fig. 10.8: Observation error probability density functions $p(\hat{v}_{m,0})$ and $p(\hat{v}_{m,22})$ estimated from real and artificial noise-free stereo training utterances for the two considered environments

Evaluating the observation mapping f defined in (10.46) for reverberant noisy utterances to compute the observation error $\hat{\mathbf{v}}_m$ (10.66) requires the availability of each utterance in its clean and noisy reverberant version as well as the corresponding noise-only signal. Since the training data of the Aurora 5 database does not contain all these three required components jointly for each utterance, we employed those observation error parameters for the enhancement of noisy utterances which were estimated on the noise-free training utterances for the corresponding environment.

The state vector $\mathbf{z}_m$ was chosen to include $L_C = 4$ successive non-reverberant feature vectors, as further increasing the value of L_C has shown to only negligibly improve the enhancement performance [24]. The number of iterations for the IEKF was set to $R = 3$.

For the enhancement of noise-free reverberant features the observation model (10.46) was simplified not to depend on $\mathbf{n}_m$ by assuming $\mathbf{n}_m = (-\infty, ..., -\infty)^T \ \forall m$. Accordingly, the state vector was set to $\mathbf{z}_m = \boldsymbol{\chi}_m$. For the enhancement of noisy reverberant speech, the noise model parameters $\boldsymbol{\mu}_\mathbf{n}$ and $\boldsymbol{\Sigma}_\mathbf{n}$ were estimated from the first and last 15 frames of each utterance.

10.6.3 Recognition Results for Feature Enhancement

The actual feature enhancement was performed with two sets of observation noise model parameters, $\boldsymbol{\mu}_\mathbf{v}$ and $\boldsymbol{\Sigma}_\mathbf{v}$, which had been obtained from either real or artificial stereo noise-free training data as described in the previous section. The covariance matrix $\boldsymbol{\Sigma}_{\mathrm{v}}$ was constrained to be diagonal, which we found to considerably improve the performance of the enhancement. The corresponding recognition accuracies for all considered reverberation and noise scenarios are given in Tables 10.5 and 10.6, respectively.

Table 10.5: Word accuracies [%] for feature enhancement using observation noise model parameters estimated from real noise-free stereo training data

SNR [dB] \ Test environment	Office	Living room
∞	98.03	95.22
15	92.51	85.78
10	83.69	74.76
5	66.31	56.20
0	40.42	33.66

Table 10.6: Word accuracies [%] for feature enhancement using observation noise model parameters estimated from artificial noise-free stereo training data

SNR [dB] \ Test environment	Office	Living room
∞	97.91	95.25
15	93.00	86.44
10	84.86	75.84
5	68.10	56.95
0	42.29	33.05

First, considering the noise-free conditions, it can be seen from each of the first rows of the tables that the results for both kinds of stereo training data are similar. Compared to the baseline results without feature enhancement given in Tables 10.2 about 76 % and 71 % of the errors caused by reverberation have been recovered for the office and living room environment, respectively. This is a considerable improvement on the performance of the ETSI SFE and AFE.

From the results for the noisy conditions it can be seen that the relative reduction of the word error rate (WER) compared to the clean case decreases on average with decreasing SNR and increasing reverberation time, while for the living room environment at 0 dB SNR only an averaged relative WER reduction of about 26 % is obtained.

This effect can be mainly attributed to the fact that the noise model (10.32) becomes more important for decreasing SNR. Using a rather coarse noise model in the form of a single Gaussian prevents an accurate capturing of the non-stationarity of the noise. Further, the observation model becomes more inaccurate for increasing reverberation, which can be seen from the observation error variances in Fig. 10.7.

The performance of the proposed feature enhancement is distinctly superior to that of the ESTI AFE for the conditions with a high SNR. For low SNRs of about

0 dB, however, where the distortion is mainly given by noise rather than by reverberation, the AFE outperforms the proposed approach. It is noticeable that the noise model parameters estimated on artificial noise-free reverberant training data always lead to slightly better results on average.

Finally, it should be emphasized that the proposed algorithm has been shown in [23] to be suited for real-time processing on today's personal computers. Further, it has been shown for the conditions without noise that the algorithm is robust against estimation errors in the reverberation time and that its performance is superior to that of CMLLR and model adaptation.

10.7 Conclusion

In this chapter an approach for the enhancement of jointly reverberant and noisy speech features has been presented, based on a Bayesian framework. The technique is aimed at estimating the mean and covariance of the posterior distribution of clean LMPSCs given all previously observed corresponding reverberant and noisy LMPSCs. The estimation is performed by carrying out suboptimal inference using the IMM approach. It exploits a priori knowledge about the speech and noise feature trajectories given in the form of an a priori model as well as knowledge about the relation between the reverberant noisy and clean features in the form of an observation model. The observation model is based on a simplified RIR model with only two parameters, RIR energy and reverberation time, which can quite easily be estimated from the reverberant signal.

The great advantage of the approach is that no modification of the speech recognition back-end is required, since the enhancement is carried out in the feature domain. Further, the proposed approach is of moderate computational complexity and is suitable for online processing as it entails a latency of only a few frames.

The performance of the enhancement algorithm was measured by means of the achieved accuracy on the connected digits recognition task of the Aurora 5 database under different reverberation and noise conditions. For the reverberation conditions without noise, up to 76 % of the errors caused by reverberation had been recovered. In jointly reverberant and noisy environments the performance decreased on average with decreasing SNR and increasing reverberation time, where for the living room at 0 dB SNR only a relative WER reduction of about 26 % was obtained.

One of the most critical components of the enhancement approach is the proposed noise model assuming the stochastic process underlying the noise feature vector trajectory to be Gaussian, white and stationary. The neglect of any non-stationarity, however, introduces errors into the observation model, which become more significant with decreasing SNR. Moreover, the proposed method for the estimation of the observation error model parameters does not consider the influence of noise, which further reduces the observation model accuracy. It is left open for future research to develop approaches to consider these aspects.

References

1. Avargel, Y., Cohen, I.: On multiplicative transfer function approximation in the short-time Fourier transform domain. IEEE Signal Processing Letters **14**(5), 337–340 (2007)
2. Bar-Shalom, Y., Li, X.R., Kirubarajan, T.: Estimation with Applications to Tracking and Navigation: Theory, Algorithms, and Software. Wiley, New York (2001)
3. Couvreur, L., Couvreur, C.: Blind model selection for automatic speech recognition in reverberant environments. Journal of VLSI Signal Processing **36**(2/3), 189–203 (2004)
4. Delcroix, M., Hikichi, T., Miyoshi, M.: On the use of lime dereverberation algorithm in an acoustic environment with a noise source. Proc. of IEEE International Conference on Acoustics, Speech and Signal Processing (ICASSP) **1**, I–I (2006)
5. Delcroix, M., Nakatani, T., Watanabe, S.: Static and dynamic variance compensation for recognition of reverberant speech with dereverberation preprocessing. IEEE Transactions on Audio, Speech, and Language Processing **17**(2), 324–334 (2009)
6. Dempster, A.P., Laird, N.M., Rubin, D.B.: Maximum likelihood from incomplete data via the em algorithm. Journal of the Royal Statistical Society: Series B (Statistical Methodology) **39**(1), 1–38 (1977)
7. Droppo, J., Acero, A.: Noise robust speech recognition with a switching linear dynamic model. In: Proc. of IEEE International Conference on Acoustics, Speech and Signal Processing (ICASSP), vol. 1, pp. I–953–6 vol.1 (2004)
8. ETSI: ETSI standard document, Speech Processing, Transmission and Quality Aspects (STQ); Distributed speech recognition; Advanced front-end feature extraction algorithm; Compression algorithms, ETSI ES 202 050 V1.1.5 (2007-01)
9. ETSI: ETSI standard document, Speech Processing, Transmission and Quality Aspects (STQ); Distributed speech recognition; Front-end feature extraction algorithm; Compression algorithms, ETSI ES 201 108 V1.1.3 (2003-09)
10. Gales, M.J.F.: Maximum likelihood linear transformations for HMM-based speech recognition. Computer Speech and Language **12**(2), 75–98 (1998)
11. Gales, M.J.F., Woodland, P.C.: Mean and variance adaptation within the MLLR framework. Computer Speech and Language **10**(4), 249–264 (1996)
12. Gannot, S., Moonen, M.: Subspace methods for multi-microphone speech dereverberation. EURASIP Journal on Applied Signal Processing **11**, 1074–1090 (2003)
13. Gürelli, M., Nikias, C.: EVAM: an eigenvector-based algorithm for multichannel blind deconvolution of input colored signals. IEEE Transactions on Signal Processing **43**(1), 134–149 (1995)
14. Habets, E.: Single- and multi-microphone speech dereverberation using spectral enhancement. Ph.D. thesis, Technische Universiteit Eindhoven (2007)
15. Hermansky, H., Morgan, N.: RASTA processing of speech. IEEE Transactions on Speech and Audio Processing **2**(4), 578–589 (1994)
16. Hirsch, H.: Aurora-5 experimental framework for the performance evaluation of speech recognition in case of a hands-free speech input in noisy environments. Tech. rep., Niederrhein University of Applied Sciences (2007)
17. Hirsch, H.G., Finster, H.: The simulation of realistic acoustic input scenarios for speech recognition systems. In: Proc. of Annual Conference of the International Speech Communication Association (Interspeech), pp. 2697–2700 (2005)
18. Hirsch, H.G., Finster, H.: A new approach for the adaptation of HMMs to reverberation and background noise. Speech Commununication **50**(3), 244–263 (2008)
19. Julier, S.J., Jeffrey, Uhlmann, K.: Unscented filtering and nonlinear estimation. In: Proceedings of the IEEE, pp. 401–422 (2004)
20. Kennedy, R., Radlovic, B.: Iterative cepstrum-based approach for speech dereverberation. In: Proc. of International Symposium on Signal Processing and its Applications (ISSPA), vol. 1, pp. 55–58 vol.1 (1999)
21. Kingsbury, B.E.D., Morgan, N.: Recognizing reverberant speech with RASTA-PLP. In: Proc. of IEEE International Conference on Acoustics, Speech and Signal Processing (ICASSP), vol. 2, pp. 1259–1262 (1997)

22. Kinoshita, K., Delcroix, M., Nakatani, T., Miyoshi, M.: Suppression of late reverberation effect on speech signal using long-term multiple-step linear prediction. IEEE Transactions on Audio, Speech, and Language Processing **17**(4), 534–545 (2009)
23. Krueger, A., Haeb-Umbach, R.: Model-based feature enhancement for reverberant speech recognition. IEEE Transactions on Audio, Speech, and Language Processing **18**(7), 1692–1707 (2010)
24. Krueger, A., Leutnant, V., Haeb-Umbach, R., Marcel, A., Bloemer, J.: On the initialisation of dynamic models for speech features. In: Proc. of ITG Fachtagung Sprachkommunikation (2010)
25. Langhans, T., Strube, H.: Speech enhancement by nonlinear multiband envelope filtering. In: Proc. of IEEE International Conference on Acoustics, Speech and Signal Processing (ICASSP), vol. 7, pp. 156–159 (1982)
26. Lebart, K., Boucher, J., Denbigh, P.: A new method based on spectral subtraction for speech dereverberation. Acta Acustica United with Acustica **87**, 359–366(8) (2001)
27. Leggetter, C.J., Woodland, P.C.: Maximum likelihood linear regression for speaker adaptation of continuous density hidden Markov models. Computer Speech and Language **9**(2), 171–185 (1995)
28. Löllmann, H.W., Vary, P.: Low delay noise reduction and dereverberation for hearing aids. In: EURASIP Journal on Advances in Signal Processing (2009)
29. Murphy, K.: Switching Kalman filters. Tech. rep., U.C. Berkeley (1998)
30. Neely, S.T., Allen, J.B.: Invertibility of a room impulse response. Journal of the Acoustical Society of America **66**(1), 165–169 (1979)
31. Qian, S., Chen, D.: Discrete Gabor transform. IEEE Transactions on Signal Processing **41**(7), 2429–2438 (1993)
32. Ratnam, R., Jones, D., O'Brien W.D., J.: Fast algorithms for blind estimation of reverberation time. IEEE Signal Processing Letters **11**(6), 537–540 (2004)
33. Ratnam, R., Jones, D.L., Wheeler, B.C., O'Brien, W.D., Lansing, C.R., Feng, A.S.: Blind estimation of reverberation time. Journal of the Acoustical Society of America **114**(5), 2877–2892 (2003)
34. Raut, C.K., Nishimoto, T., , Sagayama, S.: Model adaptation by state splitting of HMM for long reverberation. In: Proc. of Annual Conference of the International Speech Communication Association (Interspeech) (Sep 2005)
35. Rosenberg, A.E., Lee, C.H., Soong, F.K.: Cepstral channel normalization techniques for HMM-based speaker verification. In: Proc. of International Conference on Spoken Language Processing (ICSLP), pp. 1835–1838 (1994)
36. Sehr, A., Kellerman, W.: A new concept for feature-domain dereverberation for robust distant-talking ASR. In: Proc. of IEEE International Conference on Acoustics, Speech and Signal Processing (ICASSP), vol. 4, pp. IV–369–IV–372 (2007)
37. Subramaniam, S., Petropulu, A., Wendt, C.: Cepstrum-based deconvolution for speech dereverberation. IEEE Transactions on Speech and Audio Processing **4**(5), 392–396 (1996)
38. Unoki, M., Sakata, K., Furukawa, M., Akagi, M.: A speech dereverberation method based on the MTF concept in power envelope restoration. Acoustical Science and Technology **25**(4), 243–254 (2004)
39. Wu, M., Wang, D.: A two-stage algorithm for one-microphone reverberant speech enhancement. IEEE Transactions on Audio, Speech, and Language Processing **14**(3), 774–784 (2006)
40. Yegnanarayana, B., Mahadeva Prasanna, S., Sreenivasa Rao, K.: Speech enhancement using excitation source information. In: Proc. of IEEE International Conference on Acoustics, Speech and Signal Processing (ICASSP), vol. 1, pp. I–541–I–544 (2002)
41. Yegnanarayana, B., Murthy, P.: Enhancement of reverberant speech using LP residual signal. IEEE Transactions on Speech and Audio Processing **8**(3), 267–281 (2000)
42. Yoshioka, T., Nakatani, T., Miyoshi, M.: Integrated speech enhancement method using noise suppression and dereverberation. IEEE Transactions on Audio, Speech, and Language Processing **17**(2), 231–246 (2009)

43. Young, S.J., Evermann, G., Gales, M.J.F., Hain, T., Kershaw, D., Moore, G., Odell, J., Ollason, D., Povey, D., Valtchev, V., Woodland, P.C.: The HTK Book, version 3.4. Cambridge University Engineering Department, Cambridge, UK (2006)
44. Zhang, Z., Furui, S.: Piecewise-linear transformation-based hmm adaptation for noisy speech. Speech Commununication **42**(1), 43–58 (2004)

Part IV
Applications: Multiple Speakers and Modalities

Chapter 11
Evidence Modeling for Missing Data Speech Recognition Using Small Microphone Arrays

Marco Kühne, Roberto Togneri and Sven Nordholm

Abstract During the last decade microphone array processing has emerged as a powerful tool for increasing the noise robustness of automatic speech recognition (ASR) systems. Typically, microphone arrays are used as preprocessors that enhance the incoming speech signal prior to recognition. While such traditional approaches can lead to good results, they usually require large numbers of microphones to reach acceptable performance in practice. Furthermore, important information, such as uncertainty estimates and energy bounds, are often ignored as speech recognition is conventionally performed only on the enhanced output of the array. Using the probabilistic concept of evidence modeling this chapter presents a novel approach to robust ASR that aims for closer integration of microphone array processing and missing data speech recognition in reverberant multi-speaker environments. The output of the array is used to estimate the probability density function (pdf) of the hidden clean speech data using any information which may be available before and after array processing. The chapter discusses different types of evidence pdfs and shows how these models can be used effectively during HMM decoding.

Marco Kühne
The University of Western Australia, 35 Stirling Highway, Crawley, WA 6009, Australia,
e-mail: marcokuehne@gmx.de

Roberto Togneri
The University of Western Australia, 35 Stirling Highway, Crawley, WA 6009, Australia,
e-mail: Roberto.Togneri@uwa.edu.au

Sven Nordholm
Curtin University of Technology, GPO Box U1987, Perth, WA 6845, Australia,
e-mail: S.Nordholm@curtin.edu.au

D. Kolossa and R. Haeb-Umbach (eds.), *Robust Speech Recognition of Uncertain or Missing Data: Theory and Applications*, DOI 10.1007/978-3-642-21317-5_11,
© Springer-Verlag Berlin Heidelberg 2011

11.1 Introduction

Solving the so-called 'cocktail party problem' [9] has been a particular challenge for researchers and engineers alike. While people with normal hearing possess the remarkable ability to focus on and segregate a particular voice or sound from a mixture of voices, reverberation and background noise, no technical system has been developed to date that is capable of human-like performance in such environments. Most automatic speech recognition (ASR) systems are trained on clean, anechoic speech and tend to produce errors when confronted with reverberant speech mixtures. According to the study of Lippmann [26], machine error rates are often more than an order of magnitude higher than those of human listeners in both quiet and noisy environments. Lippmann also points out that the inability of modern machine recognizers to distinguish between individual voices and non-speech sounds represents a fundamental weakness that prevents these systems from performing in real-life situations.

It therefore comes as no surprise that one promising approach toward solving this problem consists in mimicking the spatial hearing mechanism of the human auditory system. In spatial hearing, sounds and voices are distinguished based on the spatial location of their sources. In principle, this mechanism can be reproduced by spatially sampling the auditory scene through the placement of a number of physically distributed microphones, called a microphone array. Spatial directivity of the array is then achieved by filtering and combining the individual microphone recordings such that a desired signal is enhanced while at the same time reducing the level of directional and ambient background noise.

Besides improving the speech input quality in noisy environments, microphone arrays offer the distinct advantage of placing fewer constraints on the user. With their help, speech input can be acquired remotely, making the speech acquisition process a truly hands-free experience, without the user being required to wear a headset or to speak directly into a close-talking microphone. However, the use of multiple channels also introduces some disadvantages, which mainly come in the form of acoustic losses due to the speaker's distance from the array as well as increased hardware and processing requirements. Driven by the continued advances in computing technology and the availability of increasingly powerful processors, computational resources no longer present a major obstacle, and microphone arrays are starting to appear more frequently in applications that require noise robust speech recognition.

Figure 11.1 shows a typical setup for a microphone array-driven speech recognition system. In this approach, the array serves as a preprocessor with the objective of producing an enhanced waveform T of the target source. A sequence of feature vectors x is then extracted from T and passed to a conventional Hidden Markov Model (HMM) recognition engine for decoding. Examples for this type of processing are the systems based on beamforming [30, 34] and independent component analysis [27].

The major advantage of such an architecture is that speech enhancement and speech recognition can be treated independently. On the downside, this approach

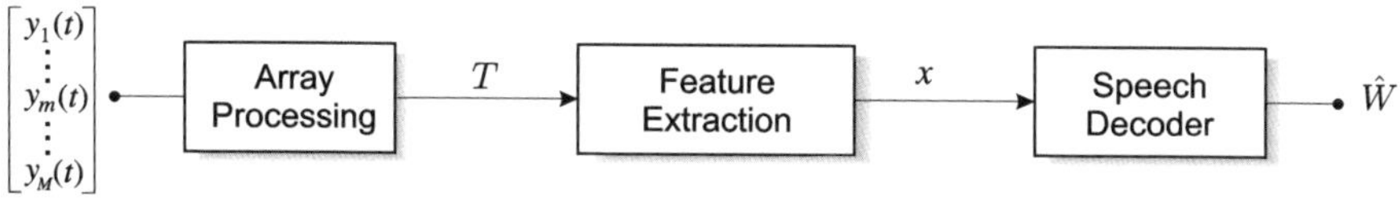

Fig. 11.1: Conventional approach for microphone array-based robust speech recognition

places high requirements on the level of enhancement that needs to be achieved by the array. A large number of microphones is usually necessary for an effective suppression of interfering speakers and background noise. Previous studies also reported that gains in speech recognition accuracy were only marginal despite the fact that the signal-to-noise ratio (SNR) had been substantially improved [14, 39].

In the field of computational auditory scene analysis (CASA) [7], researchers have pursued a different strategy that aims at closer integration of preprocessing and speech recognition. In this approach, spatial information is used to localize and select spectro-temporal areas that are representative of the target speech signal. More specifically, interaural level and intensity differences between paired microphones are utilized for the estimation of a time-frequency mask. Each element of the mask indicates whether the energy of the target speaker is masked by the noise or not. No actual spatial filtering or waveform reconstruction is performed. Instead, the reliability mask along with the acoustic features are passed to a missing data recognizer [10] that is adapted to cope with missing or unreliable features. Examples of this technique are the binaural segregation models proposed in [13, 35, 38] and the multi-microphone systems described in [18, 20].

In a further study by McCowan et al. [31], time-frequency masking was employed during a postprocessing step to improve the speech recognition performance of a small microphone array. The authors proposed utilizing the beamformer output not for recognition but for marking the reliability of each feature component in a missing data recognition framework. More specifically, a soft missing data mask was constructed using the energy difference between the noisy input and the enhanced beamformer output. Kolossa and Orglmeister [15] proposed a similar technique for BSS systems, in which mask estimation is based on the ratio of the demixed signal energies instead.

In this chapter we present a comprehensive overview of a microphone array-driven robust speech recognition system which the authors developed in [23] and [22]. Building upon the concept of evidence modeling,[1] our approach unifies the above-mentioned ideas of array-based speech enhancement and reliable feature selection under one framework. Figure 11.2 shows the building blocks of our approach, where the conventional 'Feature Extraction' module has been replaced with the new 'Evidence Model Estimation' block. The motivation behind this replacement is to use the array for gathering as much information as possible about the

[1] The probabilistic concept of evidence modeling is a generalization of the missing data paradigm and was developed by Morris et al. [33] in order to quantify the level of uncertainty when dealing with noisy input data.

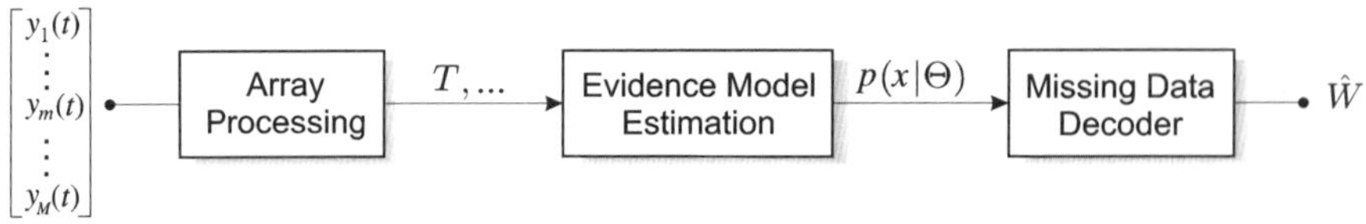

Fig. 11.2: Evidence modeling approach for microphone array based robust speech recognition (adopted from [22], © 2010 IEEE)

acoustic scene and then to transform that knowledge into a statistical description in the form of a probability density function (pdf). The pdf $p(x|\Theta)$ is known as the data evidence model [33] and is input to a missing data recognizer for decoding. It expresses the degree of belief into a possible observation x representing the hidden clean data value. The great variety of statistical distributions makes it possible to tailor the shape of this pdf to the information at hand. In our work, the array input and output are both utilized to estimate the parameter set Θ of the evidence pdf. Instead of important side information that might be available after array processing being ignored, it can now be incorporated into the recognition framework through the pdf estimation process. In this way speech recognition performance can be improved substantially relative to the conventional approach, even when only a relatively small number of microphones is available.

The remainder of this chapter is organized as follows: Section 11.2 reviews the concept of evidence modeling and discusses its implications for the likelihood computation in HMM speech recognition systems. Section 11.3 demonstrates how the parameters of the chosen evidence model can be estimated with the help of a multichannel blind source separation (BSS) front-end. Section 11.4 compares the ASR performance of various types of evidence models for a multi-source reverberant connected digit task. The section also contains a general discussion, in which the limitations and shortcomings as well as future research directions are briefly outlined. Section 11.5 concludes the chapter with a short summary of the main findings.

11.2 Robust HMM Likelihood Computation Using Evidence Models

It is common practice to model the HMM output probability distributions as Gaussian mixture models (GMMs) with diagonal covariances for each mixture component. In this case, the emission likelihood can be computed by

$$p(\mathbf{x}_k|\Lambda) = \sum_{r=1}^{G} g_r \prod_{i=1}^{D} \mathcal{N}(x_{ki}; \mu_{ri}, \sigma_{ri}^2), \tag{11.1}$$

where $\mathbf{x}_k = (x_{k1}, \ldots, x_{kD})^T$ is a deterministic feature vector of dimension D at time frame k, $g_r \in [0,1]$ is the weight of the r-th mixture, G denotes the total number of mixtures and $\mathcal{N}(x;\mu,\sigma^2)$ is a univariate Gaussian,

$$\mathcal{N}(x;\mu,\sigma^2) = \frac{1}{\sqrt{2\pi\sigma^2}} \exp\left(-\frac{(x-\mu)^2}{2\sigma^2}\right), \tag{11.2}$$

with mean μ and variance σ^2. The parameter set for the diagonal GMM (11.1) is completely defined by the mean vectors, variance vectors and mixture weights from all component densities, e.g., $\Lambda = \{g_r, \boldsymbol{\mu}_r, \boldsymbol{\sigma}_r^2 | r = 1,\ldots,G\}$ with $\boldsymbol{\mu}_r = (\mu_{r1},\ldots,\mu_{rD})^T$ and $\boldsymbol{\sigma}_r^2 = (\sigma_{r1}^2,\ldots,\sigma_{rD}^2)^T$. We hereby assume that the components of Λ are learned during model training with clean data and are assumed to be free of any uncertainty.

It is well known that acoustic mismatch due to different training and testing conditions can severely degrade recognition performance. In a noisy environment, some entries in $\mathbf{x}_k$ may be uncertain or even completely missing such that (11.1) cannot or should not be evaluated. We overcome this problem by resorting to the concept of evidence modeling [33], where feature vector components are treated as stochastic random variables, each associated with its own pdf, e.g., $x_{ki} \sim p(x_{ki}|\Theta_{ki})$. The pdf's parameter set Θ_{ki} hereby represents all available data evidence, such as noisy and enhanced observations or any information about the range of possible feature values. The major advantage of such a probabilistic framework is that decisions are now based on a more complete picture of the available evidence. Besides the noise-corrected data values, also the precision with which these values could be restored by the preprocessor is taken into consideration [33].

Different from the conventional enhancement-recognition approach, evidence modeling requires a modification to the HMM likelihood evaluation procedure. Morris et al. [33] have shown that the emission likelihood for an uncertain feature can be determined with the help of the evidence model by replacing $\mathcal{N}(x_{ki};\mu_{ri},\sigma_{ri}^2)$ in (11.1) with the following expectation integral[2]:

$$\mathrm{E}\left[\mathcal{N}(x_{ki};\mu_{ri},\sigma_{ri}^2) | p(x_{ki}|\Theta_{ki})\right] = \int_{-\infty}^{\infty} \mathcal{N}(x_{ki};\mu_{ri},\sigma_{ri}^2) p(x_{ki}|\Theta_{ki}) \, dx_{ki}. \tag{11.3}$$

This amounts to evaluating $\mathcal{N}(x_{ki};\mu_{ri},\sigma_{ri}^2)$ over all possible feature values x_{ki}, weighted by the corresponding data evidence $p(x_{ki}|\Theta_{ki})$. Note that the application of this rule minimizes the 'Bayes risk' and leads to optimal classification with missing or uncertain data [33].

A wide range of statistical distributions are available to express the confidence in each possible data value. For instance, if the only information available is the interval in which the clean value is known to reside, then this may be expressed by the uniform distribution,

$$\mathcal{U}(x;a,b) = \frac{1}{b-a} \mathbb{1}_{[a,b]}(x), \tag{11.4}$$

[2] Strictly speaking this is only an approximation, in which the missing speech prior $p(x_{ki})$ was assumed to be constant or very flat [33].

where a,b with $b > a$ specify the distribution boundaries and $\mathbb{1}_{[a,b]}(x)$ is the usual indicator function equal to 1 if $x \in [a,b]$, and 0 otherwise. On the other hand, the rather unrealistic expectation that the array processing was able to completely remove any mismatch between the training and testing environments is best modeled by the Dirac-Delta distribution. Using a sufficiently regular function $f(x)$, the distribution $\delta(x-\hat{\mu};\hat{\mu})$ can be defined over the following integral (sifting property):

$$\int_{-\infty}^{\infty} f(x)\delta(x-\hat{\mu};\hat{\mu})\,dx = f(\hat{\mu}). \tag{11.5}$$

In other words, the Dirac-Delta distribution simply 'sifts out' the value of the function $f(x)$ at the point where the argument of the delta function vanishes. Because the probability mass is concentrated at a single point, this pdf indicates full confidence in the (enhanced) feature value $\hat{\mu}$ being identical with the true clean data value. Other elementary distributions are the Gaussian pdf, as defined in (11.2), or its bounded counterpart

$$\mathscr{B}(x;\hat{\mu},\hat{\sigma}^2,\alpha,\beta) = \frac{\mathscr{N}(x;\hat{\mu},\hat{\sigma}^2)}{\Phi(\beta;\hat{\mu},\hat{\sigma}^2)-\Phi(\alpha;\hat{\mu},\hat{\sigma}^2)}\mathbb{1}_{[\alpha,\beta]}(x), \tag{11.6}$$

where α and β specify the lower and upper truncation points and $\Phi(z;\hat{\mu},\hat{\sigma}^2) = \int_{-\infty}^{z}\mathscr{N}(z;\hat{\mu},\hat{\sigma}^2)dz$ is the cumulative distribution function (CDF) of $\mathscr{N}(x;\hat{\mu},\hat{\sigma}^2)$. In its most general form, $p(x_{ki}|\Theta_{ki})$ may be represented as a mixture pdf provided that the integral in (11.3) yields a closed form solution [33]. Table 11.1 shows an overview of the most popular evidence models found in the missing data ASR literature.

With respect to equation (11.3), the HMM emission likelihoods are calculated differently, depending on the choice of $p(x|\Theta)$. In the following, we briefly comment on the decoding rules of two evidence models: the Dirac-Delta pdf and the Bounded-Gauss-Uniform mixture pdf.

Having only the single parameter $\hat{\mu}_{ki}$ makes the Dirac-Delta distribution the most simple evidence model. In this case, the solution of the expectation integral yields

$$\mathrm{E}\Big[\mathscr{N}(x_{ki};\mu_{ri},\sigma_{ri}^2)|p_\delta(x_{ki}|\Theta_{ki})\Big] = \mathscr{N}(\hat{\mu}_{ki};\mu_{ri},\sigma_{ri}^2), \tag{11.7}$$

which is identical to the conventional likelihood computation with certain data. Most conventional enhancement-recognition approaches implicitly use this model by passing only point estimates in the form of the enhanced array output $\hat{\mu}_{ki}$ to the recognizer. While this results in a fast likelihood evaluation, noise robustness remains very low due to the deficiencies associated with any kind of practical noise removal technique. After all, perfect source separation might be impossible to achieve because of theoretical and physical limitations [1].

With a total of seven parameters, the Bounded-Gauss-Uniform mixture pdf is the most complex evidence model in Table 11.1. Its decoding rule is given by [22]:

Table 11.1: Popular evidence models used in noise robust speech recognition.

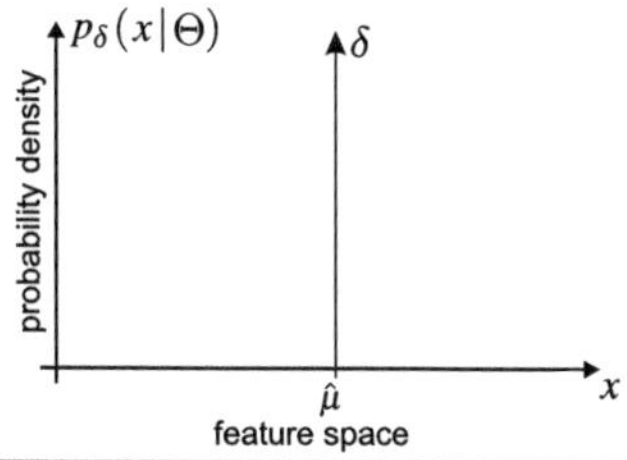

Dirac-Delta pdf:

$$p_\delta(x|\Theta) = \delta(x - \hat{\mu}; \hat{\mu})$$

Parameterset:

$$\Theta = \{\hat{\mu}\}$$

References:

[27, 34]

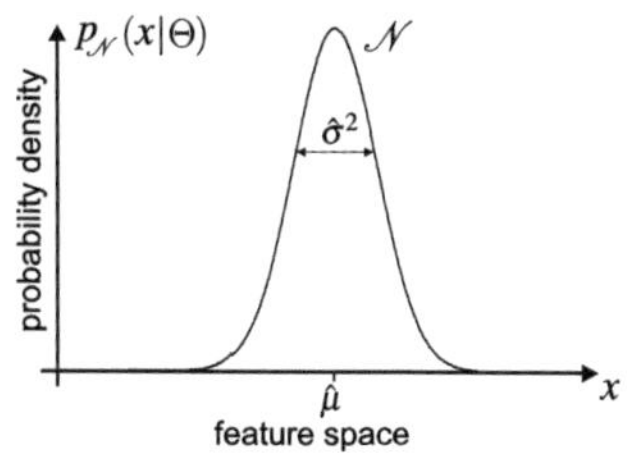

Gaussian pdf:

$$p_\mathcal{N}(x|\Theta) = \mathcal{N}(x; \hat{\mu}, \hat{\sigma}^2)$$

Parameterset:

$$\Theta = \{\hat{\mu}, \hat{\sigma}^2\}$$

References:

[4, 6, 11, 16, 41]

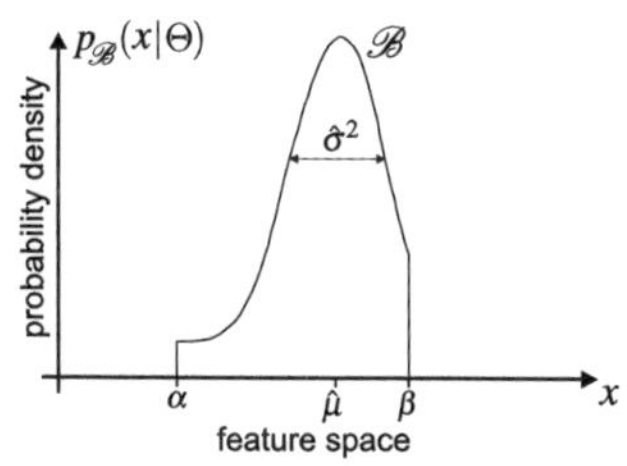

Bounded-Gaussian pdf:

$$p_\mathcal{B}(x|\Theta) = \mathcal{B}(x; \hat{\mu}, \hat{\sigma}^2, \alpha, \beta)$$

Parameterset:

$$\Theta = \{\hat{\mu}, \hat{\sigma}^2, \alpha, \beta\}$$

References:

[22]

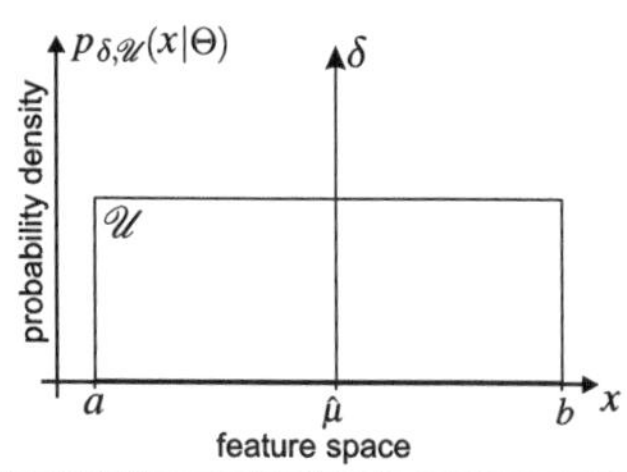

Dirac-Uniform mixture pdf:

$$p_{\delta,\mathcal{U}}(x|\Theta) = w\delta(x - \hat{\mu}; \hat{\mu}) + (1 - w)\mathcal{U}(x; a, b)$$

Parameterset:

$$\Theta = \{\hat{\mu}, w, a, b\}$$

References:

[5, 10, 13, 32, 35]

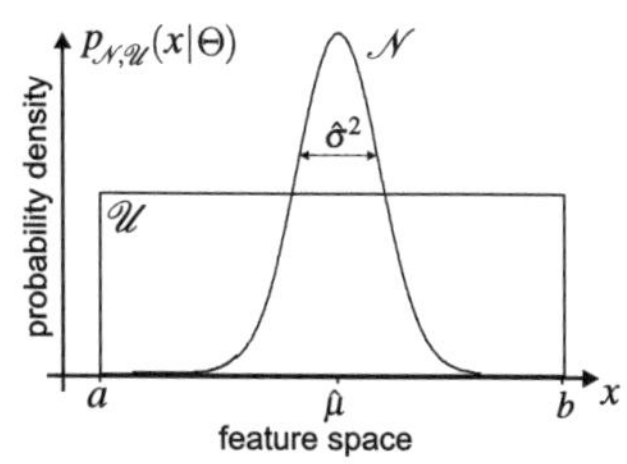

Gauss-Uniform mixture pdf:

$$p_{\mathcal{N},\mathcal{U}}(x|\Theta) = w\mathcal{N}(x; \hat{\mu}, \hat{\sigma}^2) + (1 - w)\mathcal{U}(x; a, b)$$

Parameterset:

$$\Theta = \{\hat{\mu}, \hat{\sigma}^2, w, a, b\}$$

References:

[19, 22]

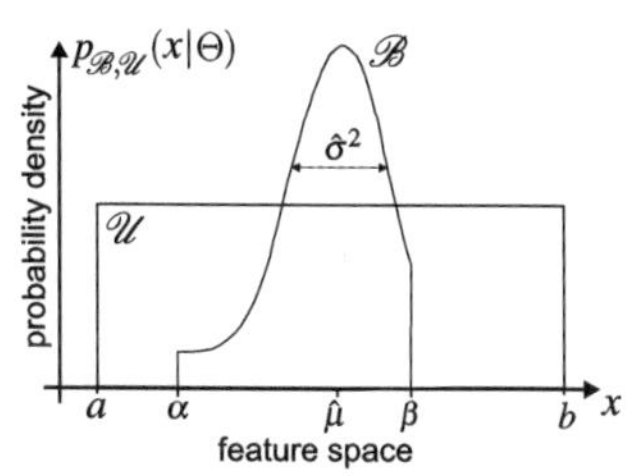

Bounded-Gauss-Uniform mixture pdf:

$$p_{\mathcal{B},\mathcal{U}}(x|\Theta) = w\mathcal{B}(x; \hat{\mu}, \hat{\sigma}^2, \alpha, \beta) + (1 - w)\mathcal{U}(x; a, b)$$

Parameterset:

$$\Theta = \{\hat{\mu}, \hat{\sigma}^2\alpha, \beta, w, a, b\}$$

References:

[22]

$$\begin{aligned}
\mathrm{E}\Big[\mathcal{N}(x_{ki};\mu_{ri},\sigma_{ri}^2)|p_{\mathcal{B},\mathcal{U}}(x_{ki}|\Theta_{ki})\Big] = \\
w_{ki}\frac{\mathcal{N}(\hat{\mu}_{ki};\mu_{ri},\hat{\sigma}_{ki}^2+\sigma_{ri}^2)}{\Phi(\beta_{ki};\hat{\mu}_{ki},\hat{\sigma}_{ki}^2)-\Phi(\alpha_{ki};\hat{\mu}_{ki},\hat{\sigma}_{ki}^2)}\int_{\alpha_{ki}}^{\beta_{ki}}\mathcal{N}(x_{ki};\tilde{\mu},\tilde{\sigma}^2)dx_{ki} \\
+(1-w_{ki})\frac{1}{b_{ki}-a_{ki}}\int_{a_{ki}}^{b_{ki}}\mathcal{N}(x_{ki};\mu_{ri},\sigma_{ri}^2)dx_{ki}
\end{aligned} \tag{11.8}$$

with

$$\tilde{\mu} = \frac{\hat{\mu}_{ki}\sigma_{ri}^2+\mu_{ri}\hat{\sigma}_{ki}^2}{\hat{\sigma}_{ki}^2+\sigma_{ri}^2} \text{ and } \tilde{\sigma}^2 = \frac{\hat{\sigma}_{ki}^2\sigma_{ri}^2}{\hat{\sigma}_{ki}^2+\sigma_{ri}^2}.$$

In comparison with (11.7), this new decoding rule provides a number of features for dealing with time-varying distortions in the spectral feature space. First, it realistically takes into account the shortcomings of the array processing with the help of the mixture weight w_{ki}. The purpose of w_{ki} is to indicate if speech enhancement was deemed successful or if distortions were too severe to be corrected by the preprocessing. Depending on the value of w_{ki}, the likelihood computation is then biased toward either the Bounded-Gaussian (first term in (11.8)) or the Uniform mixture contribution (second term in (11.8)). This ability considerably relaxes the performance requirements on the preprocessing, which is particularly important for small microphone arrays, where the enhancement capabilities are limited. Second, imperfections in the restored feature $\hat{\mu}_{ki}$ are also taken into consideration through the uncertainty parameter $\hat{\sigma}_{ki}^2$. The effect of $\hat{\sigma}_{ki}^2$ is an adaptive variance broadening of the recognizer's speech models, which helps to prevent scoring in the tails of the distributions for uncertain observations. Third, the bounded nature of spectral energy features is properly reflected through the integration limits α_{ki}, β_{ki} and a_{ki}, b_{ki} in both mixture score contributions. These energy bounds provide an effective mechanism to include counter-evidence by penalizing all speech models with insufficient spectral energy [10]. It is worth pointing out that the decoding rules for all other pdfs listed in Table 11.1 can be derived from (11.8) by appropriate choice of the model parameters. For example, when $\hat{\sigma}_{ki}^2 = 0$ and $\alpha_{ki} \to -\infty, \beta_{ki} \to \infty$, we have

$$\begin{aligned}
\mathrm{E}\Big[\mathcal{N}(.)|p_{\mathcal{B},\mathcal{U}}(.)\Big] &= w_{ki}\mathcal{N}(\hat{\mu}_{ki};\mu_{ri},\sigma_{ri}^2) \\
&+(1-w_{ki})\frac{1}{b_{ki}-a_{ki}}\int_{a_{ki}}^{b_{ki}}\mathcal{N}(x_{ki};\mu_{ri},\sigma_{ri}^2)dx_{ki},
\end{aligned} \tag{11.9}$$

which is the decoding rule of the popular Dirac-Uniform mixture model $p_{\delta,\mathcal{U}}(x|\Theta)$, known as *soft bounded marginalization* [5, 13]. Another well-known example is the Gaussian observation pdf $p_{\mathcal{N}}(x|\Theta)$, which has foremost been used in the field of *uncertainty decoding* [4, 11]. In this case, the likelihood computation simplifies to

$$\mathrm{E}\Big[\mathcal{N}(x_{ki};\mu_{ri},\sigma_{ri}^2)|p_{\mathcal{B},\mathcal{U}}(x_{ki}|\Theta_{ki})\Big] = \mathcal{N}(\hat{\mu}_{ki};\mu_{ri},\sigma_{ri}^2+\hat{\sigma}_{ki}^2), \tag{11.10}$$

which one obtains from (11.8) when $w_{ki} = 1$ and $\alpha_{ki} \to -\infty, \beta_{ki} \to \infty$. In summary, the new Bounded-Gauss-Uniform mixture pdf constitutes an important generalization which consolidates the advantages of previous evidence models into one HMM state likelihood formula.

After having discussed the modifications required for the HMM likelihood evaluation with evidence models, we now turn our attention to the question of how the parameters of these pdfs can be estimated in a practical application scenario.

11.3 Microphone Array-Based Evidence Model Parameter Estimation

This section provides a description of one possible methodology that can be used for the evidence pdf estimation in reverberant multi-source environments. We start by describing a multi-channel BSS system, which the authors developed in a previous study on reverberant speech separation [23]. This particular front-end combines adaptive beamforming and time-frequency masking for separating all source signals blindly from the mixture observations. Our system is in line with recent advances in frequency domain BSS and follows the overall architecture of current state-of-the-art separation algorithms [3, 14, 36]. The section concludes with an example illustrating how the BSS outcomes can be utilized for the evidence pdf estimation in the spectral feature space [22].

11.3.1 BSS Front-End

Consider N speech sources in a reverberant enclosure impinging on a uniform linear microphone array (ULA) made up of M identical, omni-directional sensors with inter-element spacing d. We assume that N and M as well as the sensor spacing d are known and that d is chosen such that no spatial aliasing occurs.[3]

The main architecture and building blocks of our BSS front-end are illustrated in Fig. 11.3. The algorithm consists of four processing steps, each of which we describe in the following paragraphs in more detail.

[3] Here, we only deal with the even- and over-determined case ($N < M$) and do not consider the more difficult under-determined BSS problem, where $N > M$.

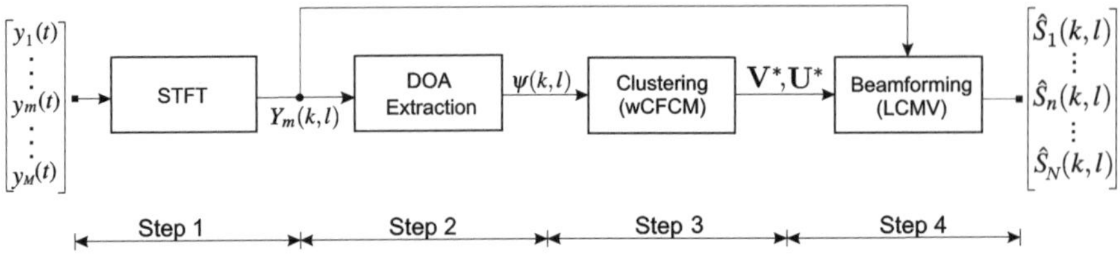

Fig. 11.3: The BSS system developed in [23], which uses a combination of adaptive beamforming and time-frequency masking to separate N sources from M mixture recordings (adopted from [22], © 2010 IEEE)

Step 1 — Short time spectral analysis

The first step converts the time domain signals $y_m(t)$ into their short-time Fourier-transform (STFT) representation

$$Y_m(k,l) = \sum_{\tau=-L/2}^{L/2-1} \mathrm{win}(\tau) y_m(\tau + k\tau_0) e^{-\iota l \omega_0 \tau}, \quad m = 1,\ldots,M \tag{11.11}$$

where $k \in \{0,\ldots,K-1\}$ is a time frame index, $l \in \{0,\ldots,L-1\}$ is a frequency bin index, $\mathrm{win}(\tau)$ is an L-point Hamming window and τ_0 and ω_0 denote the chosen STFT grid resolution parameters. This step is motivated by the fact that source separation becomes much easier in the STFT domain, where the original convolutive BSS problem can be approximated by its simpler instantaneous counterpart [1]. Furthermore, working in the frequency domain has the additional advantage that speech signals are more sparse here than in their waveform representation [45].

Step 2 — Spatial feature extraction

The second processing step extracts spatial features, so-called direction of arrival (DOA) values [2, 3], using the sound mixture recordings from the microphone array. More specifically, the DOA value at point (k,l) for microphone pair (Y_i, Y_j) is computed as

$$\psi(k,l) = -\frac{1}{l\omega_0 d_{ij} c^{-1}} \arg\left[\frac{Y_i(k,l)}{Y_j(k,l)}\right], \quad i,j \in \{1,\ldots,M\}, i \neq j \tag{11.12}$$

where d_{ij} denotes the distance between microphones i and j and c is the propagation velocity of sound. In our implementation, only DOA values extracted from the microphone pair with the biggest spacing are passed on to the following clustering step.[4] This is motivated by the observation that the longer the distance d_{ij} between a sensor pair, the better the DOA localization performance [42].

Step 3 — Fuzzy clustering of spatial features

The DOA feature set is then divided into N clusters by automatically grouping observations that originate from the same spatial location. A plethora of clustering techniques exist to perform this task, ranging from 'hard' to 'soft' data partitioning strategies. Here, we resorted to fuzzy clustering in order to reflect the localization uncertainty in a reverberant data set through a soft partitioning. Each cluster is represented by a prototype vector, called centroid, $\mathbf{V} = (v_1, \ldots, v_N)^T \in \mathbb{R}^N$, and a partition matrix $\mathbf{U} = [u_n(k,l)] \in \mathbb{R}^{N \times K \times L}$ indicating the degree $u_n(k,l)$ to which a data point $\psi(k,l)$ belongs to the n-th cluster.

In our implementation, we relied on the weighted contextual fuzzy c-means (wCFCM) cluster algorithm [23], which has been shown to perform superior to conventional fuzzy clustering on reverberant speech mixtures. The wCFCM algorithm considers the following minimization problem:

$$\min_{(\mathbf{U},\mathbf{V})} \Bigg\{ J_q(\mathbf{U},\mathbf{V}) = \sum_{n=1}^{N} \sum_{\forall (k,l)} u_n(k,l)^q w(k,l) \underbrace{|\psi(k,l) - v_n|^2}_{E_n(k,l)} + \frac{\beta}{2} \sum_{n=1}^{N} \sum_{\forall (k,l)} u_n(k,l)^q \underbrace{\sum_{\substack{\forall (k',l') \\ \in \mathrm{N}_{(k,l)}}} \sum_{\substack{n'=1, \\ n' \neq n}}^{N} u_{n'}(k',l')^q}_{C_n(k,l)} \Bigg\} \tag{11.13}$$

subject to the constraints

$$\begin{aligned} &\text{(i)} && \sum_{n=1}^{N} u_n(k,l) = 1, && \forall (k,l), \\ &\text{(ii)} && \sum_{\forall (k,l)} u_n(k,l) > 0, && \forall n, \\ &\text{(ii)} && u_n(k,l) \in [0,1], && \forall (k,l), n. \end{aligned}$$

Here, $q \in (1,\infty)$ is a constant controlling the softness of the cluster memberships and $w(k,l)$ is a user-specified weight indicating the reliability of observation $\psi(k,l)$. The parameter β controls the trade-off between minimizing the (weighted) Euclidean distances $E_n(k,l)$ in the DOA feature space and biasing the solution toward homogeneous membership masks with the help of the added regularization term in the wCFCM cost function $J_q(\mathbf{U},\mathbf{V})$ in (11.13). This penalty is minimized when the

[4] Using the SPIRE algorithm [42], spatial aliasing is dealt with by restoring the aliased phase difference values for microphone pairs where $d_{ij} > c/f_s$. f_s denotes the sampling frequency.

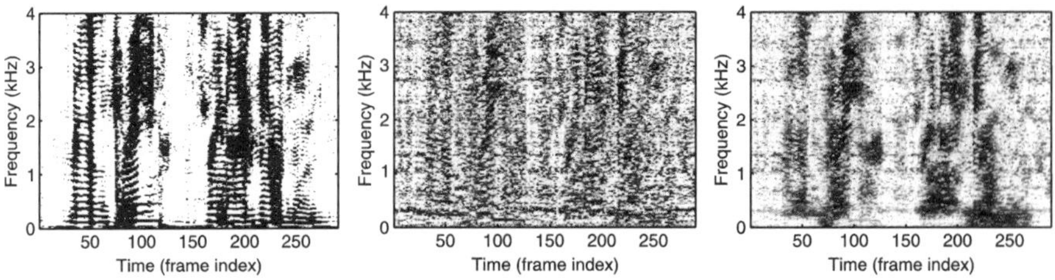

(a) oracle mask (b) FCM (c) wCFCM

Fig. 11.4: Segmentation results for a reverberant mixture of two speakers. Lighter areas represent lower and darker areas represent higher membership levels (adopted from [23], © 2010 Elsevier)

membership value for a particular cluster is large and the membership values for the other clusters in a local TF neighborhood $\mathrm{N}_{(k,l)}$ are small. The net result is a smoothing effect that causes neighboring points to have similar memberships in the same cluster. The value of β needs to be determined prior to the clustering, for example, by trial-and-error or cross-validation [23]. Likewise, the reader is referred to [21, 45], where various techniques for finding suitable observation weights $w(k,l)$ are discussed.

Starting with randomly initialized membership values, the wCFCM algorithm is implemented as an alternating optimization scheme and iterates between updates for the centroids,

$$v_n = \frac{\sum_{\forall(k,l)} u_n(k,l)^q w(k,l)\psi(k,l)}{\sum_{\forall(k,l)} u_n(k,l)^q w(k,l)}, \qquad \forall n \tag{11.14}$$

and memberships,

$$u_n(k,l) = \left[\sum_{j=1}^{N}\left(\frac{w(k,l)E_n(k,l)+\beta C_n(k,l)}{w(k,l)E_j(k,l)+\beta C_j(k,l)}\right)^{\frac{1}{q-1}}\right]^{-1}, \forall n,k,l \tag{11.15}$$

until a convergence criterion is met. Convergence is considered to be obtained when the difference between successive partition matrices is less than some predefined threshold. The final cluster centroids $\mathbf{V}^*$ represent estimates of the source DOAs, and the corresponding partition matrix $\mathbf{U}^*$ can be interpreted as a collection of N soft time-frequency masks. Note that wCFCM defaults to the conventional fuzzy c-means (FCM) algorithm if $\beta = 0$ and $w(k,l) = 1\,\forall(k,l)$.

Figure 11.4 shows an example of the segmentation results of fuzzy clustering on a reverberant speech mixture. As is evident from Fig. 11.4(b), the isolated membership assignment in FCM is highly vulnerable to noise in the DOA feature set. When compared to the "ground truth" in Fig. 11.4(a), the FCM mask is more speckled and contains many misclassifications. In contrast, the wCFCM result in Fig. 11.4(c) is much smoother and less speckled due to the inclusion of neighborhood information in the form of $C_n(k,l)$ during membership estimation.

Step 4 — Spatial filtering

While time-frequency masking on its own is a powerful tool for BSS, beamforming can further improve the separation quality for signals that have overlapping frequency content but originate from different spatial locations. Therefore, the segmentation results of the clustering are now used to compute the spatial filter weights of N adaptive beamformers, one for each detected source.

The estimate of the n-th source spectrum is obtained by the operation

$$\hat{S}_n(k,l) = \mathbf{b}_n^*(l)^{\mathrm{H}}\mathbf{Y}(k,l), \tag{11.16}$$

where $\mathbf{b}_n^*(l)$ denotes the optimum beamformer weights for the n-th source and the vector $\mathbf{Y}(k,l) = [Y_1(k,l),\ldots,Y_M(k,l)]^{\mathrm{T}}$ consists of all mixture observations. We implemented the Linearly Constrained Minimum Variance (LCMV) beamformer [28], for which the optimal filter weights are given by

$$\mathbf{b}_n^*(l) = \mathbf{R}_n^{-1}(l)\mathbf{A}(l)\left(\mathbf{A}(l)^{\mathrm{H}}\mathbf{R}_n^{-1}(l)\mathbf{A}(l)\right)^{-1}\boldsymbol{\delta}_n. \tag{11.17}$$

$\mathbf{R}_n(l)$ is the noise-plus-interference correlation matrix, $\mathbf{A}(l) = [\mathbf{a}_1(l),\ldots,\mathbf{a}_N(l)]$ is the constraint matrix containing the steering vectors

$$\mathbf{a}_n(l) = \begin{bmatrix} 1 & e^{-\iota l\omega_0 d_{21} c^{-1}\psi_n} & \ldots & e^{-\iota l\omega_0 d_{M1} c^{-1}\psi_n}\end{bmatrix}^{\mathrm{T}} \tag{11.18}$$

and $\boldsymbol{\delta}_n = (\delta_{n_1},\ldots,\delta_{n_N})^{\mathrm{T}}$ is the constraint response vector with

$$\delta_{n_i} = \begin{cases} 1, & \text{if } i = n \\ 0, & \text{otherwise.} \end{cases} \tag{11.19}$$

The LCMV beamformer preserves the desired signal while minimizing contributions to the output due to interfering signals and noise arriving from directions other than the direction of interest. In statistically optimum beamforming, $\mathbf{b}_n^*(l)$ is chosen based on the second-order statistics of the data received at the microphone array.

In practice, the true statistics for $\mathbf{R}_n(l)$ and $\mathbf{A}(l)$ are unknown and need to be derived from the available data. As proposed in [8], we determine suitable estimates $\hat{\mathbf{R}}_n(l)$ and $\hat{\mathbf{A}}(l)$ for both quantities by utilizing the estimated centers and memberships of the preceding clustering step. More specifically, $\hat{\mathbf{A}}(l)$ was obtained by replacing ψ_n in (11.18) with the estimated cluster center v_n and $\hat{\mathbf{R}}_n(l)$ was estimated via

$$\hat{\mathbf{R}}_n(l) = \frac{\sum_{k=0}^{K-1}\rho_n(k,l)\mathbf{Y}(k,l)\mathbf{Y}(k,l)^{\mathrm{H}}}{\sum_{k=0}^{K-1}\rho_n(k,l)}, \tag{11.20}$$

where the weights $\rho_n(k,l) = 1 - u_n(k,l)$ specify the degree to which the target source is deemed inactive.[5]

[5] A suitable estimation procedure for $\hat{\mathbf{R}}_n(l)$ in the under-determined BSS case is given in [8].

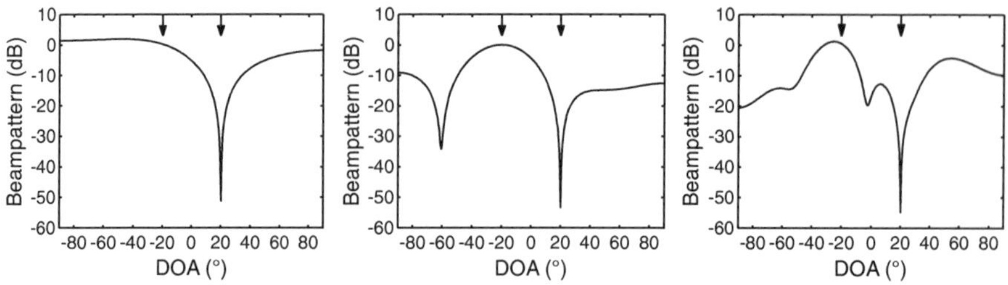

(a) 625 Hz (b) 1875 Hz (c) 3750 Hz

Fig. 11.5: Sample LCMV beampatterns at a low (a), mid (b) and high (c) frequency. The two arrows at the top of the figures mark the true DOA angles of the target and interference, located at -20° and 20°, respectively

Fig. 11.5 shows three LCMV beampatterns at different frequencies for a ULA of six microphones and a two-source scenario. The filter weights were estimated using the outcome of the wCFCM algorithm. As desired, deep nulls of up to minus 55 dB are placed at the DOA angle of the interfering source while maintaining a distortionless response in the look direction.

With this example we conclude the brief step-by-step description of our multi-channel BSS front-end. For more detailed information regarding performance assessments or parameter choices the interested reader is referred to [23].

11.3.2 Evidence PDF Estimation

Before speech recognition can take place the BSS separation results need to be converted to a representation that matches that of the acoustic models of the recognizer. For the missing data framework spectral features are preferred because contrary to the more common mel frequency cepstral coefficients (MFCCs) they prevent a smearing of spectrally local distortions over all feature vector components [10, 43].

Next, we give an example how the evidence model pdf $p_{\mathscr{B},\mathscr{U}}(x|\Theta)$ can be estimated in the log-spectral feature domain. Figure 11.6 shows a schematic of the employed estimation procedure [22], which is briefly described in the following paragraphs. In order to simplify processing, we assume that the BSS separation results are available in the form of two magnitude spectra T and I, where the former denotes the estimated target signal used for recognition and the latter represents an estimate of the noise intrusion.[6] Furthermore, let $\mathscr{M}$ be the estimated fuzzy time-frequency mask marking the dominant points for the target speaker T.

We start with the estimation of the mixture component $\mathscr{B}(x;\hat{\mu},\hat{\sigma}^2,\alpha,\beta)$. The mean value $\hat{\mu}$ is computed as the beamformed target estimate $T(k,l)$ converted to the FBANK [46] feature space. Using the Hidden Markov Model Toolkit's (HTK's)

[6] A priori knowledge was utilized in our study to select the recognition target among the separation results.

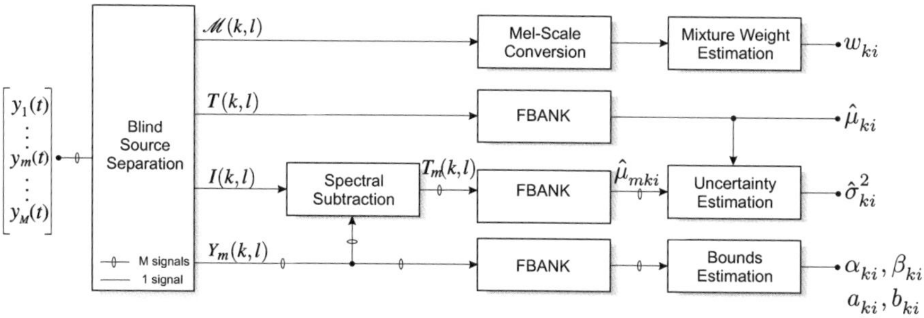

Fig. 11.6: Heuristic procedure for estimating the parameters of the Bounded-Gauss-Uniform mixture evidence model in the FBANK feature domain (adopted from [22], © 2010 IEEE)

[46] definition, this is implemented as

$$\hat{\mu}_{ki} = \log\left(\max\left\{\sum_l \lambda_i(l)T(k,l), \kappa\right\}\right), \quad i = 1, \dots, B, \tag{11.21}$$

where $\kappa = 1$ is the mel floor constant, $\lambda_i(l)$ is the i-th triangular filter of the mel filter bank and B denotes the total number of channels. Next, we consider the variance parameter $\hat{\sigma}^2_{ki}$ associated with $\mathscr{B}$. Because our BSS front-end currently does not provide uncertainty values directly, we estimate $\hat{\sigma}^2_{ki}$ indirectly with the help of a heuristic spectral subtraction (SS) scheme [22]. Our SS method utilizes the mixture observations $Y_m(k,l)$ as well as the interference estimate $I(k,l)$ to construct an additional target estimate $T_m(k,l)$ at each of the $m = 1, \dots, M$ microphones (see Fig. 11.6). Our assumption here is that the confidence in the BSS output $T(k,l)$ can be deemed high if $T(k,l)$ is similar to the SS estimates $T_m(k,l)$, $m = 1, \dots, M$, and low otherwise. After conversion of $T_m(k,l)$ to the mel frequency domain, we estimate the feature uncertainty by averaging the (weighted) square errors between the BSS-based estimate $\hat{\mu}_{ki}$ and each SS-based estimate $\hat{\mu}_{mki}$ over all microphones:

$$\hat{\sigma}^2_{ki} = \frac{1}{M}\sum_{m=1}^{M} \gamma_{ki}\left(\hat{\mu}_{mki} - \hat{\mu}_{ki}\right)^2, \qquad i = 1, \dots, B. \tag{11.22}$$

The empirical weighting factor $\gamma_{ki} = \{1 + \exp[\gamma_0[\varsigma_{ki} - \gamma_1)]\}^{-2}$ with SNR estimate $\varsigma_{ki} = 10\log_{10}(\sum_l \lambda_i(l)T^2(k,l) / \sum_l \lambda_i(l)I^2(k,l))$ followed a sigmoid-like function whose parameters $\gamma_0 = 0.25$ and $\gamma_1 = 3.0$ were determined on a small development set of ten utterances [22]. As reported elsewhere [6], the use of γ_{ki} in (11.22) helps to bias the variances $\hat{\sigma}^2_{ki}$ toward zero in high SNRs while retaining higher uncertainty values in mid and low SNRs.

We now turn to the estimation of the energy bounds associated with $\mathscr{B}$ and $\mathscr{U}$.[7] This is done by considering the smallest and largest values the clean FBANK value

[7] We assume here that the truncation points of the bounded Gaussian $\mathscr{B}$ are identical with those of the uniform distribution $\mathscr{U}$, e.g., $\alpha \equiv a$ and $\beta \equiv b$.

can take, given the noisy mixture observations. According to (11.21), the lower bound can be found by realizing that if the target emits no energy the clean FBANK value is $\log(\kappa) = \log(1) = 0$. Hence, the lower energy bounds are set to

$$\alpha_{ki} = a_{ki} = 0, \qquad i=1,\ldots,B. \tag{11.23}$$

If, on the other hand, there is no interference and all energy was emitted by the target speaker then the clean value is identical with the observed energy at the microphone. Hence, the upper bounds are set to

$$\beta_{ki} = b_{ki} = \log\left(\max\left\{\sum_l \lambda_i(l)\tilde{Y}(k,l), \kappa\right\}\right), \qquad i=1,\ldots,B \tag{11.24}$$

where $\tilde{Y}(k,l)$ denotes microphone array observation with the largest magnitude.

Finally, the mixture weight w_{ki} of $p_{\mathscr{B},\mathscr{U}}(x|\Theta)$ is obtained by converting the mask of the target speaker $\mathscr{M}_{kl}$ from its linear STFT frequency resolution to the perceptual mel frequency scale [18]. We use the same triangular filter weights λ_{il} as in (11.21) and compute the mixture weights as

$$w_{ki} = \frac{\sum_l \lambda_i(l)\mathscr{M}(k,l)}{\sum_l \lambda_i(l)} \qquad i=1,\ldots,B. \tag{11.25}$$

11.4 Speech Recognition Experiments

In this section we present some experimental evidence in support of the missing data theory and its application to noise robust ASR. This evidence mainly stems from our recent study [22], where we investigated several aspects of evidence modeling in the context of microphone array-based reverberant speech recognition. For the benefit of the reader, we briefly review the evaluation protocol and experimental setup before presenting some selected results that highlight the potential of evidence modeling for succeeding in challenging cocktail party-like scenarios. The section concludes with a general discussion, where we comment on limitations and shortcomings of the study and point out future research directions.

11.4.1 Room Simulation

Reverberant sound mixtures were created by simulating sound propagation in small-room acoustics with the help of an image-source model [24]. Figure 11.7 shows the corresponding room layout with details of the room's dimensions and the locations of the speakers and microphones. Target and noise sources were placed at horizontal angles of minus 20° and 20° to the broadside of a six-element ULA positioned in the middle of the room. Four male target speakers ('ah','ar','at','be') from the TIDIGIT corpus [25] were mixed at an SNR level of 0 dB and various reverberation levels with another male speaker ('DR2_MCEM0_SA1') from the TIMIT database [12]. Each of resulting test sets consisted of 240 speech mixtures containing a total of 785 digits.

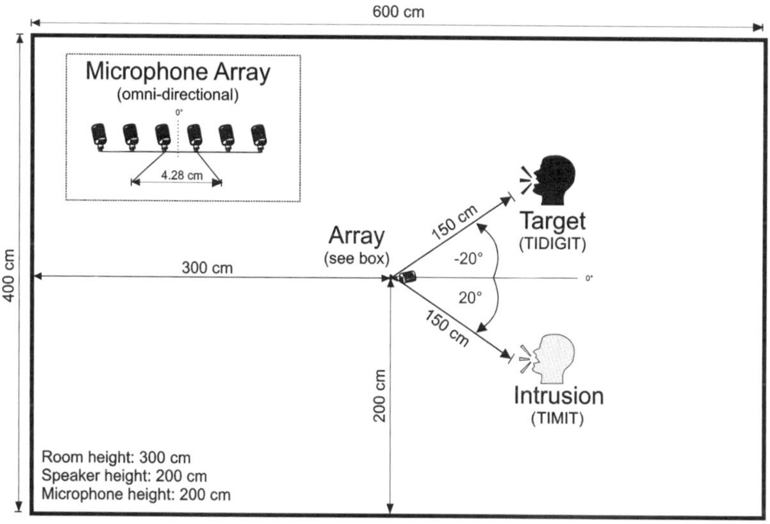

Fig. 11.7: Room layout for experimental evaluation (adopted from [22], © 2010 IEEE)

11.4.2 Model Training

HTK [46] was used to train 11 word HMMs ('1'–'9','oh','zero'), each with eight emitting states and two silence models ('sil','sp') with three and one state. All HMMs followed standard left-to-right models without skips using continuous Gaussian densities with diagonal covariance matrices and ten mixture components. Model training was performed on clean, anechoic signals using the male subset of the TIDIGIT corpus. Acoustic features were extracted using a 25 ms Hamming window and a 10 ms frame shift.

11.4.3 Performance Measures

Speech recognition performance was assessed by two standard measures obtained from HTK's alignment tool HResults [46]. The correctness score is defined as

$$\mathrm{COR} = \frac{\mathrm{NUM} - \mathrm{DEL} - \mathrm{SUB}}{\mathrm{NUM}} \times 100, \tag{11.26}$$

where NUM is the total number of tokens in the test set and DEL and SUB denote the deletion and substitution errors, respectively. The second performance measure, the accuracy score, is defined as

$$\mathrm{ACC} = \frac{\mathrm{NUM} - \mathrm{DEL} - \mathrm{SUB} - \mathrm{INS}}{\mathrm{NUM}} \times 100 \tag{11.27}$$

and in contrast to (11.26) also considers insertion errors, denoted here by INS. Hence, the accuracy score is considered the more representative performance measure among the two.

11.4.4 Results

The first experiment was concerned with the question of which type of evidence model is best in terms of speech recognition performance. Figure 11.8(a) and (b) show the correctness and accuracy scores achieved by six different evidence models of increasing complexity when the room reverberation time RT_{60}[8] was fixed at 300 ms. It can be seen that the more flexible the shape of the evidence pdf, the better the recognition performance. Note that the conventional 'BSS-only' approach is represented here by the Dirac evidence model, which utilized only the point estimates of the BSS output for recognition. In comparison with the other distributions the Dirac model is by far the least robust evidence pdf, lagging behind the leading Bounded-Gauss-Uniform mixture model with a gap of 35% for the correctness score and 60% for the accuracy measure. Furthermore, all single component evidence pdfs achieved disappointingly low accuracy scores because of the high number of insertion errors these models produced. The two-component mixture models seem to provide a much higher robustness in this regard. We attribute this to their superior ability to ignore spectro-temporal areas where the array preprocessor was unsuccessful in restoring the clean speech signal.

The second investigation studied the impact of the room reverberation time on the recognition accuracy. Figure 11.9 shows the performance of an MFCC-CMN baseline scoring on the sound mixture and the BSS output as well as the results of the missing data decoder using the Dirac-, Gauss- and Bounded-Gauss-Uniform mixture pdfs. The prefix 'a posteriori' indicates that the evidence pdf parameters were estimated using the array-based BSS algorithm. This is in contrast to the 'a priori' evidence models, where mixture weights and feature variances were obtained using 'oracle' knowledge of the clean signals [11, 17]. They are used here as a means to illustrate the upper performance limit for the missing data decoder.

[8] RT_{60} is defined as the time required for the reflections of a direct sound to decay by 60 dB following sound offset.

Several observations can be made regarding the different performances of the tested recognition paradigms. First, the severity of the acoustic mismatch problem is reflected by the abysmal recognition accuracy of the cepstral 'Mixtures' baseline. Clearly, state-of-the-art features alone provide no effective means of protection in the presence of both convolutive and additive noise. Second, the conventional approach of first enhancing and then recognizing the preprocessed speech signal performs well only for the ideal case of echo-free speech mixtures. In the absence of sound reflections, our spatial BSS algorithm has little difficulty in restoring the target signal with sufficient quality. An impressive 90% absolute improvement in recognition accuracy was achieved by the cepstral 'BSS-only' baseline in this case. However, because of the limitations inherent in spatial source separation, our LCMV beamformer cannot deal well with reverberation. Consequently, the recognition performance of the 'BSS-only' baseline dropped significantly once sound reflections came into play. At the highest tested reverberation time of 600 ms the 'BSS-only' system was only slightly better than the 'Mixture' baseline. Third, the proposed combination of BSS and evidence modeling was consistently superior to 'BSS-only' processing. All three a posteriori evidence models achieved substantial gains in recognition accuracy for all room reverberation times. At best, the performance gap between the leading a posteriori evidence model and the 'BSS-only' baseline was as wide as 40 percentage points. As before, the best results were obtained by the Bounded-Gauss-Uniform mixture model followed by the Gauss-Uniform and Dirac-Uniform mixture pdfs. Finally, the highest noise robustness of all systems was shown by the a priori evidence models. While the Gauss- and Bounded-Gauss-

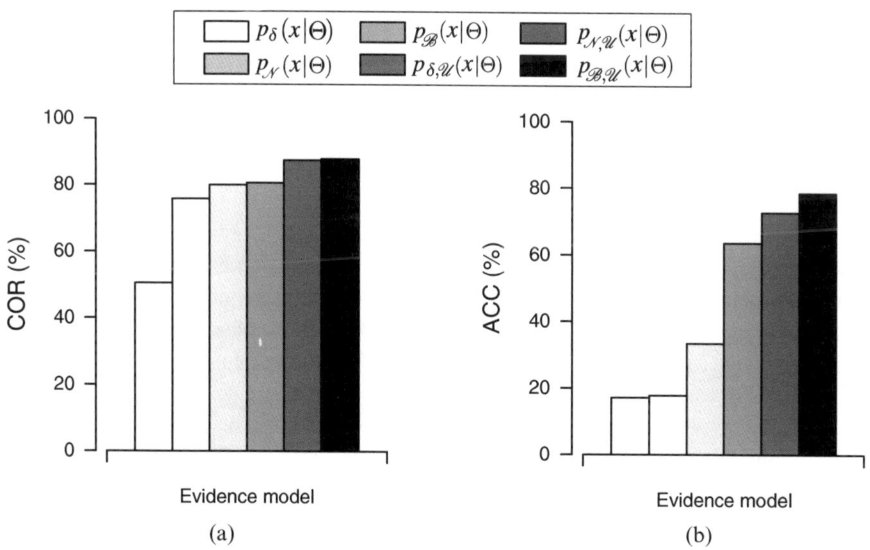

Fig. 11.8: Recognition performance of six evidence models in the presence of an interfering male speaker ($RT_{60} = 300\,\text{ms}$)

Uniform mixture pdfs remained close to ceiling performance, the Dirac-Uniform mixture model showed a moderate susceptibility to reverberation. We explain this with the model's simple 'data is clean or noisy' assumption, which becomes rather implausible at higher reverberation times due to the increasing amount of convolutive distortions. Performance nevertheless remained at a very high level, demonstrating the strong potential of evidence modeling for noise robust ASR.

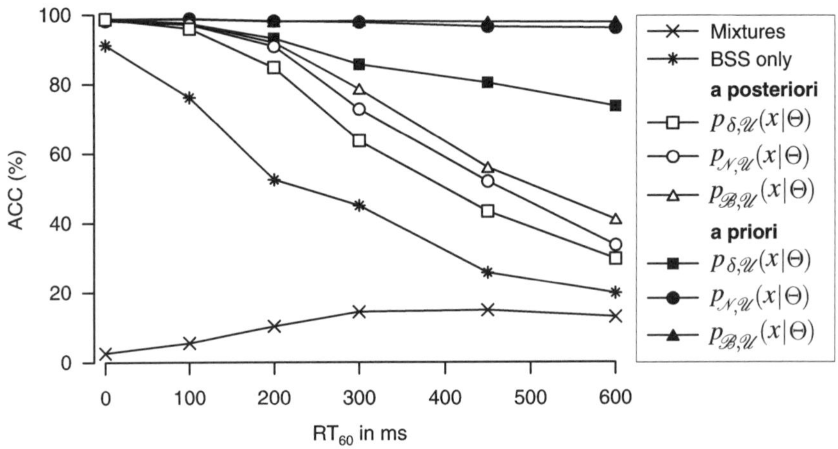

Fig. 11.9: Recognition performance for different room reverberation times in the presence of an interfering male speaker

11.4.5 General Discussion

The most compelling argument for the use of array-based evidence modeling is the higher noise robustness which can be obtained even when only a small number of microphones are available. The key idea in this chapter was to make use of the enhanced array output whenever possible while detecting and ignoring unreliable data when distortions were uncorrectable by the array preprocessor. Our results suggest that this idea is best implemented in the spectral feature space using two-component mixture pdfs, where the decision whether speech enhancement was successful or not is controlled by the value of the missing data reliability mask. In this way, the enhancement requirements for the preprocessor are considerably relaxed because it is no longer necessary for the array to produce accurate estimates of $\hat{\mu}$ and $\hat{\sigma}^2$ in spectro-temporal regions where the target is masked by interference or background noise. The results presented here and elsewhere [19, 22] have shown that the more flexible the model is in terms of its pdf shape, the better is the speech recognition accuracy. Performance gains were most evident for the more challenging setups with higher reverberation times and lower spatial separation of the sources [22].

In particular, the Gauss-Uniform and Bounded-Gauss-Uniform mixture models are strong candidates for microphone array applications that require noise robust speech recognition.

Before concluding this chapter, we would like to comment on some points that influence the performance of our system.

First of all, the speech recognizer's accuracy is driven by the quality of the BSS separation algorithm. Currently, our front-end relies exclusively on spatial information for source segregation and as a consequence performance suffers in echoic environments. With increasing level of reverberation, the extracted localization cues become more and more unreliable, making an accurate DOA and time-frequency mask estimation for the fuzzy clustering algorithm quite difficult. Furthermore, it is well known that the blind estimation of the LCMV beamformer weights is highly susceptible to errors. Any mismatch in the steering vector and the jammer correlation matrix can lead to severe performance degradation due to the phenomenon of target cancellation [28]. Future work needs to address these issues by incorporating robust beamforming schemes as well as additional knowledge sources into the BSS front-end. For example, recent work [37, 44] has shown that pitch and harmonicity cues may still provide robust information for source segregation despite the presence of strong sound reflections. Extending the fuzzy clustering algorithm to multi-dimensional feature sets is therefore one possible extension in this regard. By equipping the BSS front-end with the ability to deal with more challenging situations, it can be expected that, in turn, speech recognition performance will also increase substantially.

Another crucial factor that determines how successful our system performs in practice is the accuracy with which the parameters of the evidence pdf are estimated. In this regard, the heuristic multi-channel estimation technique developed in [22] may serve as a proof-of-concept for demonstrating the potential of our approach. Additional performance gains may be achieved by integrating evidence pdf estimation and speech feature extraction into one theoretically consistent framework. This, for example, includes uncertainty propagation from the BSS front-end to the ASR back-end [16] and feature marginalization using tighter integration bounds [40].

A closely related issue is the choice of the feature space in which the evidence modeling takes place. Spectral features possess a number of characteristics which make them attractive for noise robust ASR. Firstly, the spectral feature space is physically interpretable and constitutes the domain of choice for many source segregation and speech enhancement techniques. Secondly, the spectral feature space offers the highest level of flexibility in terms of the statistical distributions that are available to the evidence model estimation algorithm for transforming the collected information into a probabilistic description. It is instructive to point out that each transformation step in the feature extraction chain is afflicted with an increase in the uncertainty of that transformed value. For example, the mixing of all spectral bands that occurs when applying an orthogonalization transform leads to a smearing of the noise across all orthogonalized coefficients [43]. Marginalizing unreliable feature components with the help of evidence mixture models is therefore best implemented in the spectral feature domain. Furthermore, no effective bounds for the clean feature value are known in the cepstral feature space, thus restricting the choice of possible evidence models to the class of unbounded pdfs. And thirdly, working in the spectral feature space easily facilitates an exchange of information between top-down and bottom-up processing. There is ample evidence in the psychoacoustic and neurophysiological literature that supports the use of higher-level expectations in human auditory processing [29]. Considering the framework of evidence modeling, the use of top-down information could assist the decoder with the detection and correction of inconsistencies between learned expectations and incoming bottom-up 'evidence'.

However, the commitment to spectral feature representations is not without its shortcomings. Arguably, the biggest drawback is the high correlation between the components of a spectral feature vector, rendering the diagonal GMM covariance assumption rather implausible. In this sense, our use of a triangular mel filter bank with overlapping filters, as described in Section 11.3.2, further contributes to this problem, particularly with the usual feature decorrelation step missing. While this has not been an issue for the small vocabulary task considered here, we expect it to gain in importance for more complex recognition scenarios that require a larger number of speech models. In order to address this problem, future research needs to identify ways in which less correlated spectral coefficients can be extracted from the high frequency resolution supplied by the BSS front-end.

Lastly, it is worth noting that evidence modeling is not limited to multi-channel cocktail party processing. It is our hope that more recently proposed evidence models, such as the Bounded-Gauss-Uniform mixture pdf, will also find widespread use in monaural missing data approaches, where the parameters need to be derived from single-channel enhancement techniques.

11.5 Summary

Robust classification in cocktail party scenarios still presents a formidable challenge to current state-of-the-art speech recognizers. This chapter has presented a microphone array-based approach to this problem embedded into the probabilistic concept of evidence modeling. Several aspects of the approach were discussed, including evidence model selection, HMM likelihood computation and pdf parameter estimation with the help of a multi-channel BSS front-end. Experimental evaluation showed that the proposed framework outperformed the conventional enhancement-recognition approach by a substantial margin and that the more flexible the shape of the evidence pdf, the higher the obtained noise robustness.

Future research should be directed at improving the BSS algorithm in reverberant conditions and integrating top-down information into the decision process. Ultimately, speech recognition and source separation should work together under a unified framework, enabling the overall system to exploit all available information sources for optimal performance in a variety of environmental conditions. Only when we depart from the passive statistical pattern matching framework that is dominant in current ASR technology and begin the transition to more active decision making paradigms can we expect to succeed in our quest for truly robust speech recognition systems.

Acknowledgements This research was partly funded by the Australian Research Council (ARC) grant no. DP1096348.

References

1. Araki, S., Mukai, R., Makino, S., Nishikawa, T., Saruwatari, H.: The fundamental limitation of frequency domain blind source separation for convolutive mixtures of speech. IEEE Transactions on Speech and Audio Processing **11**(2), 109–116 (2003)
2. Araki, S., Sawada, H., Mukai, R., Makino, S.: DOA estimation for multiple sparse sources with normalized observation vector clustering. In: IEEE International Conference on Acoustics, Speech and Signal Processing. Toulouse, France (2006)
3. Araki, S., Sawada, H., Mukai, R., Makino, S.: Underdetermined blind sparse source separation for arbitrarily arranged multiple sensors. Signal Processing **87**(8), 1833–1847 (2007)
4. Arrowood, J.: Using observation uncertainty for robust speech recognition. Ph.D. thesis, Georgia Institute of Technology (2003)
5. Barker, J., Josifovski, L., Cooke, M., Green, P.: Soft decisions in missing data techniques for robust automatic speech recognition. In: 6th International Conference of Spoken Language Processing. Beijing, China (2000)
6. Benìtez, M., Segura, J., Ramìrez, J., Rubio, A.: Including uncertainty of speech observations in robust speech recognition. In: 8th International Conference on Spoken Language Processing. Jeju Island, Korea (2004)
7. Bregman, A.: Auditory Scene Analysis. MIT Press, Cambridge MA (1990)
8. Cermak, J., Araki, S., Sawada, H., Makino, S.: Blind speech separation by combining beamformers and a time frequency binary mask. In: International Workshop on Acoustic Echo and Noise Control. Paris, France (2006)
9. Cherry, E.: Some experiments on the recognition of speech, with one and with two ears. Journal of the Acoustical Society of America **25**(5), 975–979 (1953)
10. Cooke, M., Green, P., Josifovski, L., Vizinho, A.: Robust automatic speech recognition with missing and unreliable acoustic data. Speech Communication **34**(3), 267–285 (2001)
11. Deng, L., Droppo, J., Acero, A.: Dynamic compensation of HMM variances using the feature enhancement uncertainty computed from a parametric model of speech distortion. IEEE Transactions on Speech and Audio Processing **13**(3), 412–421 (2005)
12. Garofolo, J., Lamel, L., Fisher, W., Fiscus, J., Pallett, D., Dahlgren, N., Zue, V.: TIMIT acoustic-phonetic continuous speech corpus. Tech. rep., Linguistic Data Consortium (1993)
13. Harding, S., Barker, J., Brown, G.: Mask estimation for missing data speech recognition based on statistics of binaural interaction. IEEE Transactions on Audio, Speech, and Language Processing **14**(1), 58–67 (2006)
14. Kolossa, D., Klimas, A., Orglmeister, R.: Separation and robust recognition of noisy, convolutive speech mixtures using time-frequency masking and missing data techniques (2005)
15. Kolossa, D., Orglmeister, R.: Nonlinear postprocessing for blind speech separation. In: 5th International Conference on Independent Component Analysis and Signal Separation. Granada, Spain (2004)
16. Kolossa, D., Sawada, H., Astudillo, R., Orglmeister, R., Makino, S.: Recognition of convolutive speech mixtures by missing feature techniques for ICA. In: Asilomar Conference on Signals, Systems and Computers. Pacific Grove, CA (2006)
17. Kühne, M., Pullella, D., Togneri, R., Nordholm, S.: Towards the use of full covariance models for missing data speaker recognition. In: IEEE International Conference on Acoustics, Speech and Signal Processing. Las Vegas, USA (2008)
18. Kühne, M., Togneri, R., Nordholm, S.: Mel-spectrographic mask estimation for missing data speech recognition using short-time-Fourier-transform ratio estimators. In: IEEE International Conference on Acoustics, Speech and Signal Processing. Honolulu, USA (2007)
19. Kühne, M., Togneri, R., Nordholm, S.: Adaptive beamforming and soft missing data decoding for robust speech recognition in reverberant environments. In: Interspeech. Brisbane, Australia (2008)
20. Kühne, M., Togneri, R., Nordholm, S.: Time-frequency masking: Linking blind source separation and robust speech recognition. In: F. Milhelič, J. Žibert (eds.) Speech Recognition: Techniques, Technologies and Applications, pp. 61–80. In-Tech Open Access Publisher (2008)

21. Kühne, M., Togneri, R., Nordholm, S.: Robust source localization in reverberant environments based on weighted fuzzy clustering. IEEE Signal Processing Letters **16**(2), 85–88 (2009)
22. Kühne, M., Togneri, R., Nordholm, S.: A new evidence model for missing data speech recognition with applications in reverberant multi-source environments. IEEE Transactions on Audio, Speech and Language Processing, in press (2010)
23. Kühne, M., Togneri, R., Nordholm, S.: A novel fuzzy clustering algorithm using observation weighting and context information for reverberant blind speech separation. Signal Processing **90**(2), 653–669 (2010)
24. Lehmann, E., Johansson, A.: Prediction of energy decay in room impulse responses simulated with an image-source model. Journal of the Acoustical Society of America **124**(1), 269–277 (2008)
25. Leonard, R.: A database for speaker-independent digit recognition. In: IEEE International Conference on Acoustics, Speech, and Signal Processing. San Diego, CA (1984)
26. Lippmann, R.: Speech recognition by machines and humans. Speech Communication **22**(1), 1–15 (1997)
27. Low, S., Togneri, R., Nordholm, S.: Spatio-temporal processing for distant speech recognition. In: IEEE International Conference on Acoustics, Speech, and Signal Processing. Montreal, Canada (2004)
28. Malonakis, D., Ingle, V., Kogon, S.: Statistical and Adaptive Signal Processing. McGraw Hill (2000)
29. McAdams, S.: Recognition of Auditory Sound Sources and Events. Thinking in Sound: The Cognitive Psychology of Human Audition. Oxford University Press (1993)
30. McCowan, I.A., Marro, C., Mauuary, L.: Robust speech recognition using near-field superdirective beamforming with post-filtering. In: IEEE International Conference on Acoustics, Speech and Signal Processing. Istanbul, Turkey (2000)
31. McCowan, I.A., Morris, A., Bourlard, H.: Improving speech recognition performance of small microphone arrays using missing data techniques. In: 7th International Conference on Spoken Language Processing. Denver, USA (2002)
32. Morris, A.: Data utility modelling for mismatch reduction. In: Workshop on Consistent & Reliable Acoustic Cues for sound analysis. Aalborg, Denmark (2001)
33. Morris, A., Barker, J., Bourlard, H.: From missing data to maybe useful data: Soft data modelling for noise robust ASR. In: WISP. Stratford-upon-Avon, England (2001)
34. Omologo, M., Matassoni, M., Svaizer, P.: Speech recognition with microphone arrays. In: M. Brandstein, D. Ward (eds.) Microphone arrays, pp. 331–353. Springer (2001)
35. Palomäki, K., Brown, G., Wang, D.: A binaural processor for missing data speech recognition in the presence of noise and small-room reverberation. Speech Communication **43**(4), 361–378 (2004)
36. Roman, N., Srinivasan, S., Wang, D.: Speech recognition in multisource reverberant environments with binaural inputs. In: IEEE International Conference on Acoustics, Speech and Signal Processing. Toulouse, France (2006)
37. Roman, N., Wang, D.: Pitch-based monaural segregation of reverberant speech. Journal of the Acoustical Society of America **120**(1), 458–469 (2006)
38. Roman, N., Wang, D., Brown, G.: Speech segregation based on sound localization. Journal of the Acoustical Society of America **114**(4), 2236–2252 (2003)
39. Seltzer, M.: Microphone array processing for robust speech recognition. Ph.D. thesis, Carnegie Mellon University, Pittsburgh, USA (2003)
40. Srinivasan, S., Roman, N., Wang, D.: Exploiting uncertainties for binaural speech recognition. In: IEEE International Conference on Acoustics, Speech, and Signal Processing. Honolulu, USA (2007)
41. Stouten, V., Van hamme, H., Wambacq, P.: Accounting for the uncertainty of speech estimates in the context of model-based feature enhancement. In: International Conference on Spoken Language Processing. Jeju Island, Korea (2004)
42. Togami, M., Sumiyoshi, T., Amano, A.: Stepwise phase difference restoration method for sound source localization using multiple microphone pairs. In: IEEE International Conference on Acoustics, Speech and Signal Processing. Honolulu, USA (2007)

43. de Veth, J., de Wet, F., Cranen, B., Boves, L.: Acoustic features and a distance measure that reduces the impact of training-set mismatch in ASR. Speech Communication **34**(1-2), 57–74 (2001)
44. Wu, M., Wang, D.: A one-microphone algorithm for reverberant speech enhancement. In: IEEE International Conference on Acoustics, Speech, and Signal Processing, vol. 1, pp. 892–895. Hong Kong, China (2003)
45. Yilmaz, Ö., Rickard, S.: Blind separation of speech mixtures via time-frequency masking. IEEE Transactions on Signal Processing **52**(7), 1830–1847 (2004)
46. Young, S., Evermann, G., Gales, M., Hain, T., Kershaw, D., Liu, X., Moore, G., J., O., Ollason, D., Povey, D., Valtchev, V., Woodland, P.: The HTK Book. Cambridge University Engineering Department (2006)

Chapter 12
Recognition of Multiple Speech Sources Using ICA

Eugen Hoffmann, Dorothea Kolossa and Reinhold Orglmeister

Abstract In meetings or noisy public places, often a number of speakers are active simultaneously and the sources of interest need to be separated from interfering speech in order to be robustly recognized. Independent component analysis (ICA) has proven to be a valuable tool for this purpose. However, under difficult environmental conditions, ICA outputs may still contain strong residual components of the interfering speakers. In such cases, time-frequency masking can be applied to the ICA outputs to reduce the remaining interferences. In order to remain robust against possible resulting artifacts and loss of information, treating the processed speech feature vector as a random variable with time-varying uncertainty, rather than as deterministic, is a helpful strategy. This chapter shows the ways of improving recognition of multiple speech signals based on nonlinear postprocessing, applied together with uncertainty-based decoding techniques.

12.1 Introduction

In order for speech recognition to perform well in arbitrary, noisy environments, it is of special importance to suppress interfering speech, which poses significant problems for noise reduction algorithms due to the overlapping spectra and nonstationarity. In such cases, blind source separation can often be of great value, since it is applicable to any set of signals that is at least statistically independent and non-Gaussian, which is a fairly mild requirement. Blind source separation itself can, however, profit significantly from additional nonlinear postprocessing, in order to suppress speech or noise which remains in the separated components. Such nonlinear postprocessing functions have been shown to result in SNR improvements in excess of 10 dB, e.g., in [40].

Electronics and Medical Signal Processing Group, TU Berlin, Einsteinufer 17, 10587 Berlin.
e-mail: eugen.hoffmann.1@tu-berlin.de, dorothea.kolossa@gmx.de, reinhold.orglmeister@tu-berlin.de

D. Kolossa and R. Haeb-Umbach (eds.), *Robust Speech Recognition of Uncertain or Missing Data: Theory and Applications*, DOI 10.1007/978-3-642-21317-5_12,
© Springer-Verlag Berlin Heidelberg 2011

However, while the results of source separation are greatly improved by nonlinear postprocessing, speech recognition results often suffer from artifacts and loss of information due to such postprocessing. In order to compensate for these losses and artifacts and to obtain results exceeding those of ICA alone, we suggest the use of uncertainty-of-observation techniques for the subsequent speech recognition. This allows for the utilization of a feature uncertainty estimate, which can be derived considering the suppressed components of target speech, and will be described in more detail in Section 12.3. From such an uncertain description of the speech signal in the spectrum domain, uncertainties need to be made available also in the feature domain in order to be used for recognition. This can be achieved by uncertainty propagation, which converts an uncertain description of speech from the spectrum domain, where ICA takes place, to the feature domain of speech recognition, as described in Chapter 2. After this uncertainty propagation, recognition can take place under observation uncertainty, using uncertainty-of-observation techniques.

The entire process is vitally dependent on the appropriate estimation of uncertainties. Results given in Section 12.4.7 show that when the exact uncertainty in the spectrum domain is known, recognition results with the suggested approach are far in excess of those achievable by ICA alone. Also, a realistically computable uncertainty estimate is given, and the experiments and results in Section 12.4 show that with this practically available uncertainty measure, significant improvements of recognition performance can be attained for noisy and reverberant room recordings.

The presented method is closely related to other works that consider observation vectors as uncertain for decoding purposes, most often for noisy speech recognition [16, 18, 29, 32], but in some cases also for speech recognition in multi-talker conditions, as, for example, [10, 41], or [47] in conjunction with speech segregation via binary masking (see, e.g., [9, 49]).

The main novelty in comparison with the above techniques is the use of independent component analysis in conjunction with uncertainty estimation and with a piecewise approach to transforming uncertainties to the feature domain of interest. This allows for the suggested approach to combine the strengths of independent component analysis and soft time-frequency masking, and to be still used with a wide range of feature parameterizations. Corresponding results are shown here for MFCC coefficients, but the discussed uncertainty transformation approach also generalizes well to the ETSI advanced front-end, as shown in [48], and has been successfully used for time-frequency masking of ICA results in conjunction with RASTA-PLP features as well in [6].

12.2 Blind Source Separation

Blind source separation (BSS) is a technique of recovering the source signals of interest using only observed mixtures when both the mixing parameters and the sources are unknown. Due to a large number of applications, for example, in medical and speech signal processing, BSS has gained great attention. In the following chapter, we will discuss the application of BSS for acoustic signals observed in a real environment, i.e., convolutive mixtures of multiple speakers recorded under mildly noisy and reverberant distant-talking conditions.

In recent years, this problem has been widely studied and a number of different approaches have been proposed [1–3]. Many existing unmixing methods of acoustic signals are based on Independent Component Analysis (ICA) in the frequency domain, where the convolutions of the source signals with the room impulse response are reduced to multiplications with the corresponding transfer functions. So for each frequency bin, an individual instantaneous ICA problem can be solved in order to obtain the unmixed sources in the frequency domain [3].

Alternative methods include adaptive beamforming, which is closely related to independent component analysis when information-theoretic cost functions are applied [8], and sparsity-based methods that utilize amplitude-delay histograms [9, 10] or grouping cues typical of human stream segregation [11]. Here, independent component analysis has been chosen due to its inherent robustness to noise and its ability to handle strong reverberation by frequency-by-frequency optimization of the cost function.

12.2.1 Problem Formulation

This section provides an introduction to the problem of blind separation of acoustic signals.

At first, a general situation will be considered. In a reverberant room, N acoustic signals $\mathbf{s}(t) = [s_1(t), \ldots s_N(t)]$ are simultaneously present, where t represents the discrete time index. The vector of the source signals $\mathbf{s}(t)$ is recorded with M microphones placed in the room, so that an observation vector $\mathbf{x}(t) = [x_1(t), \ldots x_M(t)]$ results. Due to the time delay and the signal reflections, the resulting mixture $\mathbf{x}(t)$ is the result of a convolution of the source signal vector $\mathbf{s}(t)$ with unknown filter matrices $\{\underline{\mathbf{a}}_1 \ldots \underline{\mathbf{a}}_K\}$ where $\underline{\mathbf{a}}_k$ is the k-th ($k \in [1 \ldots K]$) $M \times N$ matrix with filter coefficients and K is the filter length.[1] This problem can be summarized by

$$\mathbf{x}(t) = \sum_{k=0}^{K-1} \underline{\mathbf{a}}_{k+1} \mathbf{s}(t-k) + \mathbf{n}(t). \tag{12.1}$$

The term $\mathbf{n}(t)$ denotes the additive sensor noise. Now the problem is to find filter matrices $\{\underline{\mathbf{w}}_1 \ldots \underline{\mathbf{w}}_{K'}\}$ so that by applying them to the observation vector $\mathbf{x}(t)$ the source signals can be estimated via

$$\hat{\mathbf{s}}(t) = \mathbf{y}(t) = \sum_{k'=0}^{K'-1} \underline{\mathbf{w}}_{k'+1} \mathbf{x}(t-k') \tag{12.2}$$

[1] In the following work, only the situation with $M \geq N$ is considered. For $M < N$, the so-called underdetermined case, see [9].

with K' as the filter length. In other words, for the estimated vector $\mathbf{y}(t)$ and the source vector $\mathbf{s}(t)$, $\mathbf{y}(t) \approx \mathbf{s}(t)$ should hold.

This problem is also known as the cocktail party problem. A common way to deal with the problem is to reduce it to a set of the instantaneous source separation problems, for which efficient approaches exist.

For this purpose, the time-domain observation vectors $\mathbf{x}(t)$ are transformed into a frequency domain time series by means of the short-time Fourier transform (STFT)

$$\mathbf{X}(\Omega,\tau) = \sum_{t=-\infty}^{\infty} \mathbf{x}(t) w(t-\tau R) e^{-j\Omega t}, \tag{12.3}$$

where Ω is the angular frequency, τ represents the frame index, $w(t)$ is a window function (e.g., a Hanning window) of length N_{FFT}, and R is the shift size, in samples, between successive windows [12]. Transforming Eq. (12.1) into the frequency domain reduces the convolutions to multiplications with the corresponding transfer functions, so that for each frequency bin an individual instantaneous problem

$$\mathbf{X}(\Omega,\tau) \approx \mathbf{A}(\Omega)\mathbf{S}(\Omega,\tau) + \mathbf{N}(\Omega,\tau) \tag{12.4}$$

arises. $\mathbf{A}(\Omega)$ is the mixing matrix in the frequency domain, $\mathbf{S}(\Omega,\tau) = [S_1(\Omega,\tau), \ldots, S_N(\Omega,\tau)]$ represents the source signals, $\mathbf{X}(\Omega,\tau) = [X_1(\Omega,\tau), \ldots, X_M(\Omega,\tau)]$ denotes the observed signals, and $\mathbf{N}(\Omega,\tau)$ is the frequency domain representation of the additive sensor noise. In order to reconstruct the source signals, the unmixing matrix $\mathbf{W}(\Omega) \approx \mathbf{A}^{+}(\Omega)$ is derived[2] using a complex-valued unmixing algorithm, so that

$$\mathbf{Y}(\Omega,\tau) = \mathbf{W}(\Omega)\mathbf{X}(\Omega,\tau) \tag{12.5}$$

can be used for obtaining estimated sources in the frequency domain. Here, $\mathbf{Y}(\Omega,\tau) = [Y_1(\Omega,\tau), \ldots, Y_N(\Omega,\tau)]$ is the time-frequency representation of the unmixed outputs.

12.2.2 ICA

Independent Component Analysis (ICA) is an approach that can help us find optimal unmixing matrices $\mathbf{W}$. The main idea is to obtain statistical independence of the output signals, which is mathematically defined in terms of probability densities. The components of the vector $\mathbf{Y}$ are statistically independent if and only if the joint probability distribution function $f_{\mathbf{Y}}(\mathbf{Y})$ is equal to the product of the marginal distribution functions of each signal Y_i:

$$f_{\mathbf{Y}}(\mathbf{Y}) = \prod_i f_{Y_i}(Y_i). \tag{12.6}$$

[2] $\mathbf{A}^{+}(\Omega)$ denotes the pseudoinverse of $\mathbf{A}(\Omega)$.

The process of finding the unmixing matrix $\mathbf{W}$ is now composed of two steps:

- the definition of a contrast function $\mathscr{J}(\mathbf{W})$, which is a quantitative measure of the statistical independence of all components in $\mathbf{Y}$, and
- the minimization of $\mathscr{J}(\mathbf{W})$ so that

$$\hat{\mathbf{W}} \stackrel{!}{=} \arg\min_{\mathbf{W}} \mathscr{J}(\mathbf{W}). \tag{12.7}$$

At this point, the definition of the contrast function $\mathscr{J}(\mathbf{W})$ is the key to the problem solution. For this purpose, it is possible to focus on different aspects of statistical independence, which results in the large number of ICA algorithms that have been proposed during the last few decades [2]. The most common approaches use one of the following characteristics of independent signals:

- The higher-order cross-statistic tensor of independent signals is diagonal, so $\mathscr{J}(\mathbf{W})$ is defined as the sum of the off-diagonal elements of, e.g., the fourth-order cross-cumulant (JADE algorithm [13]).
- Each independent component remains independent in time, so the cross correlation matrix $\mathbf{C}(\tau) = \mathrm{E}[\mathbf{Y}(t)\mathbf{Y}(t+\tau)^T]$ remains diagonal, i.e.,

$$\mathbf{C}(\tau) = \begin{pmatrix} \mathrm{R}_{Y_1Y_1}(\tau) & 0 & \cdots & 0 \\ 0 & \mathrm{R}_{Y_2Y_2}(\tau) & \cdots & 0 \\ \vdots & \vdots & \ddots & \vdots \\ 0 & 0 & \cdots & \mathrm{R}_{Y_NY_N}(\tau) \end{pmatrix} \tag{12.8}$$

  for each τ (SOBI algorithm [14]).
- The mutual information $\mathrm{I}(\mathbf{X}, g(\mathbf{Y}))$, with $\mathrm{g}(\cdot)$ as a nonlinear function, achieves its maximum when the components of $\mathbf{Y}$ are statistically independent. This assumption leads to the solution

$$\hat{\mathbf{W}} \stackrel{!}{=} \arg\max_{\mathbf{W}} \mathrm{H}(\mathrm{g}(\mathbf{Y})), \tag{12.9}$$

  where $H(g(\mathbf{Y}))$ is the joint entropy of $g(\mathbf{Y})$. This is known as the information maximization approach [15].
- If the distributions of the independent components are non-Gaussian, the search for maximal independence results in a search for maximal non-Gaussianity [17]. In this case, the negentropy

$$\mathscr{J}(\mathbf{W}) = \mathrm{H}(\mathbf{Y}_{\mathrm{Gauss}}) - \mathrm{H}(\mathbf{Y}) \tag{12.10}$$

  plays the role of the cost function, where $\mathbf{Y}_{\mathrm{Gauss}}$ is a vector-valued Gaussian random variable of the same mean and covariance matrix as $\mathbf{Y}$.

The last assumption leads to an approach proposed by Hyvärinen et al. [17], also known as the FastICA algorithm. The problem of this approach is the computation of the negentropy. The calculation according to Eq. (12.10) would require an estimate of the probability density functions in each iteration, which is computationally very costly. Therefore, the cost function $\mathcal{J}(\mathbf{W})$ is approximated using a nonquadratic function G:

$$\mathcal{J}(\mathbf{W}) \propto (\mathrm{E}[\mathrm{G}(\mathbf{Y})] - \mathrm{E}[\mathrm{G}(\mathbf{Y}_{\mathrm{Gauss}})])^2. \tag{12.11}$$

Using the approximation of the negentropy from Eq. (12.11), the update rule for the unmixing matrix $\mathbf{W}$ in the i-th iteration can be derived as [1]

$$\tilde{\mathbf{W}}_i = \langle \mathrm{g}(\mathbf{Y})\mathbf{X}^H \rangle - \Lambda \mathbf{W}_{i-1}, \tag{12.12}$$

$$\mathbf{W}_i = \left(\tilde{\mathbf{W}}_i \tilde{\mathbf{W}}_i^T\right)^{-1/2} \tilde{\mathbf{W}}_i \tag{12.13}$$

where Λ is a diagonal matrix with $\lambda_{ii} = \langle \mathrm{g}'(Y_i) \rangle$ and $\langle \cdot \rangle$ denotes the mean value. As for the function $\mathrm{g}(\cdot)$, the derivative of the function $\mathrm{G}(\cdot)$ from the Eq. (12.10) will be chosen, so setting

$$\mathrm{G}(x) = -\exp\left(-\frac{|x|^2}{2}\right) \tag{12.14}$$

the function $\mathrm{g}(\cdot)$ becomes

$$\mathrm{g}(x) = x \exp\left(-\frac{|x|^2}{2}\right) \tag{12.15}$$

and

$$\mathrm{g}'(x) = (1 - x^2) \exp\left(-\frac{|x|^2}{2}\right). \tag{12.16}$$

Ideally, these approaches will result in independent output signals in each frequency bin.

In order to obtain the complete spectra of all unmixed sources, it is additionally necessary to correctly sort the outputs, since their ordering after solving instantaneous ICA problems for each frequency is arbitrary and may vary from frequency bin to frequency bin. This so-called permutation problem can be solved in a number of ways and will be discussed in the following section.

12.2.3 Permutation Correction

Due to the principle of the ICA algorithms, it is highly unlikely that we will obtain a consistent ordering of the recovered signals for different frequency bins. In the case of frequency domain source separation, this means that the ordering of the outputs may change in each frequency bin. In order to obtain correctly estimated source signals in the time domain, however, all separated frequency bins have to be put in one consistent order. This problem is also known as the permutation problem.

There exist several classes of algorithms giving a solution for the permutation problem. Approaches presented in [4], [19], and [20] try to correct permutations by considering the cross-statistics (such as cross-correlation or cross-cumulants) of the spectral envelopes of adjacent frequency bins. In [21], algorithms were proposed that make use of the spectral distance between neighboring bins and try to make the impulse response of the mixing filters short, which corresponds to smooth transfer functions of the mixing system in the frequency domain. The algorithm proposed by Kamata et al. [22] solves the problem using the continuity in power between adjacent frequency components of the same source. A similar method was presented by Pham et al. [23]. Baumann et al. [24] proposed a solution which works by comparing the directivity patterns resulting from the estimated demixing matrix in each frequency bin. Similar algorithms were presented in [25], [26] and [27]. In [28], it was suggested to use the direction of arrival (DOA) of source signals implicitly estimated by the ICA unmixing matrices **W** for the problem solution. The approach in [30] exploits the continuity of the frequency response of the mixing filter. A similar approach was presented in [31] using the minimum of the L_2-norm of the resulting mixing filter and in [33] using the minimum distance between the filter coefficients of adjacent frequency bins. In [34], the authors suggest using the cosine between the demixing coefficients of different frequencies as a cost function for the problem solution. Sawada et al. [35] proposed an approach using basis vector clustering of the normalized estimated mixing matrices. In [36] the permutation problem was solved using a maximum likelihood ratio criterion between the adjacent frequency bins.

However, with a growing number of independent components, the complexity of the solution grows. This is true not only because of the factorial increase of permutations to be considered, but also because of the degradation of the ICA performance. Therefore, not all of the approaches mentioned above perform equally well for an increasing number of sources.

In all following work, permutations have been corrected by maximizing the likelihood-ratio criterion described in [36]. The correction algorithm from [36] was expanded for the case of more than two extracted channels. In order to solve the permutation problem, for each frequency bin a correction matrix $\hat{\mathbf{P}}(\Omega)$,

$$\hat{\mathbf{P}}(\Omega) \stackrel{!}{=} \arg \min_{k=1\ldots K} \prod_{i,j \in \left\{ \mathbf{P}^k_{ij}(\Omega)=1 \right\}} \gamma_{ij}(\Omega), \tag{12.17}$$

has to be found, where $\mathbf{P}^k(\Omega)$ is the k-th among K possible permutation matrices, the parameter $\gamma_{ij}(\Omega)$ is

$$\gamma_{ij}(\Omega) = \frac{1}{T} \sum_{\tau} \frac{|Y_i(\Omega,\tau)|}{\beta_j(\tau)} \tag{12.18}$$

and

$$\beta_j(\tau) = \frac{1}{N}\sum_{\Omega}\left|Y_j(\Omega,\tau)\right|. \tag{12.19}$$

In this case, $\beta_j(\tau)$ is a scaling parameter of the signal envelope. $\beta_j(\tau)$ allows us to consider the scaling of signals in permutation correction, so that the likelihood of an unmixed source at a given frequency will be weighted with the averaged magnitude of the current frame.

12.2.4 Postmasking

Even after source separation, in the majority of real-world cases, the extracted independent components are still corrupted by residual noise and interference, especially in reverberant environments. The residual disturbances are assumed to be a superposition of the other independent components and the background noise. Therefore, the quality of the recovered source signals often leaves room for improvement, which can be attained in a wide range of scenarios by applying a soft time-frequency-mask to the ICA outputs. While an ideal postprocessing function is impossible to obtain realistically, approximations to it are already advantageous. One such approximation, mask estimation based on ICA results, has been proposed and shown to be successful, for both binary and soft masks; see, e.g., [35, 37, 40].

In this section, a brief review of four types of time-frequency masking algorithms, namely

- amplitude-based masks
- phase-based masks
- interference-based masks and
- two-stage noise suppression algorithm-based masks

will be given. The data flow of the whole application is shown in Figure 12.1.

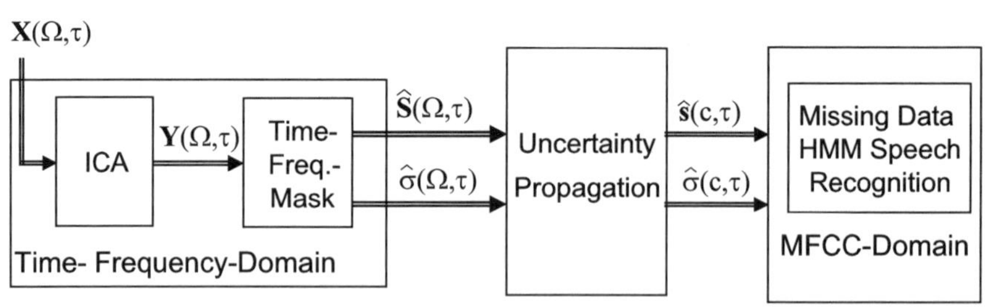

Fig. 12.1: Block diagram with data flow

12.2.4.1 Amplitude-Based Mask Estimation

The time-frequency mask is calculated by comparing the local amplitude ratios between the output signal $Y_i(\Omega,\tau)$ and all other $Y_j(\Omega,\tau)$. With an additional sigmoid transition point T, the mask can be used to block all time-frequency components of Y_i which are not at least T dB above all other estimated sources in that time-frequency point. This corresponds to computing [37]

$$M_i(\Omega,\tau) = \Psi\left(\log\left(|Y_i(\Omega,\tau)|^2\right) - \max_{\forall j \neq i} \log\left(|Y_j(\Omega,\tau)|^2\right) - \frac{T}{10}\right) \tag{12.20}$$

and applying a sigmoid nonlinearity Ψ defined by

$$\Psi(x) = \frac{1}{1+\exp(-x)}. \tag{12.21}$$

An example of the resulting masks is shown in Fig. 12.2.

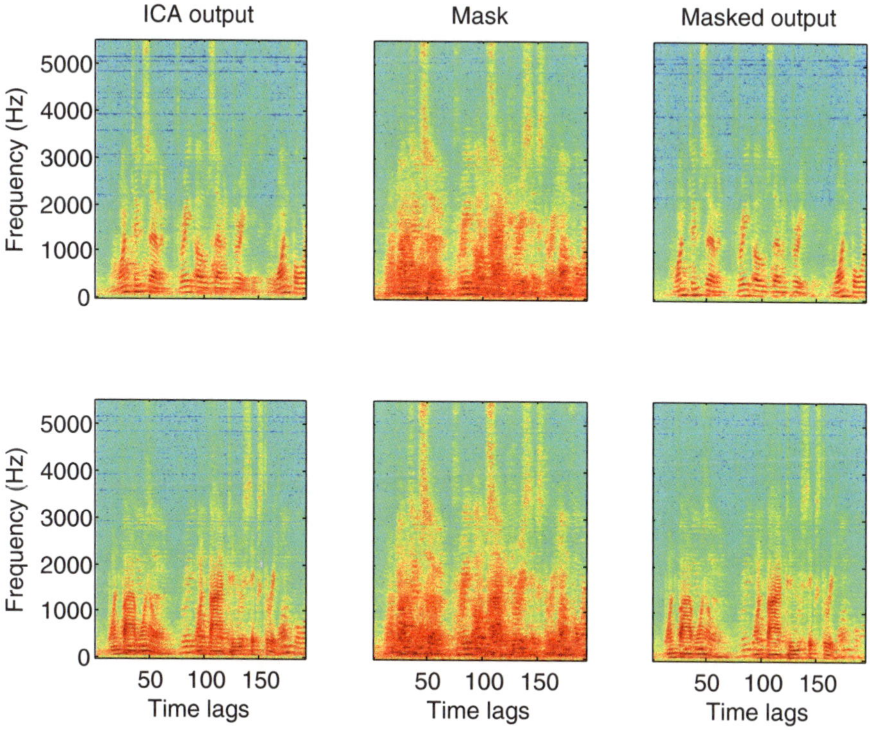

Fig. 12.2: Effect of amplitude mask for the case of $M = N = 2$. The spectrograms of 1) the output signals $Y_1(\Omega,\tau)$ and $Y_2(\Omega,\tau)$ obtained only with ICA (left column), 2) the estimated masks $M_1(\Omega,\tau)$ and $M_2(\Omega,\tau)$ (middle column), and 3) the output signals $\hat{S}_1(\Omega,\tau)$ and $\hat{S}_2(\Omega,\tau)$ obtained by a combination of ICA and T-F masking calculated with Eq. (12.20)

12.2.4.2 Phase Angle-Based Mask Estimation

The source separation performance of ICA can also be seen from a beamforming perspective. When the unmixing filters learned by ICA are viewed as frequency-variant beamformers, it can be shown that successful ICA effectively places zeros in the directions of all interfering sources [39]. Therefore, the zero directions of the unmixing filters should be indicative of all source directions. Thus, when the local direction of arrival (DOA) is estimated from the phase of any one given time-frequency bin, this should give an indication of the dominant source in this bin. This is the principle underlying phase-based time-frequency masking strategies.

Phase-based post-masking of ICA outputs was introduced in [35]. This method considers closeness of the phase angle $\vartheta_i(\Omega,\tau)$ between a column of the mixing matrix $\mathbf{a}_i(\Omega)$ and the observed signal $\mathbf{X}(\Omega,\tau)$ calculated in the whitened space, with $\mathbf{V}(\Omega) = \mathbf{R}^{-1/2}(\Omega)$ as the whitening matrix that is obtained from the signal autocorrelation $\mathbf{R}(\Omega) = \langle \mathbf{X}(\Omega,\tau)\mathbf{X}(\Omega,\tau)^H \rangle$. ^{H} denotes the conjugate or Hermitian transpose. The phase angle is given by

$$\vartheta_i(\Omega,\tau) = \arccos \frac{\left|\mathbf{b}_i^H(\Omega)\mathbf{Z}(\Omega,\tau)\right|}{\|\mathbf{b}_i(\Omega)\| \, \|\mathbf{Z}(\Omega,\tau)\|}, \tag{12.22}$$

where $\mathbf{Z}(\Omega,\tau) = \mathbf{V}(\Omega)\mathbf{X}(\Omega,\tau)$ are whitened samples and $\mathbf{b}_i(\Omega) = \mathbf{V}(\Omega)\mathbf{a}_i(\Omega)$ is the basis vector i in the whitened space. Then, the mask is calculated by

$$M_i(\Omega,\tau) = \frac{1}{1+\exp(g(\vartheta_i(\Omega,\tau) - \vartheta_T))} \tag{12.23}$$

where ϑ_T and g are parameters specifying the sigmoid transition point and steepness, respectively. Figure 12.3 shows exemplary results.

12.2.4.3 Interference-Based Mask Estimation

Interference-based mask estimation is introduced in detail in [40]. The main idea is to detect the time-frequency points in the separated signals where the source signal and the interference are dominant, assuming them to be sparse in the time-frequency domain. The mask is estimated by

$$M_i(\Omega,\tau) = \frac{1}{1+\exp(g(\tilde{\mathbf{S}}_i(\Omega,\tau) - \lambda_s))} \times \left(1 - \frac{1}{1+\exp(g(\tilde{\mathbf{N}}_i(\Omega,\tau) - \lambda_n))}\right), \tag{12.24}$$

where λ_s, λ_n and g are parameters specifying the threshold points and the steepness of the sigmoid function and $\tilde{\mathbf{S}}_i(\Omega,\tau)$ and $\tilde{\mathbf{N}}_i(\Omega,\tau)$ are speech and noise dominance measures given by

$$\tilde{\mathbf{S}}_i(\Omega,\tau,R_\Omega,R_\tau) = \frac{\left\|\Phi(\Omega,\tau,R_\Omega,R_\tau)(Y_i(\Omega,\tau) - \sum_{m\neq i} Y_m(\Omega,\tau))\right\|}{\left\|\Phi(\Omega,\tau,R_\Omega,R_\tau)\sum_{m\neq i} Y_m(\Omega,\tau)\right\|} \tag{12.25}$$

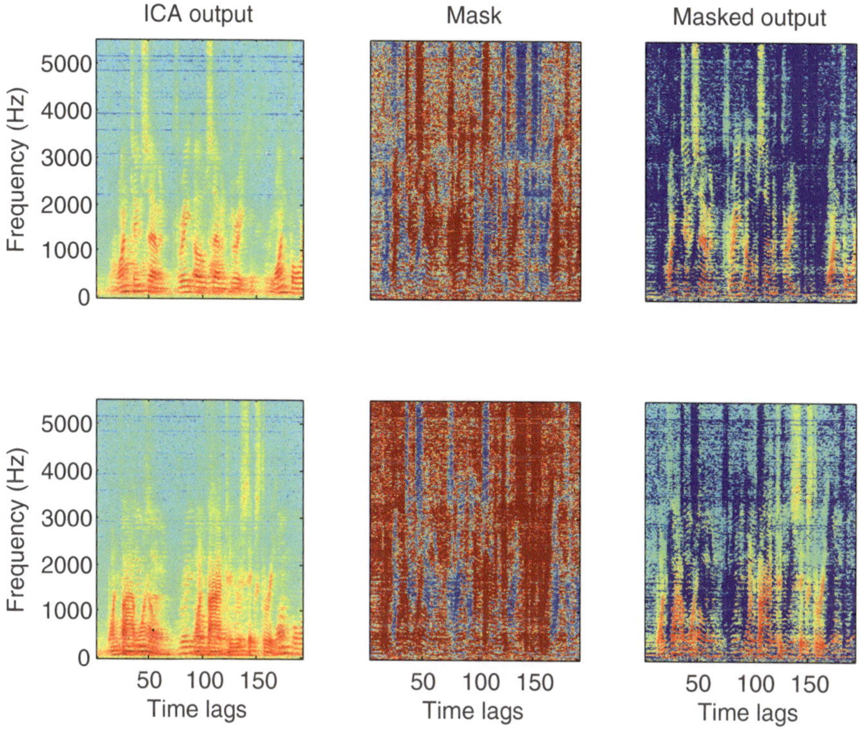

Fig. 12.3: Effect of phase mask for the case of $M = N = 2$. The spectrograms of 1) the output signals $Y_1(\Omega,\tau)$ and $Y_2(\Omega,\tau)$ obtained only with ICA (left column), 2) the estimated masks $M_1(\Omega,\tau)$ and $M_2(\Omega,\tau)$ (middle column), and 3) the output signals $\hat{S}_1(\Omega,\tau)$ and $\hat{S}_2(\Omega,\tau)$ obtained by a combination of ICA and T-F masking calculated with Eq. (12.23)

and

$$\tilde{\mathbf{N}}_i(\Omega,\tau,R_\Omega,R_\tau) = \frac{\left\|\Phi(\Omega,\tau,R_\Omega,R_\tau)(Y_i(\Omega,\tau) - \sum_{m\neq i} Y_m(\Omega,\tau))\right\|}{\|\Phi(\Omega,\tau,R_\Omega,R_\tau)Y_i(\Omega,\tau)\|}. \tag{12.26}$$

Here, $\|\cdot\|$ denotes the Euclidean norm operator and

$$\Phi(\Omega,\tau,R_\Omega,R_\tau) = \begin{cases} \mathcal{W}(\Omega-\Omega_0,\tau-\tau_0,R_\Omega,R_\tau), & |\Omega-\Omega_0| \leq R_\Omega, |\tau-\tau_0| \leq R_\tau, \\ 0 & \text{otherwise} \end{cases} \tag{12.27}$$

utilizes a two-dimensional window function $\mathcal{W}(\Omega-\Omega_0,\tau-\tau_0,R_\Omega,R_\tau)$ of the size $R_\Omega \times R_\tau$ (e.g., a two-dimensional Hanning window) [40]. This mask tends to result in a very strong suppression of interferences, as can be gleaned from its visual impression in Fig. 12.4.

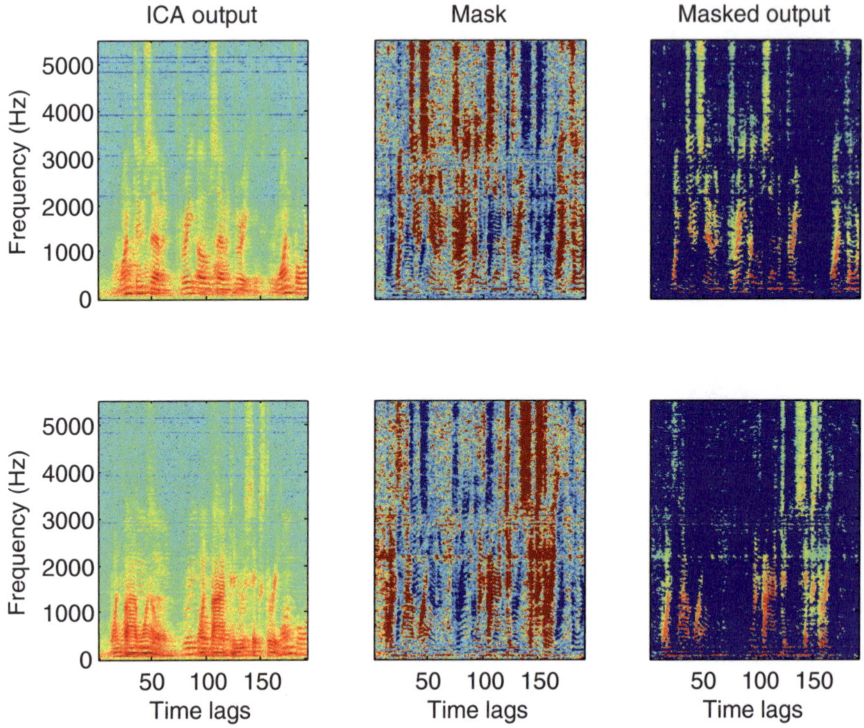

Fig. 12.4: Effect of interference mask for the case of $M = N = 2$. The spectrograms of 1) the output signals $Y_1(\Omega,\tau)$ and $Y_2(\Omega,\tau)$ obtained only with ICA (left column), 2) the estimated masks $M_1(\Omega,\tau)$ and $M_2(\Omega,\tau)$ (middle column), and 3) the output signals $\hat{S}_1(\Omega,\tau)$ and $\hat{S}_2(\Omega,\tau)$ obtained by a combination of ICA and T-F masking calculated with Eq. (12.24)

12.2.4.4 Two-Stage Noise Suppression

As an alternative criterion for masking, residual interference in the signal may be estimated and the mask may be computed as an MMSE estimator of the clean signal. For this purpose, the following signal model is assumed:

$$\mathbf{Y}(\Omega,\tau) = \mathbf{S}(\Omega,\tau) + \mathbf{N}(\Omega,\tau), \tag{12.28}$$

where the clean signal $\mathbf{S}(\Omega,\tau)$ is corrupted by a noise component $\mathbf{N}(\Omega,\tau)$, the remaining sum of the interfering signals and the background noise. The estimated clean signals are obtained by

$$\hat{\mathbf{S}}(\Omega,\tau) = \mathbf{M}_{SE}(\Omega,\tau)\mathbf{Y}(\Omega,\tau), \tag{12.29}$$

where $\mathbf{M}_{SE}(\Omega,\tau)$ is the amplitude estimator gain. For the calculation of the gain $\mathbf{M}_{SE}(\Omega,\tau)$ in Eq. (12.34), different speech enhancement algorithms can be used. In [42] the method by McAulay and Malpass [43] has been used. In the following, we use the log spectral amplitude estimator (LSA) as proposed by Ephraim and Malah [38].

In case of the LSA estimator, first the a posteriori SNR $\gamma_i(\Omega,\tau)$ and the a priori SNR $\xi_i(\Omega,\tau)$ are defined by

$$\gamma_i(\Omega,\tau) = \frac{|Y_i(\Omega,\tau)|^2}{\lambda_{D,i}(\Omega,\tau)} \tag{12.30}$$

and

$$\xi_i(\Omega,\tau) = \alpha\xi_i(\Omega,\tau-1) + (1-\alpha)\frac{\lambda_{X,i}(\Omega,\tau)}{\lambda_{D,i}(\Omega,\tau)}, \tag{12.31}$$

where α is a smoothing parameter that controls the trade-off between the noise reduction and the transient distortions [45], $Y_i(\Omega,\tau)$ is the i-th ICA output, $\lambda_{D,i}(\Omega,\tau)$ is the noise power and $\lambda_{X,i}(\Omega,\tau)$ is the approximate clean signal power. With these parameters, the log spectral amplitude estimator is given by

$$\mathbf{M}_{SE}(\Omega,\tau) = \frac{\xi(\Omega,\tau)}{1+\xi(\Omega,\tau)} \exp\left(\int_{t=\nu(\Omega,\tau)}^{\infty} \frac{e^{-t}}{t} dt\right) \tag{12.32}$$

and

$$\nu(\Omega,\tau) = \left(\frac{\xi(\Omega,\tau)}{1+\xi(\Omega,\tau)}\right)\gamma(\Omega,\tau), \tag{12.33}$$

with $\xi(\Omega,\tau)$ denoting the local a priori SNR.

According to [44], this approach can be generalized by using additional information for calculation of speech presence probabilities. The speech presence probability $\mathbf{p}(\Omega,\tau)$ can then be used to modify the spectral gain function

$$\mathbf{M}(\Omega,\tau) = \mathbf{M}_{SE}(\Omega,\tau)^{\mathbf{p}(\Omega,\tau)}\mathbf{G}_{min}^{(1-\mathbf{p}(\Omega,\tau))}, \tag{12.34}$$

where $\mathbf{G}_{min}$ is a spectral floor constant [44, 45]. Since the probability functions are not known, the masks from Sections 12.2.4.1–12.2.4.3 can be used at this point as an approximation. Considering $\mathbf{p}(\Omega,\tau) = \mathbf{M}(\Omega,\tau)$ from Eq. (12.24) as the approximate speech presence probability, we estimate the noise power $\lambda_{D,i}(\Omega,\tau)$ as

$$\begin{aligned}\lambda_{D,i}(\Omega,\tau) = {} & p_i(\Omega,\tau)\lambda_{D,i}(\Omega,\tau-1) \\ & + (1-p_i(\Omega,\tau))\left[\alpha_D\lambda_{D,i}(\Omega,\tau-1) + (1-\alpha_D)|Y_i(\Omega,\tau)|^2\right]\end{aligned} \tag{12.35}$$

with α_D a smoothing parameter and the approximate clean signal power $\lambda_{X,i}(\Omega,\tau)$,

$$\lambda_{X,i}(\Omega,\tau) = (|Y_i(\Omega,\tau)|\, p_i(\Omega,\tau))^2. \tag{12.36}$$

The effect of the two-stage mask is again a strong interference suppression; however, the spectral distortion is reduced compared to that of the interference mask. This can also be observed from the associated spectrographic representation in Fig. 12.5.

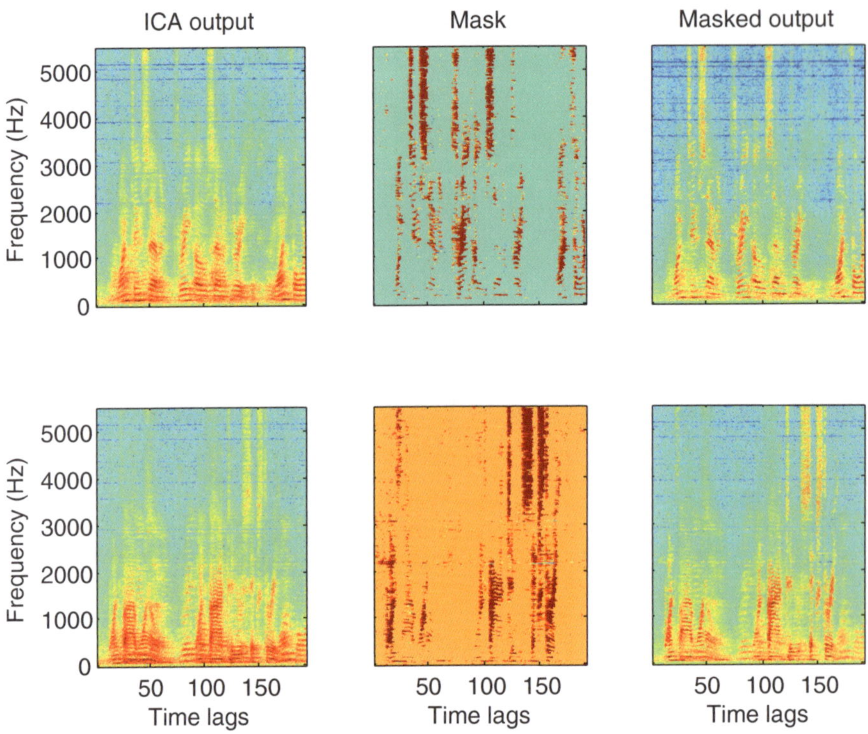

Fig. 12.5: Effect of two-stage noise suppression for the case of $M = N = 2$. The spectrograms of 1) the output signals $Y_1(\Omega,\tau)$ and $Y_2(\Omega,\tau)$ obtained only with ICA (left column), 2) the estimated masks $M_1(\Omega,\tau)$ and $M_2(\Omega,\tau)$ (middle column), and 3) the output signals $\hat{S}_1(\Omega,\tau)$ and $\hat{S}_2(\Omega,\tau)$ obtained by a combination of ICA and T-F masking calculated with Eq. (12.34)

12.3 Uncertainty Estimation

Because of the use of time-frequency masking, a part of the information of the original signal is often eliminated along with the interfering sources. To compensate for this lack of information, each masked estimated source is considered as uncertain and described in the form of a posterior distribution of each Fourier coefficient of the clean signal $S_k(\Omega,\tau)$ given the available information, as described in more detail in [6].

Estimating the uncertainty in the spectrum domain has clear advantages when contrasted with uncertainty estimation in the domain of speech recognition, since much intermediate information about the signal and noise processes as well as the mask is known in this phase of signal processing, but is generally not available in the further steps of feature extraction. This has motivated a number of studies on spectrum domain uncertainty estimation, most recently, for example, [47] and [48]. In contrast to other methods, the suggested strategy possesses two advantages: It does not need a detailed spectrum domain speech prior, which may require a large number of components or may incur the need for adaptation to the speaker and environment; and it gives a computationally very inexpensive approximation that is applicable to both binary and soft masks.

The model used here for this purpose is the complex Gaussian uncertainty model [50]:

$$p(S_k(\Omega,\tau)|\hat{S}_k(\Omega,\tau)) = \frac{1}{\pi\sigma^2(\Omega,\tau)} \exp\left(-\frac{|S_k(\Omega,\tau)-\hat{S}_k(\Omega,\tau)|^2}{\sigma^2(\Omega,\tau)}\right), \tag{12.37}$$

where the mean is set equal to the Fourier coefficient obtained from post-masking $\hat{S}_k(\Omega,\tau)$ and the variance $\sigma^2(\Omega,\tau)$ represents the lack of information, or uncertainty. In order to determine σ^2, two alternative procedures were used.

12.3.1 Ideal Uncertainties

Ideal uncertainties describe the squared difference between the true and the estimated signal magnitude. They are computed as

$$\sigma_T^2 = \left||S_k(\Omega,\tau)| - |\hat{S}_k(\Omega,\tau)|\right|^2, \tag{12.38}$$

where S_k is the reference signal. However, these ideal uncertainties are available only in experiments where a reference signal has been recorded. Thus, the ideal results may only serve as a perspective of what the suggested method would be capable of if a very high quality error estimate were already available.

12.3.2 Masking Error Estimate

In practice, it is necessary to approximate the ideal uncertainty estimate using values that are actually available. Since much of the estimation error is due to the time-frequency mask, in further experiments such a masking error was used as the single basis of the uncertainty measure.

This uncertainty due to masking can be computed by

$$\sigma_E^2 = \alpha\left||\hat{S}_k(\Omega,\tau)| - |Y_k(\Omega,\tau)|\right|^2. \tag{12.39}$$

If $\alpha = 1$, this error estimate would assume that the time-frequency mask leads to missing signal information with 100% certainty. The value should be lower to reflect the fact that some of the masked time-frequency bins contain no target speech information at all. To obtain the most suitable value for α, the following expression was minimized:

$$\alpha = \arg\min_{\tilde{\alpha}} (\sigma_E(\tilde{\alpha}) - \sigma_T)^2. \tag{12.40}$$

In order to avoid adapting parameters to each of the test signals and masks, this minimization was carried out only once and only for a mixture not used in testing, at which point stereo data was also necessary in order to compute σ_T according to (12.38). After averaging over all mask types, the same value of α was used in all experiments and for all datasets. This optimal value was $\alpha = 0.71$.

12.3.3 Uncertainty Propagation

Once the clean speech features and their uncertainties have been estimated in the STFT domain, the uncertain features need to be made available in that feature domain where speech recognition takes place. In all subsequent experiments and results, this ASR feature domain was the mel frequency cepstrum.

Therefore, after uncertainty estimation, an additional step of uncertainty propagation was necessary, as is also shown in Fig. 12.1. For this purpose, the estimated speech signal $\hat{S}_k(\Omega, \tau)$ and its variance $\sigma^2(\Omega, \tau)$ are interpreted as mean and variance of a complex Gaussian-distributed random variable. Then, the effect that subsequent MFCC feature extraction stages have on these random variables can be determined. This uncertainty propagation was carried out as described in detail in Chapter 3, and its outputs are the approximate mean and variance of the uncertain speech features after they have been nonlinearly transformed to the mel frequency cepstrum domain.

12.4 Experiments and Results

12.4.1 Recording Conditions

For the evaluation of the proposed approaches, different real room recordings were used. In these recordings, audio files from the TIDigits database [46] were used and mixtures with up to three speakers were recorded in a mildly reverberant ($T_R \approx$ 160 ms) and noisy lab room at TU Berlin. The distances L_i between loudspeakers and microphones were varied between 0.9 and 3 m.

The setup is shown schematically in Figure 12.6 and the experimental conditions are summarized in Table 12.1.

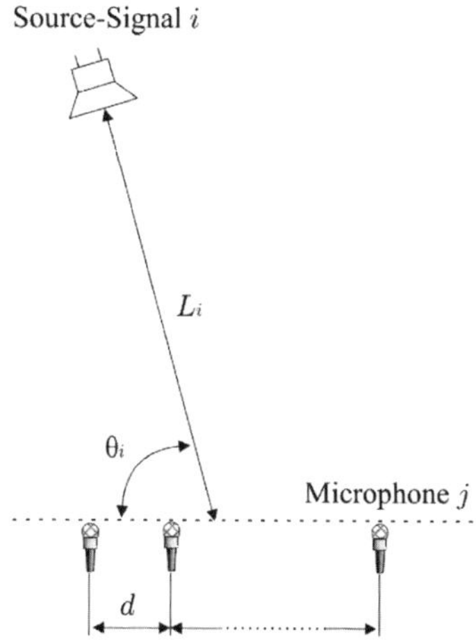

Fig. 12.6: Experimental setup used in recording of mixtures. The distance d between adjacent microphones was 3 cm for all recordings

Table 12.1: Mixture description

Mixture	Mix. 1	Mix. 2	Mix. 3	Mix. 4	Mix. 5
Number of speakers N	2	3	2	2	3
Number of microphones M	2	3	2	2	3
Speaker Codes	ar,ed	pg,ed,cp	fm,pg	cp,ed	fm,ga,ed
Distance between speaker i and array center	$L_1 = L_2 =$ 2.0 m	$L_1 = L_2 =$ $L_3 = 0.9$ m	$L_1 = 1.0$ m $L_2 = 3.0$ m	$L_1 = L_2 =$ 0.9 m	$L_1 = L_2 =$ $L_3 = 0.9$ m
Angular position of the speaker i (as shown in Fig. 12.6)	$\theta_1 = 75°$ $\theta_2 = 165°$	$\theta_1 = 30°$ $\theta_2 = 80°$ $\theta_3 = 135°$	$\theta_1 = 50°$ $\theta_2 = 100°$	$\theta_1 = 50°$ $\theta_2 = 115°$	$\theta_1 = 40°$ $\theta_2 = 60°$ $\theta_3 = 105°$

12.4.2 Parameter Settings

The algorithms were tested on the room recordings, which were first transformed to the frequency domain at a resolution of $N_{\mathrm{FFT}} = 1,024$. For calculating the STFT, the signals were divided into overlapping frames using a Hanning window with an overlap of $3/4 \cdot N_{\mathrm{FFT}}$. For the BSS, the FastICA algorithm (Eqs. (12.12)–(12.13)) with the nonlinearity $g(\cdot)$ from Eq. (12.15) and $g'(\cdot)$ from Eq. (12.16) was used.

The parameter settings for the different evaluated time-frequency masks are summarized in Table 12.2.

Table 12.2: Parameter settings

Algorithm	Settings
Amplitude based mask	$T = 0$ dB
Phase angle based mask	$\vartheta_T = 0.3 \cdot \pi$ $g = 50$
Interference based mask	$\lambda_s = 1.3$ $\lambda_n = 1.3$ $g = 20$
Two stage noise suppression	$\alpha = 0.7$ $\alpha_D = 0.3$ $\mathbf{G}_{min} = 0.1$

12.4.3 Performance Measures

For determining of the performance of ICA and time-frequency masking, the signal-to-interference ratio (SIR) was used as a measure of the separation performance and the signal-to-distortion ratio (SDR) as a measure of the signal quality. The SIR improvement ΔSIR is obtained from

$$\Delta\mathrm{SIR}_i = 10\log_{10}\frac{\sum_n y_{i,s_i}^2(n)}{\sum_{j\neq i}\sum_n y_{i,s_j}^2(n)} - 10\log_{10}\frac{\sum_n x_{i,s_i}^2(n)}{\sum_{j\neq i}\sum_n x_{i,s_j}^2(n)}, \tag{12.41}$$

and the SDR is computed by

$$\mathrm{SDR}_i = 10\log_{10}\frac{\sum_n x_{k,s_i}^2(n)}{\sum_n (x_{k,s_i}(n) - \alpha y_{i,s_i}(n-D))^2}, \tag{12.42}$$

where y_{i,s_j} is the i-th separated signal with only the source signal s_j active, and x_{k,s_j} is the observation obtained by microphone k, again when only s_j is active. α and D are parameters for phase and amplitude which are chosen automatically to optimally compensate for the difference between y_{i,s_j} and x_{k,s_j}.

12.4.4 Separation Results

All the mixtures from Table 12.1 were separated with the FastICA algorithm and subsequently the time frequency masking from Sections 12.2.4.1–12.2.4.4 was performed using parameter settings as shown in Section 12.4.2. For each result, the performance was calculated using Eqs. ((12.41))–((12.42)). Table 12.3 shows the results of the applied methods.

As can be seen in Table 12.3, the best SIR improvements were achieved by the two-stage approach. Still, the results of all time-frequency masks depend on the performance of the preceding BSS algorithm, which in turn depends on the experimental setup. As can be seen, the best BSS results were generally achieved when the microphones were placed near the source signals. Thus, given a low ICA performance for large microphone distances (in terms of SIR and SDR), a stronger signal distortion should be expected from subsequent masking as well. Furthermore, the higher the SIR improvement gained with a time-frequency mask, the lower the SDR value tends to become. The consequence of this for speech recognition will be discussed further in Section 12.4.7.

Table 12.3: Experimental results (mean value of output ΔSIR/SDR in dB)

Scenario	none	Amplitude	Phase	Interference	2-stage
Mix. 1	3.48 / 5.13	6.35 / 3.98	4.93 / **4.38**	8.43 / 2.48	**8.57** / 2.84
Mix. 2	9.06 / 4.23	11.99 / **4.10**	13.76 / 3.86	16.88 / 2.68	**17.25** / 2.87
Mix. 3	6.14 / 6.33	11.20 / 5.39	9.11 / **5.88**	14.11 / 3.78	**14.14** / 4.14
Mix. 4	8.24 / 8.68	14.56 / 7.45	11.32 / **7.91**	**19.04** / 4.88	18.89 / 5.33
Mix. 5	3.93 / 2.92	5.24 / 2.41	6.70 / **2.66**	9.31 / 0.84	**9.55** / 1.11

12.4.5 Model Training

The HMM speech recognizer was trained with the HTK toolkit [51]. The HMMs were trained at the phoneme-level with six-component mixture of Gaussian emitting probabilities and a conventional left-right structure. The training data was mixed and comprised of the 114 speakers of the TI-DIGITS clean speech database along with the room recordings for speakers sa and rk used for adaptation. The speakers that had been used for adaptation were removed from the test set. The features were mel frequency cepstral coefficients (MFCCs) with deltas and accelerations, which were postprocessed with cepstral mean subtraction (CMS) for further reduction of convolutive effects.

12.4.6 Recognition of Uncertain Data

In the following experiments, the clean cepstrum domain speech features $s_k(c,\tau)$ are assumed to be unavailable, with only an estimate $\hat{s}_k(c,\tau)$ and its associated uncertainty or variance $\sigma^2(c,\tau)$ as the available data according to Fig. 12.1.

In the recognition tests, we compare three strategies to deal with this uncertainty.

- All estimated features $\hat{s}_k(c,\tau)$ are treated as reliable observations and recognized by conventional likelihood evaluation. This is labeled *no Uncertainty* in all following tables.

- Uncertainty decoding is used, as described in [16]. This will be labeled *Uncertainty Decoding (UD).*
- Modified imputation according to [41] is employed, which will be denoted by *Modified Imputation (MI).*

The implementation that was used for the experiments with both considered uncertainty-of-observation techniques is also described in more detail in Chapter 13 of this book.

12.4.7 Recognition Results

Table 12.4 shows the baseline result, attained after some adaptation to the reverberant room environment, as well as the word error rate on the noisy mixtures and on the ICA output signals. Here, the word error rate is computed via

$$\mathrm{WER} = 100\frac{D+S+I}{N}, \tag{12.43}$$

with D as the number of deletions, S as the number of substitutions, I as the number of insertions, and N as the number of reference output tokens; and error rates are computed over all five scenarios.

Table 12.4: Word error rate (WER) for reverberant data, noisy mixtures and ICA results

	reverberant data	mixtures	ICA output
WER	9.28	72.55	26.34

The recognition results in Tables 12.5 and 12.7 are achieved with true squared errors used as uncertainties. As can be seen here, all considered masks lead to a greatly reduced average word error rate under these conditions. However, since only uncertainties estimated from the actual signal should be considered, Tables 12.6 and 12.8 show the error rate reductions that can easily be attained in practice by setting the uncertainty to a realistically available estimate as described in Section 12.3.2.

In each of the tables, the numbers in parentheses give the word error rate reduction relative to that of the ICA outputs, which are achieved by including time-frequency masking with observation uncertainties.

It is visible from the results that modified imputation clearly tends to give the best results for true uncertainties, whereas uncertainty decoding is the superior strategy for the estimated uncertainty that was tested here. This is indicative of a high sensitivity of modified imputation to uncertainty estimation errors.

However, since good uncertainty estimation is vital in any case for optimal performance of uncertain feature recognition, it will be interesting to further compare the performance of both uncertainty-of-observation techniques in conjunction with

Table 12.5: Word error rate (WER) for true squared error used as uncertainty. The relative error rate reduction in percent is given in parentheses

	none	2-stage	Phase	Amplitude	Interference
Mixtures	72.55	n.a.	n.a.	n.a.	n.a.
ICA, no Uncertainty	26.34	55.53	91.79	29.91	96.92
Modified Imputation	n.a.	**11.48 (56.4)**	13.84 (47.5)	16.42 (37.7)	16.80 (36.2)
Uncertainty Decoding	n.a.	12.35 (53.1)	18.59 (29.4)	16.50 (37.4)	22.20 (15.7)

Table 12.6: Word error rate (WER) for estimated uncertainty. The relative error rate reduction in percent is given in parentheses

	none	2-stage	Phase	Amplitude	Interference
ICA, no Uncertainty	26.34	55.53	91.79	29.91	96.92
Modified Imputation	n.a.	24.78 (5.9)	20.87 (20.8)	26.53 (-0.01)	23.22 (11.9)
Uncertainty Decoding	n.a.	20.79 (21.1)	**19.95 (24.3)**	23.41 (11.1)	21.55 (18.2)

more precise uncertainty estimation techniques, which are an important target for future work.

As for mask performance, the lowest word error rate with estimated uncertainty values is quite clearly achieved for the phase masking strategy. This corresponds well to the high SDR that has been achieved with this strategy in Table 12.3.

On the other hand, the lowest word error rate for ideal uncertainties is almost always reached using the two-stage mask. Again, it is possible to draw a conclusion by comparing with Table 12.3, which now shows that best performance is apparently possible when the largest interference suppression is reached, i.e., when the ΔSIR takes on its largest values.

Also, a more detailed analysis of results is provided in Tables 12.7 and 12.8, which each show the word error rates separately for each mixture recording. Here, it can be seen how the quality of source separation influences the overall performance gain from the suggested approach. For the lower-quality separation observable in mixtures #1 and #5, the relative performance gains are clearly lower than average, especially for estimated uncertainties. In contrast, the mean performance improvement for the best separated mixtures #2 and #4 is 36.9% for estimated uncertainties with uncertainty decoding and phase masking, and 61.6% for true squared errors with uncertainty decoding and the two-stage mask.

A special case is presented by mixture #3. Here, the separation results are comparatively good; however, due to the large microphone distance of 3 m, the recognition performance is not ideal. Thus, the rather small performance improvement for estimated uncertainties in this case can also be understood from the fact that much of the recognition error is likely due to mismatched conditions, and hence cannot be expected to be overly influenced by uncertainties derived only from the value of the time-frequency mask.

Table 12.7: Detailed results (WER) for true squared error used as uncertainty

Algorithm	none	Amplitude	Phase	Interference	2-stage
Mix. 1					
no Unc.	31.54	34.65	92.12	96.89	53.73
MI		21.16	**13.90**	18.26	15.77
UD		22.20	20.54	22.61	17.84
Mix. 2					
no Unc.	18.38	18.86	94.45	96.83	48.65
MI		11.57	9.35	11.25	8.87
UD		11.89	10.78	15.21	**7.45**
Mix. 3					
no Unc.	29.93	33.67	88.03	96.76	52.62
MI		18.45	17.71	21.45	**12.22**
UD		16.21	23.94	25.94	14.71
Mix. 4					
no Unc.	14.73	22.86	90.77	97.14	55.60
MI		10.99	9.01	9.45	6.37
UD		9.89	10.33	13.85	**5.27**
Mix. 5					
no Unc.	35.95	39.58	91.99	96.98	65.11
MI		20.09	19.03	23.26	**13.90**
UD		21.45	27.04	32.02	16.47

12.5 Conclusions

We have discussed the application of ICA for the recognition of multiple, overlapping speech signals, which have been recorded with distant-talking microphones in noisy and mildly reverberant environments. Independent component analysis can segregate these multiple sources also in such realistic environments, and can thus lead to significantly improved robustness of automatic speech recognition.

In order to gain more performance even when the ICA outputs still contain residual interferences, the use of time-frequency masking has proved beneficial. However, it improves results only in conjunction with uncertainty-of-observation techniques, in which case a further 24% relative reduction of word error rate has been shown possible, on average, for datasets with two or three simultaneously active speakers.

Even greater error rate reductions, about 39% when averaged over all tested time-frequency masks and decoding strategies, have been achieved for ideal uncertainties, i.e., when the true squared estimation error is utilized as the feature uncertainty. This indicates the need for further work on reliable uncertainty estimation as a step for greater robustness with respect to highly instationary noise and interference. This should also include an automatized parameter adaptation for the uncertainty compensation, e.g., via an EM-style unsupervised adaptation.

Another important target for further work is the source separation itself. As the presented experimental results have shown, overall performance both of time-

Table 12.8: Detailed results (WER) for estimated uncertainty

Algorithm	none	Amplitude	Phase	Interference	2-stage
Mix. 1					
no Unc.	31.54	34.65	92.12	96.89	53.73
MI		32.99	24.90	27.80	29.25
UD		27.59	**23.86**	24.48	24.07
Mix. 2					
no Unc.	18.38	18.86	94.45	96.83	48.65
MI		19.81	14.58	16.32	18.70
UD		16.16	**11.89**	13.79	12.36
Mix. 3					
no Unc.	29.93	33.67	88.03	96.76	52.62
MI		32.92	**24.19**	32.67	35.91
UD		27.68	26.18	27.93	27.68
Mix. 4					
no Unc.	14.73	22.86	90.77	97.14	55.60
MI		15.16	11.21	11.21	11.87
UD		12.75	**9.01**	10.55	9.45
Mix. 5					
no Unc.	35.95	39.58	91.99	96.98	65.11
MI		32.18	**28.55**	29.00	29.46
UD		32.02	**28.55**	30.51	30.06

frequency masking and of subsequent uncertain recognition depends strongly on the quality of preliminary source separation by ICA. Thus, more successful source separation in strongly reverberant environments would be of great significance for attaining the best overall results from the suggested approach.

References

1. A. Hyvärinen, J. Karhunen, E. Oja, Independent Component Analysis, New York: John Wiley, 2001.
2. A. Mansour and M. Kawamoto, "ICA papers classified according to their applications and performances," in *IEICA Trans. Fundamentals*, vol. E86-A, no. 3, pp. 620–633, March 2003.
3. M. S. Pedersen, J. Larsen, U. Kjems and L. C. Parra, "Convolutive blind source separation methods", in *Springer Handbook of Speech Processing and Speech Communication*, pp. 1065–1094, Springer Verlag Berlin Heidelberg, 2008.
4. J. Anemüller and B. Kollmeier, "Amplitude modulation decorrelation for convolutive blind source separation", in *Proc. ICA 2000*, Helsinki, pp. 215–220, 2000.
5. L. Deng, J. Droppo and A. Acero, "Dynamic compensation of HMM variances using the feature enhancement uncertainty computed from a parametric model of speech distortion", in *IEEE Trans. Speech and Audio Processing*, vol. 13, no. 3, pp. 412–421, May 2005.
6. D. Kolossa, R. F. Astudillo, E. Hoffmann and R. Orglmeister, "Independent Component Analysis and Time-Frequency Masking for Speech Recognition in Multitalker Conditions", in *EURASIP J. on Audio, Speech, and Music Processing*, vol. 2010, article ID 651420, 2010.

7. D. Kolossa, A. Klimas and R. Orglmeister, "Separation and robust recognition of noisy, convolutive speech mixtures using time-frequency masking and missing data techniques", in *Proc. Workshop on Applications of Signal Processing to Audio and Acoustics (WASPAA)*, pp. 82–85, Oct. 2005.
8. K. Kumatani, J. McDonough, D. Klakow, P. Garner, and W. Li, "Adaptive beamforming with a maximum negentropy criterion," in *Proc. HSCMA*, 2008.
9. O. Yilmaz and S. Rickard, "Blind separation of speech mixtures via time-frequency masking," *IEEE Trans. Signal Processing*, vol. 52, no. 7, pp. 1830–1847, July 2004.
10. M. Kühne, R. Togneri, and S. Nordholm, "Time-frequency masking: Linking blind source separation and robust speech recognition," in *Speech Recognition, Technologies and Applications*. I-Tech, 2008.
11. G. Brown and M. Cooke, "Computational auditory scene analysis," *Computer Speech and Language*, vol. 8, pp. 297–336, 1994.
12. J. B. Allen and L. R. Rabiner, "A unified approach to short-time Fourier analysis and synthesis," Proc. IEEE, vol. 65, pp. 1558–1564, Nov. 1977.
13. J.-F. Cardoso and A. Souloumiac, "Blind beamforming for non-Gaussian signals," *Radar and Signal Processing, IEEE Proceedings*, F 140(6), pp. 362-370, Dec. 1993.
14. A. Belouchrani, K. Abed Meraim, J.-F. Cardoso and E. Moulines, "A blind source separation technique based on second order statistics," in *EEE Trans. on Signal Processing*, vol. 45(2), pp. 434-444, 1997.
15. A. Bell and T. Sejnowski, "An information-maximization approach to blind separation and blind deconvolution," in *Neural Computation,*, vol. 7, pp. 1129–1159, 1995.
16. L. Deng and J. Droppo and A. Acero, "Dynamic compensation of HMM variances using the feature enhancement uncertainty computed from a parametric model of speech distortion," in *IEEE Trans. Speech and Audio Processing*, vol. 13, pp. 412–421, 2005.
17. A. Hyvärinen and E. Oja. A fast fixed-point algorithm for independent component analysis. in *Neural Computation*, vol. 9, pp. 1483–1492, 1997.
18. T. Kristjansson and B. Frey. Accounting for uncertainty in observations: A new paradigm for robust automatic speech recognition, in *Proc. ICASSP*, 2002.
19. C. Mejuto, A. Dapena and L. Castedo, "Frequency-domain infomax for blind separation of convolutive mixtures", in *Proc. ICA 2000*, pp. 315–320, Helsinki, 2000.
20. N. Murata, S. Ikeda, and A. Ziehe, "An approach to blind source separation based on temporal structure of speech signals," *Neurocomputing*, vol. 41, no. 1–4, pp. 1–24, Oct. 2001.
21. L. Parra, C. Spence and B. De Vries, "Convolutive blind source separation based on multiple decorrelation." in *Proc. IEEE NNSP workshop*, pp. 23–32, Cambridge, UK, 1998.
22. K. Kamata, X. Hu, and H. Kobatake, "A new approach to the permutation problem in frequency domain blind source separation," in *Proc. ICA 2004*, pp. 849–856, Granada, Spain, September 2004.
23. D.-T. Pham, C. Servière, and H. Boumaraf, "Blind separation of speech mixtures based on nonstationarity" in *IEEE Signal Processing and Its Applications*, Proceedings of the Seventh International Symposium, pp. 73–76, 2003.
24. W. Baumann, D. Kolossa and R. Orglmeister, "Maximum likelihood permutation correction for convolutive source separation," in *ICA 2003*, pp. 373–378, 2003.
25. S. Kurita, H. Saruwatari, S. Kajita, K. Takeda, and F. Itakura, "Evaluation of frequency-domain blind signal separation using directivity pattern under reverberant conditions," in *ICASSP2000*, pp. 3140–3143, 2000.
26. M. Ikram and D. Morgan, "A beamforming approach to permutation alignment for multichannel frequency-domain blind speech separation," in *ICASSP02*, pp. 881–884, 2002.
27. N. Mitianoudis and M. Davies, "Permutation alignment for frequency domain ICA using subspace beamforming methods", in *Proc. ICA 2004*, LNCS 3195, pp. 669–676, 2004.
28. H. Sawada, R. Mukai, S. Araki, S. Makino, "A robust approach to the permutation problem of frequency-domain blind source separation," in *Proc. ICASSP*, vol. V, pp. 381–384, Apr. 2003.

29. V. Stouten and H. Van hamme and P. Wambacq, "Application of minimum statistics and minima controlled recursive averaging methods to estimate a cepstral noise model for robust ASR," in *Proc. ICASSP*, vol. 1, May 2006.
30. D.-T. Pham, C. Servière, and H. Boumaraf, "Blind separation of convolutive audio mixtures using nonstationarity," in *Proc. ICA2003*, pp. 981–986, 2003.
31. P. Sudhakar, and R. Gribonval, "A sparsity-based method to solve permutation indeterminacy in frequency-domain convolutive blind source separation," in *Independent Component Analysis and Signal Separation: 8th International Conference, ICA 2009, Proceedings*, Paraty, Brazil, March 2009.
32. M. Van Segbroeck and H. Van hamme, "Robust speech recognition using missing data techniques in the prospect domain and fuzzy masks," in *Proc. ICASSP*, pp. 4393–4396, 2008.
33. W. Baumann, and B.-U. Khler, and D. Kolossa, and R. Orglmeister, "Real time separation of convolutive mixtures." in: *Independent Component Analysis and Blind Signal Separation: 4th International Symposium, ICA 2001, Proceedings*, San Diego, USA, 2001.
34. F. Asano, S. Ikeda, M. Ogawa, H. Asoh, and N. Kitawaki, "Combined approach of array processing and independent component analysis for blind separation of acoustic signals," in *IEEE Trans. Speech Audio Proc.*, vol. 11, no. 3, pp. 204–215, May 2003.
35. H. Sawada, S. Araki, R. Mukai and S. Makino, "Blind extraction of a dominant source from mixtures of many sources using ICA and time-frequency masking," in *ISCAS 2005*, pp. 5882–5885, May 2005.
36. N. Mitianoudis, and M. E. Davies, "Audio source separation of convolutive mixtures." in: *IEEE Transactions on Audio and Speech Processing*, vol 11(5), pp. 489-497, 2003.
37. D. Kolossa and R. Orglmeister, "Nonlinear post-processing for blind speech separation," in *Proc. ICA (LNCS 3195)*, Sep. 2004, pp. 832-839.
38. Y. Ephraim and D. Malah, "Speech enhancement using a minimum mean square error log-spectral amplitude estimator," *IEEE Trans. Acoust., Speech, Signal Processing*, vol. ASSP-33, pp. 443–445, Apr. 1985.
39. S. Araki, S. Makino, Y. Hinamoto, R. Mukai, T. Nishikawa, and H. Saruwatari, "Equivalence between frequency-domain blind source separation and frequency-domain adaptive beamforming for convolutivemixtures," in *EURASIP Journal on Applied Signal Processing*, vol. 11, p. 1157–1166, 2003.
40. E. Hoffmann, D. Kolossa and R. Orglmeister, "A batch algorithm for blind source separation of acoustic signals using ICA and time-frequency masking," in *Proc. ICA (LNCS 4666)*, Sep. 2007, pp. 480–488.
41. D. Kolossa, A. Klimas and R. Orglmeister, "Separation and robust recognition of noisy, convolutive speech mixtures using time-frequency masking and missing data techniques," in *IEEE Workshop on Applications of Signal Processing to Audio and Acoustics (WASPAA)*, pp. 82–85, New Paltz, NY, 2005.
42. E. Hoffmann, D. Kolossa, and R. Orglmeister, "A soft masking strategy based on multichannel speech probability estimation for source separation and robust speech recognition", *In: Proc. WASPAA*, New Paltz, NY, 2007.
43. R. J. McAulay and M. L. Malpass, "Speech enhancement using a soft-decision noise suppression filter," *IEEE Trans. ASSP-28*, pp. 137–145, Apr. 1980.
44. I. Cohen, "On speech Enhancement under signal presence uncertainty," *International Conference on Acoustic and Speech Signal Processing*, pp. 167–170, May 2001.
45. Y. Ephraim and I. Cohen, "Recent advancements in speech enhancement", *The Electrical Engineering Handbook*, CRC Press, 2006.
46. R. G. Leonard, "A database for speaker-independent digit recognition", *Proc. ICASSP 84*, Vol. 3, p. 42.11, 1984.
47. S. Srinivasan and D. Wang, "Transforming binary uncertainties for robust speech recognition", in *IEEE Trans. Audio, Speech and Language Processing, IEEE Transactions on Speech and Audio Processing* vol. 15, pp. 2130–2140, 2007.
48. R. F. Astudillo, D. Kolossa, P. Mandelartz and R. Orglmeister, "An uncertainty propagation approach to robust ASR using the ETSI advanced front-end", *IEEE Journal of Selected Topics in Signal Processing*, vol. 4, pp. 824–833, 2010.

49. G. Brown and D. Wang, "Separation of speech by computational auditory scene analysis", *Speech Enhancement*, eds. J. Benesty, S. Makino and J. Chen, Springer, pp. 371–402, 2005.
50. R. F. Astudillo, D. Kolossa and R. Orglmeister, "Propagation of statistical information through non-linear feature extractions for robust speech recognition", in *Proc. MaxEnt*, 2007.
51. S. Young, G. Evermann, T. Hain, D. Kershaw, G. Moore, J. Odell, D. Ollason, D. Povey, V. Valtchev, P. Woodland, "The HTK Book (for HTK Version 3.4)", Cambridge University Engineering Department, 2006.

Chapter 13
Use of Missing and Unreliable Data for Audiovisual Speech Recognition

Alexander Vorwerk, Steffen Zeiler, Dorothea Kolossa, Ramón Fernandez Astudillo, Dennis Lerch

Abstract Under acoustically distorted conditions, any available video information is especially helpful for increasing recognition robustness. However, an optimal strategy for integrating audio and video information is difficult to find, since both streams may independently suffer from time-varying degrees of distortion. In this chapter, we show how missing-feature techniques for coupled HMMs can help us fuse information from both uncertain information sources. We also focus on the estimation of reliability for the video feature stream, which is obtained from a linear discriminant analysis (LDA) applied to a set of shape- and appearance-based features. The approach has resulted in significant performance improvements under strongly distorted conditions, while, in conjunction with stream weight tuning, being lower-bounded in performance by the best of the two single-stream recognizers under all tested conditions.

Alexander Vorwerk
Electronics and Medical Signal Processing (EMSP), Technische Universität Berlin (TU Berlin), Einsteinufer 17, 10587 Berlin, e-mail: alexander.vorwerk@tu-berlin.de

Steffen Zeiler
EMSP, TU Berlin, e-mail: s.zeiler@ee.tu-berlin.de

Dorothea Kolossa
EMSP, TU Berlin, e-mail: dorothea.kolossa@gmx.de

Ramón Fernandez Astudillo
EMSP, TU Berlin, e-mail: rast79@gmail.com

Dennis Lerch
EMSP, TU Berlin, e-mail: dennis_lerch@gmx.de

D. Kolossa and R. Haeb-Umbach (eds.), *Robust Speech Recognition of Uncertain or Missing Data: Theory and Applications*, DOI 10.1007/978-3-642-21317-5_13,
© Springer-Verlag Berlin Heidelberg 2011

13.1 Introduction

The inclusion of the visual modality can significantly improve the robustness of automatic speech recognition systems. This can be achieved by using HMMs or, more generally, graphical models. Examples of successful applications have been described in [15, 22, 32, 33]. However, both the audio signals and the video images may suffer from noise or artifacts, and the video features are furthermore subject to misdetections of the face and mouth. Thus, for optimal performance, the audiovisual recognition system should be capable of considering each of the modalities as uncertain. This is of special interest here, because unreliable segments of either audio or video data can be compensated for at least in part by the other modality, making missing or uncertain feature recognition especially attractive for multimodal data.

In one such strategy, uncertainty compensation has previously been employed in [36], using uncertainty decoding to deal with unreliable features. This approach is used here as well, and it is compared with both binary uncertainty and *modified imputation*, an uncertain recognition strategy that is proving even more beneficial in the considered context.

This chapter is organized as follows. At first, Section 13.2 will introduce the use of coupled HMMs in audiovisual speech recognition and will describe the changes that are necessary to accommodate missing or uncertain feature components. Next, in Sections 13.3 and 13.4, an overview of strategies for the calculation of audio and video features will be presented. In Section 13.5 an audiovisual recognition system is described. This includes the presentation of the audiovisual speech recognizer JASPER, which will be used for all subsequent experiments. This system is based on and allows for asynchronous streams as long as synchrony is again achieved at word boundaries. Subsequently, the feature extraction and uncertainty estimation are presented. Afterwards, the strategy utilized for multistream missing-feature recognition is described. Section 13.6 deals with an efficient implementation of the presented recognizer and also shows the results for this system on the GRID database, a connected word small-vocabulary audiovisual database. All results and further implications are discussed in Section 13.7.

13.2 Coupled HMMs in Audiovisual Speech Recognition

Using a number of streams of audio and/or video features may significantly increase robustness and performance of automatic speech recognition [30, 33]. In audiovisual speech recognition (AVSR), usually two streams of feature vectors are available, which are denoted by $x_a(t)$ for the acoustical and $x_v(t)$ for the visual feature vector at time t. Both streams are not necessarily synchronized. Technical influences, like differing sampling rates or other recording conditions, may be compensated for by a variety of measures, e.g., synchronous sampling or interpolation. However, other causes of asynchronicities are based in the speech production process itself, in which variable delays between articulator movements and voice production lead to time-varying lags between video and audio. Such lags may have a duration of up to 120 ms, which corresponds to the duration of up to an entire phoneme [27].

In order to allow for such delays, various model designs have been used, e.g., multistream HMMs, coupled HMMs (CHMMs), product HMMs and independent HMMs [32]. These differ especially in the degree of required synchrony between modalities from the one extreme of independent HMMs, where both feature streams can evolve with no coupling whatsoever, to the other extreme of multistream HMMs, in which a state-wise alignment is necessary or at least implicitly assumed. Coupled HMMs allow for both streams to have lags or evolve at different speeds, with the obligation that they must again be synchronous at all word boundaries. Since this introduces some constraints but does not force unachievable frame-by-frame alignment, they present a reasonable compromise and have been used for all the following work.

13.2.1 Two-Stream Coupled HHM for Word Models

In coupled hidden Markov models, both feature vector sequences are retained as separate streams. As generative models, CHMMs can describe the probability of both feature streams jointly as a function of a set of two discrete, hidden state variables, which evolve analogously to the single state variable of a conventional HMM.

Thus, CHMMs have a two-dimensional state $\mathbf{q}$ which is composed of an audio and a video state, q_a and q_v, respectively, as can be seen in Fig. 13.1.

Each sequence of states through the model represents one possible alignment with the sequence of observation vectors. To evaluate the likelihood of such an alignment, each state pairing is connected by a transition probability, and each state is associated with an observation probability distribution.

The transition probability and the observation probability can both be composed from the two marginal HMMs. Then, the coupled transition probability becomes

$$\begin{aligned} \mathrm{p}\Big(q_a(t+1) = j_a, q_v(t+1) = j_v | q_a(t) = i_a, q_v(t) = i_v\Big) \\ = a_a(i_a, j_a) \cdot a_v(i_v, j_v), \end{aligned} \tag{13.1}$$

where $a_a(i_a, j_a)$ and $a_v(i_v, j_v)$ correspond to the transition probabilities of the two marginal HMMs, the audio-only and the video-only single-stream HMMs, respectively.

For the observation probability, both marginal HMMs could equally be composed to form a joint output probability

$$\mathrm{p}(\mathbf{x}|\mathbf{i}) = b_a(x_a|i_a) \cdot b_v(x_v|i_v). \tag{13.2}$$

Here, $b_a(x_a|i_a)$ and $b_v(x_v|i_v)$ denote the output probability distributions for the two single streams. In this case, there is a fairly large degree of independence between the two models, and coupling takes place only insofar as both streams are con-

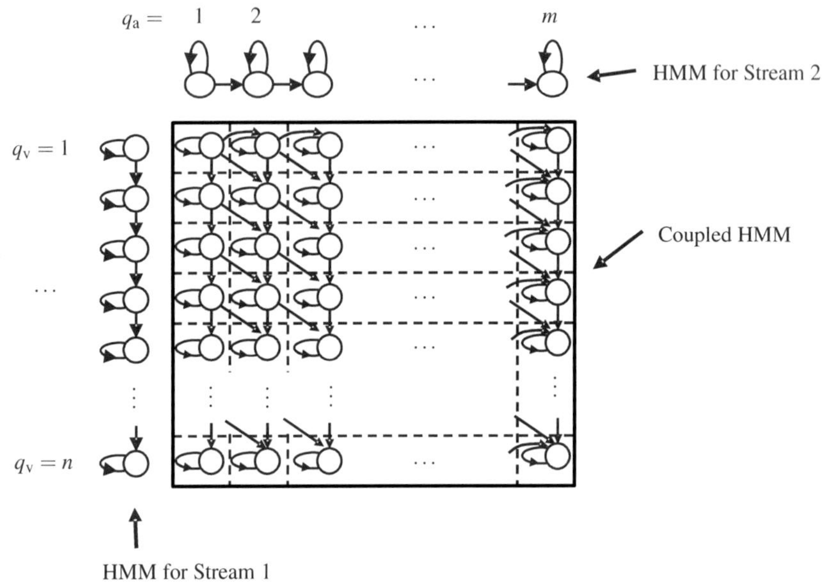

Fig. 13.1: A coupled HMM consists of a matrix of interconnected states, which each correspond to the pairing of one audio and one video HMM state, q_a and q_v, respectively

strained to be generated by the same word model, i.e., stream synchronicity is enforced at word boundaries by the CHMM structure.

However, such a formulation does not allow us to take into account the different reliabilities of audio and video streams. Therefore, Eq. (13.2) is commonly modified by an additional stream weight γ as follows[1]:

$$\mathrm{p}(\mathbf{x}|\mathbf{i}) = b_a(x_a|i_a)^{\gamma} \cdot b_v(x_v|i_v)^{1-\gamma}. \tag{13.3}$$

This approach, described in more detail in [33], is also adopted in the audiovisual speech recognizer (JASPER) that is presented in Section 13.5.4.

13.2.2 Missing Features in Coupled HMMs

Recognition of speech in noisy or otherwise difficult conditions can greatly profit from so-called *missing-feature approaches*. In these methods, those features within a signal that are dominated by noise or overly distorted by other detrimental effects are considered "missing," and subsequently disregarded in speech recognition [5].

[1] Note that the resulting function does not correspond to a probability density as such, since it does not adhere to the normalization condition unless $\gamma = 1$. However, we will still refer to these non-normalized functions with a $\mathrm{p}(\cdot)$ for simplicity of notation.

While binary uncertainties are beneficial in many situations, allowing for real-valued feature uncertainties leads to a Bayesian framework for speech recognition, which can utilize reliability information in a more precise and flexible way, as described in detail in Chapter 2 of this book. This can also allow for feature uncertainties to be estimated in one suitable domain, e.g., in the short-time Fourier transform (STFT) domain, and then propagated to the recognition domain of interest. Appropriate techniques for such an *uncertainty propagation* have been summarized in Chapter 3 and form the basis of the *uncertain feature recognition* we use here as the second alternative.

In the following section, a short overview of the utilized missing and uncertain feature techniques is given and the integration of both uncertain feature streams in a *missing-feature coupled HMM* is described.

13.2.2.1 Missing Feature Theory

In real-world applications, some parts of a speech signal often are occluded by noise or interference, while the visual modality may suffer from varying lighting conditions or misdetected faces or mouths. In these cases, missing feature theory may help the recognizer focus only on features that are considered reliable. Both continuous-valued and binary methods exist to incorporate such uncertainties in the recognition process. In either case, following the notation from Chapter 2, we assume that the clean observation vector $\mathbf{x}(t)$ is not directly observable, but that, rather, only a noisy or otherwise distorted vector $\mathbf{y}(t)$ is accessible.

In the realm of binary uncertainties, two main approaches can be distinguished, marginalization and imputation [39]. In both cases, a binary mask is needed, which marks reliable and unreliable regions in the feature domain. In the following, the uncertainty value for the audio feature stream is denoted by $u_\mathrm{a}(k,t)$ and is determined both feature-wise and frame-wise. Where the video stream is concerned, $u_\mathrm{v}(t)$ denotes a frame-wise uncertainty value.

In a standard HMM with M Gaussian mixture components, for a given state q the output probability is usually computed by

$$\mathrm{b}\big(\mathbf{x}(t)\big) = \sum_{m=1}^{M} w_m \cdot \mathcal{N}\big(\mathbf{x}(t), \boldsymbol{\mu}_{q,m}, \boldsymbol{\Sigma}_{q,m}\big) \tag{13.4}$$

with $\boldsymbol{\mu}_{q,m}$ and $\boldsymbol{\Sigma}_{q,m}$ as the mean and the covariance matrix of mixture m and w_m as the mixture weight.

Utilizing marginalization to recognize uncertain audio features as described above, the calculation of the output probability is modified to

$$\mathrm{b}\big(\mathbf{y}(t)\big) = \sum_{m=1}^{M} w_m \cdot \mathcal{N}\big(\mathbf{y}^r(t), \boldsymbol{\mu}^r_{q,m}, \boldsymbol{\Sigma}^r_{q,m}\big). \tag{13.5}$$

In Eq. (13.5), $\mathbf{y}^r(t)$ describes a reduced feature vector, which only contains those components k that are reliable at the given time, i.e., for which the uncertainty value $u_\mathrm{a}(k,t)$ equals 0, so that we can assume $\mathbf{y}^r(t) = \mathbf{x}(t)$. Similarly, $\boldsymbol{\mu}^r_{q,m}$ is a reduced mean vector and $\boldsymbol{\Sigma}^r_{q,m}$ is the reduced covariance matrix, from which all rows and columns k with an uncertainty of $u_\mathrm{a}(k,t) = 1$ have been removed.

13.2.2.2 Uncertainty Decoding

Uncertainty decoding is based on computing observation probabilities by integrating over all possible values for the true feature vector $\mathbf{x}$. As shown in [9], this results in the time-varying diagonal uncertainty covariance matrix $\boldsymbol{\Sigma}_y$ becoming an additive component to the state covariance matrix $\boldsymbol{\Sigma}_{q,m}$, which is later substituted with

$$\tilde{\boldsymbol{\Sigma}}_{q,m}(t) = \boldsymbol{\Sigma}_{q,m} + \boldsymbol{\Sigma}_y \tag{13.6}$$

in the likelihood evaluation of the m-th mixture component, so

$$b_m(\mathbf{y}(t)) = \mathcal{N}\left(\mathbf{y}(t), \boldsymbol{\mu}_{q,m}, \tilde{\boldsymbol{\Sigma}}_{q,m}(t)\right). \tag{13.7}$$

13.2.2.3 Modified Imputation

Modified imputation [21] is based on the idea of computing an estimated observation vector $\hat{\mathbf{x}}$, which is a weighted average of the mean of the currently evaluated mixture m in state q, and of the corrupted feature vector $\mathbf{y}(t)$. The relative weighting is determined from the uncertainty covariance matrix $\boldsymbol{\Sigma}_y$ and the mixture's covariance matrix

$$\hat{\mathbf{x}}(t) = \left(\boldsymbol{\Sigma}_y^{-1} + \boldsymbol{\Sigma}_{q,m}^{-1}\right)^{-1} \left(\boldsymbol{\Sigma}_y^{-1}\mathbf{y}(t) + \boldsymbol{\Sigma}_{q,m}^{-1}\boldsymbol{\mu}_{q,m}\right). \tag{13.8}$$

Finally, the observation probability of mixture component m is obtained from

$$b_m(\mathbf{y}(t)) = \mathcal{N}\left(\hat{\mathbf{x}}(t), \boldsymbol{\mu}_{q,m}, \boldsymbol{\Sigma}_{q,m}\right).$$

13.2.2.4 The Missing-Feature Coupled HMM

In the case of coupled HMMs, according to Eq. (13.3) the output probability computation is factorized into two streams. For that reason, marginalization, (modified) imputation, and uncertainty decoding can also be carried out independently for each stream.

Since we will be using binary uncertainties and marginalization for the video stream, that leaves the following three options for the implementation of the observation probability evaluation:

$$p(\mathbf{y}|\mathbf{q}) = b_{a,UD}(\mathbf{y}_a|q_a)^{\gamma} \cdot b_{v,MA}(\mathbf{y}_v|q_v)^{1-\gamma} \quad \text{or} \tag{13.9}$$

$$p(\mathbf{y}|\mathbf{q}) = b_{a,MI}(\mathbf{y}_a|q_a)^{\gamma} \cdot b_{v,MA}(\mathbf{y}_v|q_v)^{1-\gamma} \quad \text{or} \tag{13.10}$$

$$p(\mathbf{y}|\mathbf{q}) = b_{a,MA}(\mathbf{y}_a|q_a)^{\gamma} \cdot b_{v,MA}(\mathbf{y}_v|q_v)^{1-\gamma}. \tag{13.11}$$

If the video feature uncertainty is computed only once for each frame and extends to the entire video feature vector at that time (see Section 13.5.2), for each frame and each HMM state $\mathbf{q} = (q_a, q_v)$ the evaluation is simplified to the expression

$$p(\mathbf{y}|\mathbf{q}) = \begin{cases} b_a(y_a|q_a)^{\gamma} \cdot b_v(y_v|q_v)^{1-\gamma} & \text{for } u_v(t) = 0, \\ b_a(y_a|q_a)^{\gamma} & \text{for } u_v(t) = 1. \end{cases} \tag{13.12}$$

13.3 Audio Features

For audio features, we compare two feature sets. On the one hand RASTA-PLP-cepstra have been chosen for their comparatively large robustness with respect to noisy and reverberant conditions. Part of this robustness is due to the RASTA-PLP's feature extraction having been designed to concentrate on features with a certain rate of change, namely that rate of change which is typical of speech. For that purpose, features are band-pass filtered. This makes them more robust to variations in room transfer function and in speaker characteristics, and to changes caused by background noise varying more quickly or slowly than speech signals [16].

On the other hand, the ETSI advanced front-end introduced in [12] was used. The features specified there comprise mel frequency cepstral coefficients with deltas and accelerations, which are optimized for achieving maximum robustness by the inclusion of both a double Wiener filter and a cepstrum normalization step as parts of the feature extraction process.

13.4 Video Features

Finding the face region and deriving significant video features is a crucial task in any audiovisual recognition process. In this section, an overview of face finding algorithms, followed by a description of possible video features, is given.

13.4.1 Face Recognition

Any video feature extraction algorithm strongly depends on robust detection of the speaker's face. A wide range of methods have been presented for that task; an overview is given in [45]. Usually, four methods are distinguished:

- Appearance-based: use of models that are based on the appearance (intensity values) of the face, e.g., Eigenfaces, classification using artificial neural networks (ANNs), naive Bayes, Independent Component Analysis (ICA).
- Feature invariant: searching for features that are invariant to changes in pose and lighting condition, e.g., skin color, texture, facial features.
- Template matching: correlation with typical templates, e.g., of the face shape.
- Knowledge-based: methods using rules based on expert knowledge, e.g., a multi-resolution approach in order to find typical face parameters in every stage according to heuristic rules.

Despite this categorization, hybrid approaches exist, i.e., a combination of the above-mentioned face finding methods is used. For example, in [17], skin tone detection is combined with a search for eyes and mouth via maps generated by knowledge-based methods that are applied to different scales of the image.

While the selection of a method depends on the given task (e.g., single speaker close-up view vs. multi-speaker video conference scenario), it has been shown that color information is useful for detection in general. In order to detect faces independently of brightness, color images are often converted from the red/green/blue (RGB) representation to a color space that separates luminance and chrominance. A very discriminative color space that has been widely used is YCbCr, which consists of one luma and two color components (e.g., [43], [40], [13]).

An example of a hybrid face finding approach including the localization of face components such as eyes, nose and mouth is described in Section 13.5.1.

13.4.2 Feature Extraction

Once the location of the face in a speaker image is known, the region of interest (ROI) from which the desired features are to be calculated has to be determined. Usually, the mouth region, sometimes extended to parts of the lower face, is extracted. Various methods have been reported for that task, such as correlation with templates [41], binarization based on color information thresholds of the lips [44] or eyes [26] and Gaussian estimations of lip color densities [37].

Based on the ROI, visual features applicable to audiovisual speech recognition have to be derived. Three basic classes of visual feature determination are distinguished: appearance-based methods, shape- or contour-based methods and a combination of them [29]. As a preprocessing step, ROI normalization to a defined size as well as mean or variance adjustments are often conducted [44]. Subsequently to feature calculation, postprocessing stages such as linear discriminant analysis (LDA) or maximum likelihood linear transform (MLLT) can significantly improve classification results [38].

13.4.2.1 Appearance-Based Methods

These methods encompass algorithms that provide visual features on an image pixel basis. This includes statistical methods like principal component analysis (PCA) as well as linear image transforms, e.g., discrete cosine and wavelet transforms (DCT and DWT, respectively). The DCT of the speaker's gray-level image is very popular and a widely used technique because of its good results [1, 26] combined with low calculation costs. A further aspect in using the mentioned transforms is the dimensionality reduction reached when choosing only high energy coefficients.

In addition to appearance-based features that are determined on single images, it has been reported that motion information may be useful in AVSR tasks (e.g., [7]). For instance, information on ROI movements of adjacent video frames can be extracted pixelwise using optical flow (OF) estimations [6]. Examples of using OF-based visual features are given in [28, 35, 42].

13.4.2.2 Shape-Based Methods

An intuitive way of exploiting visual information is using shape parameters of the mouth or a broader face region. Such parameters may be geometrical properties of the lips, such as width, height or area inside the lip contours [14], image moments or Fourier descriptors. Furthermore, statistical lip models, such as active shape models (ASMs), have been used in the community [23]. In any method using shape information, a robust detection of the contours of the lips is necessary. For that purpose, data is either hand-labeled or automatically gathered using, e.g., deformable templates (parabolas, [2]) or so-called snakes.

Active Contours (Snakes)

Active contours in an image are somewhat similar in appearance to snakes in nature and therefore often simply called snakes. Snakes are used in image processing for object segmentation and will be applied here to approximate the lip region boundary by a closed curve. This curve in two-dimensional space is approximated by a set of sorted discrete points along the curve boundaries. Each point is connected to its neighbors by a straight line. The shape of the snake is influenced by different so-called forces, with a distinction being made between *internal* and *external* forces. The internal forces represent the properties of the snake shape. The two most common internal forces (sometimes formulated as energy terms), described in detail in [18], model the elastic behavior of the active contour as that of an elastic band. First of all there is the attractive force (see Eq. (13.13)), which is calculated with k as the index of the current, $k-1$ the index of the previous and $k+1$ the index of the next point. $\mathbf{p}_k$ is the position vector, and α_k stands for a weighting factor:

$$
\begin{aligned}
\mathbf{f}_{k,\text{attractive}} &= \alpha_k \cdot (\mathbf{p}_{k-1} - \mathbf{p}_k) + \alpha_{k+1} \cdot (\mathbf{p}_{k+1} - \mathbf{p}_k) \\
&= [\alpha_k, \alpha_{k+1}] \cdot \begin{bmatrix} \mathbf{p}_{k-1} - \mathbf{p}_k \\ \mathbf{p}_{k+1} - \mathbf{p}_k \end{bmatrix} \\
&= \boldsymbol{\alpha} \cdot \tilde{\mathbf{f}}_{k,\text{attractive}} .
\end{aligned} \tag{13.13}
$$

The so-called bending force, shown in Eq. (13.14), has a smoothing effect on the contour. Sharp corners lead to high bending forces, and with great weighting factors β_k ($\beta_k \gg 1$) the snake tends pass over local structural discontinuities and avoid any acute angles.

$$
\begin{aligned}
\mathbf{f}_{k,\text{bending}} &= \beta_{k-1} (2\,\mathbf{p}_{k-1} - \mathbf{p}_{k-2} - \mathbf{p}_k) - \\
&\quad 2\,\beta_k (2\,\mathbf{p}_k - \mathbf{p}_{k-1} - \mathbf{p}_{k+1}) + \\
&\quad \beta_{k+1} (2\,\mathbf{p}_{k+1} - \mathbf{p}_k - \mathbf{p}_{k+2}) \\
&= [\beta_{k-1}, \beta_k, \beta_{k+1}] \cdot \begin{bmatrix} 2\,\mathbf{p}_{k-1} - \mathbf{p}_{k-2} - \mathbf{p}_k \\ -2\,(2\,\mathbf{p}_k - \mathbf{p}_{k-1} - \mathbf{p}_{k+1}) \\ 2\,\mathbf{p}_{k+1} - \mathbf{p}_k - \mathbf{p}_{k+2} \end{bmatrix} \\
&= \boldsymbol{\beta} \cdot \tilde{\mathbf{f}}_{k,\text{bending}} .
\end{aligned} \tag{13.14}
$$

The total internal force of every discrete point k becomes

$$
\mathbf{f}_{k,\text{int.}} = \mathbf{f}_{k,\text{attractive}} + \mathbf{f}_{k,\text{bending}} . \tag{13.15}
$$

Image contents and other peripheral effects are considered as external forces. Object boundaries (or edges in the image) are the most important external force and will usually be included in the force calculation via image gradients $\mathbf{g}_k$. Another useful force, which can be used to allow concave active contours, is the center of gravity force

$$
\mathbf{f}_{k,\text{gravity}} = \frac{\mathbf{r}_k}{|\mathbf{r}_k|} \cdot |\mathbf{r}_k|^i = \mathbf{r}_k \cdot |\mathbf{r}_k|^{i-1} , \tag{13.16}
$$

with $\mathbf{r}_k = \mathbf{b} - \mathbf{p}_k$, $\mathbf{b}$ as the position vector of the center of gravity, and i as the order of the vector norm. Thus, the total external force, with extra weighting factors μ_k and ν_k, is

$$
\mathbf{f}_{k,\text{ext.}} = \mu_k \cdot \mathbf{g}_k + \nu_k \cdot \mathbf{f}_{k,\text{gravity}} . \tag{13.17}
$$

Active contours are deformed iteratively so as to minimize the total force

$$
\mathbf{f}_{k,\text{total}} = \mathbf{f}_{k,\text{int.}} + \mathbf{f}_{k,\text{ext.}} = \boldsymbol{\alpha} \cdot \tilde{\mathbf{f}}_{k,\text{attractive}} + \boldsymbol{\beta} \cdot \tilde{\mathbf{f}}_{k,\text{bending}} + \mu_k \cdot \mathbf{g}_k + \nu_k \cdot \mathbf{f}_{k,\text{gravity}} , \tag{13.18}
$$

until either $\mathbf{f}_{k,\text{total}} \approx \mathbf{0}$ or the maximum number of allowed iterations is exceeded. In Eq. (13.18), $\boldsymbol{\alpha}$ and $\boldsymbol{\beta}$ are weighting factors, independent of the index k in the following considerations. The deformation is an iterative minimization process. To find the optimum shape of a snake, it is necessary to calculate the partial derivatives with respect to the discrete point coordinates and set them equal to zero, for which the derivation is given in [18]. The equivalence between the energy and force formulation is shown in [24]. An application can be found in Section 13.5.2.2.

13.4.2.3 Combined Shape and Appearance-Based Methods

Instead of simply concatenating appearance- and shape-based features, an approach of incorporating both sources of information using a single statistical model has been proposed, called active appearance model (AAM). In this model, intensity values of all points inside the predetermined lip contour are used to build a statistical model based on principal component analysis (for a detailed description, see [29]).

13.4.2.4 Fisher Linear Discriminant

After features have thus been computed, a dimensionality reduction is often called for. In order to find a projection of the features that is most amenable to subsequent classification, linear discriminant analysis has proved beneficial in many studies.

The Fisher Linear Discriminant (FLD) gives a projection matrix $\mathbf{W}$ that reshapes the scatter of a data set to maximize class separability, defined as the ratio of the between-class scatter matrix to the within-class scatter matrix. This projection defines features that are optimally discriminative.

Let $\mathbf{x}_i$ be one of N column vectors of dimension D. The mean of the dataset is

$$\boldsymbol{\mu}_x = \frac{1}{N}\sum_{i=1}^{N}\mathbf{x}_i. \tag{13.19}$$

For K classes $\{C_1, C_2, \ldots, C_K\}$, the mean of class k containing N_k members is

$$\boldsymbol{\mu}_{xk} = \frac{1}{N_k}\sum_{\mathbf{x}_i \in C_k}\mathbf{x}_i. \tag{13.20}$$

The between-class scatter matrix is

$$\mathbf{S}_B = \sum_{k=1}^{K} N_k(\boldsymbol{\mu}_{xk} - \boldsymbol{\mu}_x)(\boldsymbol{\mu}_{xk} - \boldsymbol{\mu}_x)^T, \tag{13.21}$$

and the within-class scatter matrix is

$$\mathbf{S}_W = \sum_{k=1}^{K}\sum_{\mathbf{x}_i \in C_k}(\mathbf{x}_i - \boldsymbol{\mu}_{xk})(\mathbf{x}_i - \boldsymbol{\mu}_{xk})^T. \tag{13.22}$$

In order to find the within-class scatter matrix $\mathbf{S}_W$, either labelled data or the covariance matrices of a trained single-mixture full covariance HMM may be used. The transformation matrix that repositions the data to be most separable is the matrix $\mathbf{W}$ that maximizes

$$\frac{\det(\mathbf{W}^T\mathbf{S}_B\mathbf{W})}{\det(\mathbf{W}^T\mathbf{S}_W\mathbf{W})}. \tag{13.23}$$

Let $\{\mathbf{w}_1, \mathbf{w}_2, \ldots, \mathbf{w}_D\}$ be the generalized eigenvectors of $\mathbf{S}_B$ and $\mathbf{S}_W$, which can be determined as the non-trivial solutions of

$$\mathbf{S}_B \mathbf{w} = \lambda \mathbf{S}_W \mathbf{w} \tag{13.24}$$

with λ as a scalar value $\neq 0$.

Then, $\mathbf{W} = [\mathbf{w}_1, \mathbf{w}_2, \ldots, \mathbf{w}_D]$ gives a projection space of dimension D. A projection space of dimension $d < D$ can be defined by using the generalized eigenvectors with the largest d eigenvalues to give $\mathbf{W}_d = [\mathbf{w}_1, \mathbf{w}_2, \ldots, \mathbf{w}_d]$. The projection of vector $\mathbf{x}$ onto the subspace of dimension d is $\mathbf{y} = \mathbf{W}_d^T \mathbf{x}$. The resultant feature vector $\mathbf{y}$ is subsequently used for recognition.

13.5 Audiovisual Speech Recognizer Implementation and Results

In this section, an implementation of a complete audiovisual recognizer is presented. After giving an overview of the used face detection method, localizing face features and deriving uncertainty measures from the calculated audio and video features, we will discuss the utilized speech recognizer and the resulting recognition rates. All experiments have been conducted on the GRID database. It contains utterances from 34 speakers recorded under strictly defined conditions, showing the speaker in a front view reading an English six-word sentence (for details, see [8]).

13.5.1 Face Detection

The detection of the face utilizes a hybrid approach and is divided into detection of the skin, fitting an ellipse and a rotation correction. Afterwards, the mouth region is determined and, if necessary, a correction of the extracted region over a video sequence takes place.

13.5.1.1 Skin Detection

According to an algorithm proposed by [31], the skin detection is carried out in the YCbCr color space. Based on training images selected randomly from the database, a Gaussian model was trained for the color components in order to eliminate the influence of lighting conditions. Using this model, the color image of the original sequence is transformed into a gray scale image. The gray values represent every pixel's likelihood of belonging to the skin region. An adaptive threshold is used to convert the gray scale into a binary image, in which the face is then detected as a region of at least a minimum size of connected skin pixels with characteristic holes (mouth, eyes, eyebrows) in it.

13.5.1.2 Ellipse Fitting and Rotation Correction

Because the face may be partly occluded and the image may also contain other skin regions besides the face, the binary skin mask is usually not the exact face region. In order to cope with that, an ellipse with the same balance point and moment of inertia as the binary face region is searched for. From that ellipse and the location of characteristic holes in the skin region that mark the position of eyes and mouth, an angle of rotation of the face is elicited. This angle is used to rotate the face back into a vertical position, so that all further processing steps can assume an upright face.

13.5.1.3 Mouth Region Extraction

The aim of this processing step is the robust determination of a rectangular mouth region. At first, an edge filter is applied to the face region, followed by a summation to find horizontal lines that belong to the eyebrows, eyes, nostrils, mouth and chin. Only along those lines that may represent the location of the mouth is the correlation with a mouth template carried out. Starting from the point of highest correlation, the color image is searched for edges in order to find the upper and lower lip, which, after addition of a resolution-dependent number of pixels, mark the boundaries for the ROI rectangle.

Sequence Processing

When seeking the mouth in a sequence of images, one can assume that the change of location from one frame to another will be limited to a certain value according to the frame rate and resolution. In order to speed up the process of tracking the ROI, the mouth region found in one image is presented as starting point to the search engine for the next frame until a threshold, basically a certain number of frames, is reached. After that, a complete run, i.e., a reinitialization, of the mouth localization algorithm is started.

Postprocessing

It has been found that the dimension of the extracted mouth region changes over the course of a sequence to an undesired degree, especially in reinitialization frames. Since this will decrease the ability to train reliable models based on visual features, lip corners are determined for all frames of a sequence. This is done using repeated morphological filter operations on a color-corrected version of the ROI (similarly to [25]). All detected mouth corner locations are then checked using geometrical preconditions. Based on valid locations, a mean position of the mouth over the entire sequence is determined that will be used to replace wrongly detected ROIs. Furthermore, all valid mouth corners are used to recalculate the size of the ROI which they were extracted from. As a consequence, all mouth regions of a sequence will only show smooth changes in size over frames. Before the calculation of visual features, the regions of interest are resized to a 64 x 64 pixel image.

13.5.2 Video Features and Their Uncertainties

For video features, DCT coefficients and active contours are used. To compute the uncertainties of video features, frame-wise methods are used that provide an uncertainty measure for both features. These uncertainties are combined afterwards using a simple decision function to be incorporated into the recognition process as binary uncertainties.

13.5.2.1 DCT Uncertainty

Based on the mouth regions extracted as described above, coefficients of a two-dimensional discrete cosine transform according to Eq. (13.25) are calculated with $0 \leq k \leq M-1, 0 \leq l \leq N-1$ and α_k, α_l as normalization constants according to the size of the ROI. Beforehand, the color image of the mouth region is transformed to gray level:

$$DCT_{kl} = \alpha_k \alpha_l \sum_{m=0}^{M-1} \sum_{n=0}^{N-1} ROI_{mn} \cos \frac{\pi(2m+1)k}{2M} \cos \frac{\pi(2n+1)l}{2N}. \tag{13.25}$$

In order to obtain an uncertainty measure for the DCT video features, a speaker-dependent mouth model is trained on 20 hand-labelled sequences, which were selected to contain at least 30 wrongly estimated mouth regions in total. From these sequences, intact mouth regions are learned separately from a model combining both non-mouth and cropped-mouth regions. The model consists of a single Gaussian probability density function (pdf) of the same first 64 DCT coefficients which are also used for HMM training. Therefore, no additional feature extraction needs to take place.

Subsequently, a frame will be counted as reliable if the log-likelihood p_{m} of the trained mouth model,

$$p_{\mathrm{m}} = \log \frac{1}{\sqrt{(2\pi)^D |\boldsymbol{\Sigma}_{\mathrm{m}}|}} \mathrm{e}^{-\frac{1}{2}(\mathbf{y}_{\mathrm{v}}(t)-\boldsymbol{\mu}_{\mathrm{m}})' \boldsymbol{\Sigma}_{\mathrm{m}}^{-1} (\mathbf{y}_{\mathrm{v}}(t)-\boldsymbol{\mu}_{\mathrm{m}})}, \tag{13.26}$$

with mean $\boldsymbol{\mu}_{\mathrm{m}}$ and covariance matrix $\boldsymbol{\Sigma}_{\mathrm{m}}$ exceeds that of a non-mouth model

$$p_{\mathrm{n}} = \log \frac{1}{\sqrt{(2\pi)^D |\boldsymbol{\Sigma}_{\mathrm{n}}|}} \mathrm{e}^{-\frac{1}{2}(\mathbf{y}_{\mathrm{v}}(t)-\boldsymbol{\mu}_{\mathrm{n}})' \boldsymbol{\Sigma}_{\mathrm{n}}^{-1} (\mathbf{y}_{\mathrm{v}}(t)-\boldsymbol{\mu}_{\mathrm{n}})} \tag{13.27}$$

with parameters $\boldsymbol{\mu}_\mathrm{n}$ and $\boldsymbol{\Sigma}_\mathrm{n}$. Thus, the uncertainty $u_{\mathrm{v,DCT}}(t)$ of all DCT features in frame t is given by

$$u_{\mathrm{v,DCT}}(t) = \begin{cases} 0 \text{ for } p_\mathrm{m} \geq p_\mathrm{n}, \\ 1 \text{ otherwise.} \end{cases} \tag{13.28}$$

A typical example of the resultant labeling of video frames can be seen in Fig. 13.2.

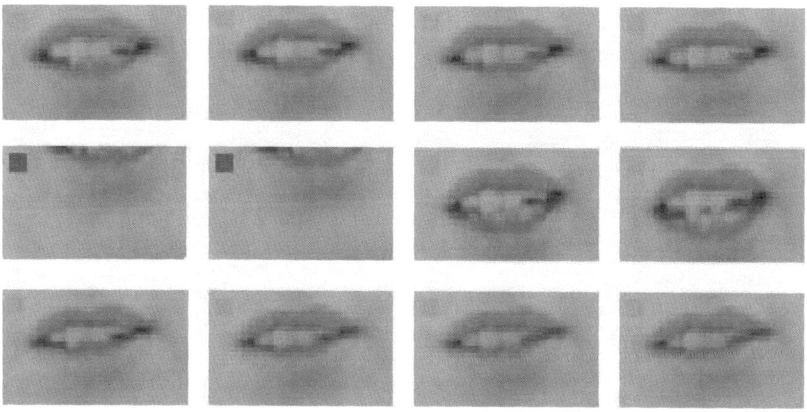

Fig. 13.2: Mouth regions and their associated labels. A light square means that the frame has been counted as reliable; otherwise, a dark square is shown

13.5.2.2 Active Contour Uncertainty

In AVSR, active contours (a.k.a. snakes) consist of a certain number of coordinate pairs that, in an ideal case, mark exactly the location of the lip contour in the region of interest. From these contour points, it is possible to derive different criteria that provide information on the uncertainty of the calculated snake. Such measures are of a geometrical kind, e.g., an enclosed area, or utilize shape information such as subsets of points that describe the upper and lower lip curves. Criteria for both characteristics are described in this section.

Snake Calculation

All of the above-mentioned forces (see Sec. 13.4.2.2) are used with optimal weighting parameters $\boldsymbol{\alpha}, \boldsymbol{\beta}, \mu_k, \nu_k$ according to Eq. (13.18) to obtain the outer lip curve. Due to the iterative curve fitting process, it is necessary to have an initial curve. For good segmentation results, it is recommended that the initial snake outline the desired lip region in each frame as closely as possible. Since the coordinates of the mouth corners are already known from previous processing steps, we have initialized the snake as an ellipse with a major axis equal to the mouth corner distance and a minor axis equal to half the mouth region height. In every step of the iteration, the coordinates of the discrete points localised initially in the mouth corners are held constant and only the other points are adjusted. The optimal weighting parameters and further investigations can be found in [24].

The calculation of the image gradients is done by a differentiated three-dimensional Gaussian filter with standard deviation of $\sqrt{1.5}$ pixels in every dimension. In order to calculate the gradients in boundary image regions, the last pixel values are replicated according to the filter size.

One of the arising problems is a considerable shift of the discrete points over the iteration process. The uniformly distributed points along the shape at the beginning are finally concentrated in regions indicated by the highest gradients. From this, it follows that in these regions the discrete points approximate the lip shape very well but in other sectors too little accuracy is reached. A modified gradient approach, which guarantees a constant relative distance between neighboring points, solves the problem. In principle, it uses a projection of the image gradients onto a line perpendicular to the secant connecting the two neighbor points $k-1$ and $k+1$ (see Fig. 13.3).

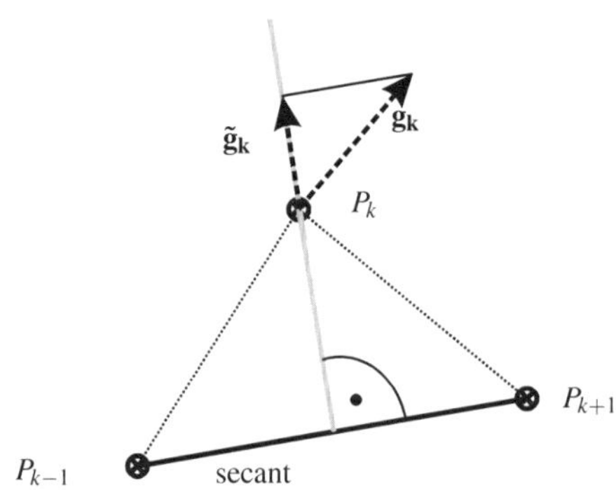

Fig. 13.3: Principle of the gradient projection. $\tilde{\mathbf{g}}_\mathbf{k}$ is the used projected gradient at the point of interest

The coordinates of the discrete contour points are finally used for calculating the video features.

Surface Area

The enclosed surface area of a snake may help us identify miscalculations. For example, a very low surface area value should not appear given a valid region of interest. Taking only the area into account is not robust enough to tolerable changes in the original size of the region of interest. Therefore, this ROI size (rS) is incorporated into the area criterion in two ways, as can be seen in Eq. (13.29). The first measure pC is derived from the product of the surface area (sA) and the number of points of the original ROI, whereas the second measure sC is a weighted sum of normalized inputs. For both measures, a threshold is calculated that is subsequently used to denote an active contour as an uncertain candidate:

$$\begin{aligned} pC &= sA \cdot rS \\ sC &= \frac{2}{3} sA_{\text{norm}} \cdot \frac{1}{3} rS_{\text{norm}}. \end{aligned} \tag{13.29}$$

Curve Parameters

The active contour shape may deliver further information on its validity. In order to recognize inappropriate curve parameters, a model has to be trained from labeled data. For this work, Artificial Neural Networks (ANNs) have been trained on coefficients of fifth-order polynomials that have been fitted to the coordinates of the upper and lower lips. All calculated snakes are classified using these ANNs, so that two additional criteria regarding uncertainty of the active contour are available.

13.5.2.3 Video Feature Uncertainty Combination for Decoding

By rule-based combination of all four criteria described above, a binary uncertainty flag is derived for the snake features, as shown in Fig. 13.4. Black dots in the corners of the images show validity criteria being met, whereas diamonds, squares and pentagrams indicate a miscalculated snake. Gray bars crossing an image stand for an uncertainty $u_{\text{v,AC}} = 1$ for this particular frame, while all other images have an uncertainty $u_{\text{v,AC}} = 0$.

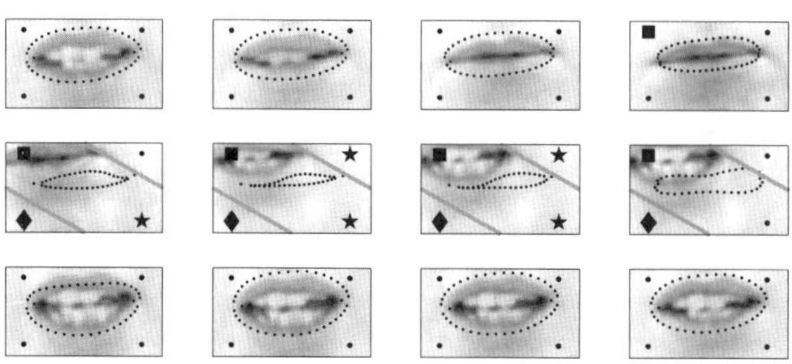

Fig. 13.4: Active contour uncertainty. Black dots near the corners indicate valid snakes. Black squares and diamonds mark invalid snakes according to surface area criteria, whereas pentagrams stand for invalid upper and lower lips. Gray bars visualize the combined hard decision of an invalid frame

13.5.3 Audio Features and Their Uncertainties

13.5.3.1 Rasta-PLP

For the RASTA-PLP cepstrum, 12 coefficients and their first and second derivatives were used, which were obtained from a power spectrum of the speech signal using a window size of 25 ms with 15 ms overlap. RASTA filtering takes place in the log-Bark spectrum, using the transfer function

$$H(z) = \frac{0.2z^4 + 0.1z^3 - 0.1z^{-3} - 0.2z^{-4}}{z^3(z - 0.94)} \tag{13.30}$$

given in [16]. Subsequently, the LPC-cepstrum is obtained from the RASTA-filtered log spectrum, using an LPC model of order 12. For these computations, the Rastamat toolbox was used, which is available from [11]. Finally, first- and second-order time derivatives were appended to the 12 RASTA-PLP cepstral coefficients, which further improved robustness.

13.5.3.2 ETSI Advanced Front-End

The feature extraction for the ETSI advanced front-end follows the standard described in [12]. It uses 13 mel frequency cepstral coefficients and the delta and acceleration parameters, which form a 39-dimensional feature vector. In it, the zero-th cepstral coefficient is replaced by the logarithm of the frame energy.

13.5.3.3 Uncertainties

For audio feature uncertainties, numerous approaches exist to estimate either binary or continuous-valued uncertainties. However, in most cases, binary uncertainty values are only used for spectra and log mel spectra, which are not very robust to variations in the speaker, the environment or the background noise. In contrast, a number of more elaborate mechanisms for estimating continuous uncertainty values in the feature domain have been developed over the past years; see, e.g., [3, 9]. In the following, we will be considering both types of uncertainty.

For binary uncertainties, which are easy to compute but less robust than ideally achievable, a simple strategy is to compare each feature vector component $y_a(k,t)$ at time t to the level of the estimated background noise n_a. These two vectors are compared for each of the feature components, and a reliability decision is made by

$$u_a(k,t) = \begin{cases} 0 \text{ for } n_a(k,t) < \alpha \cdot y_a(k,t), \\ 1 \text{ otherwise,} \end{cases} \tag{13.31}$$

i.e., a feature is deemed unreliable if the value of the background noise feature exceeds the value of α times the observed signal feature. This binary feature uncertainty will be used in the following as the uncertainty of the RASTA-PLP cepstrum only. Its application for the ETSI advanced front-end has not proven advantageous, possibly due to the integrated denoising stages of the ETSI-AFE, which would require a more advanced decision function for binary uncertainties.

As a second, continuous-valued uncertainty strategy, we will be considering the ETSI advanced front-end. Since this feature extraction already contains a Wiener filter, an estimate of the estimator's error variance is available also within the system. The continuous-valued uncertainties have been obtained from the Wiener filter estimator error variance in the complex spectrum domain [4] and propagated to the ETSI-MFCC back-end via the uncertainty propagation techniques described in Chapter 3, Section 6.

13.5.4 Audiovisual Recognition System

The JASPER (JAVA AUDIOVISUAL SPEECH RECOGNIZER) System, developed for robust single- and multi-stream speech recognition, can combine audio recognition and lipreading with the help of audiovisual CHMMs. It is based on a flexible token passing architecture applicable to a wide range of statistical speech models.

The system allows for a tight integration of the MATLAB and JAVA environments and capabilities, with an interface that lets preprocessing and feature extraction be carried out in MATLAB, whereas model training and recognition take place in JAVA.

JASPER is based on an abstract model in which connected word recognition is viewed as a process of passing tokens around a transition network [46]. Within this network, each vocabulary element is represented by a word model. These word models are statistical descriptions of the evolution of the feature stream over time within the associated words. Word models may be realized as conventional HMMs, coupled or product HMMs, or even templates, or a range of graphical models, which is an advantage of the abstraction inherent in the implemented token passing approach. Figure 13.5 shows an example of a possible word net structure for the GRID database.

In addition to the word models, further network elements are link nodes and non-terminal nodes. Link nodes provide the connections between non-terminal nodes and word models and associate a linked list of possible word alternatives with all tokens passing through them. Non-terminal nodes allow the grouping of nodes into different hierarchical levels. The highest level in the network is the non-terminal node that represents the entire language model. The elements on the lowest level are the word models.

The recognition process starts with a single token entering the top-level syntax. Every time step is split into two half steps. At first, all link nodes propagate their tokens down the hierarchy, until at the lowest level the word models are reached. The actual calculations of word model log-likelihoods given the observed features happen in the second phase. At the end of the second phase, all link nodes collect tokens

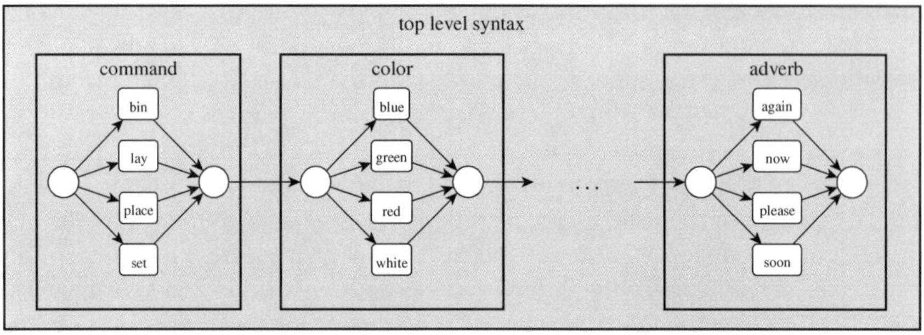

Fig. 13.5: An example of a word network for recognition of the GRID grammar described in [8]. Link nodes are depicted by circles, non-terminal nodes by bold black rectangles and word models are shown as rounded rectangles

from their incoming connections and build a connected list of the n best tokens. Tokens are ranked by a score, corresponding to their accumulated log-likelihood, and a global, adaptive threshold is used for efficient pruning. This process is iterated until a complete observation sequence has been processed and the outgoing link node of the top-level syntax contains the token with the highest score given the model and the observed data.

13.6 Efficiency Considerations

Coupled HMMs involve, in general, a significantly increased search space when compared to one-dimensional or multistream HMMs. Therefore, computational efficiency is of some significance.

Profiling the search reveals it to be dominated in a large number of cases by the computation of output distributions. In the case of missing or uncertain features, this becomes of even greater significance, since the computation of uncertain feature likelihoods causes an increase in computational demands. Therefore, the following section will describe some considerations that have proven valuable for efficient likelihood evaluations in CHMMs, both without and with missing data techniques.

13.6.1 Gaussian Density Computation

The likelihood computation for an observation vector $\mathbf{y}$ using a multivariate normal distribution

$$\mathscr{N}(\mathbf{y} \mid \boldsymbol{\mu}, \boldsymbol{\Sigma}) = \frac{1}{\sqrt{(2\pi)^D |\boldsymbol{\Sigma}|}} \exp\left(-\frac{1}{2} (\mathbf{y} - \boldsymbol{\mu})^{\mathsf{T}} \boldsymbol{\Sigma}^{-1} (\mathbf{y} - \boldsymbol{\mu}) \right) \tag{13.32}$$

with mean vector $\boldsymbol{\mu}$ and covariance matrix $\boldsymbol{\Sigma}$ is prone to overflow errors because of the limited dynamic range of common floating-point number formats. To prevent overflows, the log-likelihood

$$\log(\mathscr{N}) = -\log\left(\sqrt{(2\pi)^D|\boldsymbol{\Sigma}|}\right) - \frac{1}{2}(\mathbf{y}-\boldsymbol{\mu})^{\mathsf{T}}\boldsymbol{\Sigma}^{-1}(\mathbf{y}-\boldsymbol{\mu}) \tag{13.33}$$

is normally used instead.

For an efficient implementation, a transformation into a quadratic form,

$$\log(\mathscr{N}) = s + \mathbf{u}^{\mathsf{T}}\mathbf{y} - (\mathbf{V}\mathbf{y})^{\mathsf{T}}(\mathbf{V}\mathbf{y}), \tag{13.34}$$

is helpful. After sorting terms, (13.33) gives

$$\begin{aligned}\log(\mathscr{N}) &= -\frac{1}{2}\Big(\log(2\pi)D + \log(|\boldsymbol{\Sigma}|)\Big) - \frac{1}{2}\Big(\mathbf{y}^{\mathsf{T}}\boldsymbol{\Sigma}^{-1}\mathbf{y} - 2\boldsymbol{\mu}^{\mathsf{T}}\boldsymbol{\Sigma}^{-1}\mathbf{y} + \boldsymbol{\mu}^{\mathsf{T}}\boldsymbol{\Sigma}^{-1}\boldsymbol{\mu}\Big)\\ &= -\frac{1}{2}\Big(\log(2\pi)D + \log(|\boldsymbol{\Sigma}|) + \boldsymbol{\mu}^{\mathsf{T}}\boldsymbol{\Sigma}^{-1}\boldsymbol{\mu}\Big) + \boldsymbol{\mu}^{\mathsf{T}}\boldsymbol{\Sigma}^{-1}\mathbf{y} - \frac{1}{2}\mathbf{y}^{\mathsf{T}}\boldsymbol{\Sigma}^{-1}\mathbf{y}.\end{aligned} \tag{13.35}$$

By means of the Cholesky decomposition, the inverse covariance matrix $\boldsymbol{\Sigma}^{-1}$ is factored into a product of upper triangular matrices $\mathbf{U}^{\mathsf{T}}\mathbf{U}$, so for the computation of the Mahalanobis distance, we have

$$\mathbf{y}^{\mathsf{T}}\boldsymbol{\Sigma}^{-1}\mathbf{y} = \mathbf{y}^{\mathsf{T}}\mathbf{U}^{\mathsf{T}}\mathbf{U}\mathbf{y} = \Big(\mathbf{U}\mathbf{y}\Big)^{\mathsf{T}}\Big(\mathbf{U}\mathbf{y}\Big). \tag{13.36}$$

This matrix $\mathbf{U}$ can also be written as the matrix square root $\mathbf{U} = \boldsymbol{\Sigma}^{-\frac{1}{2}}$. Inserting this into (13.35) yields

$$\begin{aligned}\log(\mathscr{N}) = &-\frac{1}{2}\Big(\log(2\pi)D + \log(|\boldsymbol{\Sigma}|) + \boldsymbol{\mu}^{\mathsf{T}}\boldsymbol{\Sigma}^{-1}\boldsymbol{\mu}\Big)\\ &+ \boldsymbol{\mu}^{\mathsf{T}}\boldsymbol{\Sigma}^{-1}\mathbf{y} - \frac{1}{2}\Big(\boldsymbol{\Sigma}^{-\frac{1}{2}}\mathbf{y}\Big)^{\mathsf{T}}\Big(\boldsymbol{\Sigma}^{-\frac{1}{2}}\mathbf{y}\Big).\end{aligned} \tag{13.37}$$

After equating with (13.34), this gives the required coefficients:

$$s = -\frac{\log(2\pi)D + \log(|\boldsymbol{\Sigma}|) + \boldsymbol{\mu}^{\mathsf{T}}\boldsymbol{\Sigma}^{-1}\boldsymbol{\mu}}{2}, \tag{13.38}$$

$$\mathbf{u} = \boldsymbol{\mu}^{\mathsf{T}}\boldsymbol{\Sigma}^{-1}, \quad \mathbf{V} = \frac{\boldsymbol{\Sigma}^{-1/2}}{\sqrt{2}}. \tag{13.39}$$

These can be used for efficient evaluation of full covariance Gaussian mixture models.

13.6.1.1 Diagonal Model Implementations

Independently of the structure of the covariance matrix, the likelihood of a multivariate observation $\mathbf{y}$ always depends on the Mahalanobis distance

$$D_M(\mathbf{y}) = \sqrt{(\mathbf{y}-\boldsymbol{\mu})^{\mathsf{T}}\boldsymbol{\Sigma}^{-1}(\mathbf{y}-\boldsymbol{\mu})} \tag{13.40}$$

for given mean $\boldsymbol{\mu}$ and covariance $\boldsymbol{\Sigma}$. The following paragraph describes the efficient likelihood computation for diagonal covariance matrices and the necessary extensions for methods using observations with uncertainties.

13.6.2 Conventional Likelihood

For diagonal covariance matrices

$$\boldsymbol{\Sigma} = \begin{pmatrix} \sigma_1^2 & 0 & \dots & 0 \\ 0 & \sigma_2^2 & \dots & 0 \\ \vdots & \vdots & \ddots & \vdots \\ 0 & 0 & \dots & \sigma_D^2 \end{pmatrix}, \tag{13.41}$$

with the variances σ_i^2 on the main diagonal, the equations for the log-likelihood computations simplify to

$$\log(\mathscr{N}) = -\frac{1}{2}\left(\log(2\pi)D + \log\left(\prod_{i=1}^{D}\sigma_i^2\right) + \sum_{i=1}^{D}\frac{\boldsymbol{\mu}_i^2}{\sigma_i^2}\right) + \sum_{i=1}^{D}\frac{\boldsymbol{\mu}_i}{\sigma_i^2}y_i - \sum_{i=1}^{D}\frac{1}{2\sigma_i^2}y_i^2 \tag{13.42}$$

and the calculation can be expressed as a dot product

$$\left(s, \frac{\boldsymbol{\mu}_1}{\sigma_1^2}, \cdots, \frac{\boldsymbol{\mu}_D}{\sigma_D^2}, -\frac{1}{2\sigma_1^2}, \cdots, -\frac{1}{2\sigma_D^2}\right) \cdot \left(1, y_1, \cdots, y_D, y_1^2, \cdots, y_D^2\right)^{\mathsf{T}} \tag{13.43}$$

of the parameter vector and an augmented observation vector [10]. This is especially helpful for parallel implementations and has been used to obtain faster results by extending JASPER to CASPER.[2]

[2] CUDA AUDIOVISUAL SPEECH RECOGNIZER; this is the JASPER system with all log-likelihood computations done in CUDA; see [20].

13.6.3 Uncertainty Decoding

In uncertainty decoding, the diagonal matrix of feature uncertainties $\boldsymbol{\Sigma}_y$ is used to obtain a modified covariance matrix for computing

$$p(\mathbf{y}, \boldsymbol{\mu}, \boldsymbol{\Sigma}, \boldsymbol{\Sigma}_y) = \mathcal{N}\left(\mathbf{y} \mid \boldsymbol{\mu}, \tilde{\boldsymbol{\Sigma}}\right). \tag{13.44}$$

Here, the model covariance $\boldsymbol{\Sigma}$ is substituted by $\tilde{\boldsymbol{\Sigma}} = \boldsymbol{\Sigma} + \boldsymbol{\Sigma}_y$.

Because the entries of the covariance matrix are thus modified by the uncertainties, the factor s from Eq. (13.38) has to be recalculated in each frame by

$$s = -\frac{1}{2}\left(\log(2\pi)D + \log\left(\prod_{i=1}^{D} \tilde{\sigma}_i^2\right)\right). \tag{13.45}$$

Otherwise, the considerations from Section 13.6.2 apply.

13.6.4 Modified Imputation

In contrast to uncertainty decoding, modified imputation adapts the observation vector $\mathbf{y}$ with uncertainties $\boldsymbol{\Sigma}_y$,

$$p(\mathbf{y}, \boldsymbol{\mu}, \boldsymbol{\Sigma}, \boldsymbol{\Sigma}_y) = \mathcal{N}\left(\hat{\mathbf{x}}, \boldsymbol{\mu}, \boldsymbol{\Sigma}\right), \tag{13.46}$$

according to

$$\hat{x}_i = \frac{y_i \sigma_{yi}^{-1} + \mu_i \sigma_i^{-1}}{\sigma_{yi}^{-1} + \sigma_i^{-1}}. \tag{13.47}$$

Here, μ_i and σ_i denote the i-th component of the model mean and the square root of the i-th diagonal element of the covariance matrix for the currently evaluated state, respectively. σ_{yi} is the i-th component of the feature uncertainty and $\hat{x}$ is determined elementwise for each dimension $1 \leq i \leq D$.

The following simplifications are useful:

$$\left(\frac{\hat{x}-\mu}{\sigma}\right)^2 = \frac{1}{\sigma^2}\left(\frac{\frac{y}{\sigma_y}+\frac{\mu}{\sigma}}{\frac{1}{\sigma_y}+\frac{1}{\sigma}}-\mu\right)^2 \tag{13.48}$$

$$= \frac{1}{\sigma^2}\left(\frac{\frac{y}{\sigma_y}+\frac{\mu}{\sigma}-\frac{\mu}{\sigma_y}-\frac{\mu}{\sigma}}{\frac{1}{\sigma_y}+\frac{1}{\sigma}}\right)^2 \tag{13.49}$$

$$= \frac{1}{\sigma^2}\left(\frac{\frac{y}{\sigma_y}-\frac{\mu}{\sigma_y}}{\frac{1}{\sigma_y}+\frac{1}{\sigma}}\right)^2 \tag{13.50}$$

$$= \frac{1}{\sigma^2}\left(\frac{y-\mu}{1+\frac{\sigma_y}{\sigma}}\right)^2 \tag{13.51}$$

$$= \frac{(y\quad\mu)^2}{(\sigma+\sigma_y)^2}. \tag{13.52}$$

This last expression can be used to compute the Mahalanobis distance more efficiently than can the original definition of modified imputation.

13.6.5 Binary Uncertainties

For binary uncertainties, only the reliable subvector of $\mathbf{y}$, composed of only the reliable components of $\mathbf{y}$ and denoted by $\mathbf{y}_r$, is used. If full covariances are needed, the considerations from Section 13.6.2 apply; otherwise, computation of a reduced inner product according to

$$(\tilde{s}, \boldsymbol{\mu}_r) \cdot (1, \mathbf{y}_r)^{\mathsf{T}} \tag{13.53}$$

will give an implementation that is potentially suitable for utilizing the vector lanes of SIMD architectures. However, it is first necessary in each frame to recompute the scaling factor s via $\frac{1}{\sqrt{(2\pi)^d \det(\Sigma_q+\Sigma_y)}}$, with d as the number of reliable components in $\mathbf{y}_r$, and to obtain the reliable subvector of $\boldsymbol{\mu}$, a process which is hard to parallelize over time because the number of reliable components tends to change quickly.

13.6.6 Overview of Uncertain Recognition Techniques

An overview of all uncertainty-of-observation techniques is shown in Table 13.1. In the last line of Table 13.1, the binary uncertainty pdf is only computed for reliable features $y_i \in \mathbf{y}_r$ and is set to 1 otherwise.

Since these computations need to take place for each of the GMM mixture components and at each time, they parallelize well in principle. An overview of possible architectures for exploiting this fact using graphics processors with Nvidia's *Compute Unified Device Architecture* (CUDA) [34] is given in [10] for audio-only and

Table 13.1: Mahalanobis distance and likelihood computation for diagonal covariance models in the standard implementation, and using uncertainty decoding, modified imputation, and binary uncertainties

Likelihood Evaluation	Mahalanobis Distance	Scaling Factor s
diagonal covariance	$\sqrt{\frac{(y-\mu)^2}{\sigma^2}}$	s is unchanged an can be precomputed offline
uncertainty decoding	$\sqrt{\frac{(y-\mu)^2}{\sigma^2+\sigma_y^2}}$	s changes and has to be updated
modified imputation	$\sqrt{\frac{(y-\mu)^2}{(\sigma+\sigma_y)^2}}$	s is unchanged an can be precomputed offline
binary uncertainties	$\sqrt{\frac{(y_r-\mu_r)^2}{\sigma_r^2}}$	s changes and has to be updated

in [20] for audiovisual speech recognition. A more detailed discussion of parallel kernels for missing-feature probability computations is given in [19].

13.6.7 Recognition Results

The system has been tested in two setups. In the first, continuous-valued uncertainties were used. Here, for the audio feature extraction, the ETSI advanced front-end was used, since its integrated Wiener filter yields the posterior error variance of spectrum domain features. This value is useful directly for modified imputation or uncertainty decoding once it has been propagated to the MFCC domain used in the ETSI standard as described in Chapter 3. Results for this setup are shown in Table 13.2.

Table 13.2: Recognition results for clean video and additive babble noise

SNR	-30 dB	-20 dB	-10 dB	0 dB	10 dB	20 dB	30 dB
Video only	83.1	83.1	83.1	83.1	83.1	83.1	83.1
Audio only	18.4	27.4	71.0	96.9	98.1	98.1	98.3
Conventional AVSR	84.1	85.6	92.1	97.8	**98.2**	**98.3**	**98.4**
AVSR + MI	84.4	**86.7**	**93.5**	**98.0**	**98.2**	**98.3**	**98.4**
AVSR + UD	**84.5**	86.5	93.1	**98.0**	**98.2**	**98.3**	98.3

As can be seen, there are consistent, however, rather small, improvements in accuracy over the entire range of SNRs. In contrast, in the case where video data is also distorted by additive noise, the improvement due to uncertain recognition becomes more noticeable, in the form of a much more graceful degradation of performance for low SNRs. Here, modified imputation outperforms uncertainty decoding, as will be visible in the following set of results, in Table 13.3.

Table 13.3: Recognition results for noisy video and additive babble noise

SNR	-30 dB	-20 dB	-10 dB	0 dB	10 dB	20 dB	30 dB
Video only	44.1	44.1	44.1	44.1	44.1	44.1	44.1
Audio only	18.4	27.4	71.0	96.9	98.1	98.1	98.2
Conventional AVSR	45.5	48.0	77.7	97.6	**98.2**	**98.3**	**98.4**
AVSR + MI	45.6	**53.0**	**92.3**	**98.0**	**98.2**	**98.3**	**98.4**
AVSR + UD	**45.7**	52.7	92.1	97.9	98.1	**98.3**	**98.4**

Tables 13.4 and 13.5 show the results when no stream weight adaptation is carried out, but, rather, the same stream weight is used over the entire range of SNRs.[3] Here, it is noticeable that is is still possible to retain a performance above that of the video model, as long as an SNR greater than about minus 20 dB is given, indicating the usefulness of the audio stream even under severely distorted conditions. Also, on average the performance is clearly in excess compared to that of conventional AVSR for noisy video data, showing the helpfulness of missing data techniques for the integration of multiple uncertain streams.

However, since the use of stream adaptation still results in clearly the best performance, it is likely that the best overall system design will rely on a combination of both optimizations, adaptive stream weighting and uncertain recognition.

Table 13.4: Recognition results for clean video and additive babble noise without stream weight tuning

SNR	-30dB	-20dB	-10dB	0dB	10dB	20dB	30dB
Video only	**83.1**	**83.1**	83.1	83.1	83.1	83.1	83.1
Audio only	18.4	27.4	71.0	96.9	**98.1**	**98.1**	**98.3**
Conventional AVSR	30.0	50.5	86.7	95.5	96.6	96.6	96.9
AVSR + MI	26.6	51.8	**89.5**	**97.0**	97.6	98.0	98.1
AVSR + UD	32.5	56.2	89.4	95.7	96.6	96.8	97.0

Results for Videos with Missing Frames

For videos where the main source of error lies in missing frames or frames with misdetected mouths, binary uncertainties are the natural choice. In the following set of experiments, binary uncertainties have been used for the audio features as well, where they were estimated as described in Section 13.5.3.3.

[3] For each of the methods, the stream weights were chosen for best average performance over the SNR range from minus 10 dB to 30 dB.

Table 13.5: Recognition results for noisy video and additive babble noise without stream weight tuning

SNR	-30dB	-20dB	-10dB	0dB	10dB	20dB	30dB
Video only	**44.1**	44.1	44.1	44.1	44.1	44.1	44.1
Audio only	18.4	27.4	71.0	**96.9**	98.1	98.1	98.2
Conventional AVSR	19.9	42.7	66.3	92.6	98.0	98.1	98.0
AVSR + MI	23.1	**49.8**	**80.1**	96.1	**98.3**	**98.2**	**98.3**
AVSR + UD	21.1	45.0	70.9	94.4	98.1	98.1	98.1

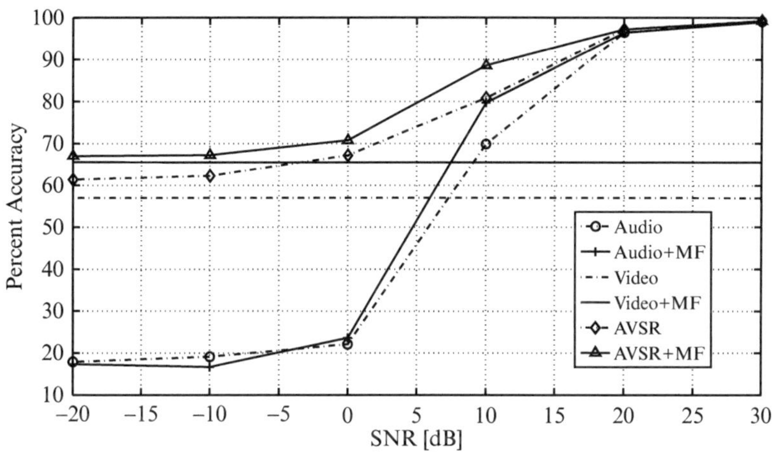

Fig. 13.6: Recognition results with purely binary uncertainties, applied both on the audio and video features

In the case of binary uncertainties, it can be seen that despite the very simple strategy, missing-feature-based recognition clearly outperforms conventional AVSR, especially in the area around 10 dB. This is a rather typical result of many such experiments, and it reflects the fact that missing data techniques are apparently most successful in aiding stream integration when both sources of information have roughly equivalent contributions to the final result.

13.7 Conclusion

The use of missing data techniques can help increase the robustness of audiovisual speech recognition. In all tested cases and scenarios, using both binary and continuous-valued uncertainties, the best performance was achieved with the help of missing data techniques. When the stream weights are adjusted appropriately, the performance is also, in all cases, lower-bounded by the best performance of either audio-only or video-only speech recognition.

The best performance gains are generally observed when both audio and video signal have some mismatch with the model; otherwise, it can be seen that the second modality is very well able to compensate for the shortcomings of the first when the stream weights are chosen well. However, it can be argued that in most cases where it will be needed, audiovisual speech recognition is confronted with both types of mismatch in general, since video features are not yet as independent to changes in environment and speaker as would be desirable, and since distorted audio conditions will be the major reason for including the computationally expensive visual modality.

For future work, it will be interesting to use missing data techniques in conjunction with model adaptation, which has not been tried in this context but has been shown to be a successful strategy for dealing with reverberant data in Chapter 9 of this book.

In order for audiovisual speech recognition to remain computationally feasible, it is also important to consider implementation strategies which make use of the fast-growing potential of parallel processing. This is another target of future work: to make real-time human-machine interaction with large vocabularies feasible with the loose stream integration that CHMMs can offer in conjunction with the missing data techniques discussed here.

References

1. Ahmad, N., Datta, S., Mulvaney, D., Farooq, O.: A comparison of visual features for audiovisual automatic speech recognition. In: Acoustics 2008, Paris, pp. 6445–6448 (2008). DOI 10.1121/1.2936016
2. Aleksic, P.S., Williams, J.J., Wu, Z., Katsaggelos, A.K.: Audio-visual speech recognition using MPEG-4 compliant visual features. EURASIP Journal on Applied Signal Processing **11**, 1213–1227 (2002)
3. Astudillo, R.F., Kolossa, D., Orglmeister, R.: Uncertainty propagation for speech recognition using RASTA features in highly nonstationary noisy environments. In: Proc. ITG (2008)
4. Astudillo, R.F., Kolossa, D., Orglmeister, R.: Accounting for the uncertainty of speech estimates in the complex domain for minimum mean square error speech enhancement. In: Proc. Interspeech (2009)
5. Barker, J., Green, P., Cooke, M.: Linking auditory scene analysis and robust ASR by missing data techniques. In: Proceedings WISP 2001 (2001)
6. Barron, J., Fleet, D., Beauchemin, S.: Performance of optical flow techniques. International Journal of Computer Vision **92**, 236–242 (1994). URL `citeseer.ist.psu.edu/barron92performance.html`
7. Cetingl, H., Yemez, Y., Erzin, E., Tekalp, A.: Discriminative analysis of lip motion features for speaker identification and speech-reading. Image Processing, IEEE Transactions on **15**(10), 2879–2891 (2006). DOI 10.1109/TIP.2006.877528
8. Cooke, M., Barker, J., Cunningham, S., Shao, X.: An audio visual corpus for speech perception and automatic speech recognition. Acoustical Society of America Journal **120**, 2421–2424 (2006). DOI 10.1121/1.2229005
9. Deng, L., Droppo, J., Acero, A.: Dynamic compensation of HMM variances using the feature enhancement uncertainty computed from a parametric model of speech distortion. IEEE Trans. Speech and Audio Processing **13**(3), 412–421 (2005)
10. Dixon, P.R., Oonishi, T., Furui, S.: Harnessing graphics processors for the fast computation of acoustic likelihoods in speech recognition. Comput. Speech Lang. **23**(4), 510–526 (2009). DOI http://dx.doi.org/10.1016/j.csl.2009.03.005
11. Ellis, D.P.W.: PLP and RASTA (and MFCC, and inversion) in Matlab. `http://www.ee.columbia.edu/~dpwe/resources/matlab/rastamat/` (2005). Online web resource, last checked: 01 July 2010
12. ETSI: Speech processing, transmission and quality aspects (STQ); distributed speech recognition; front-end feature extraction algorithm; compression algorithms, ETSI ES 202 050 v1.1.5 (2007-01) (January 2007)
13. Gejguš, P., Šperka, M.: Face tracking in color video sequences. In: SCCG '03: Proceedings of the 19th Spring Conference on Computer Graphics, pp. 245–249. ACM, New York, NY, USA (2003). DOI http://doi.acm.org/10.1145/984952.984992
14. Goecke, R.: A stereo vision lip tracking algorithm and subsequent statistical analyses of the audio-video correlation in Australian English. Ph.D. thesis, Australian National University, Canberra, Australia (2004). URL `citeseer.ist.psu.edu/goecke04stereo.html`
15. Gowdy, J., Subramanya, A., Bartels, C., Bilmes, J.: DBN based multi-stream models for audio-visual speech recognition. In: Proc. ICASSP, vol. 1, pp. I–993–6 vol.1 (2004). DOI 10.1109/ICASSP.2004.1326155
16. Hermansky, H., Morgan, N.: RASTA processing of speech. Speech and Audio Processing, IEEE Transactions on **2**(4), 578–589 (1994). DOI 10.1109/89.326616
17. Hsu, R.L., Abdel-Mottaleb, M., Jain, A.K.: Face detection in color images. IEEE Transactions on Pattern Analysis and Machine Intelligence **24**, 696–706 (2002)
18. Kass, M., Witkin, A., Terzopoulos, D.: Snakes: Active contour models. International Journal of Computer Vision **1**(4), 321–331 (1988). URL `http://www.springerlink.com/content/q7g93335q86604x6/fulltext.pdf`
19. Kolossa, D., Astudillo, R.F., Zeiler, S., Vorwerk, A., Lerch, D., Chong, J., Orglmeister, R.: Missing feature audiovisual speech recognition under real-time constraints. Accepted for publication in ITG Fachtagung Sprachkommunikation (2010)

20. Kolossa, D., Chong, J., Zeiler, S., Keutzer, K.: Efficient manycore CHMM speech recognition for audiovisual and multistream data. Accepted for publication in Proc. Interspeech (2010)
21. Kolossa, D., Klimas, A., Orglmeister, R.: Separation and robust recognition of noisy, convolutive speech mixtures using time-frequency masking and missing data techniques. In: Proc. WASPAA, pp. 82–85 (2005). DOI 10.1109/ASPAA.2005.1540174
22. Kratt, J., Metze, F., Stiefelhagen, R., Waibel, A.: Large vocabulary audio-visual speech recognition using the Janus speech recognition toolkit. In: DAGM-Symposium, pp. 488–495 (2004)
23. Lan, Y., Theobald, B.J., Ong, E.J., Bowden, R.: Comparing visual features for lipreading. In: Int. Conf. on Auditory-Visual Speech Processing (AVSP2009). Norwich, UK (2009)
24. Lerch, D.: Audiovisuelle Spracherkennung unter Berücksichtigung der Unsicherheit von visuellen Merkmalen. Diploma thesis, TU Berlin, dennis_lerch@gmx.de (2009)
25. Lewis, T.W., Powers, D.M.W.: Lip feature extraction using red exclusion. In: VIP'00: Selected Papers from the Pan-Sydney Workshop on Visualisation, pp. 61–67. Australian Computer Society, Inc., Darlinghurst, Australia, Australia (2001)
26. Lucey, P.J., Dean, D.B., Sridharan, S.: Problems associated with current area-based visual speech feature extraction techniques. In: AVSP 2005, pp. 73–78 (2005). URL http://eprints.qut.edu.au/12847/
27. Luettin, J., Potamianos, G., Neti, C.: Asynchronous stream modelling for large vocabulary audio-visual speech recognition. In: Proc. ICASSP (2001)
28. Mase, K., Pentland, A.: Automatic lip-reading by optical flow analysis. Trans. Systems and Computers in Japan **22**, 67–76 (1991)
29. Matthews, I., Cootes, T., Bangham, J., Cox, S., Harvey, R.: Extraction of visual features for lipreading. IEEE Transactions on Pattern Analysis and Machine Intelligence **24**(2), 198–213 (2002). DOI 10.1109/34.982900
30. Metze, F.: Articulatory features for conversational speech recognition. Ph.D. thesis, Universität Fridericiana zu Karlsruhe (2005)
31. Naseem, I., Deriche, M.: Robust human face detection in complex color images. IEEE International Conference on Image Processing, ICIP 2005. **2**, 338–341 (2005). DOI 10.1109/ICIP.2005.1530061
32. Nefian, A., Liang, L., Pi, X., Liu, X., Murphy, K.: Dynamic Bayesian networks for audio-visual speech recognition. EURASIP Journal on Applied Signal Processing **11**, 1274–1288 (2002)
33. Neti, C., Potamianos, G., Luettin, J., Matthews, I., Glotin, H., Vergyri, D., Sison, J., Mashari, A., Zhou, J.: Audio-visual speech recognition. Tech. Rep. WS00AVSR, Johns Hopkins University, CLSP (2000). URL citeseer.ist.psu.edu/neti00audiovisual.html
34. NVIDIA Corporation: NVIDIA CUDA Compute Unified Device Architecture Programming Guide (2007)
35. Otsuki, T., Ohtomo, T.: Automatic lipreading of station names using optical flow and HMM. Technical report of IEICE. HIP **102**(473), 25–30 (2002). URL http://ci.nii.ac.jp/naid/110003271904/en/
36. Papandreou, G., Katsamanis, A., Pitsikalis, V., Maragos, P.: Adaptive multimodal fusion by uncertainty compensation with application to audiovisual speech recognition. Audio, Speech, and Language Processing, IEEE Trans. **17**(3), 423–435 (2009). DOI 10.1109/TASL.2008.2011515
37. Patterson, E.K., Gurbuz, S., Tufekci, Z., Gowdy, J.N.: Moving-talker, speaker-independent feature study, and baseline results using the CUAVE multimodal speech corpus. EURASIP J. Appl. Signal Process. **2002**(1), 1189–1201 (2002). DOI http://dx.doi.org/10.1155/S1110865702206101
38. Potamianos, G., Neti, C., Luettin, J., Matthews, I.: Audio-visual automatic speech recognition: An overview. In: E. Vatikiotis-Bateson, P. Perrier (eds.) Issues in Visual and Audio-Visual Speech Processing. MIT Press (2004)
39. Raj, B., Stern, R.: Missing-feature approaches in speech recognition. Signal Processing Magazine, IEEE **22**(5), 101–116 (2005)

40. Schwerdt, K., Crowley, J.L.: Robust face tracking using color. In: Proc. of 4th International Conference on Automatic Face and Gesture Recognition, pp. 90–95. Grenoble, France (2000). URL `citeseer.ist.psu.edu/schwerdt00robust.html`
41. Shdaifat, I., Grigat, R.R., Lütgert, S.: Recognition of the German visemes using multiple feature matching. In: B. Radig, S. Florczyk (eds.) Lecture Notes in Computer Science, Pattern Recognition, vol. 2191/2001, pp. 437–442. Springer-Verlag Berlin Heidelberg (2001)
42. Tamura, S., Iwano, K., Furui, S.: A robust multi-modal speech recognition method using optical-flow analysis. In: Multi-Modal Dialogue in Mobile Environments, ISCA Tutorial and Research Workshop (ITRW). ISCA, Kloster Irsee, Germany (2002)
43. Vezhnevets, V., Sazonov, V., Andreeva, A.: A survey on pixel-based skin color detection techniques. In: Proc. Graphicon, pp. 85–92. Moscow, Russia (2003). URL `citeseer.ist.psu.edu/vezhnevets03survey.html`
44. Wang, X., Hao, Y., Fu, D., Yuan, C.: ROI processing for visual features extraction in lipreading. In: 2008 International Conference on Neural Networks and Signal Processing, pp. 178–181. IEEE (2008)
45. Yang, M.H., Kriegman, D.J., Ahuja, N.: Detecting faces in images: A survey. IEEE Transactions on Pattern Analysis and Machine Intelligence **24**(1), 34–58 (2002)
46. Young, S., Russell, N., Thornton, J.: Token passing: A simple conceptual model for connected speech recognition systems. Tech. Rep. CUED/FINFENG /TR.38, Cambridge University Engineering Department (1989)

Index

A

A priori model 24, 194, 266
  joint, noise and speech 196
  noise 195, 267
  speech 195, 266
  training 212
Acoustic echo canceler (AEC) 163, 164, 170, 172, 173
Acoustic model 11, 13, 20, 121, 190
  adaptation 102
  training 166
Adaptive training 116
Additive noise 130, 144, 160
Algonquin
  algorithm 71, 72
  model 68, 70–72, 76, 83, 86, 88, 90, 94
Audiovisual speech recognition (AVSR) 345, 346, 352, 363, 371
Aurora 61, 159, 162, 165, 171, 180
  Aurora 2 database 210
  Aurora 4 database 211
  Aurora 5 database 281
Automatic speech recognition (ASR) 157
Autoregressive model 266

B

Bayesian decision rule 11, 189
Bayesian predictive classification (BPC) 84, 85
Beamforming 305, 321, 328
Blind source separation 319–321, 328
Bounded MAP estimation 129, 137, 138, 140

C

Cepstral mean normalization 129, 147
Cepstral mean subtraction (CMS) 213
Classifier compensation 128, 129, 133, 149–151
Clean feature posterior 15, 19, 21, 194
Clean speech posterior 107
Cluster-based reconstruction 129, 135, 138, 139, 141, 145, 146, 150–152
Cocktail party problem 322
Codeword-Dependent Cepstral Normalisation (CDCN) 21
Complex Gaussian uncertainty model 37–40, 43, 53, 55, 57, 333, 334
Conditional Bayesian estimation 187, 193
Conditional independence 12, 190
Connected digits recognition 281
Correlation-based reconstruction 129, 135, 136, 141, 145, 146
Coupled hidden Markov model (CHMM) 346–348, 350, 363, 364, 372

D

Decision rule
  Bayesian 69, 84, 189
Denoising 257
Dereverberation *see* Reverberation, 229, 257
Deutlichkeit value (D50) 242
Direction of arrival (DOA) 302, 325, 328
Discrete cosine transform (DCT) 25, 114
Discriminative training 119, 121
Domain-based compensation 105
Dynamic variance compensation 228, 231

D. Kolossa and R. Haeb-Umbach (eds.), *Robust Speech Recognition of Uncertain or Missing Data: Theory and Applications*, DOI 10.1007/978-3-642-21317-5,
© Springer-Verlag Berlin Heidelberg 2011

E

EM algorithm 117, 237
ETSI front-end feature extraction
  advanced front-end (AFE) 38, 40, 52, 53, 61, 177, 178, 180–182, 213, 282, 351, 362, 363, 369
  standard front-end (SFE) 197, 213, 262, 281
Evidence modeling 295
Expectation maximization (EM) algorithm 195, 267

F

False reliables 170
Feature compensation 68–71, 76, 82, 83, 86, 88, 91–93, 129, 130, 133, 135, 145, 149, 152
Feature enhancement 159, 176, 180, 182, 257, 264, 277–280
Feature extraction 197
Finite impulse response (FIR) 164, 172, 173
Fragment decoding 153
Fuzzy clustering 303

G

Gaussian mixture model (GMM) 15, 24, 38, 72, 88, 110, 188, 266, 365
Gaussian-dependent imputation 158, 161
Generalized pseudo Bayesian (GPB) 206, 208

H

Hands-free scenario 261
Harmonic decomposition 164, 178
Hidden Markov model (HMM) 189
  training 211

I

Imputation 19, 349, 350
  modified 20, 61, 338, 350, 367–370
Independent component analysis (ICA) 320–322, 340, 352
Inference 26, 205, 264, 276
  clean feature posterior 194
  scheme 194
Interacting multiple model (IMM) 206, 277

J

Joint uncertainty decoding (JUD) 21, 70, 76, 115

K

Kalman filter
  extended (EKF) 27, 208
  iterated extended (IEKF) 207, 209, 276, 278
  unscented (UKF) 27, 276

L

Language model 11, 109, 190
Likelihood 14, 113
Linear discriminant analysis (LDA) 166, 352, 355
Linear dynamic model (LDM) 195
Logarithmic mel power spectral coefficient (LMPSC) 263, 274
LVCSR 227, 247

M

Marginalization 19, 128, 133–135, 144, 145, 147, 149–153, 193, 349, 350
Mask estimation 132, 149, 161, 170, 177, 180, 326–328, 330
  oracle mask 163, 165, 171
  semi-oracle mask 159, 163, 170, 172, 182
  soft mask 19
  SVM mask 165, 171, 181, 182
  vector quantization (VQ) mask 164, 171, 177, 178, 182
Maximum likelihood linear regression (MLLR) 102
  constrained (CMLLR) 103
    noisy (NCMLLR) 103
    predictive 116
Mel filter bank 199
Mel frequency cepstral coefficient (MFCC) 25, 42, 262, 320, 334, 369
Microphone array 301
Minimum mean squared error (MMSE) 19, 107, 264, 277
Mismatch function 104
  dynamic parameter 106
  static parameter 104
Missing data 18, 296
Missing data mask 158, 160
Missing data techniques (MDT) 158, 160, 370, 371
  assumptions 158, 183
  imputation 158
  imputation bounds 161, 176, 180–182
Missing feature 18
Missing-feature coupled HMM 350

MMSE
  estimate 37, 71–73, 82, 88, 89, 134, 143, 330
  estimator 128, 141, 151
  posterior variance 37, 38
  STSA estimator 38, 39, 44, 89
Model compensation 68, 69, 75, 76, 84–88, 91
Multi-condition training 159, 166, 167, 170, 176, 182
Mutual information discriminant analysis (MIDA) 166, 178

N

Negentropy 323
Noise adaptive training (NAT) 89–93, 95
Noise robustness 187

O

Observation error 275
Observation model 25, 194, 197, 268
  phase-insensitive 199
  phase-sensitive 198

P

Parallel model combination (PMC) 110
  data-driven 112
Particle filter 28
Permutation correction 324
Phase factor 197, 202
  central moments 202, 204
  characteristic function 203
  distribution 198
Phase-based compensation 105
Phase-insensitive observation model 199
Phase-sensitive distortion model 76, 77, 80, 82–84, 86–88, 94
Phase-sensitive observation model 187, 198
Probabilistic optimal filtering (POF) 21
PROSPECT features 158, 161, 166, 167
Pseudo-Monte Carlo method 41, 45

R

RASTA-PLP 46, 351, 362, 363
Real-world recordings 158, 160, 167, 177, 182, 320, 334
Resource management database 144
Retraining 92, 167
Reverberant speech recognition 227, 257
Reverberation 159, 171, 173, 182, 257
Reverberation time 23, 271
Rice distribution 43
Room impulse response (RIR) 172, 174, 175, 261, 268, 271, 321
  model 271

S

Spectral subtraction 82, 83, 88–91, 107, 147, 159, 176, 177, 181, 182
Spectrogram 130, 160
Spectrographic mask 129, 150, 153
  binary 132, 141, 152
  estimation 132, 148, 151
  oracle 148, 149
  soft 129, 133, 135, 151
Speech recognition 189
Speech separation 301
SpeechDat-Car 158, 162
SPEECON 158, 162
SPHINX 144, 151
SPLICE 21, 70, 88–91, 93, 109, 233
State-based imputation 128, 133, 134, 144, 145, 147, 149
Static and dynamic variance adaptation (SDVA) 235
  Covariance model 235–236
  Digit experiment 239–246
  DVA 239
  Large vocabulary experiment 247–249
  Model parameter estimation 237–239
  SDVA 239
  SVA 238
Statistical independence 323
Support vector machine (SVM) 165
Switching linear dynamic model (SLDM) 24, 188, 195, 267

T

Taylor series 199
Time-frequency masking 305, 326, 328, 332, 338, 340
  binary 320
  soft 320

U

Uncertainty decoding 8, 17, 61, 62, 68–71, 76, 84, 191, 228, 233, 338, 346, 350, 367, 369
Uncertainty propagation 35–39, 42, 57, 58, 61, 62, 320, 334
Unscented transform 41, 42, 45, 112

V

Vector Taylor series (VTS) 21, 70, 71, 76, 83, 86–89, 91–93, 110, 111, 115
Viterbi algorithm 190
Voice activity detector (VAD) 164, 165, 178

W

Wall Street Journal task 247
Wiener filter 89

IMMUNOLOGY SERIES

Editor-in-Chief
NOEL R. ROSE
Professor and Chairman
Department of Immunology and Infectious Diseases
The Johns Hopkins University
School of Hygiene and Public Health
Baltimore, Maryland

European Editor
ZDENEK TRNKA
Basel Institute for Immunology
Basel, Switzerland

1. Mechanisms in Allergy: Reagin-Mediated Hypersensitivity
   Edited by Lawrence Goodfriend, Alec Sehon and Robert P. Orange
2. Immunopathology: Methods and Techniques
   Edited by Theodore P. Zacharia and Sidney S. Breese, Jr.
3. Immunity and Cancer in Man: An Introduction
   Edited by Arnold E. Reif
4. *Bordetella pertussis:* Immunological and Other Biological Activities
   J.J. Munoz and R.K. Bergman
5. The Lymphocyte: Structure and Function (in two parts)
   Edited by John J. Marchalonis
6. Immunology of Receptors
   Edited by B. Cinader
7. Immediate Hypersensitivity: Modern Concepts and Development
   Edited by Michael K. Bach
8. Theoretical Immunology
   Edited by George I. Bell, Alan S. Perelson, and George H. Pimbley, Jr.
9. Immunodiagnosis of Cancer (in two parts)
   Edited by Ronald B. Herberman and K. Robert McIntire
10. Immunologically Mediated Renal Diseases: Criteria for Diagnosis and Treatment
    Edited by Robert T. McCluskey and Giuseppe A. Andres
11. Clinical Immunotherapy
    Edited by Albert F. LoBuglio
12. Mechanisms of Immunity to Virus-Induced Tumors
    Edited by John W. Blasecki
13. Manual of Macrophage Methodology: Collection, Characterization, and Function
    Edited by Herbert B. Herscowitz, Howard T. Holden, Joseph A. Bellanti, and Abdul Ghaffar
14. Suppressor Cells in Human Disease
    Edited by James S. Goodwin
15. Immunological Aspects of Aging
    Edited by Diego Segre and Lester Smith
16. Cellular and Molecular Mechanisms of Immunologic Tolerance
    Edited by Tomáš Hraba and Milan Hašek

CELLULAR ONCOGENE ACTIVATION